Springer Texts in Statistics

Advisors:
Stephen Fienberg   Ingram Olkin

Springer Texts in Statistics

(continued after index)

Harold R. Lindman

Analysis of Variance in Experimental Design

With 31 Figures

Springer-Verlag
New York Berlin Heidelberg London Paris
Tokyo Hong Kong Barcelona Budapest

Harold R. Lindman
Department of Psychology
Indiana University
Bloomington, IN 47401
USA

Mathematics Subject Classifications: 62J10, 62K

Library of Congress Cataloging-in-Publication Data
Lindman, Harold R.
    Analysis of variance in experimental design/Harold R. Lindman.
        p. cm. — (Springer texts in statistics)
    Includes bibliographical references and index.
    ISBN 0-387-97571-3 (alk. paper)
    1. Analysis of variance. I. Title. II. Series.
QA279.L573   1991
519.5′38—dc20   91-17298

Printed on acid-free paper.

Production managed by Francine Sikorski; manufacturing supervised by Robert Paella.
Photocomposed copy prepared using LaTeX.
Printed and bound by R.R. Donnelley & Sons, Harrisonburg, VA.
Printed in the United States of America.

9 8 7 6 5 4 3 2 1

ISBN 0-387-97571-3 Springer-Verlag New York Berlin Heidelberg
ISBN 3-540-97571-3 Springer-Verlag Berlin Heidelberg New York

Preface

This is a general text on the analysis of variance. It is aimed at serious students in research with a general background in statistics but with little additional knowledge of mathematics (e.g., graduate students in the biological or social sciences). It begins with applications rather than with abstract concepts such as linear models (which can put off the nonmathematical reader), and the emphasis throughout is on *intelligent* application of the methods; that is, enough theory (including linear models) is developed to enable the reader to choose procedures and to interpret results appropriately.

The text is relatively complete, containing most of the information one needs to perform analyses of variance and related procedures. Thus, although it is designed as a text, it can also serve as a reference. It contains some material (e.g., some multiple comparisons in Chapter 4, and some material on mixed models in Chapter 15) not found previously in general texts on the subject.

It begins at a relatively simple level, but gradually increases in difficulty and generality. One who studies the entire text should not only become knowledgeable about analysis of variance, but should also learn more about statistics in general.

The first eight chapters should be covered in order. Each chapter depends on, and in turn adds to, the material in the previous chapters. Beginning with Chapter 9, more flexibility is possible. A good basic course would cover Chapters 1 through 10. Chapter 11 is somewhat complicated and is not essential for most research.

Chapters 12 through 14 cover multivariate models, including multivariate analysis of variance and analysis of covariance. These are becoming increasingly important in research, largely because computers are now available to do such analyses. However, the computer is a mixed blessing; with the calculations made easy, multivariate analyses are sometimes done without adequate thought as to the nature of the data and the goals of the research. Chapters 12 through 14 give advice on when and how to apply multivariate methods, as well as describe the methods themselves.

Chapter 15 presents an almost completely general linear model for analyzing variance. It gives a general theory covering nearly all of the analyses in the rest of the text. It is also the only chapter with a relatively rigorous, theorem-oriented approach. It is there mainly for those who are curious about the deeper mathematical foundations of the analysis of variance. Those who are interested primarily in applications can ignore it.

There is little emphasis on computers in this text. To begin with, any stress on a particular computer program is likely to make the text dated as soon as the program is revised or superseded by some other program. In addition, I believe that the details of using a program are comparatively easy; it is much more difficult to design a good piece of research, do *appropriate* analyses, and understand the results at the end. Knowing how to use a statistical program does not make one a statistician any more than knowing how to use a word processor makes one a writer. Finally, knowing the calculations, we can better understand and evaluate computer programs designed to do those calculations.

Accordingly, I believe that at least some of the exercises provided should be done by hand with the aid of a pocket calculator or, perhaps, a spreadsheet computer program. (No exercises are given for the final chapters; multivariate analyses of variance are too difficult to do by hand.) However, two popular statistical packages are described in some detail in Appendixes C and D. Each appendix is organized according to the individual chapters. Thus, after reading a given chapter, you can immediately refer to the section of the appendix that relates to that chapter. Alternatively, you can read the entire appendix after studying the rest of the text.

My thanks for assistance in writing this text go primarily to the many students who have suffered through "preprinted" versions while the text was being written. I am grateful not only for their patience but also for their excellent suggestions for improvements and their diligence in finding errors.

Bloomington, Indiana   HAROLD R. LINDMAN

Contents

Acknowledgments for Permissions

Springer-Verlag wishes to thank the publishers listed below for their copyright permission and endorsement to use their previously published material in this book. Their invaluable help in this matter has made the publication of this volume possible.

Figures 1.1, 1.2, 3.1, 3.2, 5.1, 5.2, 7.3, 10.1 to 10.8, 11.1 to 11.8, and 14.1 to 14.3 have been reproduced with the kind permission of W.H. Freeman and Company from *Analysis of Variance in Complex Experimental Designs* by Harold R. Lindman.

Table 2.2 has been reproduced with the kind permission of the Royal Statistical Society.

Table 2.3 has been reproduced with the kind permission of the Institute of Mathematical Statistics.

Table 2.9 has been reproduced with the kind permission of J. Wiley and Sons.

Table 2.11 has been reproduced with the kind permission of the Institute of Mathematical Statistics.

The examples of the output and syntax from SAS presented in Appendix C have been reproduced with the kind permission of SAS Institute, Inc.

The examples of the output and syntax from SPSS presented in Appendix D have been reproduced with the kind permission of SPSS, Inc.

1

Review of Statistical Concepts

This text is written for those who have already had an intermediate level, noncalculus course in statistics. In this chapter we will review certain basic concepts and cover some fine points that may have been overlooked in earlier study. This chapter will also introduce the special notation used in the book, and my own statistical biases.

Statistics is, in fact, a rather controversial subject. Although there is wide agreement on the principles of statistics, there is considerable disagreement on the application of these principles. My concern in this book is with knowledgeable application of the principles of statistics, so certain controversial subjects will necessarily be considered. An explanation of my own biases may help clarify my position on these issues.

As to the special notation, unfortunately, much statistical notation is not standardized. Although the notation for new concepts will be introduced with the concepts, it seems simplest to present the basic statistical notation used in this book at an early point. (But see the symbol list at the end of the book.)

Probability Theory

MATHEMATICAL CONCEPTS OF PROBABILITY

Mathematically, probability theory is concerned with assigning numbers to events in such a way that those numbers represent, in some sense, how likely each event is to occur. To do this plausibly, the numbers must have certain properties. The basic properties are: that the probability (Pr) of an impossible event is zero; that no probability is greater than one; and that if two events, A and B, are mutually exclusive (if they cannot *both* occur), then $Pr(A \text{ or } B) = Pr(A) + Pr(B)$. From these basic properties, the other important properties are derived.

APPLIED PROBABILITY THEORY

In the basic mathematical theory, neither events nor probabilities are given any practical definition. That task is left to applied probability theory and statistics. The most common definition limits the assignment of probabilities to the outcomes of *experiments*. An experiment is defined technically as a set of acts that result in one of a group of possible events, and that can in principle be repeated an infinite number of times under identical conditions. By this definition, the outcomes of flipping a coin have probabilities associated with them because the coin can be flipped over and over again under conditions that, for all practical purposes at least, are identical. The outcomes of a boxing match, on the other hand, do not have probabilities associated with them because there is no way in which a boxing match can be repeated under even remotely identical conditions.

For this definition, usually called the *relative frequency* definition, the probability of an event is the limiting relative frequency of the event as the number of repetitions of the experiment approaches infinity. All of the basic theory of statistics in this text was originally developed with the relative frequency interpretation of probability.

A different applied definition, rejected by most statisticians until recently, but now growing in acceptance, is the *subjective* (or *personalist*) definition. It holds that a probability need not be tied to any particular relative frequency of occurrence, but that it is a measure of an individual's belief about the likelihood of the occurrence of an event. According to this view, if a person has a belief about the likelihood of each possible outcome of a boxing match, the possible outcomes of the boxing match have probabilities. The probability of an event may therefore be different for different people, since people can differ in their beliefs about the outcomes of boxing matches. Moreover, the probability need not be related in any rigid way to any relative frequency, even if one exists, although the mathematics of probability theory assert that under certain circumstances beliefs are influenced by relative frequencies.

Detailed discussions (including the philosophical bases) of each view of probability are not within the scope of this text. Nevertheless, these opposing views reflect differences in the general approach to statistics. Those who hold to relative frequency have usually had a more rigid approach to statistical inference, setting up specific rules that must be followed exactly to obtain valid statistical results. For example, one cannot assign a probability to the assertion that a null hypothesis is true. Since a null hypothesis is not the outcome of an infinitely repeatable experiment, it is considered improper to assign it a probability. From this is derived the practice of unqualified acceptance or rejection of a null hypothesis on the basis of a predetermined significance level.

Personalists, on the other hand, freely talk about the probability that a null hypothesis is true, but refuse to unqualifiedly accept or reject it on the basis of inconclusive evidence. As a general rule, the personalist uses statistics more freely, feeling that correct interpretation is more important than exact procedure. I am a personalist, and although the mathematics in this text have all been developed within the framework of relative frequency theory, much of the discussion on interpretation and proper procedure is colored by my bias for freer use of the procedures with intelligent interpretation of the results.

Summary Values

It is common practice to characterize distributions by single numbers that represent certain important properties of the distributions. The most commonly used such numbers are measures of central tendency: the mean (or expected value), the median, and the mode. Other important measures are the measures of dispersion: usually the range, interquartile range, mean absolute deviation, and standard deviation, or its square, the variance.

SKEWNESS AND KURTOSIS

Two other important summary measures in this text are *skewness* and *kurtosis*. They are best defined in terms of the third and fourth *central moments*. For a value of r, μ'_r, the rth central moment is defined as the expected value of the rth power of the deviations from the mean:

$$\mu'_r = \underline{E}[(X - \mu)^r] = \Sigma_i (X_i - \mu)^r Pr(X_i).$$

The first central moment is zero, and the second is the variance.

To find the skewness, we divide the third central moment by the cube of the standard deviation:

$$Sk = \mu'_3/\sigma^3.$$

To find the kurtosis we take the fourth central moment, divide it by the fourth power of the standard deviation, and subtract three from the result:

$$Ku = (\mu'_4/\sigma^4) - 3.$$

The 3 plays an important role and cannot be neglected. It serves to standardize the kurtosis measure so that the kurtosis of the normal distribution is zero.

Kurtosis differs from skewness primarily in that kurtosis measures the overall tendency toward a long tail at either end; skewness measures the degree to which this tendency differs at the two ends. The skewness can take on any value between plus and minus infinity, although it will seldom be greater than plus or minus two. The possible range of the kurtosis is limited by the skewness:

$$(Sk)^2 - 2 \leq Ku \leq \infty.$$

The absolute lower limit for the kurtosis is -2. The normal distribution has a kurtosis of 0, and the rectangular distribution has a kurtosis of -1.2. Figures 1.1 and 1.2 illustrate distributions differing in skewness and kurtosis.

Notation

Most of the mathematical terms in this book follow standard mathematical notation. However, there are some differences, and not everyone is equally familiar with mathematical terms, so we will review the basic system.

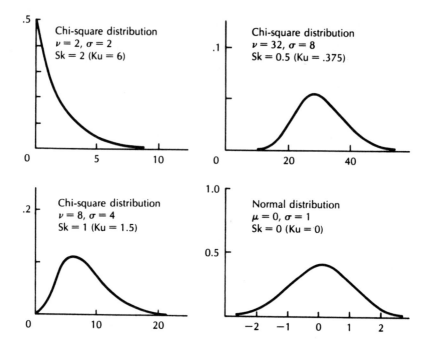

FIGURE 1.1. Distributions varying in skewness.

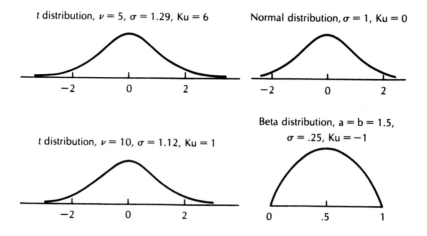

FIGURE 1.2. Distributions varying in kurtosis.

First, both Greek and Roman letters are used as symbols. With few exceptions, Greek letters are used exclusively to represent properties (parameters) of populations. In general, then, Greek letters represent values that can be estimated but not known. Two exceptions are the letters α and β, used to represent the significance level and the probability of a Type II error, respectively. Other exceptions will occur occasionally in the text; it will always be clear when an exception occurs. The following Greek letters are used in the text:

α (alpha) β (beta) γ (gamma) δ (delta)
ϵ (epsilon) θ (theta) λ (lambda) μ (mu)
ν (nu) π (pi) ρ (rho) σ (sigma)
τ (tau) ϕ (phi) χ (chi) ψ (psi)
ω (omega)

Roman letters are used to represent sample values, i.e., statistics. For example, s^2 stands for the variance of a sample, which we will define as

$$s^2 = \Sigma(X_i = \overline{X})^2/n,$$

while σ^2 represents the corresponding population variance. There is an important exception to this practice. When statistics are used to estimate population values, an estimate is represented by the symbol for the particular population parameter, but the symbol wears a "hat." For example, $\hat{\sigma}^2$ stands for the unbiased estimate of the population variance. Specifically,

$$\hat{\sigma}^2 = ns^2/(n-1). \tag{1.1}$$

Subscripts $(i, j, k, l, \ldots)$ are used to denote different values of quantities of the same type. For example, suppose that we have data for 25 subjects in each of four schools in each of three towns. Then X_{ijk} might represent the score of the kth subject in the jth school in the ith town. The number of possible values of each subscript is indicated by the corresponding capital letter. Thus, in the example just given, I would be the total number of towns studied $(I = 3)$, and J would be the number of schools in each town $(J = 4)$.

However, we will use a different convention for the number of subjects. Subjects are usually randomly sampled, and n traditionally represents the number of subjects in a single random sample. We will usually use n to designate the size of such a random sample (in our example, $n = 25$).

In the experiments discussed in this text, several different samples will be taken from different groups. In the above study, for example, we have taken 12 samples (4 schools in each of 3 towns) of 25 each, for a grand total of $12 \times 25 = 300$ scores. We will let N represent the total number of scores in the entire experiment (in the example, $N = 300$).

Because sample means are very important, we will have a special method of designating them. A bar over a symbol will indicate that a mean is intended, and the subscript over which we have averaged will be replaced by a dot. Thus,

$\overline{X}_{ij.}$ will represent the average, over the 25 subjects (subscript k, replaced by a dot), for the jth school in the ith town. Or,

$$\overline{X}_{ij.} = (1/n)\Sigma_{k=1}^{n}X_{ijk}.$$

If we next averaged over the schools, we would have, for the ith town,

$$\begin{aligned}\overline{X}_{i..} &= (1/J)\Sigma_{j=1}^{J}\overline{X}_{ij.} \\ &= (1/nJ)\Sigma_{j=1}^{J}\Sigma_{k=1}^{n}X_{ijk}.\end{aligned}$$

Finally, the grand mean of all the scores would be:

$$\begin{aligned}\overline{X}_{...} &= (1/I)\Sigma_{i=1}^{I}\overline{X}_{i..} \\ &= (1/N)\Sigma_{i=1}^{I}\Sigma_{j=1}^{J}\Sigma_{k=1}^{n}X_{ijk}.\end{aligned}$$

For computational purposes, we will sometimes work with totals instead of means. The total, or sum, of a set of scores will be represented by t with appropriate subscripts. Thus,

$$\begin{aligned}t_{ij.} &= \Sigma_{k=1}^{n}X_{ijk} = n\overline{X}_{ij.}, \\ t_{i..} &= \Sigma_{j=1}^{J}t_{ij.} \\ &= \Sigma_{j=1}^{J}\Sigma_{k=1}^{n}X_{ijk} = nJ\overline{X}_{i..}, \\ t_{...} &= \Sigma_{i=1}^{I}t_{i..} = \Sigma_{i=1}^{I}\Sigma_{j=1}^{J}\Sigma_{k=1}^{n}X_{ijk} = N\overline{X}_{...}.\end{aligned}$$

Important Distributions

This text requires an understanding of several different distributions. In this section, we will review these distributions and discuss their properties. We will also illustrate the notation to be used for representing distributions, and we will show how the distributions relate to each other.

NORMAL DISTRIBUTION

The normal distribution should be familiar already; it is the "bell-shaped curve" of elementary statistics. It is completely specified by its mean and variance. We will write

$$X \sim N_{(\mu,\sigma^2)}$$

to indicate that X has a normal distribution with mean μ and variance σ^2. (The symbol $\sim$ should be read "is distributed as .") The *standard normal* distribution has a mean of zero and a variance of 1. Standard normal variables are usually represented with the letter Z, i.e.,

$$Z \sim N_{(0,1)}.$$

All of the other distributions we will be concerned with are derived from the standard normal distribution.

CHI-SQUARE

The chi-square distribution with ν degrees of freedom is defined to be the distribution of the sum of ν independent squared random variables, each having the standard normal distribution. If a set of ν independent random variables, Z_i, all have a standard normal distribution, then

$$\Sigma_{i=1}^{\nu} Z_i^2 \sim \chi^2_{(\nu)}$$

(i.e., the sum of the squares has a chi-square distribution with ν degrees of freedom).

The mean of the chi-square distribution is ν, and the variance is 2ν. It has positive skewness and positive kurtosis, with both approaching zero as ν approaches infinity. By the central limit theorem, the chi-square distribution is approximately normal for large ν. Figure 1.1 shows some examples of the chi-square distribution.

The chi-square distribution is important primarily because, when the population is normally distributed, it gives us the distribution of the sample variance, s^2. Specifically,

$$ns^2/\sigma^2 \sim \chi^2_{(n-1)}.$$

If we write this in terms of the best estimate of the population variance (Eq. 1.1), we have

$$(n-1)\hat{\sigma}^2/\sigma^2 \sim \chi^2_{(n-1)}. \tag{1.2}$$

Another important property of the chi-square distribution is that the sum of *independent* random variables, each having a chi-square distribution, also has a chi-square distribution. The degrees of freedom of the sum are equal to the sum of the degrees of freedom of the individual chi-squares. Specifically, given a set of independent random variables, X_i, each having a chi-square distribution with ν_i degrees of freedom,

$$\Sigma X_i \sim \chi^2_{\Sigma(\nu_i)}.$$

NONCENTRAL CHI-SQUARE

A related distribution, less commonly discussed in statistics texts, is the *noncentral chi-square* distribution. The noncentral chi-square distribution with ν degrees of freedom and noncentrality parameter ϕ is the sum of ν independent normally distributed random variables, each of which has a variance of 1, but a mean $\neq 0$. It thus differs from the chi-square distribution in the means of the variables that are squared and summed. Given a set of independent random variables,

$$X_i \sim N_{(\mu_i,1)},$$

then

$$\Sigma_{i=1}^{\nu} X_i^2 \sim \chi^2_{(\nu,\phi)},$$

where $\phi^2 = \Sigma_i \mu_i^2$. The mean of the noncentral chi-square distribution is $\nu + \phi^2$, and the variance is 2ν. The noncentral chi-square distribution is important mainly because the noncentral F distribution (to be described later) is derived from it.

t DISTRIBUTION

Let Z and X be independent random variables,

$$Z \sim N_{(0,1)}$$
$$X \sim \chi^2_{(\nu)}.$$

Then

$$Z/(X/\nu)^{1/2} \sim t_{(\nu)};$$

that is, the ratio has a t distribution with ν degrees of freedom. The t distribution is symmetric with a mean of zero and a variance of $\nu/(\nu - 2)$. The kurtosis is positive, decreasing to zero as ν approaches infinity, when the t distribution approaches the normal. Some of the distributions in Figure 1.2 are examples of the t distribution.

The t distribution is important because it is central to the commonly used t test. The t distribution and the t test, however, should not be confused—the one is a probability distribution, the other is a statistical test that makes use of that distribution.

The t test takes advantage of the unique statistical properties of the normal distribution. In particular, if the population distribution is normal, then several things are true. First, the sample mean is normally distributed with a mean of μ and a variance of σ^2/n, so that

$$(\overline{X}_. - \mu)n^{1/2}\sigma \sim N_{(0,1)}.$$

Second is Equation 1.2. Third, the sample mean and variance are independent, allowing us to take ratios to get

$$(\overline{X}_. - \mu)n^{1/2}/\hat{\sigma} \sim t_{(n-1)}.$$

The t test for a single sample tests the null hypothesis H_0: $\mu = \mu^*$ (for a specified μ^*) by substituting μ^* for μ and comparing the result with the tabled values of the t distribution.

Similarly, the t test for the difference between the means of two groups takes advantage of the fact that the sum of independent variables having chi-square distributions has a chi-square distribution. The test assumes that the two groups being compared have equal population variances.

F DISTRIBUTION

The F distribution is defined as the ratio of two independent chi-squares, each divided by its degrees of freedom. If X_1 and X_2 are independent random variables distributed as chi-square with ν_1 and ν_2 degrees of freedom, respectively, then

$$(X_1/\nu_1)/(X_2/\nu_2) \sim F_{(\nu_1,\nu_2)}$$

(i.e., the ratio has an F distribution with ν_1 degrees of freedom in the numerator and ν_2 degrees of freedom in the denominator).

The mean of F is $\nu_2/(\nu_2-2)$, which is approximately one if ν_2 is large. The formula for the variance is complicated, but when ν_2 is large, it is approximately $2/\nu_1 + 2/\nu_2$.

One useful property of the F distribution is that if

$$X \sim F_{(\nu_1,\nu_2)},$$

then

$$1/X \sim F_{(\nu_2,\nu_1)};$$

that is, if X has an F distribution, then its reciprocal also has an F distribution with the degrees of freedom reversed. This property allows us to table only the upper percentage points of the F distribution. The lower tail can be found from the upper tail of the reciprocal of F.

Another important property of the F distribution is that both the chi-square and the t distribution are special cases of it. If

$$X \sim F_{(\nu,\infty)},$$

then

$$\nu X \sim \chi^2_{(\nu)}.$$

If

$$X \sim t_{(\nu)},$$

then

$$X^2 \sim F_{(1,\nu)}.$$

Consequently, any table of the F distribution also contains, implicitly, tables of the chi-square and t distributions.

The F distribution can theoretically be used to test the null hypothesis that two different populations have the same variance. Under that null hypothesis, and the assumption that the population distributions are normal, the ratio of the sample variances has the F distribution. However, the assumption of normal population distributions is critical to the test; if that assumption is violated, the test is not valid, no matter how large the samples may be. Because the populations can seldom safely be assumed to be precisely normal, the test is seldom valid. The real importance of the F distribution is the subject of the rest of this book.

NONCENTRAL F

Just as there is a noncentral version of the chi-square distribution, so there is a noncentral version of the F distribution. In fact, the two noncentral distributions are very closely related; noncentral F is defined like central F except that the variable in the numerator has a noncentral rather than a central chi-square distribution. (The variable in the denominator has a central chi-square distribution.) The noncentral F will be important for some of the power calculations discussed in later chapters.

TABLES AND INTERPOLATION

Appendix B contains tables of the normal, chi-square, t, and F distributions. Values that are not tabled can be found by special interpolation methods (simple linear interpolation usually does not work too well).

One easy and fairly accurate way to interpolate using graph paper is illustrated in Figure 1.3. Here we find the significance level of an F ratio obtained in Chapter 2. The obtained F is 4.40 with $\nu_1 = 2, \nu_2 = 12$. From the F table we find and plot all the tabled F ratios versus their significance levels for 2 and 12 degrees of freedom. We then draw a smooth curve through these points. Finally, we read from the graph that $F = 4.40$ has a significance level of about .04 (the correct value is .038).

The graph in Figure 1.3 is on semi-log paper. Ordinary graph paper will do, but the curve drawn on semi-log paper is usually closer to a straight line. (Generally, the straighter the curve, the easier it is to draw and read.) If obtainable, "normal-curve" paper (with values on the Y axis spaced so that the normal integral plots as a straight line) is usually even better. If both degrees of freedom are small, say, less than five, log-log paper may work best.

The fact that semi-log paper gives nearly straight-line plots suggests a way of interpolating algebraically when graph paper is unavailable or inconvenient. We interpolate linearly in the logarithms of the α values instead of in the α values themselves. For the problem above, we first find from the tables that for $F = 4.4, .05 > \alpha > .025$. Then we find the F values, $F_{.05(2,12)} = 3.885$, and $F_{.025(2,12)} = 5.096$. (In this notation, the first subscript is the α level.) Next, we solve for k in the equation

$$4.4 = 3.885k + 5.096(1 - k),$$

or $k = .575$. Finally, we calculate

$$\log(\alpha) = \log(.025) + .575[\log(.05) - \log(.025)] = -1.429.$$

Taking the antilogarithm, we get $\alpha \simeq .037$.

Accurate interpolation between tabled degrees of freedom is best accomplished by linear interpolation between their reciprocals. To facilitate this, degrees of freedom listed in the tables in Appendix B are mostly in multiples of either 60 or 120.

For example, if we want to find the F needed for $\alpha = .05$, with 3 and 14 degrees of freedom, we will find that values are listed for 12 and 15 degrees of freedom, but not for 14 degrees of freedom. We first solve for k in the equation

$$120/14 = (120/12)k + (120/15)(1 - k),$$

or $8.57 = 10k + 8(1 - k), k = .285$. Then we have $F_{\alpha(3,14)} = .285F_{\alpha(3,12)} + (1 - .285)F_{\alpha(3,15)}$. For $\alpha = .05, F_{.05(3,14)} = .285(3.490) + .715(3.287) \simeq 3.345$.

Tables of the noncentral F and chi-square distributions are not commonly found in statistics texts. Adequate tables of these distributions would be impractically large because of the number of parameters involved. We will not

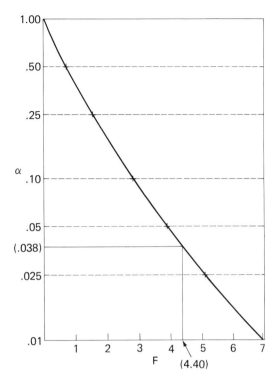

FIGURE 1.3. Interpolating in the F distribution.

use the noncentral chi-square distribution at all, and we will use the noncentral F only for calculating the power of statistical tests. For this purpose, the noncentral F distribution can be approximated by a central F distribution, and this approximation can be used to evaluate, approximately, its probabilities.

If we want the tail probability of a noncentral F with ν_1 and ν_2 degrees of freedom and noncentrality parameter ϕ, we calculate

$$F' = [\nu_1/(\nu_1 + \phi^2)]F,$$
$$\nu'_1 = (\nu_1 + \phi^2)^2/(\nu_1 + 2\phi^2),$$

and look up F' in an ordinary table of central F (e.g., Table A.5) with ν'_1 degrees of freedom in the numerator and ν_2 degrees of freedom in the denominator.

always an unbiased estimate of σ_e^2. The other, based on the differences *between the means* of the groups, is unbiased if and only if the null hypothesis is true. If the null hypotheses is true, these estimates should be approximately equal to each other, and their ratio should be approximately equal to one. If the null hypothesis is false, we can expect their ratio to be very different from one. When making an F test, therefore, we are in the unusual position of using the ratio of two variances to test for a difference between means.

Analysis: Equal Sample Sizes

For simplicity, we will first develop the F test for equal sample sizes; we will then generalize it to unequal sample sizes.

To see how the t test is extended to an F test comparing two variances, we must first digress to an estimation problem. Suppose we have a sample of only two scores (X_1 and X_2) from a normal population, and we wish to estimate the variance of that population. The mean of a sample of two is

$$\overline{X} = (X_1 + X_2)/2,$$

so the variance is

$$s^2 = \{[X_1 - (X_1 + X_2)/2]^2 + [X_2 - (X_1 + X_2)/2]^2\}/2.$$

With a little algebra, this can be shown to be

$$s^2 = (X_1 - X_2)^2/4. \tag{2.2}$$

From Equation 1.1, the unbiased estimate of the population variance is then

$$\hat{\sigma}^2 = 2s^2 = (X_1 - X_2)^2/2.$$

The formula for the t test with equal sample sizes can be written

$$t = [(\overline{X}_{1.} - \overline{X}_{2.})(n/2)^{1/2}/\hat{\sigma}_e,$$

where $\hat{\sigma}_e^2$ is an unbiased estimate of the variance of the two groups. If we square the formula for the numerator of the t ratio, assuming equal sample sizes, we get $(\overline{X}_{1.} - \overline{X}_{2.})^2(n/2)$, which is just n times the unbiased estimate of variance that we would get if we regarded $\overline{X}_{1.}$ and $\overline{X}_{2.}$ as single scores from the same population instead of as the sample means from two different populations. Under what circumstances can we assume that $\overline{X}_{1.}$ and $\overline{X}_{2.}$ are two scores from the same population? Consider the assumptions necessary for the t test:

$$\overline{X}_{1.} \sim N_{(\mu_1, \sigma_3^2/n)},$$
$$\overline{X}_{2.} \sim N_{(\mu_2, \sigma_e^2/n)}.$$

If $\mu_1 = \mu_2 = \mu$ (i.e., if the null hypothesis is true), then $\overline{X}_{1.}$ and $\overline{X}_{2.}$ are identically distributed, and for all practical purposes we can regard them as a sample of two scores from the same population. Furthermore, if the null hypothesis is true, then the square of the numerator of the t test is an estimate of n times the variance of $\overline{X}_{1.}$ and $\overline{X}_{2.}$. But the variance of $\overline{X}_{1.}$ and $\overline{X}_{2.}$ is σ_e^2/n, so the squared numerator of the t test is an unbiased estimate of σ_e^2.

The squared denominator of the t ratio is another estimate of σ_e^2, the average of the separate variance estimates from each sample. Thus, the t statistic squared is the ratio of two variance estimates; furthermore, this ratio has an F distribution with one degree of freedom in the numerator according to the discussion in Chapter 1.

The generalization to three or more groups is straightforward. We find two estimates of σ_e^2, the first being the average of the estimates from the individual groups and the second being based on the variance of the group means. Under the null hypothesis that the population means of the groups are all equal, the ratio of the two estimates will have an F distribution. In the next two sections we will find these two estimates of σ_e^2.

MEAN SQUARE WITHIN

For the first estimate, we consider each of the three samples separately. For each sample, we find the usual unbiased estimate of the population variance:

$$\hat{\sigma}_i^2 = [\Sigma_j(X_{ij} - \overline{X}_{i.})^2]/(n-1)$$

where $\overline{X}_{i.}$ is the mean of the sample of scores from group A_i, and n is the number of scores in each sample.

It can be shown that the best overall unbiased estimate of σ_e^2 in this situation is the average of the separate estimates obtained from the three samples:

$$MS_w = (1/I)\Sigma_i\hat{\sigma}_i^2. \tag{2.3}$$

The symbol MS_w refers to *mean square within*; it is so named because it is the mean of the squared errors within the samples.

We now consider how MS_w is distributed. According to Equation 1.2,

$$(n-1)\hat{\sigma}_i^2/\sigma_e^2 \sim \chi_{(n-1)}^2.$$

In addition, each $\hat{\sigma}_i^2$ is independent of the others because they are from independent random samples. Therefore, if we define

$$
\begin{aligned}
SS_w &= \Sigma_i(n-1)\hat{\sigma}_i^2 \\
&= (n-1)\Sigma_i\hat{\sigma}_i^2 \tag{2.4}\\
&= I(n-1)MS_w \quad \text{(from Eq. 1.3)}
\end{aligned}
$$

(SS_w refers to *sum of squares within*), then (from Eq. 1.2)

$$SS_w/\sigma_e^2 = [\Sigma_i(n-1)\hat{\sigma}_i^2]/\sigma_e^2$$

is a sum of chi-square variables, each with $(n-1)$ degrees of freedom. Consequently, from the theory given in Chapter 1,

$$SS_w/\sigma_e^2 \sim \chi^2_{(N-I)}.$$

Because MS_w is equal to SS_w divided by its degrees of freedom (see Eq. 2.4), it follows that MS_w/σ_e^2 is a chi-square variable divided by its degrees of freedom. (Remember from Chapter 1 that both the numerator and denominator of an F ratio consist of chi-square variables divided by their degrees of freedom.) MS_w will be the denominator of the F ratio.

MEAN SQUARE BETWEEN

According to normal distribution theory, the mean of each sample is normally distributed with a mean equal to the population mean and a variance equal to the population variance divided by n:

$$\overline{X}_{i.} \sim N_{(\mu_i, \sigma_e^2/n)}.$$

Thus, under the null hypothesis, for all groups, A_i

$$\overline{X}_{i.} \sim N_{(\mu, \sigma_e^2/n)}.$$

This means that all the $\overline{X}_{i.}$ are normally distributed with the same mean and variance. We can treat them in effect as if they themselves were a random sample of I scores from the same population. We can estimate the variance of this population by calculating the sample variance in the usual way. We must remember, however, that there are only I treatment groups, so there are only I scores in the sample. The estimate of the variance is

$$\hat{\sigma}_e^2/n = [\Sigma_i(\overline{X}_{i.} - \overline{X}_{..})^2]/(I-1)$$

where $\overline{X}_{..} = (1/N)\Sigma_i\Sigma_j X_{ij}$ is the grand mean of all of the scores.

Multiplying the above estimate by n gives us MS_{bet}, an unbiased estimate of σ_e^2:

$$MS_{bet} = n(\hat{\sigma}_e^2/n) = [n\Sigma_i(\overline{X}_{i.} - \overline{X}_{..})^2]/(I-1). \qquad (2.5)$$

Finally, just as we found it useful to define SS_w for calculational purposes, we shall define *sum of squares between* as

$$SS_{bet} = (I-1)MS_{bet} = n\Sigma_i(\overline{X}_{i.} - \overline{X}_{..})^2.$$

Because MS_{bet} is effectively a variance estimate from a sample of size I,

$$SS_{bet}/\sigma_e^2 = (I-1)MS_{bet}/\sigma_e^2 \sim \chi^2_{(I-1)},$$

and MS_{bet}/σ_e^2, like MS_w/σ_e^2, is a chi-square variable divided by its degrees of freedom.

MEAN SQUARE TOTAL

A third estimate also plays a role in later discussions. MS_{bet} is an estimate of σ_e^2 if and only if the null hypothesis is true. However, it is not the best estimate. The best estimate is found by treating the data as though they were all from the same population. We would then assume that the sample contained N scores, and the best estimate of σ_e^2 would be

$$MS_t = [\Sigma_i \Sigma_j (X_{ij} - \overline{X}_{..})^2]/(N-1).$$

The symbol MS_t refers to the *mean square total*, as it is based on the totality of the scores. We can now define the *sum of squares total*:

$$SS_t = (N-1)MS_t.$$

MS_t and SS_t are seldom used in practice to estimate σ_e^2 (they require that H_0 be true), but they play an important role in other calculations. For example, it is easy to show that

$$SS_t = SS_w + SS_{bet}.$$

THE F TEST

In general, MS_{bet} is the worst estimate of the three. It has two serious disadvantages. First, it is based only on the variability of the means of the samples, ignoring the variability of the scores within a sample. Therefore, all of the information that can be gained from the variability of the scores within each sample (the variability that is used explicitly in calculating MS_w) is lost when MS_{bet} is used as an estimate of σ_e^2. Second, the estimate requires H_0 to be true.

Nevertheless, these two disadvantages become advantages when we set out to test the null hypothesis stated earlier. Let us consider the ratio $F = MS_{bet}/MS_w$ under the assumptions we have made so far, including the null hypothesis (Eq. 2.1). Under these assumptions, MS_w and MS_{bet} are independent random variables. Their independence derives from the fact that MS_w depends only on the variances of the samples, whereas MS_{bet} depends only on their means. The mean and variance of a sample from a normal population are independent, so MS_w and MS_{bet} are also independent. Moreover, each is a chi-square variable divided by its degrees of freedom. Thus, their ratio, MS_{bet}/MS_w, has an F distribution with $(I-1)$ degrees of freedom in the numerator and $(N-I)$ degrees of freedom in the denominator. Notice, however, that MS_{bet} is a valid variance estimate only if the null hypothesis is true. Therefore, the ratio is distributed as F only if the null hypothesis is true. This suggests that we can test the null hypothesis by comparing the obtained ratio with that expected from the F distribution. The procedure is the same as in a t test, except that an F distribution is used instead of a t distribution. If the obtained ratio varies greatly from the expected ratio (which, for an F distribution, is approximately one because MS_{bet} and MS_w are unbiased estimates of the same quantity and therefore should be about equal), we can conclude that the null hypothesis is false. For our example, $F_{(2,12)} = 12.6/2.867 = 4.40$. According to the F table, the probability of obtaining a ratio this large or larger, with two and twelve degrees of freedom, is smaller than .05 but not as small as .025. By using the graphic method described in Chapter 1, we can determine that the p value is in fact about .037.

Before we can reject the null hypothesis with confidence, however, two important questions must be answered. This first is whether a one-tailed or two-tailed test is appropriate. If a one-tailed test is appropriate, we can reject the null hypothesis at any α level $\geq .037$. If it is not, we must consider the other tail of the distribution when calculating the p value.

The other question involves the importance of the assumptions for the F test. In addition to the null hypotheses that the population means are equal, the F test requires that all scores be sampled randomly and independently from normally distributed populations with the same variance, σ_e^2. If the obtained value of F deviates significantly from its expected value, we can conclude with some confidence that one of the assumptions is false. However, until we have examined the issues more carefully, we cannot confidently reject the null hypothesis. The large F ratio may be due to nonnormal populations, unequal population variances, or nonrandom sampling. The next two sections will take up these issues.

EXPECTED VALUES OF MEAN SQUARE WITHIN AND BETWEEN

The question of whether a one-tailed or two-tailed test is appropriate is most easily answered by considering the expected values of MS_w and MS_{bet}. Under the assumptions discussed above, including that of the null hypothesis, MS_w and MS_{bet} both have the same expected value, σ_e^2, because they are both unbiased estimates of the population variance. However, the expected value of MS_w is σ_e^2 even if the null hypothesis is false; based only on the variances of the samples, MS_w is independent of any assumptions we might make about the group means.

On the other hand, the expected value of MS_{bet} depends in a very definite way on the true means of the population from which the samples were drawn. To derive the expected value of MS_{bet}, we must first introduce the concept of a *linear model*, and we can best introduce the concept by deriving an appropriate linear model for the data under discussion.

Linear Models. Let us begin by letting μ_i represent the true mean of the ith treatment population. (Notice that μ_i represents the mean of the *population*, not the sample.) Then each score can be represented as

$$X_{ij} = \mu_i + \epsilon_{ij} \tag{2.6}$$

where

$$\epsilon_{ij} \sim N_{(0,\sigma_e^2)}.$$

This is just another way of stating the assumptions made at the beginning of this chapter, but it shows that each score can be thought of as a random deviation from the population mean. Notice, now, that in Equation 2.6, X_{ij} is expressed as the sum of two variables. Whenever an observed value is expressed as a linear combination (in this case, a simple sum) of two or more other variables, we have a *linear model*. Linear models will become increasingly important as we consider more and more complicated experimental designs. They provide us with a means of separating out the various influences affecting a given score. In this very simple case, for example, the linear model allows us to distinguish between the constant effect due to the group from which a score is drawn (μ_i) and a random component due to the random selection of the scores from that population (ϵ_{ij}).

The linear model of Equation 2.6 is easier to use if it is expanded slightly. To expand it, we first define

$$\mu = \Sigma_i \mu_i / I$$

and

$$\alpha_i = \mu_i - \mu. \tag{2.7}$$

That is, the *grand mean*, μ, is the average of the population means, and α_i is the difference between each group mean and the grand mean. The linear model then becomes

$$X_{ij} = \mu + \alpha_i + \epsilon_{ij} \tag{2.8}$$

where, because of the definition of the α_i in Equation 2.7,

$$\Sigma_i \alpha_i = 0.$$

Because $\mu_i = \mu + \alpha_i$ in terms of this model, the null hypothesis (Eq. 2.1) can be restated as

$$H_0 : \alpha_i = 0, \quad \text{for all} \quad i.$$

The quantities μ and α_i can be regarded as concrete though unobservable values. As concrete quantities, they can be estimated from the data. The best unbiased estimates of them are

$$\begin{aligned} \hat{\mu} &= \overline{X}_{..} \\ \hat{\alpha}_i &= \overline{X}_{i.} - \overline{X}_{...} \end{aligned}$$

Note that Equation 2.5 can now be written as

$$MS_{bet} = [n\Sigma_i \hat{\alpha}_i^2]/(I - 1).$$

The estimates ($\hat{\alpha}_i$) can help us interpret the differences among the groups. For example, for the data in Table 2.1,

$$\begin{aligned} \hat{\alpha}_1 &= 4.0 - 3.4 = 0.6, \\ \hat{\alpha}_2 &= 1.6 - 3.4 = -1.8, \\ \hat{\alpha}_3 &= 4.6 - 3.4 = 1.2. \end{aligned}$$

From these we see that the most deviant group is A_2, whose mean is much smaller than that of either A_1 or A_3. This deviation is the main contributor to the large MS_{bet}, and therefore to the significant F. Although the deviation of A_2 is only half again as great as that of A_3, the *squared* deviation is more than twice as large. Consequently, group A_2 contributes more than twice as much to MS_{bet} and F.

Expected Value of MS_{bet}. We can use Equation 2.8 to derive the expected value of MS_{bet}. The derivation is rather long and complicated; the final result is

$$\underline{E}(MS_{bet}) = \sigma_e^2 + n\Sigma_i\alpha_i^2/(I-1). \tag{2.9}$$

The first thing to note from this equation is that if the null hypothesis ($\alpha_i = 0$ for all i) is true, then $\underline{E}(MS_{bet}) = \sigma_e^2$, as we stated earlier. If the null hypothesis is false, a term must be added to σ_e^2 to find $\underline{E}(MS_{bet})$. Moreover, the added term will always be positive because it is a positive constant times the sum of the squares of the α_i. Therefore, if the null hypothesis is false, $\underline{E}(MS_{bet})$ will be larger than $\underline{E}(MS_w)$, and the F ratio will tend to be larger than one. An obtained F smaller than one cannot be due to a false null hypothesis; it must be due to the violation of one or more of the other assumptions discussed above, to calculating errors, or to pure chance. It may be due to any of these causes, but an F smaller than one cannot be grounds for rejecting the null hypothesis. Consequently, in an analysis of variance, the F test is always one-tailed. In our example, the actual obtained p value is .037, and we arrive at that figure without considering the other tail of the F distribution. However, we must still assure ourselves that this abnormally large value of F has been caused by the violation of the null hypothesis rather than by the violation of one of the other assumptions.

Robustness of F

None of the above assumptions are ever fully satisfied by real data. The underlying populations from which the samples are drawn are never exactly normally distributed with precisely equal variances. If the violation of these assumptions were likely to result in a large F, then the F test would be a poor test of the null hypothesis. We could never know whether a significant F justified rejection of the null hypothesis or was due to the violation of one of the other assumptions. A statistical test is *robust* if it is not greatly affected by such extraneous assumptions. We will examine the robustness of the F test with respect to each of the assumptions that must be made. The treatment here will necessarily be sketchy. A more thorough theoretical treatment is beyond the scope of this book (such a treatment can be found in Scheffé, 1959, Chapter 10).

Nonnormality

The most important departures from normality are likely to be either nonzero skewness or nonzero kurtosis (see Chapter 1 for formal definitions of skewness and kurtosis). These, of course, are not the only ways in which the distribution can be nonnormal, but other departures from normality are not likely to be very important. In general only the kurtosis of the distribution is likely to have any appreciable effect on F. Although MS_{bet} and MS_w are unbiased estimates of σ_e^2 no matter what the population distribution, their *variances* depend on the kurtosis of the population.

Fortunately, in most cases the effects of nonzero kurtosis can be ignored. Box and Anderson (1955) have shown that when the kurtosis is not zero, an approximate F test can still be used. In the approximate test, F is calculated in the usual way but with different degrees of freedom. The degrees of freedom are found by multiplying both the numerator and the denominator degrees of freedom, calculated in the usual way, by $1 + Ku/N$, where Ku is the kurtosis of the population. To illustrate this, consider again the example in Table 2.1. These data were actually taken from a rectangular distribution with $Ku = -1.2$. Consequently, we should have multiplied each of the degrees of freedom by $1 + (-1.2)/15 = .92$. Our new degrees of freedom would then have been (2) $(.92) \simeq 2$ and (12) $(.92) \simeq 11$. With two and eleven degrees of freedom, our p value would be about .039, which is very close to the actual value of .037. As can be seen from this example, moderate deviations from normality seldom have much effect on F.

Of course, the correction factor just discussed will have little value in practical applications because the kurtosis of the population will not be known. However, it can be used as a rough guide in estimating how large an effect nonnormality is likely to have. In most applications the kurtosis will fall somewhere between plus or minus two, although in some cases it may be as high as seven or eight (it is mathematically impossible for the kurtosis to be smaller than minus two). With such small values of kurtosis, even moderate sizes of N will usually make the correction factor so close to one that it can be ignored. Table 2.2 gives actual significance levels for a *nominal* significance level of .05 (i.e., the experimenter thinks the significance level is .05), assuming five groups of five scores each taken from populations with varying amounts of skewness and kurtosis. For most practical applications, deviations as small as those in Table 2.2 can be ignored.

We conclude, then, that the F test is robust with respect to nonnormality if N is large and is likely to be robust even if N is only moderately large.

Unequal Variances

The F test also tends to be robust with respect to the assumption of equality of variances so long as each group contains the same number of scores. The exact effects of unequal variances are difficult to evaluate. However, when the assumption of equal variances is violated, there is a tendency to reject the

TABLE 2.2. Probabilities of Type I error with the F test for the equality of means of five groups of five, each at the nominal five-percent significance level, approximated for the Pearson distribution by a correction on the degrees of freedom.

Sk^2	Ku				
	-1	-0.5	0	0.5	1
0	.053	.051	.050	.048	[a]
0.5	0.52	0.51	0.50	0.49	[a]
1	0.52	.050	.049	.048	.048

Source: Box and Andersen (1955), p. 14.
[a] The method of approximation used was unsatisfactory for these combinations of values.

hypothesis when it should be accepted. Table 2.3 shows the true significance levels associated with a nominal significance level of $\alpha = .05$ for some selected sample sizes and variance ratios. These data make it clear that some bias exists when the variances are unequal (usually the F ratio will not be quite as significant as it appears to be), but the bias is small if n is large compared to I and the variances are not too greatly different.

TESTS FOR EQUALITY OF VARIANCES

It is a common practice to test the assumption of equality of variances before calculating F. If the test for equality of variances is significant, the experimenter concludes that the F test should not be done; if it is not significant, the experimenter concludes that the F test is valid. This amounts to accepting the null hypothesis as true, and the appropriateness of accepting a null hypothesis, given nonsignificant statistical results (and an unknown value of β), has long been a subject of debate. Aside from this, however, there are other arguments against the commonly used methods of testing for equality of variances.

The three most commonly used tests are the Hartley, Cochran, and Bartlett tests. In the Hartley test the variance of each sample is calculated and the ratio of the largest to the smallest variance is compared with a set of tabled values. The Cochran test compares the largest sample variance with the average of all of the sample variances. The most sensitive test, the Bartlett test, compares the logarithm of the mean of the sample variances with the mean of their logarithms. All of these tests share a serious deficiency: they are all very sensitive to departures from normality in the population distribution. They all tend to mask existing differences in variance if the kurtosis is smaller than zero, or to exhibit nonexistent differences if the kurtosis is greater than zero.

TABLE 2.3. Effects of inequality of
variances on probability of Type I
error with F test for equality of
means at nominal .05 level.

I	n	Group variances	True α
2	7	1:2	.051
		1:5	.058
		1:10	.063
3	5	1:2:3	.056
		1:1:3	.059
5	5	1:1:1:1:3	.074
7	3	1:1:1:1:1:1:7	.12

Source: Combined data from Hsu
(1938) and Box (1954).

This deficiency is particularly serious because the F test itself is relatively insensitive to both nonnormality and inequality of variance when sample sizes are equal. If the population kurtosis is greater than zero, the experimenter may decide not to perform a perfectly valid F test. All three tests for inequality of variances are described in Winer (1971), with appropriate tables provided. However, I do not recommend their use.

Transformations on the Data. In some cases the variances of the treatment groups have a known relationship to the means of the groups. Such is the case, for example, when each score, X_{ij}, is a proportion. If the mean of a sample of proportions is close to one or zero, the variance will be small; if the mean is close to one-half, the variance will be larger. If the variances have a known relationship to the group means, it may be possible to transform the X_{ij} so that the variances of the transformed X_{ij} are independent of their means. When the X_{ij} are proportions, for example, the *arcsin transformation*, $Y_{ij} = \arcsin(X_{ij}^{1/2})$, is an appropriate transformation. If the standard deviations are approximately proportional to the means of the distributions, i.e., if the variances are approximately proportional to the squares of the means, then $Y_{ij} = \log(X_{ij})$ is appropriate. If the variances are proportional to the means of the distributions, then $Y_{ij} = 1/X_{ij}$ is appropriate. Still other transformations can be found for various other relationships between the means and variances. Moreover, these transformations often make the population distributions more nearly normal as well.

However, such transformations are usually not necessary with equal sample sizes. Furthermore, these transformations change the group means as well as the variances. When the means of the original scores are equal, the means of the transformed scores might not be, and vice versa. Often the means are changed in ways that are not intuitively meaningful. A statement about the mean of a sample of proportions is easily understood. A statement about the mean of the arcsins of the square roots of the proportions is more difficult to interpret.

However, sometimes a transformation is appropriate. For example, if each score (X_{ij}) is itself an estimate of a variance, one might be undecided whether to perform the F test on the variance estimates themselves or on their square roots, which would be estimates of standard deviations. An appropriate transformation might be $Y_{ij} = \log(X_{ij})$. This is especially appropriate when the scores are variances because the difference between the logarithm of σ^2 and the logarithm of σ (the estimated standard deviation) is only a multiplicative factor of two. Multiplication by two is a linear transformation, and linear transformations do not affect F; therefore, the F test on the logarithms of the estimated variances will be the same as the test on the logarithms of the estimated standard deviations. The test is essentially a test of the equality of the geometric means of the estimated variances (or, equivalently, of the estimated standard deviations). (The geometric mean is the antilog of the mean of the logarithms of the scores.) In this case the transformation has some intuitive justification.

A transformation may also be appropriate when the data are measured in time units, such as running times, reaction times, or number of trials required to learn a response. Such data are often highly skewed, with the variance increasing as a function of the mean. The reciprocals of the times can be interpreted as measures of speed, and they often (though not always) have more satisfactory statistical properties. A test on the reciprocals can be regarded as a test of the harmonic means of the original scores. (The harmonic mean is the reciprocal of the mean of the reciprocals of the scores.)

INDEPENDENCE OF SCORES

The F test clearly requires that each sample be random, with each score independent of all other scores. It might be assumed that a test that is robust with respect to the other assumptions will be robust with respect to this requirement as well. However, such is not the case. If the data are not independent random samples, the F ratio may be strongly affected. The direction of the effect depends on the nature of the dependence between the scores. In most cases that occur in practice, the dependencies probably tend to make MS_w smaller than it should be and F correspondingly too large (although the opposite can also occur). The obtained p-value will then be smaller than it should be. The only solution to this problem is to obtain the data in such a way that you can be reasonably certain the assumption of independence is justified.

POWER OF THE F TEST

We have found so far that the F test may be a useful test of the null hypothesis that the means of I independently sampled populations are all equal. In addition, we have found that the F test is relatively insensitive to departures from the normality and equal variance assumptions. The final problem is to determine just how sensitive it is to departures from the equality of means specified by the null hypothesis. If it is not sensitive to such departures, none of its other features can make it a useful test.

The sensitivity of F to the violation of the null hypothesis is called its *power*. The power of F is measured as the probability that the null hypothesis will be rejected when it is in fact false. In other words, the power is $1 - \beta$, where β is the probability of a Type II error.

The power of the F test depends on the distribution of the F ratio when the null hypothesis is false. This distribution is *noncental F*, and it depends importantly on four factors: the actual means of the treatment populations, the population variance, the significance level, and the number of scores in each sample. To calculate the power exactly, all four of these quantities must be known. Because the first two cannot be specified in practice, we cannot calculate the power of the F test exactly. However, some things are known. First, for many sets of population means, a given σ_e^2, and a given α, and under the assumptions used in computing F, the F test is the most powerful test one can make (a discussion of this can be found in Scheffé, 1959, Chapters 1,2). Also, it has been shown that when all groups have the same number of scores, the power of the F test is usually not greatly affected by violation of the assumptions of normality and equal variance. Thus it is a powerful practical test of the null hypothesis.

Finally, even though the power of the F test cannot be calculated exactly, it can sometimes be estimated. The estimate can then be used before the experiment to guess how large an n will be needed for a specified significance level and power. For example, suppose we are comparing five different treatment populations ($I = 5$), and we want to detect the difference between the groups if any two treatment means differ by more than six. Let us say that the significance level has been set at .05 and the desired power is .90. The population variance is estimated to be about 40. The number of scores needed per group is then to be estimated.

The effects of σ_e^2 and the population means on the power of F can be summarized in a single parameter:

$$(\phi')^2 = \Sigma_i(\mu_i - \mu)^2/(I\sigma_e^2)$$

$$= (\Sigma_i\alpha_i^2)/(I\sigma_e^2)$$

(2.10)

where α_i, in this case, refers to the α_i in the linear model of Equations 2.7 and 2.8 rather than the α level of significance. (Unfortunately, it is common practice to use the symbol α to signify both. In this text, when the symbol α

has *no subscript*, it will always refer to the α level of significance. The symbol α_i, *with a subscript*, will refer to the values in the linear model of Equation 2.8.) By comparing the expression for $(\phi')^2$ with the first term to the right of the equality in Equation 2.9, we can see why Equation 2.10 is useful.

Feldt and Mahmoud (1958) have provided charts giving the sample size needed for various values of I, α, ϕ', and $1 - \beta$.

An alternative way to calculate power is to use the noncentral F distribution directly. This can be done from tables or by the approximate method, described in Chapter 1, for calculating F. From Eq. 2.9 and the fact that $\underline{E}(MS_{bet}) = \nu + \phi^2$, we conclude that

$$\phi^2 = n(\Sigma_i \alpha_i^2)/\sigma_e^2 = nI(\phi')^2.$$

We cannot use this formula to solve for n directly, but we can use it to test various possible values of N. Suppose we guess that we need an n of about 35 for our example. We can use the formula to decide whether $n = 35$ is appropriate. We must first find the smallest value of $(\phi')^2$ that can occur when two means differ from each other by six. Because $(\phi')^2$ depends on the squares of the α_i, this smallest value will occur when one of the α_i is equal to -3, another is equal to +3, and all other α_i are equal to zero. We can then calculate $(\phi')^2$ as

$$(\phi')^2 = [(3)^2 + (-3)^2]/[(5)(40)] = 0.09.$$

In general, the minimum $(\phi')^2$ for a given maximum difference between treatment means is found by setting one α_i value to $+1/2$ the specified difference, another α_1 to $-1/2$ the difference, and all remaining α_i to zero.

For the example, $\phi^2 = (35)(5)(.09) = 15.75$ and $\nu_1 = 4$. With $n = 35$,

$$\nu_2 = 170, F_{.05(4,170)} = 2.425, F' = .4911, \nu' = 11,$$
$$\beta = Pr[F^{(11,17\text{o})} < .4911] = .09,$$

and the power is $1 - \beta = .91$, which is approximately .90. (To find β from the F table we have to invert it, obtaining $1/.4911 = 2.036$, and look that value up with $\nu_1 = 170, \nu_2 = 11$.) If the power had been much larger than .90, we would have done the calculation again with a smaller n; if it has been much smaller than .90, we would have tried a larger n. This is only an approximate sample size. However, given the other uncertainties involved in calculating power, the approximation is usually adequate for practical purposes.

Calculating Formulas

Having established the F test as an appropriate and useful test of the null hypothesis that the means of several treatment groups are equal, and having examined some of the properties of this test, we are now ready to consider the best ways of calculating F for a given set of data.

TABLE 2.4. Summary of calculations for one-way F test, equal ns; $t_{i.} = \Sigma_j X_{ij}$; $T = \Sigma_i t_{i.} = \Sigma_i \Sigma_j X_{ij}$.

	RS	SS	df	MS	F
m		T^2/N			
bet	$(1/n)\Sigma_i t_{i.}^2$	$RS_{bet} - SS_m$	$I - 1$	$SS_{bet}/(I-1)$	MS_{bet}/MS_w
w	$\Sigma_i\Sigma_j X_{ij}^2$	$RS_w - RS_{bet}$	$N - I$	$SS_w/(N-I)$	
t		$RS_w - SS_m$	$N - 1$		

It may be noticed in Table 2.1 that

$$SS_t = SS_w + SS_{bet}.$$

It can be shown that this is true in general. In fact, this is a special case (for the one-way design) of an important general principle in analysis of variance. Every analysis of variance divides SS_t into two or more independent sums of squares whose total is SS_t. Since the sums of squares into which SS_t is divided are independent, the mean squares of some of them can be the denominators of F ratios used to test others; in the one-way design, MS_w is the denominator for testing MS_{bet}.

The calculation of SS_{bet}, SS_w, and SS_t can be simplified by introducing the following terms:

$$
\begin{aligned}
t_{i.} &= n\overline{X}_{i.} = \Sigma_j X_{ij}, \\
T &= \Sigma_i t_{i.} = N\overline{X}_{..} = \Sigma_i \Sigma_j X_{ij}, \\
SS_m &= T^2/N, \\
RS_{bet} &= \Sigma_i t_{i.}^2/n, \\
RS_w &= \Sigma_i \Sigma_j X_{ij}^2.
\end{aligned}
$$

Then

$$
\begin{aligned}
SS_{bet} &= RS_{bet} - SS_m, \\
SS_w &= RS_w - SS_{bet} - SS_m = RS_w - RS_{bet}, \\
SS_t &= S_w + SS_{bet}.
\end{aligned}
$$

Each $t_{i.}$ is the total of the scores in group A_i, and T is the grand total of all of the scores in the experiment. The term RS stands for *raw sum of squares*; it refers to a sum of squares based on the squared raw data values rather than on the squares of the deviations of the values from their means. The general procedure is to first calculate the $t_{i.}$ and T. These values are then used to calculate SS_m (*sum of squares for the grand mean, RS_{bet}, and RS_w*). The calculation of the sums of squares follows. Table 2.4 summarizes this procedure. Table 2.5 illustrates the calculations for the data of Table 2.1.

TABLE 2.5. Illustration of calculations on data of Table 2.1; $t_{1.} = 20$; $t_{2.} = 8$; $t_{3.} = 23$; $T = 51$.

	RS	SS	df	MS	F	p
m		$(51)^2/15 = 173.4$				
bet	$(1/5)(20^2 + 8^2 + 23^2) = 198.6$	$198.6 - 173.4 = 25.2$	2	$25.2/2 = 12.6$	$12.6/2.867 = 4.40$	.037
w	233	$233 - 198.6 = 34.4$	12	$34.4/12 = 2.867$		
t		$233 - 173.4 = 59.6$	14			

It is useful to note two patterns in the above calculations. First, the three terms SS_m, SS_{bet}, and SS_w form a kind of hierarchy, based on the number of scores in the sum used to calculate each. SS_m is at the top of the hierarchy because it is based on T, the sum of all of the scores; next is SS_{bet}, because it is based on the t_i, each of which is the sum of the scores in a given group; SS_w is at the bottom because it is based on the raw scores, that is, they are not summed at all. Each sum of squares is found by first calculating its associated raw sum of squares, and then subtracting the sums of squares for all terms higher in the hierarchy. No subtraction is needed to find SS_m (i.e., $SS_m = RS_M$) because it is at the top.

Second, note that each raw sum of squares is found by squaring the appropriate totals (or raw scores in the case of RS_w), adding them up (if there is more than one), and then dividing by the number of scores that entered into each total. Both of these principles will be generalized to higher-way designs in later chapters.

Unequal Sample Sizes

So far, we have assumed that the same number of scores was sampled from each group. In general, it is better if the number of scores in each group is the same (for reasons that will be evident later). However, this is not always possible. Circumstances beyond our control may dictate different sample sizes in different groups. In an experiment like that in Table 2.1, for example, we may choose five students from each field of study, but some of the students may not show up to take the test. If one student failed to show up in group A_1 and two failed to show up in group A_3, the data might look like those in Table 2.6.

Of course, one might then question the validity of the samples because those who showed up may be systematically different from those who did not. However, we cannot always meet the assumptions exactly. Frequently, we have to use judgment to decide whether the departures from the assumptions are large enough to make a difference. We will assume for the purposes of the example that the effects of the departures are small.

To derive the F test with unequal sample sizes, we have to change our notation slightly. As before, we will assume that I different groups are being compared, but the number of scores in each sample will be n_i, with the subscript denoting the fact that the sample sizes are different for the different groups.

In this case, the formulas for SS_{bet} and SS_w become

$$SS_{bet} = \Sigma_i n_i (\overline{X}_{i.} - \overline{X}_{..})^2,$$

$$(2.11)$$

$$SS_w = \Sigma_i (n_i - 1)\hat{\sigma}_i^2.$$

These are simply generalizations of the formulas for equal sample sizes.

TABLE 2.6. Hypothetical data for independent samples from three groups with unequal numbers of scores.

	A_1	A_2	A_3	Sum
	3	1	5	
	5	2	5	
	6	2	2	
	1	0		
		3		
ΣX_{ij}	15	8	12	35
$\overline{X}_{i.}$	3.75	1.6	4.0	2.92
$n_i \overline{X}_{i.}^2$	56.25	12.80	48.00	117.05
ΣX_{ij}^2	71	18	54	143
$(n_i - 1)\hat{\sigma}_i^2$	14.75	5.20	6.00	25.95
$\hat{\sigma}_i^2$	4.92	1.30	3.00	
$\hat{\sigma}_i$	2.22	1.14	1.73	

	SS	df	MS	F	p
bet	14.97	2	7.483	2.60	.12
w	25.95	9	2.883		
t	40.92	11			

COMPUTATIONAL FORMULAS

The computational procedure with unequal sample sizes is the same as with equal sample sizes, except for the calculation of RS_{bet}. When calculating RS_{bet}, each t_i^2 is divided by its associated n_i, and the resulting ratios are then summed. The entire procedure is summarized in Table 2.7; the calculation for the data in Table 2.6 are shown in Table 2.8.

POWER

As with equal sample sizes, the distribution of the F ratio is noncentral F when the null hypothesis is false. However, then we cannot evaluate the power in terms of a single value of n. Instead of $(\phi')^2$, we must use the general noncentrality parameter $\phi^2 = [(\Sigma_i n_i \alpha_i^2)/(\sigma_e^2)]$.

Power calculations are seldom performed for designs with unequal sample sizes because there is a different n_i for each sample. This complicates the calculations, and we seldom plan to have unequal ns. However, they may occasionally be useful. The noncentral F can then be evaluated from tables, more elaborate charts (e.g., Scheffé, 1959), or by the approximate methods described above.

TABLE 2.7. Summary of calculations in one-way F test, unequal $ns; t_{i.} = \Sigma_j X_{ij}; T = \Sigma_i t_{i.} = \Sigma_i \Sigma_j X_{ij}$.

	RS	SS	df	MS	F
m		T^2/N			
bet	$\Sigma_i t_i^2/n_i$	$RS_{bet} - SS_m$	$I - 1$	$SS_{bet}/(I - 1)$	MS_{bet}/MS_w
w	$\Sigma_i \Sigma_j X_{ij}^2$	$RS_w - RS_{bet}$	$N - I$	$SS_w/(N - I)$	
t		$RS_w - SS_m$	$N - 1$		

ROBUSTNESS

Previously, we found that when sample sizes were equal, the F test was robust with respect to nonnormality and unequal variances. Although very little has been done to assess the effect of nonnormality with unequal sample sizes, the results appear to indicate that the effect of nonzero skewness is again negligible, and the effect of nonzero kurtosis, although slightly greater than with equal sample sizes, is still too small to be of concern in most applications.

However, the results are very different with respect to inequality of variances. Then the effect may be quite large; the direction of the effect depends on the sizes of the variances in the larger samples as compared to those in the smaller samples. This is true because, as Equation 2.11 indicates, MS_w is a weighted average of the sample variances, with the weights being equal to $(n_i - 1)$. Greater weight is placed on the variances of the larger samples, so MS_w will tend to be larger if the larger samples are from the populations with the larger variances. Because MS_w is in the denominator of F, a large MS_w will tend to result in a small F, and vice versa. Accordingly, if the larger samples are taken from the populations with the larger variances, MS_w will tend to be too large and F will tend to be too small. If, on the other hand, the larger samples are from populations with smaller variances, F will tend to be too large. Table 2.9 shows the effects of inequality of variances on the significance level of the F test on two treatment groups when the samples are large. Table 2.10 and Table 2.11 show the same results for selected values of I and n_i and for selected variances.

It is clear from these data that relatively small departures from the assumption of equal variances may have a sizable effect if the sample sizes are unequal. Moreover, the effect is not greatly reduced by increasing the sizes of the samples, so long as the ratios of the sample sizes remain the same. The safest procedure is to use equal, or nearly equal, sample sizes whenever possible.

Estimating the Effect of Unequal Variances. A rough index of the effect of unequal variances on the distribution of F can be found by noting that in this case MS_w is a weighted average of the unbiased variance estimates $(\hat{\sigma}_i^2)$ for the individual groups (Eq. 2.11). If we let σ_i^2 be the "true" variance of group

TABLE 2.8. Illustration of calculations for one-way F test (using data of Table 2.6); $t_1. = 15; t_2 = 8; t_3. - 12; T = 35$.

	RS	SS	df	MS	F	p
m		$35^2/12 = 102.08$				
bet	$15^2/4 + 8^2/5 + 12^2/3 = 117.05$	$117.05 - 102.08 = 14.97$	2	$14.97/2 = 7.483$	$7.483/2.883$ $= 2.60$	.12
w	143	$143 - 117.05 = 25.95$	9	$25.95/9 -= 2.883$		
t		$143 - 102.08 = 40.92$	11			

TABLE 2.9. Effect of unequal variances and unequal group sizes on true probability of Type I error for F test at nominal .05 level on two samples with large N.

n_1/n_2		0^a	0.2	0.5	σ_1^2/σ_2^2 1	2	5	∞^a
1	$E(F)$	1.00	1.00	1.00	1.00	1.00	1.00	1.00
	α	.050	.050	.050	.050	.050	.050	.050
2	$E(F)$	2.00	1.57	1.25	1.00	0.80	0.64	0.50
	α	.17	.12	.080	.050	.029	.014	.006
5	$E(F)$	5.00	2.60	1.57	1.00	0.64	0.38	0.20
	α	.38	.22	.12	.050	.014	.002	10^{-5}
∞	$E(F)$	∞	5.00	2.00	1.00	0.50	0.20	0.00
	α	1.00	.38	.17	.050	.006	10^{-5}	.00

Source: α values taken from Scheffe, 1959, p. 340.
^{a}Unattainable limiting cases to show bounds.

i, and

$$\sigma_w^2 = [\Sigma_i(n_i - 1)\sigma_i^2]/(N - I),$$

so that σ_w^2 is a weighted average of the true group variances, then MS_w is an unbiased estimate of σ_w^2. That is, $\underline{E}(MS_w) = \sigma_w^2$.

Similarly, we can let

$$\sigma_u^2 = [\Sigma_I \sigma_i^2]/I$$

so that σ_u^2 is the unweighted average of the group variances. It can be shown that

$$\underline{E}(MS_{bet}) = \sigma_w^2 + [(N - 1)/N][I/(I - 1)](\sigma_u^2 - \sigma_w^2). \qquad (2.12)$$

Combining these and applying some algebra, we find that

$$\underline{E}(F) \simeq \underline{E}(MS_{bet})/\underline{E}(MS_w)$$

$$\qquad (2.13)$$

$$= 1 + [(N - 1)/N][I/(I - 1)][(\sigma_u^2/\sigma_w^2) - 1].$$

We can estimate $\underline{E}(F)$ by estimating σ_w^2 and σ_u^2 and inserting these estimates into Equation 2.13. We already know that MS_w is an unbiased estimate of

TABLE 2.10. Effect of unequal variances on probability of Type I error with F test at nominal .05 level for two samples of selected sizes.

n_1	n_2		0^a	0.1	0.2	0.5	σ_1^2/σ_2^2 1	2	5	10	∞^a
15	5	$\underline{E}(F)$	3.38	2.58	2.12	1.43	1.00	0.70	0.49	0.41	0.32
		α	.32	.23	.18	.098	.050	.025	.008	.005	.002
5	3	$\underline{E}(F)$	1.88	1.66	1.50	1.22	1.00	0.82	0.68	0.62	0.56
		α	.22	.14	.10	.072	.050	.038	.031	.030	.031
7	7	$\underline{E}(F)$	1.00	1.00	1.00	1.00	1.00	1.00	1.00	1.00	1.00
		α	.072	.070	.063	.058	.050	.051	.058	.063	.072

Source: α values from Hsu, 1938.
[a] Unattainable limiting cases to show bounds.

TABLE 2.11. Effect of inequality of variances on probability of Type I error with F test at nominal .05 level.

I	σ_i^2	n_i	N	$\underline{E}(F)$	α
3	1, 2, 3	5, 5, 5	15	1.00	.056
		3, 9, 3	15	1.00	.056
		7, 5, 3	15	1.20	.092
		3, 5, 7	15	0.80	.040
3	1, 1, 3	5, 5, 5	15	1.00	.059
		7, 5, 3	15	1.35	.11
		9, 5, 1	15	1.93	.17
		1, 5, 9	15	0.60	.013
5	1, 1, 1, 1, 3	5, 5, 5, 5, 5	25	1.00	.07
		9, 5, 5, 5, 1	25	1.28	.14
		1, 5, 5, 5, 9	25	0.73	.02
7	1, 1, 1, 1, 1, 1, 7	3, 3, 3, 3, 3, 3, 3	21	1.00	.12

Source: α values from Box, 1954.

σ_w^2, and it is easy to show that

$$MS_u = (1/I)\Sigma_i \hat{\sigma}_i^2$$

is an unbiased estimate of σ_u^2. Inserting these values into Equation 2.13, we get

$$\hat{\lambda} = 1 + [(N-1)/N][I/(I-1)][(MS_u/MS_w)-1],$$

where $\hat{\lambda}$ is an approximately unbiased estimate of $\underline{E}(F)$ under the null hypothesis. This estimate is independent of the validity of the null hypothesis itself because it is based only on the variances of the individual groups and not on their means. In general, then, if $\hat{\lambda}$ is close to one, the F test is probably not greatly affected by possibly unequal variances. If it is larger than one, the obtained value of F is likely to be too large, so the obtained p value will be smaller than it should be. If it is smaller than one, the obtained F is likely to be too small, and the obtained p value will be too large. The approximate values of $\underline{E}(F)$ (Eq. 2.13) under the null hypothesis are included in Tables 2.9, 2.10, and 2.11.

An Approximate F Test. Just as there is an approximate t test when variances are unequal, so there is an approximate F test. However the approximate F test is less well studied than the approximate t; it is probably not advisable to use the approximate F unless the degrees of freedom are very large. Nevertheless, a comparison between the approximate F and the standard F can give an indication of the effect of unequal variances on the standard F.

The approximate F is

$$F^* = N(I - 1(MS_{bet}/[\Sigma_i(N-n_i)\hat{\sigma}_i^2] = F/\hat{\lambda}.$$

The degrees of freedom must be estimated; the estimates are

$$
\begin{aligned}
\hat{\nu}_1 &= [\Sigma_i(N-n_i)\hat{\sigma}_i^2]^2/[N\Sigma_i(N-2n_i)\hat{\sigma}_i^4 + (\Sigma_i n_i \hat{\sigma}_i^2)^2], \\
\hat{\nu}_2 &= [\Sigma_i(n_i-1)\hat{\sigma}_i^2]^2/[\Sigma_i(n_i-1)\hat{\sigma}_i^4].
\end{aligned}
$$

This approximate test can be used with either equal or unequal sample sizes. With equal sample sizes, $F^* = F$, but the degrees of freedom will be different. However, again the test should be used mainly as an indicator of the effect of unequal variances—the results of the test should not be taken seriously unless the sample sizes are large.

For the example in Table 2.1, $F^* = F = 4.40, \hat{\nu}_1 = 1.8, \hat{\nu}_2 = 10, p = .044$, as compared with the previously obtained $p = .037$. This suggests that unequal variances have only a small effect on the F test for these data.

For the data in Table 2.6, $F^* = .917F = 2.38, \hat{\nu}_1 = 1.8, \hat{\nu}_2 = 6.9 \simeq 7, p = .17$, as compared with the previously obtained $p = .13$. This suggests that the unequal variances have a larger effect for these data (as may be expected with unequal sample sizes), but even here there is little effect on the final conclusion.

Other Possible F Tests

You may have noticed the degrees of freedom in the F test add up to one less than the number of scores—one degree of freedom has been lost somewhere. The usual explanation is that this degree of freedom is lost in estimating the grand mean. In reality, it is not lost; it is simply not used. It can be used, however, and sometimes its use is appropriate. Consider the quantity SS_m, defined earlier:

$$SS_m = T^2/N.$$

To begin with, T is the sum of N normally distributed random variables, so it is distributed as

$$T \sim N_{(N\mu, N\sigma_e^2)}, \qquad (2.14)$$

where

$$\mu = (1/N)\Sigma_i n_i \mu_i.$$

That is, μ is the weighted average of the μ_i.

Now, consider the null hypothesis

$$H_0 : \mu = 0. \qquad (2.15)$$

According to Equation 2.14, this null hypothesis is equivalent to

$$H_0 : \underline{E}(T) = 0.$$

But this, combined with Equation 2.14, implies that

$$T^2/(N\sigma_e^2) \sim \chi_{(1)}^2$$

because it is the square of a normally distributed random variable with a mean of zero and a variance of one. Consequently, SS_m/σ_e^2 has a chi-square distribution.

Furthermore, because SS_m has only one degree of freedom, we can define

$$MS_m = SS_m/1 = SS_m,$$

and MS_m/σ_e^2 is a chi-square variable divided by its degrees of freedom. Finally, MS_m is independent of MS_w for the same reason that MS_{bet} is independent of MS_w, so under the null hypothesis (Eq. 2.15),

$$MS_m/MS_w \sim F_{(1, N-I)}.$$

This F ratio uses the "lost" degree of freedom to test the null hypothesis that the population grand mean is zero (Eq. 2.15). It should be pointed out that MS_m and MS_{bet} are also independent of each other, because MS_m is based only on the average of the sample means and MS_{bet} is based on their variance. We say that the two null hypotheses (Eqs. 2.1 and 2.15) are *independent* or *orthogonal* null hypotheses. Note, however, that both F tests use

MS_w in the denominator, so the tests of these hypotheses are not entirely independent even though the hypotheses are. (The tests are nearly independent if N is large.) If, by chance, MS_w is an underestimate or an overestimate of σ_e^2, both F tests will be affected in the same way.

The above discussions illustrate an important general principle in the analysis of variance. We noted earlier that $SS_{bet} + SS_w = SS_t$. We now note that the degrees of freedom for these two sums of squares total $N - 1$. When we analyze more complicated designs, we will find that we always begin by dividing SS_t into a number of independent sums of squares—some of which are used in the numerators of F ratios and others of which are used in the denominators—and that we correspondingly divide the total N among them, with 1 degree of freedom left over for testing the grand mean.

The test of the grand mean can be modified somewhat if we wish to test a more general null hypothesis:

$$H_0 : \mu = \mu^*,$$

where μ^* is any value specified by the experimenter. By a derivation like that above, we can calculate

$$
\begin{aligned}
MS_m^* &= SS_m^* = (T - N\mu^*)^2/N, \\
F_{(1,N-I)} &= MS_m^*/MS_w
\end{aligned}
$$

to test this null hypothesis.

The original F test on the differences between groups can also be modified in various ways. For example, to test the null hypothesis that each group has a specified population mean, μ_i^*, we calculate

$$
\begin{aligned}
SS_{bet}^* &= \Sigma_i n_i (\overline{X}_{i.} - \mu_i^*)^2, \\
MS_{bet}^* &= SS_{bet}^*/I, \\
F_{(I,N-I)} &= MS_{bet}^*/MS_w.
\end{aligned}
$$

Notice here that there are I degrees of freedom rather than $I - 1$. This test is not independent of the test of the grand mean. Notice also that if $\mu_i^* = 0$ for all i, then $SS_{bet}^* = SS_{bet} + SS_m$. This is a single test of two null hypotheses, $\mu = 0$ and $\alpha_i = 0$ for all i; accordingly it is tested by adding the sums of squares for these two tests.

Tests such as these are not often used in practice. Nevertheless, they illustrate how the degrees of freedom may be used in different ways. The next chapter will illustrate more imaginative ways of using the available degrees of freedom to make significance tests.

Proportion of Variance Accounted for

The F test really tells us very little about the differences among the groups. If it is significant, we can be reasonably confident that some differences exist between the population means of the I groups being compared. We cannot tell, however, what the differences are or whether they are large enough to be important. In this section we will derive a measure of the importance of the differences: in Chapter 3 we will develop ways of testing for specific expected differences among the means. In Chapter 4 we will discuss ways of exploring the data for differences that may not have been predicted in advance.

The question, "Are the differences large enough to be important?" can be partially answered by estimating ω_{bet}^2, the *proportion of variance accounted for* by the between variance. If ω_{bet}^2 is small, the differences between the group means are small relative to the variability of the scores within each group. Then the differences, even though they may be significant, may not be practically or scientifically important. A large value of ω_{bet}^2 implies that the differences between the groups are large relative to the variability within groups, and the overlap between the scores in the different groups is small.

To derive ω_{bet}^2, we first consider the entire population of scores from which the I groups have been drawn. That is, we think of the I different populations as combined into one large population, and we calculate the mean and variance of this composite population.

To simplify the problem somewhat, we will assume that the proportion of scores from a given group, A_i, in the population, is n_i/N, the proportion of scores that are in group A_i in the sample. Then it is easy to prove that for the *combined* population

$$\underline{E}(X_{ij}) = (1/N)\Sigma_i n_I \mu_i = \mu,$$

$$\underline{E}(X_{ij} - \mu)^2 = \underline{E}(X_{ij} - \mu_i)^2 + \underline{E}(\mu_i - \mu)^2. \tag{2.16}$$

The term $\underline{E}(\mu_i - \mu)^2$ needs some explanation because in ordinary applications of statistics the μ_i are not random variables. Remember that we randomly selected a group and then randomly selected a score from that group. Randomly selecting a group is the same as randomly selecting a value of μ_i, i.e., the population mean of the group selected. In this sense, μ_i and, consequently, $(\mu_i - \mu)$ can be regarded as random variables, and their means and variances calculated.

If we let σ_e^2 be the variance of scores within a given group, and

$$\sigma_{bet}^2 = \underline{E}(\mu_i - \mu)^2 = (1/N)\Sigma_i n_i(\mu_i - \mu)^2, \tag{2.17}$$

then, by Equation 2.16,

$$\sigma_t^2 = \underline{E}(X_{ij} - \mu)^2 = \sigma_e^2 + \sigma_{bet}^2. \tag{2.18}$$

The total variance of the scores is the sum of two variances: the variance of the *scores within each group* and the variance of the *group means about the grand mean*. If the latter is large with respect to the former, there will be relatively little overlap between the scores in the I groups. If it is small, the variability of the group means will be small compared to the random variation of the scores within each group. The ratio

$$\omega_{bet}^2 = \sigma_{bet}^2 / \sigma_t^2$$

is the proportion of the total variance that can be "accounted for" by differences between the means of the I groups.

Note that ω_{bet}^2 is a different kind of quantity from F. The quantity F is a measure of the statistical significance of the differences; the quantity ω_{bet}^2 is a measure of their practical or scientific importance. Any difference between the means, no matter how small, will be significant if the N is large enough, while an important difference may be nonsignificant if the N is small. Conversely, ω_{bet}^2, as a population parameter, is unaffected by the size of N.

CONCEPT OF ANALYSIS OF VARIANCE

The above discussion illustrates another important concept, from which the term "analysis of variance" is actually derived. We have just divided or *analyzed* (the classical definition of this term is division into parts; it is the opposite of *synthesize*) the total variance, σ_t^2, into two separate variances, σ_e^2, representing variation within groups, and σ_{bet}^2, representing variation between groups. This division corresponds to the division of SS_t into SS_w and SS_{bet}, and the corresponding division of the degrees of freedom.

This "analysis" of the total variance will be carried further in later discussions. Then σ_t^2 will be analyzed into more than two component variances, with a corresponding division of SS_t and the degrees of freedom.

ESTIMATING ω_{bet}^2

The exact value of ω_{bet}^2 depends on the population values, σ_{bet}^2 and σ_t^2. Moreover, no unbiased estimate exists.

An approximately unbiased estimate can be obtained by finding unbiased estimates of σ_{bet}^2 and σ_t^2 and taking their ratio. From Equations 2.9 and 2.18 we can see that

$$\underline{E}(MS_{bet}) = \sigma_e^2 + [N/(I-1)]\sigma_{bet}^2 \tag{2.19}$$

for equal sample sizes. A similar derivation from Equation 2.12 shows that this is also true for unequal sample sizes. We can rewrite Equation 2.19 as

$$\underline{E}(SS_{bet}) = (I-1)\sigma_e^2 + N\sigma_{bet}^2.$$

Because MS_w is an unbiased estimate of σ_e^2, a little algebra reveals that

$$\hat{\sigma}_{bet}^2 = (1/N)[SS_{bet} - (I-1)MS_w] \tag{2.20}$$

is an unbiased estimate of σ_{bet}^2. By Equation 2.18 and the fact that MS_w is an unbiased estimate of σ_e^2, it follows that

$$\hat{\sigma}_t^2 = \hat{\sigma}_{bet}^2 + MS_w = (1/N)(SS_t + MS_w) \tag{2.21}$$

is an unbiased estimate of σ_t^2. The last equality follows from some algebra plus the fact (noted earlier) that $SS_{bet} + SS_w = SS_t$.

The ratio of Equation 2.20 to Equation 2.21 is

$$
\begin{aligned}
\hat{\omega}_{bet}^2 &= \hat{\sigma}_{bet}^2/\hat{\sigma}_t^2 \\
&= [SS_{bet} - (I-1)MS_w]/(SS_t + MS_w) \\
&= (F-1)/[F - 1 + N/(I-1)].
\end{aligned}
\tag{2.22}
$$

This is not an unbiased estimate of ω_{bet}^2. It tends to underestimate ω_{bet}^2, but with large samples the bias is small. For the data in Table 2.1,

$$
\hat{\omega}_{bet}^2 = [25.2 - (2)(2.867)]/(59.6 + 2.867) = .31.
$$

For the data in Table 2.6, $\hat{\omega}_{bet}^2 = .21$.

This estimate is used in many standard computer programs. It has one minor disadvantage—although ω_{bet}^2 can never be negative, $\hat{\omega}_{bet}^2$ can. In fact, whenever $F < 1, \hat{\omega}_{bet}^2$ will be negative. When that happens, you may either report that $\hat{\omega}_{bet}^2 = 0$ or report the negative estimate, assuming that the reader can interpret the estimate correctly. In this text negative estimates will be enclosed in parentheses to indicate that they are not true estimates.

Exercises

(1.) An experimenter studying the effects of drugs on behavior of mental patients divided 30 patients randomly into six groups of five patients each. The control group (A_1) received no drugs, four experimental groups (A_2 through A_5) each received a different type of tranquilizer, and one group (A_6) received a new "psychic energizer." The data were before and after differences on a specially designed personality inventory (positive scores indicate improvement). The data are as follows:

A_1	A_2	A_3	A_4	A_5	A_6
1.3	2.5	1.0	0.5	2.7	2.2
2.5	-1.1	-2.9	-1.0	2.0	3.7
-1.0	-0.2	-1.1	3.3	3.3	2.9
0.4	1.7	-1.1	1.2	2.3	3.2
2.8	2.1	1.6	3.0	3.2	3.0

(a.) Do an analysis of variance on the above data, finding the p value and $\hat{\omega}^2_{bet}$.

(b.) Interpret the results: Which groups appear to have the smallest scores, which the largest?

(c.) Estimate the variance and standard deviation in each group and comment on the validity of the assumption of equal variances.

(d.) Is this F test robust to the assumption of equal variances? Explain.

(e.) Test $H_0 : \mu = 0$, finding the p value and $\hat{\omega}^2_{bet}$.

(f.) Estimate the power of the test in part e against $H_a : \mu = 1$. Use $\alpha = .05$, and use MS_w as your estimate of σ^2_e.

(2.) An experimenter ran five groups of rats in a T maze. Group 1 had only olfactory and tactile cues, i.e., they had no visual or auditory cues; Group 2 had only visual and tactile cues; Group 3 had only visual and olfactory cues; Group 4 had visual, olfactory, and tactile cues. Group 5 also had visual, olfactory, and tactile cues, but were given shocks when they made an error. The data below are for the trial of last error, so low scores indicate good performance.

A_1	A_2	A_3	A_4	A_5
7	5	9	6	19
8	4	11	12	6
5	4	6	8	3
9	6	8	5	12
10	3	7	11	13

(a.) Do an analysis of variance on these data, finding the p value and $\hat{\omega}_{bet}^2$.

(b.) Interpret the results: What effect does each treatment appear to have on learning?

(c.) Estimate the variance and standard deviation for each group. Comment on the validity of the assumption of equal variances in this design and the effect of its violation on the validity of the analysis of variance.

(d.) What transformation might be used to make the variances more nearly equal, but such that the transformed values would still be intuitively meaningful?

(e.) Make the transformation you suggested in part c, and do the analysis of variance on the transformed data. Compare your results with those in part a.

(3.) Refer to the data in Table 2.1. Suppose that a fourth group had also been run and the data were $A_4 : 3, 1, 2, 5, 4$.

(a.) Do an analysis of variance on the four groups, finding the p value and $\hat{\omega}_{bet}^2$.

(b.) Suppose only the first three scores were taken in Group A_4, i.e., suppose the data are $A_4 : 3, 1, 2$. Do the analysis of variance again on the four groups.

(c.) If you analyzed the data correctly, the data in part a were not significant at the .05 level, but those in part b were. By adding data, we increased the p value (in comparison with the smaller amount of data in Table 2.1) and then by removing some of the added data, we decreased it again. Explain this paradox. What conclusions can you draw for application to practical problems of experimental design?

(4.) An experimenter wants to set her power at .9 for detecting a difference as large as 10 between *any* two means. She estimates $\sigma_e^2 = 50$, and $\alpha = .05$.

(a.) How large, approximately, must n be if $I = 4$? How large must N be?

(b.) How large, approximately, must n and N be if $I = 3$? Compare these values with those in part a.

(5.) (a.) Derive Equation 2.2.

(b.) Derive Equation 2.12.

(c.) Prove that $SS_t = SS_{bet} + SS_w$.

(d.) Derive Equation 2.16.

For the remaining questions:

(a.) Do an analysis of variance on the data assuming that all of the assumptions for a one-way test are met. Find the p value and $\hat{\omega}^2_{bet}$.

(b.) Interpret the results, commenting on which differences are most likely to have contributed to the significant effect (if there was one).

(c.) Tell which, if any, of the assumptions of a one-way analysis of variance are most probably violated and comment on the effect each violation is likely to have had on the validity of the analysis.

(d.) If there are both positive and negative scores in the data, test the null hypothesis that the grand mean is zero and comment on the meaning of such a test for the particular set of data being tested.

(6.) Four males and four females were each given two forms of a short achievement test. The scores were:

| | Males | | | Females | |
	A_1	A_2		A_3	A_4
m_1	3	2	f_1	3	1
m_2	3	3	f_2	1	2
m_3	1	2	f_3	2	0
m_4	3	1	f_4	1	1

(For part a, assume that this is a one-way analysis of variance with four groups and four subjects per group.)

(7.) Mr. Gallop was curious as to whether locality was a significant factor in the political opinions of New York families. He polled three different areas within the city: the area surrounding New York University (A_1), a slum area (A_2), and a wealthy section (A_3). In each area he randomly selected three families; in each family he asked both the husband and the wife how happy they were with current efforts at school integration. Answers were scored on a 7-point scale from +3 (deliriously happy) to -3 (thoroughly nauseated). A score of 0 indicated indifference. Thus, six scores (2 from each family) were obtained in each area. The scores were:

A_1	A_2	A_3
-3	-1	2
-3	-2	3
-1	2	1
-2	2	0
1	-3	3
0	0	3

(8.) An experimenter studied reaction times of politicians at a cocktail party; subjects were sober (A_1), moderately drunk (A_2), or completely sloshed (A_3). Because the data were highly skewed (one subject was unable to react at all), the experimenter decided that it would be more legitimate to rank the reaction times and do an analysis of variance on the ranks instead of on the original reaction times. The ranks (with the subject unable to react receiving a rank of 7) were:

A_1	A_2	A_3
1	4	7
	5	6
	2	
	3	

(9.) A sociologist conducting a survey asked several college-age youths whether they liked, disliked, or felt neutral toward their parents. The people surveyed fell into the six groups shown below. Each respondent got a score of +1 if s/he liked his/her parents, -1 if s/he did not like them, and 0 if s/he felt neutral toward them. The scores were

Large College		Small College		Noncollege	
Male	Female	Male	Female	Male	Female
A_1	A_2	A_3	A_4	A_5	A_6
-1	0	1	1	0	-1
0	-1	0	1	-1	0
-1	-1	-1	1	0	0
-1	0	-1	-1	0	1
-1	1	0	1	0	0
0	-1	0	1	1	0
1	1	1	1	0	0
-1	-1	1	1	1	
0	-1	0		1	
1	-1	0		0	

3

Comparing Groups

In Chapter 2 we learned how to test for general differences among the groups in a one-way design, and how to estimate the overall size of those differences. However, none of those techniques enabled us to tell where the differences were, that is, exactly which means were significantly different from which others. In this chapter and the next we will discuss methods of finding and testing for specific differences. In this chapter we will concentrate on testing for differences that we planned to test before obtaining the data. In Chapter 4 we will discuss ways of testing for differences that were not planned, but were suggested by the data. Both kinds of techniques are useful. In a well-planned experiment there are often specific differences in which we are interested; however, we should also be aware of unexpected differences in the data.

Graphs and Figures

One of the most useful (and at the same time, most neglected) ways of comparing groups is with a graph. Statistical consultants are often asked, "What do these results mean?" If the answer is simple enough—and it usually is—a graph will make it clear. For example, Figure 3.1 is a graph of the means for the data in Table 2.1. It is clear from this figure that F is significant primarily because the scores of group A_2 tend to be much lower than those of groups A_1 and A_3, which are more nearly equal. Of course, these observed differences may be due to chance, just as the significant F may be due to chance. Nevertheless, if a real difference does exist, Figure 3.1 tells us where that difference is likely to be.

Yet comparison of group means by a graph alone can be deceptive. In order to be confident of our interpretation, we need some way of testing whether specific observed differences are due to chance.

Planned Comparisons

Frequently, we have some ideas in advance about which group means are likely to be different and which are not. Consider, for example, the design in Table 3.1

Here, four groups with eight subjects each are being compared. These are hypothetical data showing errors in a paired-associate learning task by sub-

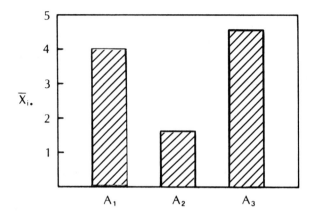

FIGURE 3.1. Group means from Table 2.1.

jects under the influence of two drugs. Group A_1 is a control group, not given any drug. Groups A_2, A_3, and A_4 are experimental groups; Group A_2 was given one drug, Group A_3 another, and Group A_4 was given both drugs. The results of an overall F test are given in the table; Figure 3.2 is a graph of the results. The estimate of ω_{bet}^2 for these data is .45.

Suppose we are really interested in answering three specific questions. On the average do the drugs have any effect on learning at all? Do subjects make more errors if given both drugs than if given only one? Do the two drugs differ in the number of errors they produce? In the remainder of this chapter we will consider issues involved in formulating and answering such questions.

AN EXAMPLE

The first question essentially asks whether the mean of group A_1 differs from the average of the means for groups A_2, A_3, and A_4. That is, we wish to test the null hypothesis

$$H_0(1) : \mu_1 = (\mu_2 + \mu_3 + \mu_4)/3.$$

TABLE 3.1. Hypothetical error data for four groups of eight subjects each.

	A_1	A_2	A_3	A_4	Total	
	1	12	12	13		
	8	6	4	14		
	9	10	11	14		
	9	13	7	17		
	7	13	8	11		
	7	13	10	14		
	4	6	12	13		
	9	10	5	14		
ΣX	54	83	69	110	316	
$\overline{X}$	6.750	10.375	8.625	13.750	9.875	
ΣX^2	422	923	663	1532	3540	
$\hat{\sigma}^2$	8.214	8.839	9.696	2.786	7.384	
	RS	SS	df	MS	F	p	$\hat{\omega}^2$
m		3120.50	1				
bet	3333.25	212.75	3	70.92	9.60	<.001	.45
w	3540.00	206.75	28	7.384			
t		419.50	31				

An equivalent way to state this hypothesis is

$$H_0(1) : \mu_1 - (1/3)\mu_2 - (1/3)\mu_3 - (1/3)\mu_4 = 0. \tag{3.1}$$

In other words, the hypothesis says that a particular linear combination of the true group means has a value of zero.

To develop a test of this hypothesis, we first find the distribution of the same linear combination of the obtained means:

$$C_1 = \overline{X}_{1.} - (1/3)\overline{X}_{2.} - (1/3)\overline{X}_{3.} - (1/3)\overline{X}_{4.}.$$

With $n = 8$, the obtained sample means are independently distributed as

$$\overline{X}_{i.} \sim N_{(\mu_i, \sigma_e^2/8)}$$

so that C_1 is normally distributed with mean and variance equal to

$$
\begin{aligned}
\underline{E}(C_1) &= \mu_1 - (1/3)\mu_2 - (1/3)\mu_3 - (1/3)\mu_4, \\
\underline{V}(C_1) &= (1 + 1/9 + 1/9 + 1/9)\sigma_e^2/8 = \sigma_e^2/6.
\end{aligned}
$$

Furthermore, according to the null hypothesis (Eq. 3.1), $\underline{E}(C_1) = 0$, and $H_0(1)$ can be restated as

$$H_0(1) : C_1 \sim N_{(0, \sigma_e^2/6)}$$

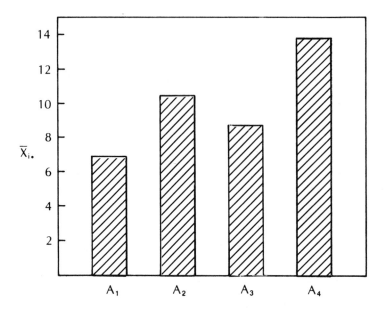

FIGURE 3.2. Group means from Table 3.1.

or, equivalently,

$$H_0(1) : (6/\sigma_e^2)^{1/2}C_1 \sim N_{(0,1)}.$$

Therefore, if the null hypothesis is true,

$$6C_1^2/\sigma_e^2 \sim \chi_{(1)}^2,$$

and

$$6C_1^2/MS_w \sim F_{(1,N-I)}.$$

We can thus use an F test on the null hypothesis. If the F is significantly larger than one, we reject the null hypothesis.

When the null hypothesis is false,

$$\underline{E}(6C_1^2) = \sigma_e^2 + 6\psi_1^2,$$

where $\psi_1 = \mu_1 - (1/3)\mu_2 - (1/3)\mu_3 - (1/3)\mu_4$, so a one-tailed test is appropriate. For the data in Table 3.1, the values actually obtained are $C_1 = -4.167, F = 14.1, p = .002$.

Equivalently, because the square root of F with one degree of freedom in the numerator is distributed as t, we could also test $H_0(1)$ with

$$t_{(N-I)} = 6C_1/MS_w^{1/2} = -3.76.$$

Under the null hypothesis this has a t distribution with $N - I$ degrees of freedom. The result is the same, of course, with either the t test (two-tailed) or the F test. A one-tailed t test could also be done.

GENERAL THEORY

The calculations discussed in the previous pages can be generalized to any linear comparison:

$$H_0(k) : \Sigma_i c_{ik} \mu_i = 0.$$

The c_{ik} are constants that define the desired comparison; for $H_0(1), c_{11} = 1$, and $c_{21} = c_{31} = c_{41} = -1/3$.

The theory for the most general case—the possibility that there may be unequal numbers of scores in the different groups—will be given first. It will then be given in a simpler form for equal sample sizes.

In general,

$$\overline{X}_{i.} \sim N_{(\mu_i, \sigma_e^2/n_i)}.$$

A linear combination $C_k = \Sigma_i c_{ik} \overline{X}_{i.}$ is then distributed as

$$C_k \sim N_{\Sigma_i c_{ik} \mu_i, \sigma_e^2 \Sigma_i c_{ik}^2/n_i)},$$

so that, under the null hypothesis,

$$C_k' = C_k/(\Sigma_i c_{ik}^2/n_i)^{1/2} \sim N_{(0, \sigma_e^2)}.$$

If we let $SS_k = (C_k')^2$, then

$$SS_k/\sigma_e^2 \sim \chi_{(1)}^2.$$

Finally, dividing by MS_w, we get

$$SS_k/MS_w \sim F_{(1, N-I)}.$$

The square root of this has a t distribution:

$$(SS_k/MS_w)^{1/2} \sim t_{(N-I)}. \tag{3.2}$$

If the null hypothesis is not true, then

$$\underline{E}(SS_k) = \sigma_e^2 + \psi_k^2/(\Sigma_i c_{ik}^2/n_i), \tag{3.3}$$

where

$$\psi_k = \Sigma_i c_{ik} \mu_i. \tag{3.4}$$

Power calculations for determining sample size, etc., can be made using non-central F as described in the previous chapter, with

$$\phi_k^2 = \psi_k^2/[\sigma_e^2 \Sigma_i (c_{ik}^2/n_i)].$$

If every group has the same number of scores, then

$$\overline{X}_{i.} \sim N_{(\mu_i, \sigma_e^2/n)},$$
$$C_k' = C_k(n/\Sigma_i c_{ik}^2)^{1/2},$$

TABLE 3.2. Tests of three planned comparisons of the data in Table 3.1.

	A_1	A_2	A_3	A_4		
$\overline{X}_{i\cdot}$	6.750	10.375	8.625	13.750		
c_{i1}	1	-1/3	-1/3	-1/3		
c_{i2}	0	-1/2	-1/2	1		
c_{i3}	0	1	-1	0		

	C	Σc^2	SS	F	t	p
$H_0(1)$	-4.167	4/3	104.2	14.1	-3.76	.002
$H_0(2)$	4.250	3/2	96.33	13.0	3.61	.002
$H_0(3)$	1.750	2	12.25	1.66	1.29	.21

and

$$E(SS_k) = \sigma_e^2 + n\psi_k^2/(\Sigma_i c_{ik}^2),$$
$$(\phi_k')^2 = \psi_k^2/(2\sigma_e^2 \Sigma_i c_{ik}^2).$$

We can now apply these formulas to the other two experimental questions asked at the beginning of this section. The first, whether subjects make more errors if given both drugs rather than only one, can be answered by testing the null hypothesis

$$H_0(2) : \mu_4 - (1/2)\mu_2 - (1/2)\mu_3 = 0. \tag{3.5}$$

The other can be answered by testing the null hypothesis

$$H_0(3) : \mu_2 - \mu_3 = 0.$$

These two tests, along with the test of $H_0(1)$, are summarized in Table 3.2. (The tabular form used there is highly recommended for clear, organized calculations.)

Given these results, we can conclude with some confidence that the drugs degrade performance and that both drugs together have a greater effect than either drug separately. However, we cannot be confident that there is a difference between the effects of the two drugs.

CALCULATIONAL SIMPLIFICATIONS

Before moving on to other theoretical issues, here are some calculational simplifications. The null hypotheses that we will discuss always specify that some linear function of the means is zero, so the hypotheses will not be changed if every c_{ik} is multiplied by the same constant. If, for example, in Equation 3.5 we multiply through by two, we will have

$$H_0(2) : 2\mu_4 - \mu_2 - \mu_3 = 0, \tag{3.6}$$

which is obviously equivalent, mathematically, to Equation 3.5. The hypotheses are equivalent, so the tests ought to be equivalent and they are. The obtained value of C_k will differ, but SS_k will be the same. For example, for Equation 3.6 we have

$$
\begin{aligned}
C_2 &= 8.5 \\
SS_2 &= (8.5)^2(8/6) = 96.33,
\end{aligned}
$$

which is the same as the value in Table 3.2. The advantage of Equation 3.6 over Equation 3.5 is that the c_{ik} are all integers, which are generally easier to work with than fractions. Most linear combinations that are of interest can be expressed in such a way that the c_{ik} are all integers; the preceding discussion shows that this can be done without changing the nature of the statistical test.

Another simplification can be achieved if, instead of finding the p value for each F, we set a significance level in advance and either accept or reject each null hypothesis accordingly. (I do not recommend this; see Chapter 1.) It is easy to show from Equation 3.2 that we should reject the null hypothesis if and only if

$$\mid C_k' \mid \ge t_{\alpha(N-I)}(MS_w)^{1/2},$$

where $t_{\alpha(N-I)}$ is the value of t needed to reach the α level of significance. Equivalently, we should reject the null hypothesis if and only if

$$SS_k \ge F_{\alpha(1,N-I)} MS_w.$$

The advantage of these simplifications is that the terms on the right need be calculated only once for all significance tests.

RELATIONSHIP TO ORDINARY t TEST

It should already be obvious that planned comparisons are similar to the ordinary t test. However, an analysis of this similarity illustrates another important general principle.

First, let us consider $H_0(3)$ (Table 3.2). If A_2 and A_3 were the only two groups studied, it would be tested by an ordinary t test. The numerator would contain C_2', and the denominator would contain an estimate of σ_e based on those two groups.

The general principle is that the denominator can contain *any* estimate of σ_e, so long as the estimate is statistically independent of C_k' and its square is an unbiased estimate of σ_e^2. If A_2 and A_3 had been the only two groups studied, then our best estimate would have been based on those two groups. Even though we have studied four groups, the estimate based on only the two groups would still be legitimate (and we shall see later that it is sometimes the most desirable). However, we maximize our power by using the *best* estimate available. The *best* estimate is based on all four groups, i.e., it is $MS_w^{1/2}$. Thus, $MS_w^{1/2}$ is not the only possible denominator term, but it is the *best*. (This point may appear to contradict a popular maxim that one must not do multiple t tests; we will discuss that prohibition later.)

Confidence Intervals for Planned Comparisons

In addition to testing the null hypothesis, it is possible to derive confidence intervals for the ψ_k. From the above derivations it can be shown that, with probability $1 - \alpha$,

$$C_k - S \leq \psi_k \leq C_k + S,$$

where

$$S = t_{\alpha(N-I)}[MS_w \Sigma_i (c_{ik}^2 / n_i)]^{1/2}.$$

For the case of equal sample sizes, this becomes

$$S = t_{\alpha(N-I)}[MS_w (1/n) \Sigma_i c_{ik}^2]^{1/2}.$$

For example, for the three null hypotheses of Table 3.2, the 95-percent confidence intervals are

$$
\begin{aligned}
-6.44 &\leq \psi_1 \leq -1.89, \\
+1.84 &\leq \psi_2 \leq +6.66, \\
-1.03 &\leq \psi_3 \leq +4.53.
\end{aligned}
$$

Necessary Assumptions and Their Importance

The same assumptions as in the ordinary F test are made in planned comparisons. The scores are assumed to be independently normally distributed with a constant variance. In addition, with one exception, the robustness properties of planned comparisons are the same as for the overall F test. Violation of the assumption of normality generally has little effect on the result of a planned comparison unless the kurtosis of the population is very different from zero, and N is small. Violation of independence, on the other hand, has very large effects, just as it does with the overall F test. Violation of the assumption of equal variances requires some discussion.

The discussion will be simplified if at first we limit it to simple comparisons between two groups, such as $H_0(3)$ above. For this case the planned comparison is like an ordinary t test. In both tests the denominator is an estimate of the population standard deviation. In the t test, however, the estimate is obtained only from the scores in the two samples being compared. In a planned comparison we take advantage of the assumption of equality of variances to use the scores in all I groups. This provides a more powerful test, as can be seen in the difference between the degrees of freedom in the two cases. In the

t test the number of degrees of freedom is $2(n-1)$ for the case of equal sample sizes; in the planned comparison the number is $(N - I) = I(n - 1)$.

When the assumption of equal variances is violated, MS_w, as we have seen, is an estimate of the weighted mean of the population variances. However, the mean of all I population variances is not an appropriate error term for comparing only two groups. The error term should be based only on the variances of the two groups being compared. If, for example, the two specific groups being compared have population variances larger than the average for all I groups, MS_w will underestimate the average for those two groups. The obtained t (or F) will then be too large and the p value will be too small. The opposite will be true if the two groups in question have population variances smaller than the average over all I groups.

More complicated comparisons produce additional difficulties. Even when all of the means are involved in a comparison, they usually are weighted differently. In testing $H_0(1)$, for example, the mean of group A_1 is weighted three times as heavily as the means of the other three groups. If the variance of the scores in Group A_1 is larger than the variances in the other three groups, the ratio will be too large. The problem exists, moreover, even when sample sizes are large and equal.

One solution is to find a transformation on the data that will make the variances in all I groups approximately the same. The advantages and disadvantages of such a transformation were discussed in Chapter 2.

APPROXIMATE F TEST

Another solution is an approximate F test. When the variances are not all equal, the expected value of c_k^2 is

$$E(C_k^2) = \Sigma_i(c_{ik}^2\sigma_i^2/n_i) + \psi_k^2,$$

where σ_i^2 is the variance of population A_i and ψ_k is defined in Equation 3.4. The null hypothesis specifies that ψ_k^2 is zero, so when the null hypothesis is true,

$$E(C_k^2) = \Sigma_i c_{ik}^2\sigma_i^2/n_i.$$

We can form an appropriate F ratio if we can find a denominator that depends only on the variances within groups and has the same expected value as C_k^2 under the null hypothesis. The obvious answer is

$$D_k = \Sigma_i c_{ik}^2\hat{\sigma}_i^2/n_i.$$

Approximately,

$$C_k^2/D_k \sim F_{(1,d_k)},$$

where d_k, the denominator degrees of freedom, is approximately

$$d_k = D_k^2/[\Sigma_i c_{ik}^4\hat{\sigma}_i^4/(n_i^3 - n_i^2)].$$

TABLE 3.3. Tests of three planned comparisons of the data in Table 3.1 (equality of variances not assumed).

	A_1	A_2	A_3	A_4
$\overline{X}_{i.}$	6.750	10.375	8.625	13.750
c_{i1}	1	-1/3	-1/3	-1/3
c_{i2}	1	-1/2	-1/2	1
c_{i3}	0	1	-1	0

	C	D	d	F	p
$H_0(1)$	-4.167	1.323	11	13.1	.005
$H_0(2)$	4.250	.9275	21	19.5	.001
$H_0(3)$	1.750	2.317	14	1.32	.27

Note, however, that d_k and the F ratio are both only approximate. The approximations hold well for large n_i; they probably hold well for moderate n_i, but not too well if the n_i are very small. A good rule of thumb is that d_k should be at least 20. However, even when $d_k < 20$, the approximate method may be preferable to the standard method when unequal variances are likely.

When the n_i are equal, the above equations reduce to

$$D_k = (1/n)\Sigma_i c_{ik}^2 \hat{\sigma}_i^2,$$
$$d_k = (n^3 - n^2)D_k^2/(\Sigma_i c_{ik}^4 \hat{\sigma}_i^4).$$

The sample variance in Table 3.1 for group A_4 is considerably smaller than those for the other three groups. Although the obtained differences could easily be due to chance, they may well lead us to suspect that the population variances are different. The results of the approximate F tests on $H_0(1)$, $H_0(2)$, and $H_0(3)$ are shown in Table 3.3. The d_k values are not as large as we would like, but the example serves to illustrate the calculations nevertheless. A comparison with Table 3.2 shows that *for these particular data* the principle difference is a loss of degrees of freedom when the variances are not assumed to be equal.

When the test is a simple comparison between the means of two groups, e.g., A_1 and A_2, the above approximation is identical to the approximate t test proposed by Satterthwaite (1946). (The square root of the approximate F is of course distributed approximately as t.) With equal sample sizes, a simple comparison between two means is identical to an ordinary t test, except for the adjustment in degrees of freedom. (This explains our earlier statement that in some cases a simple t test is preferable to a planned comparison using MS_w.)

DESIRABLE PROPERTIES OF PLANNED COMPARISONS

Planned comparisons are a very versatile tool. However, it is easy to misuse this tool, using it to conduct a kind of "fishing expedition" in which one searches for significant differences. When we use planned comparisons in this way, we run too high a risk of drawing invalid conclusions. If there are enough groups, and if we are not careful in our use of planned comparisons, there may be a very high probability of making a Type I error even though our significance level is .05 or less. Suppose, for example, we make a large number of comparisons. Although no single comparison has a high probability of being significant by chance, there is nevertheless a high probability that one or more of the set will be significant. It is similar to, say, a roulette wheel; although a zero has a low probability of occurring on any single spin of the wheel, it has a high probability of occurring at least once if the wheel is spun several times. To avoid abuse, some basic principles should be followed:

(1) Planned comparisons should be *planned*. That is, the planned comparisons should be chosen before the experiment is conducted. Moreover, they should be chosen to answer specific questions that are of particular interest. If we choose our comparisons after having viewed the data, we may be tempted to choose those comparisons that the data indicate will be significant. This will increase our probability of a Type I error.

(2) Planned comparisons should be limited in number. This avoids the "fishing expedition" problem.

(3) Planned comparisons should not have α values that are too large. Even though the number of comparisons is small, the probability that one or more will be significant is greater than α. Therefore, we should not make α too large to begin with.

It is possible to choose a significance level so low that the probability of *one or more* significant results is no higher than a preselected value. For example, if we are testing r planned comparisons and we test each at a significance level of α/r, then the probability that one or more tests will be significant by chance will be less than α. In addition, this sets the *expected number* of Type I errors equal to α.

Alternatively, we may use a conventional significance level and use judgment in interpreting the pattern of results. Less weight would then be given to an occasional significant result if it occurred along with a number of results that were nonsignificant. More weight would be given to it if it were part of an expected pattern of results.

I feel that if the comparisons have been chosen carefully to begin with, then significant results can usually be believed with significance levels as large as .05, although levels of .01 may be better if more than a few tests are to be done. Significance levels of .10 or larger should usually be avoided.

(4) When feasible, planned comparisons should be *orthogonal*. That is, each null hypothesis should be formulated independently of the others.

ORTHOGONAL HYPOTHESES

There are several advantages to formulating the null hypotheses for different planned comparisons in such a way that the truth of each null hypothesis is unrelated to the truth of any others. The three null hypotheses in our previous example were formulated in this way. If $H_0(1)$ is false, we know that drugs have some effect on learning, although $H_0(1)$ says nothing about whether the effect of taking both drugs simultaneously is different from the average of the effects of the drugs taken separately. That is, the truth of $H_0(1)$ is unrelated to the truth of $H_0(2)$ and vice versa; $H_0(1)$ involves the *average* of μ_1, μ_2, and μ_3, whereas $H_0(2)$ involves *differences* among them. The two hypotheses $H_0(1)$ and $H_0(2)$ are *independent*. In the same way, $H_0(3)$ is independent of both $H_0(1)$ and $H_0(2)$ because $H_0(1)$ and $H_0(2)$ both involve the average of μ_2 and μ_3, and $H_0(3)$ involves their difference. Planned comparisons that test independent hypotheses are said to be *orthogonal*.

Advantages of Orthogonal Tests. One important advantage of orthogonal planned comparisons is that they allow a maximum amount of information to be extracted from each test. Each new test says something new about the data. Suppose that in Table 3.2 we had also tested the null hypothesis that learning is the same in group A_4 (in which the subjects took both drugs) as in the control group, A_1; this might be wasting time on the obvious. The results of the test on $H_0(1)$ make it clear that the drugs individually have an effect on learning, so it is likely that both drugs taken together will have an effect. Instead, by testing the null hypothesis $H_0(2)$ that the mean of group A_4 is the same as the average of the means of groups A_2 and A_3, we are asking a question that is independent of that asked by $H_0(1)$.

More easily interpreted results are another advantage of orthogonal planned comparisons. Consider, for example, testing both $H_0(1)$ and the null hypothesis

$$H_0(4) : \mu_4 - \mu_1 = 0,$$

proposed in the last paragraph. It is possible to obtain data for which we could reject $H_0(4)$ but not $H_0(1)$, although the obtained means for groups A_2, A_3, and A_4 were all higher than those for group A_1. Table 3.4 contains data for which $H_0(4)$ can in fact be rejected at the .05 level, whereas $H_0(1)$ cannot be rejected even at the .25 level.

We are now in the peculiar position of having one test that says the drugs affect learning and another that says they do not! How much weight should be assigned to each test in reaching a conclusion? This entire problem can be avoided by initially formulating the null hypotheses so that they are independent. Each result can then be interpreted on its own merits, relatively independently of the other statistical tests. Note the word "relatively." Although the hypotheses may be independent, the tests themselves generally

TABLE 3.4. Hypothetical error data illustrating inconsistent results with nonindependent tests.

	A_1	A_2	A_3	A_4	Total
	5	0	10	4	
	4	8	3	5	
	4	3	5	9	
	5	9	8	10	
	8	7	4	10	
	7	5	2	7	
	9	7	5	13	
	2	7	9	14	
ΣX	44	46	46	72	208
$\overline{X}$	5.50	5.75	5.75	9.00	6.50
ΣX^2	280	326	324	736	1666

	RS	SS	df	MS	F	p
m		1352.0	1			
bet	1419.0	67.0	3	22.33	2.53	.078
w	1666.0	247.0	28	8.821		
t		314.0	31			

	A_1	A_2	A_3	A_4
$\overline{X}_{i.}$	5.50	5.75	5.75	9.00
c_{i1}	3	-1	-1	-1
c_{i4}	-1	0	0	1

	C	SS	F	p
$H_0(1)$	-4.0	10.67	1.21	.29
$H_0(4)$	3.5	49.00	5.55	0.26

are not *completely* independent because both F ratios use the same denominator, MS_w. If by pure chance MS_w is smaller than expected, all of the t ratios will then be too large, and vice versa. (However, if N is relatively large, the danger of a spuriously small or large MS_w is not very great. Thus, with a large N the t tests are nearly independent.)

A final advantage is that orthogonal tests are inherently limited in number. If there are I groups, then no more than I orthogonal planned comparisons can be done. If we wish our tests to be *contrasts*, then they are even more limited. Contrasts are planned comparisons for which $\Sigma_i c_{ik} = 0$. We will show later that contrasts are orthogonal to the grand mean, μ; they test only differences between means. The questions of interest usually involve only differences between means, so most of the planned comparisons done in actual practice are limited to contrasts. No more than $(I-1)$ orthogonal contrasts

TABLE 3.5. Tabular representation of
planned comparisons to facilitate tests
for independence.

	A_1	A_2	A_3	A_4
$H_0(1)$	1	-1/3	-1/3	-1/3
$H_0(2)$	0	-1/2	-1/2	1
$H_0(3)$	0	1	-1	0
$H_0(4)$	-1	0	0	1

can be tested. (We will show that the Ith test, orthogonal to the other $(I-1)$, must then be the test of the grand mean.)

Testing for Orthogonality. A mathematical test for orthogonality is easily derived. Comparisons on two hypotheses, $H_0(k)$ and $H_0(k')$, are orthogonal if their associated linear combinations, C_k and $C_{k'}$, are statistically independent. Because C_k and $C_{k'}$ are both normally distributed, they will be independent if and only if

$$E(C_k C_{k'}) = E(C_k)E(C_{k'}).$$

But it can be shown that this is true if and only if

$$\Sigma_i c_{ik} c_{ik'}/n_i = 0. \tag{3.7}$$

When all groups have the same number of scores, this simplifies to

$$\Sigma_i c_{ik} c_{ik'} = 0. \tag{3.8}$$

When two or more comparisons are orthogonal, their C_k values are independent and vice versa. Thus, their C'_k values (the numerators of the t ratios) are also independent.

The three hypotheses tested in Table 3.2 can now be checked for independence. Table 3.5 shows the c_{ik} values for each of the three independent hypotheses as well as for $H_0(4)$. You can readily verify that all the sums of the cross-products are zero, with the exception of the comparisons of $H_0(4)$ with $H_0(1)$ and $H_0(2)$; for these the sums of the cross-products are -4/3 and +1, respectively.

Constructing Orthogonal Tests. A general method of constructing orthogonal tests makes heavy use of matrix algebra. However, most contrasts of interest test a simple average of one set of means against another. If all sample sizes are equal, such tests are easily constructed. The basic principle is that any test on the differences among a given set of means will be orthogonal to any test involving their average.

The tests in Table 3.2 were constructed in this way. $H_0(1)$ involves the average of μ_2, μ_3, and μ_4, while $H_0(2)$ and $H_0(3)$ involve differences among these

true). Such is the case, for example, with the data in Table 3.4, where the overall F value of 2.43 is not significant but the test of $H_0(4)$ is significant. The overall F test tests all contrasts simultaneously. As such, it must be less sensitive to individual contrasts; otherwise an occasional chance difference would produce a significant result too easily.

In the past, some have argued that it is not permissible to do planned comparisons unless the overall F test is significant. As can be seen here, both the overall F test and the planned comparisons are based on the same statistical theory, and there is no reason why one should have priority over the other. In fact, the procedure advocated in the past may well prevent us from finding interesting differences in the data because the overall F test has less power than planned comparisons for testing the specific differences that are most interesting. However, remember the caution made earlier about testing only a few well-chosen hypotheses.

The overall F test remains useful when few or no planned comparisons are made. If groups A_2, A_3, and A_4 in Table 3.1 were simply groups of people given three different drugs, and if we had no a priori reason to believe that one drug would retard learning more than another, it would not be reasonable to test $H_0(3)$. There would be no better reason to test the hypothesis that group A_4 is different from groups A_2 and A_3 than there would be to test the hypothesis that A_2 is different from A_3 and A_4, or the hypothesis that A_3 is different from A_2 and A_4. Similarly, $H_0(2)$ would not be reasonably tested by a planned comparison either. Only $H_0(1)$ could be reasonably tested.

The overall F test thus remains useful when there are few or no natural planned comparisons. Although only one legitimate planned comparison is possible in the above example, an overall F test can still determine whether there are any significant differences among the means.

Combining Orthogonal Contrasts

Moreover, the relationships between contrasts and the overall F can be used to improve the tests even for the above example. Suppose groups A_1, A_3, and A_4 are given three different drugs, no one of which can be expected a priori to have any specifically different effect than the other two. One possibility would be to do an F test on the four groups and, if the test was not significant, discontinue testing. However, then a significant difference between the control group and the other three groups might be masked by the overall test. We should thus do a planned comparison on $H_0(1)$ even if the overall F test is not significant. If the planned comparison is significant, we are in the predicament described earlier; one test denies the existence of differences while the other test asserts their existence. The theory discussed above provides us with a way out of the dilemma. More specifically, instead of the overall F test, we can do two independent statistical tests: one on the null hypothesis that the scores for the control group are the same as those under drugs, i.e., $H_0(1)$, and the other on the general null hypothesis that there are no differences between the three groups receiving drugs.

Recall that SS_{bet} for the overall F test involves the sum of three statistically independent SS_k values. Any three values corresponding to three orthogonal contrasts will do. (Note that they must be contrasts, that is, the c_{ik} must sum to zero.) First we test the contrast implied by $H_0(1)$ and find it significant for the data of Table 3.1. Next we note that $H_0(2)$ and $H_0(3)$ are both independent of $H_0(1)$; therefore they would be legitimate contrasts if there were a priori reasons for testing them. Unfortunately, in the hypothetical situation we are now considering, there are no a priori reasons.

Suppose, though, that we had originally studied only groups A_2, A_3, and A_4. We would then have three groups, and our test with two numerator degrees of freedom would use the sum of the SS_k of *any* two orthogonal contrasts. Hypotheses $H_0(2)$ and $H_0(3)$ involve only differences between these latter three groups, so they would be acceptable as the two orthogonal contrasts. What all of this complicated reasoning means is that if an ordinary F test were done only on groups A_2, A_3, and A_4, the resulting SS_{bet} would be

$$SS_{A_2 A_3 A_4} = SS_2 + SS_3. \tag{3.9}$$

Of course, this can be verified:

$$
\begin{aligned}
SS_{A_2 A_3 A_4} &= (1/8)(83^2 + 69^2 + 110^2) - (1/24)(83 + 69 + 110)^2, \\
&= 2968.75 - 2860.17 = 108.58, \\
SS_2 + SS_3 &= 96.33 + 12.25 = 108.58.
\end{aligned}
$$

Thus, the general F test on the three remaining groups corresponds to the simultaneous testing of two contrasts, both of which are orthogonal to $H_0(1)$. However, when making the test, we should use the MS_w found for the F test with all four groups, rather than calculate a new MS_w based only on the three groups being tested (unless we suspect unequal variances—then the MS_w should be based only on the groups being tested). This is because MS_w is our best estimate of σ_e^2. The F test is

$$
\begin{aligned}
MS_{A_2 A_3 A_4} &= 108.58/2 = 54.29, \\
F_{(2,28)} &= 54.29/7.384 = 7.35, \\
p &= .003.
\end{aligned}
$$

We have thus made two independent tests and can reject both null hypotheses with a high degree of confidence. We can confidently conclude both that the drugs affect learning and that the drugs differ in their effects on learning.

Actually, it is not necessary to calculate the sum of squares directly. From Equation 3.9 and the fact that the values for all three contrasts sum to SS_{bet}, we have

$$SS_{A_2 A_3 A_4} = SS_{bet} - SS_1 = 212.75 - 104.17 = 108.58.$$

These results are merely a special case of a general principle. Suppose we have a set of scores from I groups and we wish to test r orthogonal contrasts.

If r is smaller than $(I-1)$, we can also make an orthogonal test of the null hypothesis that there are no differences other than the r differences already tested. We do this by finding

$$SS_{rem} = SS_{bet} - \Sigma_{k=1}^{r} SS_k.$$

Then SS_{rem} must correspond to the sum of $(I-1-r)$ additional potential contrasts, and SS_{rem}/σ_e^2 must be distributed (under the null hypothesis of no additional differences) as chi-square with $(I-1-r)$ degrees of freedom. Consequently, the test using

$$F_{(I-1-r,N-I)} = MS_{rem}/MS_w,$$

where

$$MS_{rem} = SS_{rem}/(I-1-r),$$

is independent of the r contrasts already tested.

PLANNED COMPARISONS AND THE GRAND MEAN

In the previous chapter we saw that SS_m could be used to test the null hypothesis $H_0 : \mu = 0$. Now consider the following planned comparison:

$$H_0(m) : \Sigma_i n_i \mu_i = 0. \tag{3.10}$$

This is a perfectly legitimate planned comparison, using $c_{im} = n_i$, and it can be tested in the usual way. (Notice, however, that it is not a contrast.) When we make the planned comparison, we find that

$$
\begin{aligned}
C_m &= N\overline{X}_{..}, \\
SS_m &= N\overline{X}_{..}^2.
\end{aligned}
$$

In other words, the test of the grand mean is a planned comparison.

Even though this test is seldom used, its relationship to other planned comparisons, and particularly contrasts, helps to clarify the relationship between the overall F test and planned comparisons.

It is not difficult to prove that every contrast is orthogonal to $H_0(m)$. From Equations 3.7 and 3.10 we can see that a particular planned comparison is orthogonal to $H_0(m)$ if and only if

$$\Sigma_i c_{ik} n_i / n_i = \Sigma_i c_{ik} = 0,$$

that is, if and only if the planned comparison is a contrast.

As we stated earlier, the F test is a test of the null hypothesis that all *contrasts* are zero; moreover the total number of possible contrasts is $(I-1)$. Actually, the total number of planned comparisons is I. However, if we wish our planned comparisons to be independent of the grand mean as well as of each other (i.e., to be orthogonal contrasts), there will be only $(I-1)$. The Ith planned comparison is then the test of the grand mean. (In practice, planned comparisons are almost always contrasts, but there is no reason why they must be.)

ω^2 FOR CONTRASTS

Just as for overall tests, every difference tested for by a contrast, i.e., every ψ_k, can be said to account for a proportion of the variance in the data. The proportion accounted for by a given ψ_k, i.e., ω_k^2, can be estimated in a manner similar to the estimation of ω_{bet}^2 in the overall F test.

The theory for this rests on the fact (whose derivation is too complicated to give here) that just as

$$SS_{bet} = \Sigma_k SS_k,$$

so also

$$\sigma_{bet}^2 = \Sigma_k \sigma_k^2,$$

where

$$\sigma_k^2 = \psi_k^2/(N\Sigma_i c_{ik}^2/n_i) \qquad (3.11)$$

and the ψ_k are the values of $(I-1)$ orthogonal contrasts (Eq. 3.4). Therefore, Equation 2.18 can be extended to

$$\sigma_t^2 = \sigma_e^2 + \Sigma_k \sigma_k^2,$$

and the σ_k^2 further analyze the total variance. In fact, they analyze σ_t^2 as finely as possible; σ_e^2 cannot be divided more finely, and σ_{bet}^2 is divided into elements, each of which have only one degree of freedom.

We can now let

$$\omega_k^2 = \sigma_k^2/\sigma_t^2 \qquad (3.12)$$

be the proportion of the total variance accounted for by the linear contrasts ψ_k.

For an approximately unbiased estimate, we can find from Equations 3.3 and 3.11 that

$$\hat{\sigma}_k^2 = (SS_k - MS_w)/N$$

is an unbiased estimate of σ_k^2. Consequently, from Equations 2.21 and 3.12,

$$\hat{\omega}_k^2 = (SS_k - MS_w)/(SS_t + MS_w)$$

is an approximately unbiased estimate of ω_k^2. For the tests in Table 3.2 the estimates are

$$\hat{\omega}_1^2 = .23, \quad \hat{\omega}_2^2 = .21, \quad \hat{\omega}_3^2 = .01.$$

Finally, it can be shown that

$$\Sigma_k \hat{\omega}_k^2 = \hat{\omega}_{bet}^2.$$

Therefore, these estimates can be extended to the case in which r different orthogonal contrasts are combined into one general F test. In that case the estimate of the proportion of variance accounted for, by all of the r contrasts making up the overall test, is the sum of the $\hat{\omega}_k^2$ for the r individual contrasts. The estimated proportion of between variance not accounted for by the contrasts is

$$\hat{\omega}_{rem}^2 = \hat{\omega}_{bet}^2 - \Sigma_{k=1}^r \hat{\omega}_k^2.$$

Exercises

(1.) For the data in Table 2.1:

(a.) Test the following null hypothesis, by planned comparisons, finding both the p value and $\hat{\omega}_k^2$.

$$H_0(1) : \mu_1 = \mu_2,$$
$$H_0(2) : \mu_3 = (\mu_1 + \mu_2)/2.$$

(b.) Add the following data, from a fourth group, A_4: 3, 1, 2 (these are the same as in Problem 2b, Chapter 2), and test the following two null hypotheses:

$$H_0(1) : \mu_1 = \mu_4,$$
$$H_0(2) : \mu_2 = (\mu_1 + 2\mu_4)/3.$$

(c.) Redo the analyses in parts a and b, without assuming equal variances. Which type of analysis do you think is better for these data?

(d.) Using the analysis already performed in Chapter 2, how could you have calculated the value of F for $H_0(2)$ in part a without first calculating C_2?

(2.) The following data are the group means from an experiment on three groups, with nine scores in each group:

	A_1	A_2	A_3
$\overline{X}_{i.}$	12	-6	0

(a.) Assuming that $MS_w = 100$, do a planned comparison on each of the following hypotheses, finding the p value and $\hat{\omega}_k^2$ (when appropriate):

$$H_0(1) : \quad \mu = -3,$$
$$H_0(2) : \quad 3\mu_2 = \mu_3 + 2\mu_1,$$
$$H_0(3) : \quad \mu_1 = 2\mu_3.$$

(b.) Which pairs of tests in part a are orthogonal? Which are not orthogonal?

(c.) A fourth comparison, orthogonal to $H_0(2)$ and $H_0(3)$ in part a, can be made. What would its p value be? (Hint: It is not necessary to find the comparison to determine its p value.)

(d.) Test $H_0 : \mu_1 = 0$. The mean of the third group is zero, so this would appear to be the same test as $H_0(3)$ of part a; is it? Why, or why not?

(3.) For the data in Problem 6, Chapter 2:

(a.) Assuming that the assumptions of a one-way analysis of variance are met, test the following:

$$H_0 : \quad \mu_1 + \mu_2 = \mu_3 + \mu_4,$$
$$H_0(2) : \quad \mu_1 + \mu_3 = \mu_2 + \mu_4,$$

$H_0(3)$: There are no other differences, orthogonal to the above, among the means of the four groups.

Test each as a planned comparison, finding the p value and $\hat{\omega}_k^2$.

(b.) Redo the tests in part a without assuming equal variances. Which type of analysis do you think is most appropriate in this case?

(c.) Can you find a way to test each of the above hypotheses without assuming that the two scores obtained from a single individual are independent? (Hint: With a little ingenuity, each of the first two tests can be reduced to a t test.)

(4.) For the data in Problem 8, Chapter 2:

(a.) Test $H_0(1) : \mu_1 = \mu_2$, assuming that all of the assumptions are met. Find the p value and $\hat{\omega}^2$.

(b.) Find the contrast that is statistically orthogonal to $H_0(1)$.

(5.) For the data in Problem 9, Chapter 2:

(a.) Do a planned comparison on the null hypothesis that, overall, there is no difference between the average scores of males and females, finding both the p value and $\hat{\omega}_k^2$.

(b.) Redo the test in part a without assuming equal variances. Which type of analysis do you think is most appropriate for this test?

(c.) Find another test that is logically orthogonal to the test in part a, and tell whether it is also statistically orthogonal.

(6.) For the data in Problem 1, Chapter 2:

(a.) Test the following two null hypotheses, using planned comparisons and finding both the p value and $\hat{\omega}_k^2$:

$$H_0(1) : \mu_1 = (\mu_2 + \mu_3 + \mu_4 + \mu_5)/4 = \mu_6$$

(no overall difference due to type of drug given), and

$$H_0(2) : \mu_2 = \mu_3 = \mu_4 = \mu_5$$

(no differences in the effects of the four different tranquilizers).

(Note that this last hypothesis involves more than one equality. It cannot be tested by a single planned comparison. Nevertheless, you are being asked to find a single test of this null hypothesis.)

(7.) For the data in Problem 2, Chapter 2:

(a.) Test the following null hypotheses using planned comparisons; find the p value and $\hat{\omega}_k^2$:

$$H_0(1) : \mu_4 = \mu_5$$

(shock has no effect), and

$$H_0(2) : \mu_1 = \mu_2 = \mu_3$$

(the three types of cues are equally important in learning the discrimination). (See note following Problem 6.)

$$H_0(3) : (\mu_1 + \mu_2 + \mu_3)/3 = (\mu_4 + \mu_5)/2.$$

(b.) Redo the tests in part a without assuming equal variances. Which type of test do you think is most appropriate for these data?

(c.) Are there any possible differences in the data that are not related to the three tests in part a? If so, test for them.

(8.) Prove that

(a.) Multiplying each c_{ik} in a planned comparison by a constant, $b \neq 0$, does not change the value of SS_k; and

(b.) C_k and $C_{k'}$ are uncorrelated if $\Sigma_i c_{ik} c_{ik'}/n_i = 0$.

(9.) Refer to Problem 4, Chapter 2. If we choose our n so that the power is exactly .9 under the conditions given, what is the smallest w_{bet}^2 for which the power would be .9 if

(a.) $I = 4$
(b.) $I = 3$

(Hint: Both power and w_{bet}^2 depend on $\Sigma_i \alpha_i^2$.)

4

Other Multiple Comparison Methods

A complete data analysis often requires more than a simple overall F test or a limited number of planned comparisons. Many important discoveries are "after the fact"—unanticipated relationships found in the data. Such relationships cannot be rested by planned comparisons; the choice of a comparison on the basis of apparent differences among the obtained means would introduce a strong bias in favor of rejecting the null hypothesis. Other techniques for making multiple comparisons exist, but they have very low power; in many cases, they will not find a significant difference unless it is large enough to be obvious without a test.

Nevertheless, these methods are sometimes useful. We will discuss two specific kinds of multiple comparisons. The first is *general post hoc comparisons*, methods of testing arbitrary comparisons that have not been planned in advance. The second is a set of methods for making pairwise tests of every mean against every other mean, a help in deciding which obtained means are clearly different and which are not.

The key to understanding all of these tests is the concept of a *protection level*. If a number of tests are done at an α protection level, then the probability is no more than α that one or more of the tests will be significant by chance; i.e., the probability is at least $(1 - \alpha)$ that *none* of the tests will be significant by chance. (The protection level for all tests made in the experiment is the experimentwise error rate discussed in Chapter 3.) In order to achieve this protection level, we must of course make the probability of a Type I error on any given test much less than α, and this considerably reduces the power. However, these tests do provide strong protection against Type I errors.

Post Hoc Comparisons

We will discuss three methods for making post hoc comparisons. They are the Bonferroni method, the Scheffé, or S, method (Scheffé, 1959, pp. 66–72), and the Tukey, or T, method (see Scheffé, 1959, pp. 73–75).

BONFERRONI METHOD

Sometimes we have an idea of which comparisons are likely to be interesting, before the data are gathered. With a list of r such potential comparisons, we can apply the Bonferroni method. The calculations are the same as for ordinary planned comparisons, but we use a significance level of α/r (even though the possible tests are not all orthogonal). Frequently, for example, we expect to test differences only among pairs of means. Later in this chapter we will discuss methods for comparing all pairs of means; if we want to test only a subset of these, the Bonferroni method might be appropriate.

Suppose, given the data in Table 3.1, that we had wanted to compare each experimental group against the control group, A_1, at a protection level of $\alpha = .05$. We would simply do planned comparisons on μ_1 versus μ_2, μ_1 versus μ_3, and μ_1 versus μ_4, but we would test these three differences at $\alpha = .05/3 = .0166$ instead of $\alpha = .05$. Or, equivalently, we could multiply the obtained p value by 3 before comparing it with α. For the first comparison,

$$
\begin{aligned}
H_0(1) : \mu_1 - \mu_2 &= 0, \\
C_1 &= -3.625, \\
F_{(1,28)} &= 7.12, \\
p &= 3(.013) = .039.
\end{aligned}
$$

Note that if this had been a planned comparison, the p value would have been .013.

To use the Bonferroni method we must make up a list of potential comparisons before conducting the experiment. This prevents us from testing possibly interesting differences that the data bring to our attention. Moreover, if the list is large, the Bonferroni method has very little power because α must be divided by r, the number of tests on the list, to obtain the significance level for each individual test.

THE SCHEFFÉ METHOD

The S and T methods both have the same rationale. In a sense, each extends the Bonferroni method to an unlimited list of potential post hoc contrasts. For both methods the protection level, α, is such that if there are no differences among the true means, the probability of finding *any significant contrasts at all* will be no larger than α. In other words, if there are no differences among the population means, the probability is $(1 - \alpha)$ that every single contrast that might possibly be tested will fail to be significant at the α level. The Scheffé method is presented first because it is most similar to the tests already described.

We found previously that SS_k/σ_e^2 had a chi-square distribution with one degree of freedom. The S method simply consists of treating each SS_k/σ_e^2 as if it were distributed as chi-square with $(I-1)$ degrees of freedom instead. We can make a post hoc test of any contrast that may interest us by calculating

$$F'_{(I-1,N-I)} = SS_k/[(I-1)MS_w]$$

and looking up the resulting F' in an ordinary F table with $(I-1)$ and $(N-I)$ degrees of freedom.

The method will be illustrated on the data of Table 3.1. Suppose, on inspecting the data, we note that the combined effect of the two drugs administered together appears to be greater than the sum of the effects of the two drugs administered singly. More specifically, it appears to us that

$$(\mu_4 - \mu_1) > (\mu_2 - \mu_1) + (\mu_3 - \mu_1).$$

To test this, we set up the null hypothesis

$$H_0 : (\mu_4 - \mu_1) = (\mu_2 - \mu_1) + (\mu_3 - \mu_1),$$

which is algebraically equivalent to the hypothesis

$$H_0 : \mu_4 - \mu_3 - \mu_2 + \mu_1 = 0.$$

This form of the null hypothesis is a contrast that is tested with

$$
\begin{aligned}
SS_k &= 8(13.75 - 8.625 - 10.375 + 6.75)^2/(1+1+1+1) \\
&= 4.50, \\
F'_{(3,28)} &= 4.50/[(3)(7.384)] = .20.
\end{aligned}
$$

This difference is nonsignificant, so we cannot conclude that the effect of giving both drugs simultaneously is greater than the sum of the effects of the individual drugs.

If we decide to set a significance level in advance, the calculations for the S method can be made somewhat simpler, just as with planned comparisons. We reject the null hypothesis if and only if

$$| C'_k | > [(I-1)MS_w F_{\alpha(I-1,N-I)}]^{1/2}.$$

The above results are valid only if all of the post hoc comparisons are contrasts, that is, if the c_{ik} values sum to zero. If we wish to test some comparisons that are not contrasts we can do so, but we must treat SS_k as if it has I degrees of freedom instead of $(I - 1)$. All of the above equations will then be valid if $(I - 1)$ is replaced by I wherever it appears. Note, however, that we must treat all post hoc comparisons the same way even though all but one may be contrasts. This is necessary to maintain the protection level. The power of the tests will of course be reduced accordingly.

Scheffé with Unequal Variances. It should be remembered that just as with planned comparisons the validity of post hoc comparisons depends somewhat on the validity of the assumption of equal variances. Because of the general conservatism of the tests, violations of the assumption probably have less effect than with planned comparisons. Nevertheless, if the variances are unequal, the results of post hoc comparisons may be suspect. An appropriate transformation on the data (if one exists) might make the variances more equal.

Alternatively, Brown and Forsythe (1974) have proposed an approximate Scheffé test for the case of unequal variances. Their method is simply a modification of the method described earlier for planned comparisons with unequal variances. Now, however, in keeping with the fact that we are doing post hoc tests, the F ratio is $C_k^2/[(I-1)D_k]$. The numerator degrees of freedom are $(I-1)$ as for the ordinary Scheffé test, and the denominator degrees of freedom are d_k as for the approximate planned comparison.

We will illustrate by redoing the test comparing μ_4 with μ_3, without assuming equal variances. (See Table 3.3 for the sample estimates of the means and variances.) For this test,

$$
\begin{aligned}
C &= 13.750 - 8.625 = 5.125, \\
D &= (1/8)(2.786 + 9.696) = 1.56025, \\
d &= (8^3 - 8^2)(1.56025^2)/(2.786^2 + 9.696^2) \simeq 11, \\
F'_{(3,11)} &= 5.125^2/[(3)(1.56025)] = 5.61, \\
p &= .014.
\end{aligned}
$$

We obtained a larger F ratio using the approximate method with these data because the variance of group A_4 was much smaller than the variances of the other three groups. However, our denominator degrees of freedom are now only 11. There is usually a loss of power due to the decrease in degrees of freedom.

Relationship Between S and F. The S method is related to the overall F test in an interesting and useful way. It can be proved that the p value of the overall F test is the smallest protection level (i.e., value of α) at which any S method post hoc contrast can be significant. That is, if the overall F test is just significant at the α significance level, then there is exactly one post hoc contrast that would be just significant at an α protection level (and none would be significant if α were smaller). For this contrast,

$$ c_{ik} = n_i(\overline{X}_{i.} - \overline{X}_{..}). $$

If the overall F test is not significant at the α level, then no post hoc contrast can be significant at an α protection level. (This is why α is a protection level.)

Even if a significant post hoc test exists, we may not be interested in testing it. Nevertheless, this information can influence the decision to do post hoc tests. If we are interested only in results that reach a particular protection level, and the overall F test does not reach that level of significance, then no post hoc contrast will reach it. For example, for the data of Table 3.4, no S method post hoc contrast can be significant if $\alpha < .08$.

THE TUKEY METHOD

The Tukey, or T, method of post hoc comparisons is similar in principle to the S method, but the calculations are different. It is strictly limited to tests of contrasts with equal sample sizes, and it is based on a distribution known as the *Studentized range distribution*. The Studentized range distribution is defined as follows: Let z_i be a normally distributed random variable with a mean of zero and a variance of 1, and suppose that k values of z_i are sampled randomly and independently. Let $z_{\max}$ be the largest value sampled and $z_{\min}$ be the smallest, so $z_{\max} - z_{\min}$ is the range of the z_i values in the sample. Let y, independent of the z_i, have a chi-square distribution with ν degrees of freedom. Then

$$t'_{(k,\nu)} = (z_{\max} - z_{\min})/(y/\nu)^{1/2}$$

has a Studentized range distribution with parameters k and ν. The Studentized range distribution is tabled in Appendix B.

The T method cannot be used unless all of the groups contain the same number of scores. The test is made by calculating the statistics

$$
\begin{aligned}
C''_k &= n^{1/2}C_k/[(1/2)\Sigma_i \mid c_{ik} \mid], \\
t'_{(I,N-I)} &= \mid C''_k \mid /MS_w^{1/2},
\end{aligned}
$$

and finding the significance of t' in the table of the Studentized range distribution. Notice that C''_k differs form C'_k in only one way; we divide by $1/2$ the sum of the absolute values of the c_{ik} instead of by the square root of the sum of their squares.

We will illustrate the T method by testing the null hypothesis $H_0 : 3\mu_4 - \mu_3 - \mu_2 - \mu_1 = 0$ on the data of Table 3.1. The calculations are

$$
\begin{aligned}
C'' &= 8^{1/2}(15.5)/[(1/2)(3+1+1+1)] = 14.61, \\
t'_{(4,28)} &= 14.61/7.384^{1/2} = 5.38, \\
p &< .01.
\end{aligned}
$$

Once again, if a specific significance level is chosen in advance, the calculations can be simplified by rejecting the null hypothesis if and only if

$$\mid C''_k \mid \geq t'_{\alpha(N-I)}MS_w^{1/2}.$$

The robustness of the T method is not completely known. It appears, however, to be similar to planned comparisons; it is probably robust to the assumption of normality but not to the assumptions of random sampling and equal variances.

Relationship of T Method to an Overall Test of Significance. Just as the S method is closely related to the overall F test, the T method is closely related to a different overall test. If the T method is used to test the difference between the largest and smallest of the I obtained means, the result is an overall test of the null hypothesis that there are no differences among the means of the I

populations. For the data of Table 3.1, this would be the difference between groups A_4 and A_1. The t' value of 7.28 is significant beyond the .01 level.

The theory behind this kind of overall test has been as explicitly derived as for the F test, and mathematically it is a perfectly legitimate test. However, it is rarely as powerful as the F test, so it is usually not recommended. It has more power than the F test when most of the group means are equal or nearly equal, with only a few means very different from the others.

The T method of testing for overall differences is related to individual tests by the T method in the same way that the overall F test is related to tests by the S method. The level of significance of the difference between the largest and smallest obtained means is the smallest protection level (i.e., value of α) that could be attained in any test by the T method. For the data of Table 3.4, for example, the overall T test is made on the difference between groups A_4 and A_1. The t' value of 3.33 is not significant at the .10 level; therefore we can be certain that no contrast tested by the T method would be significant at a protection level of .10.

SIMULTANEOUS CONFIDENCE INTERVALS

The theory behind post hoc tests is easily adapted to finding *simultaneous* confidence intervals. A set of $100(1 - \alpha)\%$ simultaneous confidence intervals is constructed so that the probability is $(1 - \alpha)$ that *every* confidence interval in the set will contain the true value of the corresponding parameter (i.e., the probability is only α that one or more confidence intervals will be incorrect).

The $100(1 - \alpha)\%$ confidence intervals for any set of comparisons are easily found. For example, if we use the Bonferroni method with r potential comparisons, each confidence interval is

$$C_k - S \le \psi_k \le C_k + S,$$

where

$$S = t_{\alpha/r(N-I)}[MS_w \Sigma_i(c_{ik}^2/n_i)]^{1/2}.$$

That is, we use α/r instead of α with the Bonferroni method.

For the Scheffé method,

$$C_k - S' \le \psi_k \le C_k + S',$$

where

$$S' = [(I - 1)MS_w F_{\alpha(I-1,N-I)}\Sigma_i c_{ik}^2/n_i]^{1/2}.$$

For the Tukey method,

$$C_k - T' \le \psi_k \le C_k + T',$$

where

$$T' = (MS_w/n)^{1/2} t'_{\alpha(I,N-I)}[(1/2)\Sigma_i \mid c_{ik} \mid].$$

INCREASING THE POWER OF POST HOC COMPARISONS

The power attained by post hoc comparisons depends greatly on the size of the set of possible comparisons. By limiting the set of potential post hoc comparisons, we can increase their power.

T Method. To see how this can work with the T method, consider the situation discussed previously, where A_2, A_3, and A_4 represent groups given three different drugs, with no a priori reasons for assuming that any specific drug would have a greater effect than the others. We first did a planned comparison on the difference between the control group, A_1, and the mean of the three experimental groups, A_2, A_3, and A_4. Then we did an F test on the means of the three experimental groups. Presumably, having already tested for a difference between the control and experimental groups, we now wish to focus our attention on differences among the three experimental groups. When testing these differences by the T method, we calculate t' just as before; but because only the three groups are being compared, the result can be regarded as $t'_{(3,28)}$ instead of $t'_{(4,28)}$. The increase in power can be seen in the fact that the difference between A_4 and A_2 (with a t' value of 3.51) is significant at the .05 level using $t'_{(3,29)}$, but not when using $t'_{(4,28)}$. In general, if all of the potential contrasts to be tested by the T method are limited to some subset of r groups, then the value of t' can be regarded as $t'_{(r,N-I)}$ rather than $t'_{(I,N-I)}$, usually with some increase in power.

  Note: Not all statisticians agree with this recommendation. It does not hold the experimentwise error rate to α. If you want the experimentwise error rate to be α, you will have to modify this procedure. For example, you might test the planned comparison at a significance level of $\alpha/2$, and then do the post hoc tests at the protection level of $\alpha/2$.

S Method. A similar principle holds true for the S method. If all of the potential comparisons to be made by the S method are limited to a subset of r groups, then the SS_k can be regarded as having $(r-1)$ instead of $(I-1)$ degrees of freedom. The test is then

$$F'_{(r-1,N-I)} = SS_k/[(r-1)MS_w].$$

If, as postulated above, all post hoc tests are limited to differences between the means of the three experimental groups, the difference between groups A_4 and A_2 is tested by

$$
\begin{aligned}
SS_k &= (8)(3.375)^2/2 = 45.56, \\
F'_{(2,28)} &= 45.56/[(2)(7.384)] = 3.09, \\
p &= .062.
\end{aligned}
$$

The same test, with $(I-1)$ degrees of freedom, is not significant at the .10 level. (However, to take advantage of this increase in power we must choose the r groups before seeing the data.)

For the S method, this approach can be generalized. If all of the potential post hoc contrasts are orthogonal to a given set of r' orthogonal contrasts, then the SS_k for the post hoc contrasts can be tested with $(I - r' - 1)$ instead of $(I - 1)$ degrees of freedom. In the example above, by limiting the post hoc comparisons to the three experimental groups, we made them all orthogonal to the planned comparison on $H_0(1) : \mu_1 - (1/3)\mu_2 - (1/3)\mu_3 - (1/3)\mu_4 = 0$. Because our post hoc comparisons were all independent of this one planned comparison, the SS_k had two degrees of freedom instead of three. (Once again, the r' orthogonal contrasts must be chosen before seeing the actual data.)

We can illustrate the extension of this principle to more complicated analyses by applying it to the data in Table 4.1. These are also hypothetical data, obtained as follows. To investigate students' attitudes toward school, an educator gave a questionnaire to a number of students in the first, second, and third grades. A high score on the questionnaire indicated a favorable attitude toward school. According to this school's policy, first grade pupils were divided into three classes: "slow learners" (group A_1), "average learners" (group A_2), and "fast learners" (group A_3). Children in the second and third grades were divided into two classes, but the division was random. Because the questionnaire required considerable time to administer, only four pupils from each class were tested. Table 4.1 contains the scores of the pupils from each of the seven classes. No overall F test was performed because the educator had a number of specific questions to ask by means of planned and post hoc comparisons. The mean square within was 108.1. The first question the educator wished to ask was whether the average scores in the three grades were equal. The null hypothesis was

$$H_0(1) : (\mu_1 + \mu_2 + \mu_3)/3 = (\mu_4 + \mu_5)/2 = (\mu_6 + \mu_7)/2.$$

This hypothesis is true if and only if the following null hypotheses are true (you should verify this statement):

$$H_0(1a) : \quad 4\mu_1 + 4\mu_2 + 4\mu_3 - 3\mu_4 - 3\mu_5 - 3\mu_6 - 3\mu_7 = 0,$$
$$H_0(2a) : \quad \mu_4 + \mu_5 - \mu_6 - \mu_7 = 0.$$

The SS_k for these two contrasts are

$$SS_{1a} = 1067.86,$$
$$SS_{2a} = 351.56.$$

If $H_0(1)$ is true, the sum of these two values is a chi-square with two degrees of freedom, so

$$F_{(2,21)} = [1067.86 + 351.56)/2]/108.1 = 6.57,$$
$$p = .007.$$

Thus, the educator can confidently conclude that there are differences in attitude among the three grades.

TABLE 4.1. Hypothetical data with seven groups of four subjects each.

	G_i			G_2		G_3		
	A_1	A_2	A_3	A_4	A_5	A_6	A_7	Sum
	35	41	41	31	19	31	31	
	22	38	42	20	9	24	37	
	37	73	43	31	40	26	45	
	31	50	65	25	33	41	47	
ΣX	125	202	191	107	101	122	161	1,009
$\overline{X}$	31.25	50.50	47.75	26.75	25.25	30.50	40.25	36.04
ΣX^2	4,039	10,954	9,519	2,947	3,131	3,894	6,627	41,111
$\hat{\sigma}$	6.65	15.84	11.53	5.32	13.91	7.59	6.99	

$$SS_w = 2269.75 \quad df = 21 \quad MS_w = 108.1.$$

	\multicolumn{6}{c}{Post hoc contrasts $[F'_{(4,21)}]$}

	G_1 vs. G_2	G_1 vs. G_3	G_3 vs. G_2	A_2 vs. A_1	A_3 vs. A_1	A_2 vs. A_3
C_k	17.167	7.792	9.375	19.25	16.50	2.75
F	3.27	0.67	0.81	1.71	1.26	0.03
p	0.031	0.62	0.53	0.19	0.32	1.00

The second question is whether there are attitude differences among students in different classes within the first grade:

$$H_0(2) : \mu_1 = \mu_2 = \mu_3.$$

The numerator sum of squares for this hypothesis can be found by regarding the three first-grade classes as the only three levels in the experiment; i.e., we calculate SS_{bet} on only A_1, A_2, and A_3. The result is

$$F_{(2,21)} = (867.17/2)/108.1 = 4.01$$
$$p = .034$$

Therefore, the educator can conclude with some confidence that there are differences in attitude among the pupils in the different classes of the first grade.

We do not expect to find differences between classes in the second and third grades, but there are two degrees of freedom left out of the original 6, and we might as well use them to test for these differences:

$$H_0(3) : \mu_4 - \mu_5 = 0,$$
$$F_{(1,21)} = 4.50/108.1 = .042.$$
$$H_0(4) : \mu_6 - \mu_7 = 0,$$

$$F_{(1,21)} = 190.12/108.1 = 1.76,$$
$$p = .20.$$

There appear to be few or no differences between classes in the second and third grades.

The educator would now like to make post hoc tests for specific differences between grades and between classes within the first grade. None of the tests will be between classes in the second and third grades, so the comparisons actually tested will all be independent of $H_0(3)$ and $H_0(4)$. Consequently, SS_k can be regarded as having 4 degrees of freedom (i.e., 6 - 2) instead of 6. Testing the difference between grades one and two by the S method, we get

$$H_0(1,2) : 2\mu_1 + 2\mu_2 + 2\mu_3 - 3\mu_4 - 3\mu_5 = 0,$$
$$SS_{1,2} = 1414.5,$$
$$F'_{(4,21)} = 1414.5/[(4)(108.1)] = 3.27,$$
$$p = .032.$$

If, instead, we had used 6 degrees of freedom, the result would have been:

$$F'_{(6,21)} = 1414.5/[(6)(108.1)] = 2.18,$$
$$p = .087.$$

The difference in power is obvious. The results of the remaining tests, both on differences between grades and on differences between classes, are shown in Table 4.1. The difference between grades one and two is the only significant difference.

Note: This approach increases the power, but at the cost of increasing the experimentwise error rate. Appropriate caution as discussed above should be used when applying this approach.

COMPARISON OF THE METHODS

Usually the greatest power will be achieved when a small number of tests is selected in advance and tests are made by the Bonferroni method. However, if the number of potential comparisons is large, or if we do not want to limit ourselves to a specific list, the S and T methods have more power.

For a set of contrasts, the choice between the S and T methods depends on a number of factors. First, it should be remembered that the T method is more limited as to the kinds of data and comparisons for which it can be used. For the T method to be used, all of the groups must contain the same number of scores and the comparisons must all be contrasts. If either of these requirements is not met, the T method cannot be used.

If both requirements are met, the choice of a method depends primarily on the kinds of comparisons being made. The T method was designed initially for studying simple pairwise differences between means, and it was later extended to more complex tests. The S method was designed initially to test more general kinds of comparisons. Consequently, the T method tends to be more sensitive than the S method to simple pairwise differences between means, but the S method is often more sensitive when testing more complex comparisons. If most or all of the contrasts in which we are interested are simple pairwise differences between means, the T method is probably the best choice. However, if we are also interested in testing more complex comparisons, the S method will probably have more power.

CRITIQUE OF THE S AND T METHODS

Unfortunately, there is a problem with the application of either the S or the T method to multiple comparisons. The problem lies in the fact that both methods limit the probability of making *one or more* Type I errors to α. However, they say nothing about *how many* Type I errors we might make, given that we have already made one.

For example, consider an experiment with three groups and a very large N. Suppose (unknown to us, of course) all three population means are equal. We wish to test all pairs of means by the T method, using $\alpha = .05$. Suppose that this experiment is one of those 5% in which we make at least one Type I error. What then will be the probability of making another? By some rather complicated calculations we can show that the probability of making a second Type I error, *conditional on having made the first*, is .14. (For $\alpha = .01$, the probability is .10.) Having made one Type I error, there is an unexpectedly high probability of making a second. Moreover, if the tests are not limited to pairwise differences between means, this conditional probability may be much higher. (It *can* be very close to 1.0, though that is unlikely.)

Although Type I errors are relatively rare with the S and T methods, they may occur in bunches when they do occur. One is therefore wise to be cautious about experiments in which a large number of unexpected T or S tests are significant.

CONFIDENCE INTERVALS FOR INDIVIDUAL MEANS

Scheffé (1959, p. 79) has presented a method of finding confidence intervals for the means of I groups in which the probability is $1 - \alpha$ that all of the confidence intervals will simultaneously cover the true means. The method is based on still another distribution, known as the *Studentized maximum modulus*. This distribution has been tabled, but the tables are not widely available (the interested reader is referred to Scheffé, 1959).

Pairwise Multiple Comparisons

In some experiments, we may not have any particular contrasts in mind. It may then be desirable to test every group against every other group in a set of pairwise tests. Such a procedure is employed all too often as an alternative

TABLE 4.2. Newman–Keuls (N–K), Ryan, and Welsch methods applied to data in Table 3.1. (Cells contain t' values.)

	A_1	A_3	A_2	A_4	N–K	Ryan	Welsch
			*	*	Critical Value		
A_1		1.95	3.77	7.29	3.86	3.86	3.86
				*			
A_3			1.82	5.33	3.50	3.68	3.50
				*			
A_2				3.51	2.90	3.33	3.33
A_4							

$^*\alpha < .05.$

to careful thought about the data, but sometimes it can be very useful. It is possible to use either the Scheffé or the Tukey test for pairwise comparisons, but other methods have more power.

The first such method was proposed by Newman (1930) and Keuls (1952). A variation was proposed by Duncan (1955). Their tests are no longer considered valid by most statisticians, but they are still in common use. Newer tests do not have the deficiencies of the Newman–Keuls and Duncan tests. However, because the Newman–Keuls method is still in common use, and because the more recently developed methods are variations on it, we will describe the Newman–Keuls method first.

NEWMAN–KEULS METHOD

The Newman–Keuls method, like all of the methods to be described, requires that a protection level, α, be set in advance. Each pair of means is then tested against the same level of α. The first step is to construct a table like the one in Table 4.2 for the data in Table 3.1. (Table 4.2 will be used later to illustrate the method.)

The groups are listed across the top and down the left side in order of increasing values of the obtained cell means. Each cell above the main diagonal represents a potential comparison between the means of the groups heading the corresponding row and column.

The first test is in the cell in the upper right-hand corner; it compares the two groups whose means are most different. The test is identical to the Tukey test, using the Studentized range distribution with numerator parameter I and denominator degrees of freedom $(N - I)$. If this test is not significant, we conclude that there are no significant differences among the means; we *cannot* then conduct any other tests.

If it is significant, we test the two cells adjacent to the upper right-hand corner; these cells represent the pairs of groups whose means are next furthest apart. We again calculate a Tukey statistic for each of these tests. However, we compare them with the tabled value of the Studentized range distribution having numerator parameter $(I - 1)$ instead of I. Because the first test was significant, we have already rejected the possibility that all of the means are equal. Therefore we need now consider only the possibility that $(I - 1)$ of the means might be equal.

If either of these two tests is nonsignificant, then all tests represented by cells below and/or to the left of that cell must automatically be considered nonsignificant. This requirement is absolutely essential; it imposes consistency on the tests. It prevents smaller differences from being declared significant when larger differences are not, and it keeps the protection level at the preset value of α. If either test is significant, more tests can be made.

The next tests to be made are represented by the cells adjacent to the last cells tested. Each is tested by the Tukey statistic, but now the numerator parameter is $(I - 2)$.

In general, we proceed through the table from the upper right corner toward the main diagonal, reducing the numerator parameter by one each time we move to a new diagonal. Throughout, we impose the consistency requirement: if the test represented by any given cell is not significant, then all tests in cells below and/or to the left of that cell are automatically nonsignificant as well. The testing continues until every cell has been either tested or declared nonsignificant by the consistency requirement.

The calculations for the data in Table 3.1 are in Table 4.2. The chosen protection level is $\alpha = .05$. The first test is between μ_1 and μ_4. The obtained value of t' for this comparison is 7.29, larger than the value of 3.86 needed for the Studentized range statistic at the .05 level with $I = 4$ and $\nu = 28$. If this test had not been significant, we would have stopped. Because it is significant, we proceed to the two tests on the next diagonal, testing μ_1 against μ_2 and μ_3 against μ_4. The t' values for these two tests are 3.77 and 5.33, respectively. They are compared with the value of 3.50 needed for the Studentized range statistic with $I = 3$ and $\nu = 28$. (Note that for this test we use $I = 3$ instead of $I = 4$.)

If either of these tests had been nonsignificant, then all tests below and/or to the left of the corresponding cell would also have been nonsignificant. However, both are significant, so we proceed to the three tests on the last diagonal. The obtained t' values of 1.95, 1.82, and 3.51 are compared with the value of 2.90 needed for the Studentized range statistic at the .05 level with $I = 2$ and $\nu = 28$. Only one of these, the comparison between μ_2 and μ_4, is significant. We conclude that μ_1 is significantly different from μ_2 and μ_4, and that μ_4 is significantly different from μ_2 and μ_3.

The Newman–Keuls test has two main problems. The first is that if one Type I error is made, there is a higher than expected probability of making another. In fact, with $\alpha = .05$, if a Type I error is made on the first test, the probability of a second Type I error may be higher than .20.

The other problem is more serious. The Newman–Keuls method provides a protection level of α only if all cell means are equal to each other. If some pairs of means are equal, while others are not, there may be a much higher probability of making a Type I error. For example, consider a case with six groups where

$$\mu_1 = \mu_2 << \mu_3 = \mu_4 << \mu_5 = \mu_6.$$

In this example there are three pairs of equal means, but each pair is very different from each other pair. Let us also assume that the differences are large enough to be almost certainly detected at $\alpha = .05$. Then the Newman–Keuls method eventually reduces to three t tests, μ_1 versus μ_2, μ_3 versus μ_4, and μ_5 versus μ_6, on the last diagonal. However, if each test is a t test at the α level of significance, the probability is $1 - (1 - \alpha)^3 \simeq .14$ that one or more Type I errors will be made. This is much higher than the supposed protection level of .05.

The Duncan method is similar to the Newman–Keuls, but it gives even higher probabilities of Type I errors. Consequently, neither method can be recommended.

Variations on Newman–Keuls

A number of variations on the Newman–Keuls method have been proposed to guarantee a protection level of α in all cases.

One such proposal is attributed to Ryan (Welsch, 1977). The Ryan method follows the same procedures as the Newman–Keuls, but it adjusts the α level used when comparing the obtained statistic with the tabled value. Let I' be the numerator parameter used for finding the value in the Studentized range table. Then $I' = I$ for the first test (the cell in the upper right-hand corner), $I' = (I - 1)$ for the next two tests, and so on until $I' = 2$ for the tests closest to the diagonal. Ryan's method adjusts the significance level for each test, using $\alpha_{I'} = (I'/I)\alpha$ instead of α. Thus, as we work our way toward the main diagonal, each new set of comparisons is tested with a smaller value of $\alpha_{I'}$.

The values in Table 4.2 can be used to illustrate Ryan's method. The first test, comparing μ_1 with μ_4, is identical to the Newman–Keuls test. However, for the next two tests, we compare the obtained t' with the tabled value for the .0375 level ($\alpha_3 = 3/4\alpha = .0375$) with $I' = 3$ and $\nu = 28$. This value has to be found by interpolation between the .05 and .01 levels; the method of logarithmic interpolation described in Chapter 1 works well. The interpolated value is 3.68; this is slightly larger than the value of 3.50 needed for the Newman–Keuls but slightly smaller than the value of 3.86 that would be needed for a Tukey test. Thus, the Ryan method has less power than the Newman–Keuls but more power than a set of Tukey tests.

Both tests are significant, so we proceed to the final diagonal. Here the Studentized range value needed is the value for the .025 level ($\alpha_2 = 2/4\alpha = .025$)

with $I' = 2$ and $\nu = 28$. Again by interpolation, we obtain the comparison value of 3.33. Thus for this particular example, the Ryan method gives exactly the same results as the Newman–Keuls. Of course in other cases the Newman–Keuls may give significant results when the Ryan does not. (The reverse can never be true.) The price paid for a guaranteed protection level is a small to moderate decrease in power.

Einot and Gabriel (1975) showed that slightly more power could be achieved without losing the guaranteed protection level by using a slightly larger value of $\alpha_{I'}$. The value they proposed was $\alpha_{I'} = 1 - (1 - \alpha)^{I'/I}$. However, in most cases the difference between this value and Ryan's value is small—for our example, α_3 would be .0377 instead of .0375, and α_2 would be .0253 instead of .025—and calculations are more difficult both for finding $\alpha_{I'}$ and for interpolating in the Studentized range table.

Welsch (1977) found a way to increase the power even more. He used either Ryan's or Einot and Gabriel's value of $\alpha_{I'}$ for all values of I' except $I' = I - 1$. For $I' = I - 1$, he set $\alpha_{I'} = \alpha$. In effect this means that the first three tests are all made by the Newman–Keuls method, while the remaining tests follow the procedure of either Ryan or Einot and Gabriel. Welsch's method is illustrated in Table 4.2. It increases power somewhat, and it simplifies the calculations. However, there will tend to be more "bunching" of Type I errors. (I.e., if one Type I error is made, there is a high probability of making another one as well.)

METHODS BASED ON THE F TEST

The methods described above are all based on the Studentized range distribution. They are generally thought of as methods for determining where the differences lie, while the overall F test determines whether there are any differences at all. However, as we stated earlier, the Tukey test on the largest pairwise difference among the means is an alternative to the overall F tests. Thus, the methods described above first test whether there are any overall differences among the means (by the test on the cell in the upper right-hand corner), and then repeat this test on successively narrower ranges of means until no significant differences are found.

We can follow exactly the same procedures, using the F test instead of the Studentized range to find overall differences. We can adjust the α level by either Ryan's method or by Einot and Gabriel's. We may also use the variation proposed by Welsch. However, we calculate an F ratio instead of t'. The method is best described by an illustration. We will again use the data in Table 3.1 as our example. The calculations are in Table 4.3.

We begin by constructing the same table as for the methods described above, and we proceed to test in the same sequence. The first test, in the upper right-hand corner, is a test of the null hypothesis that there are no differences among the means. That test is the ordinary F test on all of the groups. From Table 3.1 we find that $F = 9.60$, so that value of F is entered in the upper right-hand corner of Table 4.3. Just as for the ordinary F, we

TABLE 4.3. Application of F variation on Ryan and Welsch methods to data in Table 3.1. (Cells contain F ratios.) F ratios in parentheses are not calculated if Ryan's method is used.

	A_1	A_3	A_2	A_4	Critical Value Ryan	Welsch
			*	**		
A_1		(1.90)	3.56	9.60	2.95	2.95
				**		
A_3			(1.66)	7.35	3.71	3.34
				**		
A_2				6.17	5.61	5.61
A_4						

*$\alpha < .05$ by Welsch method.
**$\alpha < .05$ by both methods.

compare it with the tabled value of 2.95 for the .05 level with $\nu_1 = 3$ and $\nu_2 = 28$. In other words, this first test is identical to the overall F test in every respect (except that α must of course be set in advance).

The second test, comparing μ_1 with μ_2, is actually a test of the null hypothesis that $\mu_1 = \mu_3 = \mu_2$ That is, it is a test of the null hypothesis that there are no differences in the *range* of means between μ_1 and μ_2. To conduct the test, we temporarily ignore group A_4 and calculate MS_{bet} from groups A_1, A_3, and A_2. However, the denominator of the F ratio is the MS_w from the overall analysis on all four groups. (That is still our best estimate of the error unless we suspect unequal variances; then we use MS_w based only on the groups being compared.) For our example, $MS_{bet} = 26.29$ and $F = 26.29/7.384 = 3.56$; that value is shown in the appropriate cell of Table 4.3.

If we are using Ryan's method, we compare the obtained value of 3.56 with the tabled value of F for the .0375 level with $\nu_1 = 2$ and $\nu_2 = 28$. ($\nu_1 = 2$ because an F test on three groups has two numerator degrees of freedom.) Using interpolation as before, we find that the critical value needed for the .0375 level is $F = 3.71$. This is larger than the obtained value, so we accept the null hypothesis, and we conduct no more tests in cells below or to the left of the one we just tested.

If we had used the method of Einot and Gabriel, our α would have been .0377 instead of .0375, but this small change would not have affected the final result. (The critical value would have been 3.70.)

Although we can no longer test μ_1 against μ_3, or μ_3 against μ_2 (because the previously tested null hypothesis was accepted), we can still test μ_3 against μ_4. We regard this as an overall test of $H_0 : \mu_3 = \mu_2 = \mu_4$. For this test we temporarily ignore group A_1, and we do an F test on groups A_3, A_2, and A_4. Then $MS_{bet} = 54.29$, and $F = 54.29/7.384 = 7.35$, which is larger than the tabled value of 3.71 for $\alpha = .0375, \nu_1 = 2, \nu_2 = 28$.

The final test compares μ_2 with μ_4. These two means are adjacent in Table 4.3, so this last test reduces to an ordinary t test. For consistency, Table 4.3 contains $t^2 = F = 6.17$ instead of t. The tabled value is 5.61 for $\alpha = .025$, with $\nu_1 = 1$ and $\nu_2 = 28$. Because the obtained F is larger than the tabled value, we reject this null hypothesis.

From this analysis, we conclude that group A_4 differs from the other three groups, but that the other three do not differ significantly from each other.

If we use Welsch's variation, the tests in the second diagonal are made at the .05 level. Then the critical value is $F = 3.34$, and we reject both null hypotheses. We then test all three cells on the final diagonal, using $\alpha = .025$. The only significant difference is between A_2 and A_4. We conclude that A_4 differs from the other three groups, but that the other three groups do not differ from each other.

Note that although these tests seem superficially similar to the Scheffé test, they are actually very different. A pairwise test using the Scheffé method would use only the two means being compared. In the F method of pairwise comparisons, a comparison of two means involves every mean between them in value as well. For this reason, the F method has more power than the Scheffé method on each individual test. The F method "pays" for this increase in power by being limited to pairwise comparisons, by being bound by the consistency rule, and by using smaller values of α in successive tests.

COMPARISON OF THE METHODS

Like the S and T methods of post hoc comparison, all of the pairwise comparison methods discussed here have a tendency to produce multiple significant results. If one Type I error occurs, the probability of another is fairly high. In fact, it is higher for pairwise comparisons than for the S and T methods. If you perceive this to be a problem with your data, you can reduce it by making all pairwise tests by either the S or the T method. It is perfectly legitimate to do this, and it reduces the probability of multiple Type I errors. However, the price is a decrease in power.

The F methods share the basic properties of the methods based on the Studentized range distribution. They guarantee a protection level of α, but when Type I errors do occur, there is a tendency for multiple Type I errors to occur. Ryan's $\alpha_{I'}$ gives less power than Einot and Gabriel's, using either F or t', but the difference is usually small, and the calculations are simpler. Welsch's method gives the most power, but there is an even greater tendency for multiple Type I errors to occur (about as great as for the Newman–Keuls test).

However, methods based on the F distribution have some advantages over those based on the Studentized range. Just as the F test usually has more power than the Studentized range as an overall test, pairwise comparisons

TABLE 4.4. Test in Table 4.3, without the assumption of equal variances. (Cells contain F ratios.)

	A_1	A_3	A_2	A_4	Ryan	Welsch
				*	Critical Value	
A_1		- - -	2.95	9.60	2.95	2.95
				*		
A_3			- - -	7.64	3.861	3.47
				*		
A_2				7.84	6.30	6.30
A_4						

$^*\alpha < .05$ by both methods.

based on F should usually have more power than those based on t'. The exception would occur when two means are very different from each other, while all of the other means are clustered together about halfway between them.

The F methods can also be applied when t' methods cannot. Methods based on the Studentized range distribution can be used only if both the variances and the sample sizes are equal. Methods based on F can be used with unequal sample sizes. They can also be used when the variances are unequal, so long as the sample sizes are equal and moderately large. In that case, however, MS_w should be based only on the groups actually being compared. Thus in the above analysis, the comparison of μ_2 and μ_4 would be an ordinary F test on the three groups, A_3, A_2, and A_4, using the MS_w based on those three groups instead of the overall MS_w. The other tests are made similarly; complete results are given in Table 4.4.

OTHER METHODS

A number of othe methods have been proposed for pairwise multiple comparisons. We will briefly describe three.

Peritz (Begun and Gabriel, 1981) has proposed a variation on the above methods; it increases the power still more, while retaining the guaranteed protection level. However, Peritz's method is very complicated. For example, for the analyses in Tables 4.2, 4.3, and 4.4, the acceptance or rejection of $H_0 : \mu_1 = \mu_3$ could depend on the outcome of the seemingly irrelevant test comparing μ_2 and μ_4. Moreover, experts disagree as to whether Peritz' variation significantly increases power.

Ryan (1960) has proposed a method that uses ordinary t tests for all of the comparisons, while still guaranteeing the protection level. He accomplishes this by letting $\alpha_{I'} = 2\alpha[I(I' - 1)]$. His method has the simplicity of using ordinary t tests, but it has less power than any of the methods described above. The method can be applied to unusual problems where data other than simple means are being compared. Suppose, for example, that we are comparing a number of product moment correlation coefficients. The method for comparing pairs of correlation coefficients is well known. (It uses Fisher's Z transformation, and it is described in most introductory statistics tests.) This method could be applied, using Ryan's adjusted significance levels, $\alpha_{I'}$, to make pairwise comparisons among a number of correlation coefficients.

Finally, Welsch (1977) has proposed a method that proceeds from the "inside out," starting with the cells on the innermost diagonal and ending with the cell in the upper right-hand corner. However, his method requires special tables that are not readily available. It is not known whether his method has greater power than the tests described here.

Reviews of many of these methods, with comparisons of their power, can be found in Einot and Gabriel (1975), and in Ramsey (1978).

Exercises

(1.) For the data in Problem 1, Chapter 2:

(a.) Do a post hoc test on the hypothesis

$$H_0 : \mu_5 = (\mu_2 + \mu_3 + \mu_4)/3,$$

assuming that the only post hoc tests the experimenter had intended to make were among the four tranquilizer groups. Do the test twice, first with the S method and then with the T method (report the p values).

(b.) Repeat the test in part a, using the S method, but do not assume equal variances. Which method is preferable for these data?

(c.) Do pairwise tests, by (1) Ryan's method, and (2) Welsch's variation, on groups A_2, A_3, A_4, and A_5, with $\alpha = .05$ (assuming that these were the only groups the experimenter had intended to test).

(d.) Repeat part c, using F tests instead of studentized range tests.

(e.) Repeat part d, but do not assume equal variances.

(2.) For the data in Problem 2, Chapter 2:

(a.) After looking at the data, the experimenter noted that the mean for group 2 was lower than the means for groups 1 and 3. See if this difference is significant, using (1) the S method and (2) the T method (report the p values).

(b.) Repeat the tests in part a, assuming that prior to running the experiment the researcher had decided to limit all post hoc tests to differences among groups A_1, A_2, and A_3.

(c.) Repeat parts a and b using the S method, but do not assume equal variances. Compare your results with those in parts a and b.

(d.) What is the smallest protection level (i.e., the smallest vlaue of α) that could possibly be obtained from the above data using the S method? Using the T method? (It may not be necessary to determine *which* contrast is most significant in order to answer the question.)

(e.) Do pairwise tests, by (1) Ryan's method, and (2) Welsch's variation, on all five groups, with $\alpha = .10$.

(f.) Repeat part e, using F tests instead of studentized range tests.

(g.) Repeat part f, but do not assume equal variances.

(3.) The T method of post hoc comparisons is used on all pairs of means in an experiment having I groups. It happens that all necessary assumptions for these tests are met, and the overall null hypothesis ($\mu_i = \mu$ for all A_i) is true. If all tests are made at the .05 level, would you expect that:

(a.) The probability that at least two tests are significant was likely to be greater than, less than, or equal to .0025? (Explain.)

(b.) The expected number of significant results was likely to be greater than, less than, or equal to .05($I - 1$)? (Explain.)

(4.) Repeat Problem 3, assuming that the tests are done by (1) Ryan's method, and (2) Welsch's variation.

(5.) Do pairwise tests on all seven groups in Table 4.1, using $\alpha = .05$ and

(a.) Ryan's method.

(b.) Welsch's variation.

(c.) Ryan's method with F tests rather than Studentized range tests.

(d.) Welsch's variation with F tests rather than Studentized range tests.

(6.) Do pairwise tests on all six groups in Problem 3-9, using $\alpha = .05$, and

(a.) Ryan's method with F tests rather than Studentized range tests.

(b.) Welsch's variation with F tests rather than Studentized range tests.

(c.) Can Ryan's method be applied using Studentized range tests? Why, or why not?

(7.) Using the kind of reasoning used in this chapter, tell how large the probability of a Type I error can be, using Neuman–Keuls, if

(a.) $\alpha = .05, I = 8$.

(b.) $\alpha = .05, I = 10$.

(c.) $\alpha = .01, I = 8$.

(d.) $\alpha = .01, I = 10$.

5

Two-Way Analysis of Variance

Often, a simple one-way analysis of variance, with or without planned comparisons, is the best way to analyze data. Sometimes, however, simple one-way models are not appropriate, and even when they are appropriate, more complicated methods may be better suited to our needs.

Consider, for example, the data in Table 5.1. Here scores were obtained from six groups of three subjects each. A simple one-way analysis of variance is significant at $p = .013$. A set of planned comparisons could also have been devised and tested.

However, some data can be understood better if they are organized in a two-way table. Such a table not only clarifies some relationships but suggests specific hypotheses for testing. In this chapter, we will discuss the tests, their assumptions, and their interpretations in detail.

Main Effects

The data in Table 5.1 are hypothetical improvement scores made by patients in two psychiatric categories (Group A_1 are schizophrenics, group A_2 are depressives) under each of three different tranquilizers, B_1, B_2, and B_3.

This design suggests the arrangement of the data in Table 5.2, where the rows are the two psychiatric categories and the columns are the three types of drugs. Each 'score' is the difference between that patient's scores on an emotional adjustment scale before and after being given the drug. We wish to ask two questions: (1) In general, is there a difference in the improvement of the schizophrenics versus that of the depressives when given tranquilizers? (2) Is there a difference in general effectiveness among the three drugs?

THE A MAIN EFFECT

The first question can be answered by a planned comparison on the data in Table 5.1. The planned comparison would be

$$H_0(A) : \mu_1 + \mu_2 + \mu_3 - \mu_4 - \mu_5 - \mu_6 = 0. \tag{5.1}$$

Following the procedure outlined in Chapter 3, we obtain for this contrast the values $F = 2.04$ and $p = .18$. Thus, there is not a significant overall difference in improvement between schizophrenics and depressives under these treatments.

TABLE 5.1. Hypothetical data with six groups and three scores per group.

	G_1	G_2	G_3	G_4	G_5	G_6	Sum
	8	8	4	10	0	15	
	4	10	6	6	4	9	
	0	6	8	14	2	12	
ΣX	12	24	18	30	6	36	126
$\overline{X}$	4	8	6	10	2	12	7
ΣX^2	80	200	116	332	20	450	1,198
$\hat{\sigma}$	4	2	2	4	2	3	

	RS	SS	df	MS	F	p	$\hat{\omega}^2$
m		882					
bet	1,092	210	5	42.00	4.75	.013	.51
w	1,198	106	12	8.833			
t		316	17				

TABLE 5.2. Data from Table 5.1, rearranged.

	B_1	B_2	B_3	Row Sum
	8	8	4	
	4	10	6	
A_1	0	6	8	
	—	—	—	
	$\Sigma = 12$	$\Sigma = 24$	$\Sigma = 18$	$\sigma = 54$
	$\overline{X} = 4$	$\overline{X} = 8$	$\overline{X} = 6$	$\overline{X} = 6$
	10	0	15	
	6	4	9	
A_2	14	2	12	
	—	—	—	
	$\Sigma = 30$	$\Sigma = 6$	$\Sigma = 36$	$\Sigma = 72$
	$\overline{X} = 10$	$\overline{X} = 2$	$\overline{X} = 12$	$\overline{X} = 8$
Column	$\Sigma = 42$	$\Sigma = 30$	$\Sigma = 54$	$\Sigma = 126$
Sum	$\overline{X} = 7$	$\overline{X} = 5$	$\overline{X} = 9$	$\overline{X} = 7$

THE B MAIN EFFECT

The second question can also be answered in terms of planned comparisons; here, however, two contrasts must be combined into a single test. Furthermore, it is not immediately obvious what those contrasts should be. The problem can be simplified by relabeling the cells of Table 5.2. We define the two rows to be the two levels of Factor A and the three columns to be the three levels of Factor B; the levels of Factor A are the two categories of patients and the levels of Factor B are the three types of drugs. We then relabel each group, or *cell*, to conform with the levels under which it is found. Thus, instead of being labeled G_1 as in Table 5.1, the first cell is labeled AB_{11} because it is in the first level of Factor A and the first level of Factor B (i.e., in the first row and column). The second cell is labeled AB_{12} because it falls in the first level (row) of Factor A and the second level (column) of Factor B. The six cells are then labeled as follows:

	B_1	B_2	B_3
A_1	AB_{11}	AB_{12}	AB_{13}
A_2	AB_{12}	AB_{22}	AB_{23}

Similarly, the true mean of the population sampled in cell AB_{ij} is designated μ_{ij}, and the obtained sample mean for that cell is $\overline{X}_{ij.}$. Finally, we need some notation to represent the averages of these values, taken over either or both of the subscripts. Accordingly, we define:

$$
\begin{aligned}
I &= \text{number of levels of Factor } A \ (I = 2 \text{ in Table 5.2}), \\
J &= \text{number of levels of Factor } B (J = 3 \text{ in Table 5.2}), \\
\mu_{i.} &= (I/J)\Sigma_j\mu_{ij} \ (= \text{the average of the } \mu_{ij}, \text{ taken over the subscript } j), \\
\mu_{.j} &= (1/I)\Sigma_i\mu_{ij} \ (= \text{the average of the } \mu_{ij}, \text{ taken over the subscript } i), \\
\mu &= (1/I)\Sigma_i\mu_{i.} = (1/J)\Sigma_j\mu_{.j} \ (= \text{the overall mean of all of the } \mu_{ij}), \\
\overline{X}_{ij.} &= (1/n)\Sigma_k X_{ijk} \ (= \text{the average within each cell}), \\
\overline{X}_{i..} &= (1/J)\Sigma_j\overline{X}_{ij.} \ (= \text{the average within a row}), \\
\overline{X}_{.j.} &= (1/I)\Sigma_i\overline{X}_{ij.} \ (= \text{the average within a column}), \\
\overline{X}_{...} &= (1/J)\Sigma_i\overline{X}_{i..} = (1/I)\Sigma_j\overline{X}_{.j.} \ (= \text{the overall mean of all the scores}).
\end{aligned}
$$

With this notation, the second question can be expressed by the null hypothesis

$$H_0(B) : \mu_{.j} = \mu, \quad \text{for all} \ \ j. \tag{5.2}$$

A test of a hypothesis such as this is called a test of the *main effect* of Factor B, or simply the *B main effect*. (Similarly, the test of $H_0(A)$ (Eq. 5.1) is referred to as a test of the *A main effect*.)

To see how $H_0(B)$ can be tested, consider the three obtained means, $\overline{X}_{.1.}$, $\overline{X}_{.2.}$, $\overline{X}_{.3.}$. Each $\overline{X}_{.j.}$ is the average of the I cell means, $\overline{X}_{ij.}$. Because each $\overline{X}_{ij.}$ is normally distributed as

$$\overline{X}_{ij.} \sim N_{(\mu_{ij},\sigma_e^2/n)},$$

it follows that each $\overline{X}_{.j.}$ is distributed as

$$\overline{X}_{.j.} \sim N_{(\mu_{.j.},\sigma_e^2/nI)}.$$

A comparison of this with Equation 5.2 shows that under $H_0(B)$ the $\overline{X}_{.j.}$ are identically distributed independent random variables. The formal identity with the test in Chapter 2 can be seen by comparing Equation 5.2 with Equation 2.1. From here on the derivation follows exactly the same steps as in Chapter 2.

Defining

$$SS_b = nI\Sigma_j(\overline{X}_{.j.} - \overline{X}_{...})^2, \tag{5.3}$$

$$MS_b = SS_b/(J-1), \tag{5.4}$$

we find that

$$SS_b/\sigma_e^2 \sim \chi^2_{(J-1)}$$

so that

$$\begin{aligned} \underline{E}(SS_b) &= (J-1)\sigma_e^2, \\ \underline{E}(MS_b) &= \sigma_e^2. \end{aligned}$$

MS_b fulfills all of the requirements for the numerator of an F test. Under the null hypothesis it is an unbiased estimate of σ_e^2, based on a chi-square variable divided by its degrees of freedom.

Therefore, under the null hypothesis,

$$MS_b/MS_w \sim F_{(J-1,N-IJ)}. \tag{5.5}$$

The denominator degrees of freedom for this F ratio are found from the fact that, in general, the degrees of freedom for MS_w are equal to the total number of scores (N) minus the total number of groups, or cells (IJ). For the data in Table 5.2, these values are

$$\begin{aligned} SS_b = (3)(2)[(7-7)^2 + (5-7)^2 + (9-7)^2] &= 48, \\ MB_b = 48/2 &= 24, \\ F_{(2,12)} = 24/8.833 &= 2.72, \\ p &= .11. \end{aligned}$$

RELATIONSHIP TO PLANNED COMPARISONS

We stated earlier that the null hypothesis $H_0(b)$ could have been tested as two simultaneous contrasts. This follows from the fact, discussed in Chapter 3, that any test of differences between means with ν_1 degrees of freedom in the numerator can be factored into ν_1 different independent contrasts. It may be instructive to illustrate how this can be done with the data in Table 5.2. $H_0(B)$

can be factored in an infinite number of ways; the following two hypotheses illustrate one of them:

$$H_0(B_1): \quad \mu_{11} + \mu_{21} - \mu_{12} - \mu_{22} = 0,$$
$$H_0(B_2): \quad \mu_{11} + \mu_{21} + \mu_{12} + \mu_{22} - 2\mu_{13} - 2\mu_{23} = 0.$$

You should test these two contrasts as an exercise. When you do, you will find that

$$SS_{B_1} + SS_{B_2} = 12 + 36 = SS_b.$$

Similarly, instead of testing $H_0(A)$ as a contrast, we could have tested it using the theory developed for testing $H_0(B)$. This follows from the fact that Equation 5.1 can be stated as

$$H_0(A): \mu_{i.} = \mu \quad \text{for all} \quad i. \tag{5.6}$$

In this form it is identical to Equation 5.1. Finally, the choice of rows and columns in Table 5.2 was arbitrary, so all of the theory for testing $H_0(B)$ will be valid for testing $H_0(A)$ if we simply reverse the roles of I and J and substitute $\overline{X}_{i..}$ for $\overline{X}_{.j.}$ throughout. When we do this, Equations 5.3, 5.4, and 5.5 become, respectively,

$$SS_a = nJ\Sigma_i(\overline{X}_{i..} - \overline{X}_{...})^2,$$
$$MS_a = SS_a/(I-1),$$
$$MS_a/MS_w \sim F_{(I-1, N-IJ)}.$$

For the data in Table 5.2, we have the following values:

$$SS_a = (3)(3)[(6-7)^2 + (8-7)^2] = 18,$$
$$MS_a = 18/1 = 18, \tag{5.7}$$
$$F_{(1,12)} = 18/8.833 = 2.04,$$

and this F ratio is identical to the one obtained in testing the contrast, Equation 5.1.

Interaction

We now have a new problem. When we tested for the main effect of Factor A, the results were not significant; when we tested for the main effect of Factor B, the results were not significant either. We are led to conclude from these tests that the data indicate no reliable differences either in the overall reactions of the two types of patients to the drugs or in the overall effect of the drugs on the patients. Yet the overall F test on all six groups was significant, indicating that reliable differences do exist. The tests on main effects have missed those differences.

An initial clue to the nature of the differences we are looking for can be found by noting that the overall F test on the six groups had five degrees

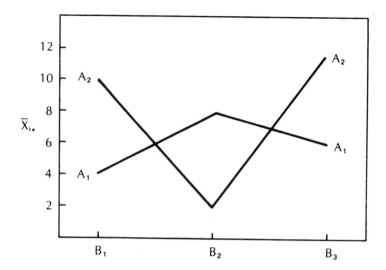

FIGURE 5.1. Cell means from Table 5.2.

of freedom, and this corresponded to a simultaneous test of five orthogonal contrasts. The test of $H_0(A)$ was one contrast; the test of $H_0(B)$ was a simultaneous test of two contrasts. Our tests so far have covered only three of the five possible orthogonal contrasts. (It is trivial to show that Equations 5.1 and 5.2 are independent null hypotheses, so all three of the contrasts already tested are orthogonal.)

Two degrees of freedom remain. We can therefore follow the theory of Chapter 3 and test for any remaining differences among the means of the six groups by calculating

$$SS_{rem} = SS_{bet} - SS_a - SS_b = 210 - 18 - 48 = 144,$$

which has two degrees of freedom. Accordingly,

$$
\begin{aligned}
MM_{rem} &= 144/2 = 72, \\
F_{(2,12)} &= 72/8.833 = 8.15, \\
p &= .006.
\end{aligned}
$$

There are significant differences among the cell means, independent of the differences tested for in the previous section.

But what are these differences? Are they meaningful? A more complete answer comes from Figure 5.1. This figure is a plot of the obtained means in Table 5.2. (Here is another example of using a figure to show important relationships in the data.) It shows clearly that the relative effectiveness of a drug depends on the type of patient to whom it is given. Or, alternatively, the two kinds of patients react differently to the three drugs. The statistician combines both of these sentences into one by saying there is an *interaction* between drugs and patients.

The Model Equation

The nature of the interaction may become clearer if we resort to the strategy, used previously, of constructing a linear model. If we let

$$\alpha_i = \mu_{i.} - \mu, \tag{5.8}$$

$$\beta_j = \mu_{.j} - \mu, \tag{5.9}$$

and X_{ijk} = the kth score in cell AB_{ij}, then

$$
\begin{aligned}
X_{ijk} &= \mu_{ij} + \epsilon_{ijk} \\
&= \mu + (\mu_{i.} - \mu) + (\mu_{.j} - \mu) + (\mu_{ij} - \mu_{i.} - \mu_{.j} + \mu) + \epsilon_{ijk} \quad (5.10) \\
&= \mu + \alpha_i + \beta_j + (\mu_{ij} - \mu_{i.} - \mu_{.j} + \mu) + \epsilon_{ijk}.
\end{aligned}
$$

It is easy to verify that $H_0(a)$ is equivalent to the null hypothesis that all α_i are zero, and that $H_0(b)$ is equivalent to the null hypothesis that all β_j are zero. Equation 5.10 shows, however, that even though α_i and β_j are both zero, the expected value of X_{ijk} may still differ from μ, the grand mean. The amount by which it can differ is defined to be

$$
\begin{aligned}
\alpha\beta_{ij} &= \mu_{ij} - \mu_{i.} - \mu_{.j} + \mu \\
&= (\mu_{ij} - \mu) - (\mu_{i.} - \mu) - (\mu_{.j} - \mu) \quad (5.11) \\
&= (\mu_{ij} - \mu) - \alpha_i - \beta_j.
\end{aligned}
$$

(The symbol $\alpha\beta_{ij}$ contains two Greek letters, but it should be read as a single symbol representing a single quantity. We will use this kind of notation generally to denote interaction terms.) Thus, the interaction, $\alpha\beta_{ij}$, is the amount by which each population cell mean differs from the value we would expect from a knowledge of the row and column means only. Inserting Equation 5.11 into Equation 5.10, we obtain the basic model for a two-way analysis of variance:

$$X_{ijk} = \mu + \alpha_i + \beta_j + \alpha\beta_{ij} + \epsilon_{ijk}. \tag{5.12}$$

Interaction Mean Square

The sum of squares for interaction was obtained above as a residual sum of squares. However, a direct method for calculating it will now be given. We will begin by reviewing the tests for main effects in light of Equation 5.10. We will then show the analogy between these tests and the test for interaction.

However, before we can review the tests for main effects, we must obtain estimates of the terms in Equation 5.10. On the basis of the definitions in Equations 5.8, 5.9, and 5.11, it is easy to show that the following sums are all identically zero:

$$\Sigma_i \alpha_i = \Sigma_j \beta_j = \Sigma_i \alpha\beta_{ij} = \Sigma_j \alpha\beta_{ij} = 0.$$

To obtain the estimates, we take the expected value of X_{ijk} in Equation 5.12. Then averaging over the appropriate subscripts and eliminating sums

TABLE 5.3. Estimates of $\mu, \alpha_i, \beta_j,$
and $\alpha\beta_{ij}$. The values in the
cells are estimates of $\alpha\beta_{ij}$. The
marginal values are estimates of
α_i (row margins), β_j (column margins), and μ (lower right corner).

	B_1	B_2	B_3	$\hat{\alpha}_i$
A_1	-2	+4	-2	-1
A_2	+2	-4	+2	+1
$\hat{\beta}_j$	0	-2	+2	7

that are zero, we obtain the following equations:

$$\begin{aligned}
E(\overline{X}_{ij.}) &= \mu + \alpha_i + \beta_j + \alpha\beta_{ij}, \\
E(\overline{X}_{i..}) &= \mu + \alpha_i, \\
E(\overline{X}_{.j.}) &= \mu + \beta_j, \\
E(\overline{X}_{...}) &= \mu.
\end{aligned}$$

Finally, from these equations we obtain the following unbiased estimates:

$$\begin{aligned}
\hat{\mu} &= \overline{X}_{...}, \\
\hat{\alpha}_i &= \overline{X}_{i..} - \overline{X}_{...}, \\
\hat{\beta}_j &= \overline{X}_{.j.} - \overline{X}_{...}, \\
\hat{\alpha\beta}_{ij} &= \overline{X}_{ij.} - \overline{X}_{i..} - \overline{X}_{.j.} + \overline{X}_{...}.
\end{aligned}$$

Table 5.3 contains the estimates of the quantities for the data in Table 5.2.

From Equation 5.8 we can see that Equation 5.6 is equivalent to the null hypothesis

$$H_0(a) : \alpha_i = 0 \text{ for all } i.$$

As can be seen from Equation 5.7, this hypothesis is tested by finding the sum of the squared estimates of the α_i and multiplying that sum by the number of scores averaged over to obtain the $\overline{X}_{i..}$ values:

$$SS_a = nJ\Sigma_i\hat{\alpha}_i^2. \tag{5.13}$$

Similarly, Equation 5.2 can be rewritten as

$$H_0(b); \beta_j = 0 \text{ for all } j$$

and tested using

$$SS_b = nI\Sigma_j\hat{\beta}_j^2 \text{ (see Eq. 5.3)}.$$

These results suggest that the hypothesis

$$H_0(ab) : \alpha\beta_{ij} = 0 \text{ for all } i, j,$$

that is, that there is no interaction, can be tested by summing the squared estimates of the $\alpha\beta_{ij}$ and multiplying the sum by n, the number of scores averaged over to obtain the $\overline{X}_{ij.}$ values:

$$SS_{ab} = n\Sigma_i\Sigma_j\alpha\hat{\beta}_{ij}^2 = n\Sigma_i\Sigma_j(\overline{X}_{ij.} - \overline{X}_{i..}\overline{X}_{.j.} + \overline{X}_{...})^2. \qquad (5.14)$$

We can see from Table 5.3 that for the data in Table 5.2

$$SS_{ab} = 3[(-2)^2 + 4^2 + (-2)^2 + 2^2 + (-4)^2 + 2^2] = 144.$$

Under the null hypothesis $H_0(ab)$

$$SS_{ab}/\sigma_e^2 \sim \chi^2_{[(I-1)(J-1)]}.$$

The total degrees of freedom for testing contrasts are $(IJ-1)$. We use $(I-1)$ of these to test the A main effect and $(J-1)$ to test the B main effect. Consequently, the degrees of freedom which remain for testing the interaction are

$$(IJ-1) - (I-1) - (J-1) = (I-1)(J-1).$$

Another interpretation of the degrees of freedom for the test of $H_0(ab)$ can be given in terms of the number of estimates that are "free to vary" (in fact, this interpretation is why the parameters of F are called "degrees of freedom"). If we consider the estimates, $\hat{\alpha}_i$, we will see that they must always sum to zero, just as the α_i sum to zero. Consequently, although the data are "free" to determine the values of the first $(I-1)$ estimates arbitrarily, the Ith estimate will always be determined from the other $(I-1)$ plus the restriction that the estimates sum to zero. We say therefore that only $(I-1)$ of the $\hat{\alpha}_i$ are "free" to vary, and the test of $H_0(a)$ has $(I-1)$ degrees of freedom. By the same reasoning, only $(J-1)$ of the $\hat{\beta}_j$ are free to vary, so the test of $H_0(b)$ has $(J-1)$ degrees of freedom. The estimates of the $\alpha\beta_{ij}$, however, must sum to zero over both the rows and columns. Consequently, only $(I-1)$ of the values in any given column are free to vary, and only $(J-1)$ are free to vary in any given row. Thus, only $(I-1)(J-1)$ of the $\alpha\hat{\beta}_{ij}$ are free to vary. It should be clear in Table 5.8 that once $\alpha\hat{\beta}_{11}$ and $\alpha\hat{\beta}_{12}$ are set, the rest of the $\alpha\hat{\beta}_{ij}$ are determined by the restriction that both the rows and columns of the matrix must sum to zero.

The test for the interaction is made by calculating

$$MS_{ab} = SS_{ab}/[(I-1)(J-1)],$$
$$F_{[(I-1)(J-1),N-IJ]} = MS_{ab}/MS_w.$$

For the data in Table 5.2 we have the following:

$$MS_{ab} = 144/2 = 72,$$
$$F_{(2,12)} = 72/8.833 = 8.15,$$
$$p = .006.$$

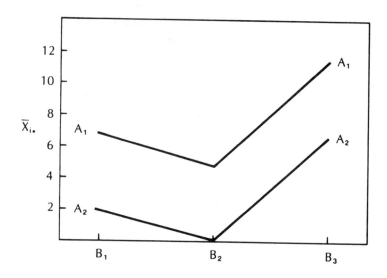

FIGURE 5.2. Sample data with both main effects but no interactions.

We have thus divided the degrees of freedom into tests of three independent null hypotheses on two main effects and an interaction. Of course, these particular tests are not necessarily the ones that should be made; the questions that we wish to ask may be better answered by a different set of tests, e.g., planned comparisons. Nevertheless, these tests are appropriate for the questions we often ask in a two-way analysis of variance.

INTERPRETING INTERACTION

The interpretation of a significant interaction is a problem commonly encountered in two-way analyses of variance. Unfortunately, no general solution can be given to the problem because the scientific rather than the statistical importance of the interaction is usually at issue. Statistically, the meaning of a significant interaction can be found easily with a graph such as Figure 5.1. A general procedure for drawing such a graph is to represent the levels of Factor B as values on the X axis and plot the cell means at those levels as values on the X axis, making a separate curve for each level of Factor A. (Alternatively, of course, the levels of A could be represented on the X axis and a curve could be drawn for each level of B.) If there were no interaction, these lines would be parallel, like those in Figure 5.2; to the extent that they are not parallel, there is an interaction. However, the importance or "meaning" of the fact that the lines are not parallel depends on the particular experiment. For example, Figure 5.1 shows that drug B_2 has a greater effect than drugs B_1 and B_3 on schizophrenics, whereas B_1 and B_3 are more effective with depressives; that is the source of the interaction in these data.

Interpreting Main Effects

Interpreting main effects in the presence of an interaction requires care. In the *absence* of an interaction, the tests on the main effects can be interpreted independently. Figure 5.2 illustrates data from the same kind of experiment as in Figure 5.1; in the data in Figure 5.2 both main effects are significant but there is no interaction. We can conclude that no matter which drug we use, schizophrenics will respond better than depressives, and that no matter which patient we use it on, drug B_3 will be most effective. As a practical matter, the hospital could stock only drug B_3 and give schizophrenics priority if the drug were in short supply.

The situation is very different, however, if there is an interaction. Again, consider Figure 5.1. Even if we rejected the null hypothesis of no B main effect, we could not conclude that drug B_3 was most effective for all patients. The hospital might be wise to stock drug B_2 for schizophrenics and drug B_3 for depressives.

Neither can we conclude from the *absence* of an A main effect that there are no differences between types of patients. Figure 5.1 shows that there are clear and definite differences in the responses to particular drugs. There is no A main effect only because these differences average out when we average responses over drugs.

Simple Effects

When there is an interaction, tests of *simple effects* are sometimes more useful than tests of main effects and interactions. Simple-effect tests are one-way analyses of variance across levels of one factor, performed separately at each level of the other factor. We might, for example, test for the B main effect separately at each level of A. The test at each level can be made as though the other level of A were not even in the experiment. The results of the simple-effects tests at each level of A are given in Table 5.4.

They show that there is a significant difference in the effects of drugs on depressives but there is no significant difference in the effects on schizophrenics. A practical result of these tests might be a decision to stock only drug B_3; it works best on depressives, and all three drugs work about equally well on schizophrenics.

We can increase the power of these tests if we assume that the variances for the two types of patients are equal. Then we can use MS_w from the overall analysis in the denominator of F, increasing the degrees of freedom from 6 to 12. The test on A_1 is still not significant, but the test on A_2 is now

$$F_{(2,12)} = 84/8.833 = 9.51,$$
$$p = .004.$$

The similarity of the separate MS_w values (8.000 and 9.667) to the overall

TABLE 5.4. Tests of simple effects on Factor B.

	RS	SS	df	MS	F	p	$\hat{\omega}^2$
		Level A_1					
m		324	1	324.0	40.5	<.001	
bet	348	24	2	12.00	1.50	.30	.10
w	396	48	6	8.00			
t		72	8				

	RS	SS	df	MS	F	p	$\hat{\omega}^2$
		Level A_2					
m		576	1	576.0	59.6	<.001	
bet	744	168	2	84.00	8.69	.017	.63
w	802	58	6	9.667			
t		226	8				

MS_w of 8.833 lends some support to the assumption of equal variances, but we should also have good a priori reasons to expect equal variances.

Simple tests can also be made across levels of A for each level of B. For this particular experiment, however, they would probably have little practical value.

Note that tests of simple effects are not orthogonal to tests of main effects and interactions. The simple effects tested above, for example, are *not* orthogonal to either the B main effect or the interaction. In fact, they use exactly the same four degrees of freedom (and thus the same variation) as the B main effect and AB interaction. The total sum of squares for the two simple effects is $SS_b + SS_{ab} = 48 + 144 = 192$.

The choice of a standard or a simple-effects analysis depends on your purposes. There is no reason why either test should take precedence over the other. However, if both are done, the cautions given in Chapter 3 about nonorthogonal multiple tests apply.

SIMPLE-EFFECTS TESTS ON MEANS

The simple-effects tests described above *are* orthogonal to both the A main effect and the grand mean. Moreover, the total of two degrees of freedom in the latter tests can be divided into two tests on single means. The two tests are on

$$H_0(A_1) : \mu_{1.} = 0,$$
$$H_0(A_2) : \mu_{2.} = 0.$$

In practical terms, they tell us whether, on the average, each type of patient benefits at all from the drugs.

These tests, in fact, have already been performed; they are the two tests on SS_m in Table 5.4. For these tests both null hypotheses are rejected, as the

TABLE 5.5. Calculational formulas for two-way analysis of variance with equal sample sizes.

	RS	SS	df
m		T^2/N	1
a	$(1/nJ)\Sigma_i t_{i.}^2$	$RS_a - SS_m$	$I - 1$
b	$(1/nI)\Sigma_j t_{.j}^2$	$RS_b - SS_m$	$J - 1$
ab	$(1/n)\Sigma_i\Sigma_j t_{ij}^2$	$RS_{ab} - SS_a - SS_b - SS_m$	$(I-1)(J-1)$
w	$\Sigma_i\Sigma_j\Sigma_k X_{ijk}^2$	$RS_w - SS_{ab} - SS_a - SS_b - SS_m$	$N - IJ$
		$= RS_w - RS_{ab}$	
t		$RS_w - SS_m$	

TABLE 5.6. Analysis of data in Table 5.2.

	RS	SS	df	MS	F	p	$\hat{\omega}^2$
m		882					
a	900	18	1	18.00	2.04	.18	.03
b	930	48	2	24.00	2.72	.11	.09
ab	1,092	144	2	72.00	8.15	.006	.39
w	1,198	106	12	8.833			
t		316	17				

tests on the A main effect and grand mean would lead us to expect (for these particular data).

This discussion has not exhausted the possibilities for specific tests. In fact, the possibilities are limited only by our ingenuity. Remember that the experiment can be regarded as a one-way design, and all of the methods in Chapters 2, 3, and 4 can be applied.

Calculational Formulas

Just as with the overall F tests, the calculations involved in finding mean squares can be simplified considerably by algebraic manipulations on Equations 5.3, 5.7, and 5.14. The resulting formulas are summarized in Table 5.5. Table 5.6 shows the calculations for the data in Table 5.2.

Note that the patterns described in Chapter 2 occur here. First we calculate raw sums of squares, and then we obtain sums of squares by subtraction. Each raw sum of squares is found by first obtaining a set of associated totals (e.g., $t_{ij}, t_{i.}, t_{.j}$), then squaring each total, and adding the squares. The final step consists of dividing by the number of scores in each total (e.g., for RS_a we divide by nJ because each $t_{i.}$ is a total of nJ scores).

To find the sums of squares we set up a hierarchy of effects. A given effect is above another in the hierarchy if its associated totals can be obtained by summing over the totals associated with the other. For example, the totals for the A main effect are the row totals, while the totals for the AB interaction are the total in individual cells. The A totals can be obtained by summing over the AB totals, so A is above AB in the hierarchy.

The complete hierarchy is

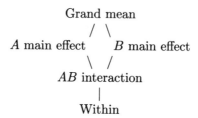

Note that A and B are at the same level. This hierarchy is said to be a *partial ordering*. To find each sum of squares we take the raw sum of squares and subtract the sums of squares of all effects that are above that effect in the hierarchy.

Expected Mean Squares

The expected value of MS_w is, of course, σ_3^2, as in the one-way analysis of variance. MS_w is the same whether the design is treated as one-way or two-way; it is simply the average of the cell variances.

Note that if we substitute $\overline{X}_{i..}$ for $\overline{X}_{i.}$ and (nj) for n in Equation 2.5 (defining SS_{bet}), the result is Equation 5.7 (defining SS_a). It follows that the expected value of MS_w is similar to that of MS_{bet}. The same holds true for MS_b because the labels A and B are arbitrarily assigned to the factors. The expected values are

$$
\begin{aligned}
\underline{E}(MS_a) &= \sigma_e^2 + nJ\Sigma_i\alpha_i^2/(I-1) \\
&- \sigma_e^2 + nJ[I/(I-1)]\sigma_a^2, \\
\underline{E}(MS_b) &= \sigma_e^2 + nI\Sigma_j\beta_j^2/(J-1) \\
&= \sigma_e^2 + nI[J/(J-1)]\sigma_b^2,
\end{aligned}
\tag{5.15}
$$

where

$$
\begin{aligned}
\sigma_a^2 &= (1/I)\Sigma_i\alpha_i^2, \\
\sigma_b^2 &= (1/J)\Sigma_j\beta_j^2.
\end{aligned}
$$

The similarity of these equations to Equation 2.9 should be obvious.

If we let

$$
\begin{aligned}
\tau_a^2 &= [I/(I-1)]\sigma_a^2, \\
\tau_b^2 &= [J/(J-1)]\sigma_b^2,
\end{aligned}
$$

then

$$E(MS_a) = \sigma_e^2 + nJ\tau_a^2,$$
$$E(MS_b) = \sigma_e^2 + nI\tau_b^2.$$

By defining τ_a^2 and τ_b^2 this way, we simplify the formulas for expected mean squares. The value of this simplification will be more apparent in later chapters. For now, notice that the multiplier of τ^2 in each of these equations is just the divisor that we used when we calculated the associated raw sum of squares (see Table 5.5).

Finding the expected value of MS_{ab} is more difficult, but the result is basically the same:

$$E(MS_{ab}) = \sigma_e^2 + \{n/[(I-1)(J-1)]\}\Sigma_i\Sigma_j\alpha\beta_{ij}^2$$
$$= \sigma_e^2 + \{nIJ/[(I-1)(J-1)]\}\sigma_{ab}^2,$$

where

$$\sigma_{ab}^2 = (1/IJ)\Sigma_i\Sigma_j\alpha\beta_{ij}^2,$$

the variance of the $\alpha\beta_{ij}$. We can now define τ_{ab}^2 in a manner analogous to τ_a^2 and τ_b above as

$$\tau_{ab}^2 = \{IJ/[(I-1)(J-1)]\}\sigma_{ab}^2$$

to obtain

$$E(MS_{ab}) = \sigma_e^2 + n\tau_{ab}^2.$$

NOTE ON ANALYZING VARIANCE

With σ_a^2, σ_b^2, and σ_{ab}^2 defined as above, we have again divided (i.e., analyzed) the total variance. More specifically, if we treat the design as a one-way design and define σ_t^2 as in Chapter 2, then

$$\sigma_t^2 = \sigma_e^2 + \sigma_a^2 + \sigma_b^2 + \sigma_{ab}^2. \tag{5.16}$$

Moreover, this corresponds to our division of SS_t and $(N-1)$:

$$SS_t = SS_w + SS_a + SS_b + SS_{ab},$$
$$N-1 = (N-IJ) + (I-1) + (J-1) + (I-1)(J-1),$$
$$= df_w + df_a + df_b + df_{ab}.$$

Note: the values we have called τ_a^2, τ_b^2, and τ_{ab}^2 are defined to be the variances by some statisticians (and some packaged statistical programs). This has the obvious advantage of simplifying the calculations. However, then Equation 5.16 no longer holds. I prefer to retain Equation 5.16 at a small expense in calculations. The τ^2 are introduced only to simplify some equations; they have no meaning in themselves.

Estimating ω^2

Because of the additivity displayed by Equation 5.16, we can define the proportion of variance accounted for by each effect:

$$\begin{aligned}
\omega_a^2 &= \sigma_a^2/\sigma_t^2, \\
\omega_b^2 &= \sigma_b^2/\sigma_t^2, \\
\omega_{ab}^2 &= \sigma_{ab}^2/\sigma_t^2.
\end{aligned}$$

To find approximately unbiased estimates of the ω^2, we must first find unbiased estimates of the σ^2. MS_w is an unbiased estimate of σ_e^2. To find the others we first solve for the associated variances, using the formulas for the expected mean squares. For the A main effect, for example, we solve for σ_a^2 in Equation 5.15 to get

$$\sigma_a^2 = [(I-1)/(nIJ)][\underline{E}(MS_w) - \sigma_e^2].$$

Then each term on the right is replaced by its sample estimate. The estimate of $\underline{E}(MS_a)$ is simply MS_a, and the estimate of σ_e^2 is MS_w, so

$$\hat{\sigma}_a^2 = [(I-1)/(nIJ)](MS_a - MS_w).$$

Similarly,

$$\begin{aligned}
\hat{\sigma}_b^2 &= [(J-1)/(nIJ)](MS_b - MS_w), \\
\hat{\sigma}_{ab}^2 &= [(I-1)(J-1)/(nIJ)](MS_{ab} - MS_w).
\end{aligned}$$

(These can be simplified somewhat by noting that $nIJ = N$.)

The estimated total variance is then the sum of the individual estimates:

$$\hat{\sigma}_t^2 = \hat{\sigma}_e^2 + \hat{\sigma}_a^2 + \hat{\sigma}_b^2 + \hat{\sigma}_{ab}^2,$$

and each $\hat{\omega}^2$ is the ratio of the appropriate estimates as before.

For the data in Table 5.2,

$$\begin{aligned}
\hat{\sigma}_a^2 &= (1/18)(18 - 8.833) = .509, \\
\hat{\sigma}_b^2 &= (2/18)(24 - 8.833) = 1.685, \\
\hat{\sigma}_{ab}^2 &= (2/18)(72 - 8.833) = 7.019, \\
\hat{\sigma}_e^2 &= 8.833, \\
\hat{\sigma}_t^2 &= 8.833 + .509 + 1.685 + 7.019 = 18.046. \\
\hat{\omega}_a^2 &= .509/18.046 = .03, \\
\hat{\omega}_b^2 &= 1.685/18.046 = .09, \\
\hat{\omega}_{ab}^2 &= 7.019/18.046 = .39.
\end{aligned}$$

POWER

For the overall F tests in Chapter 2 we found that when the null hypothesis was false, the F ratio was distributed as noncentral F. The same is true for the F ratios used in testing $H_0(a), H_0(b)$, and $H_0(ab)$. Furthermore, a simple general relationship holds between the power of a test and the expected value of the mean square in the numerator of the corresponding F ratio. In general, for *any* null hypothesis $H_0(g)$ (in the fixed-effects model) tested by the ratio MS_g/MS_w, where MS_g has ν degrees of freedom,

$$\phi_g^2 = \nu[\underline{E}(MS_g)/\sigma_e^2 - 1]. \tag{5.17}$$

For equal sample sizes, we have the general function

$$(\phi_g')^2 = \phi_g^2/nT_g, \tag{5.18}$$

where T_g is the number of cells in the corresponding table in the two-way design; i.e., $T_a = I, T_b = J, T_{ab} = IJ$.

Applying Equations 5.17 and 5.18 and letting g equal a, b, and ab in turn, we get

$$
\begin{array}{llll}
\phi_a^2 & = & nJ\sigma_a^2/\sigma_e^2, & (\phi_a')^2 & = & J\sigma_a^2/\sigma_e^2, \\
\phi_b^2 & = & nI\sigma_b^2/\sigma_e^2, & (\phi_b')^2 & = & I\sigma_b^2/\sigma_e^2, \\
\phi_{ab}^2 & = & n\sigma_{ab}^2/\sigma_e^2, & (\phi_{ab}')^2 & = & \sigma_{ab}^2/\sigma_e^2.
\end{array}
$$

Power can be estimated by the methods described previously. However, note that a different sample size may be needed to achieve the same degree of power for each of the three tests.

Robustness

The discussions in Chapters 2 and 3, on robustness of overall F tests and planned comparisons, apply to all of the tests in this chapter. In general, the tests are robust to violations of normality but not to violations of independence. In addition, the tests are robust to violations of the assumption of equal variances if each cell contains the same (or approximately the same) number of scores.

In Chapter 3 we found that, even when the sample sizes were equal, F tests on planned comparisons generally were not robust to the assumption of equal variances. Because the tests on main effects and interactions described in this chapter are composed of planned comparisons, we might suspect that they would not be robust either; fortunately, however, they are. Planned comparisons are not usually robust because in a planned comparison the means are not usually weighted equally. A planned comparison in which all of the means are weighted equally (e.g., the test on the grand mean, or Equation 5.1, where all of the weights have an absolute value of one) is robust. Few planned comparisons meet this requirement, but planned comparisons can be combined in such a way that all of the means are weighted equally in the combined test.

TABLE 5.7. Theoretical data illustrating effects of transformations on interactions (untransformed data).

		B_1	B_2	Row Sum
		84	23	
		78	27	
A_1		—	—	
		$\Sigma = 162$	$\Sigma = 50$	$\Sigma = 212$
		$\overline{X} = 81$	$\overline{X} = 25$	$\overline{X} = 53$
		26	4	
		24	0	
A_2		—	—	
		$\Sigma = 50$	$\Sigma = 4$	$\Sigma = 54$
		$\overline{X} = 25$	$\overline{X} = 2$	$\overline{X} = 13.5$
Column		$\Sigma = 212$	$\Sigma = 54$	$\Sigma = 266$
Sum		$\overline{X} = 53$	$\overline{X} = 13.5$	$\overline{X} = 33.25$

	RS	SS	df	MS	F	p	$\hat{\omega}^2$
m		8,844.5					
a	11,965	3,120.5	1	3,120.5	347	<.001	.46
b	11,965	3,120.5	1	3,120.5	347	<.001	.46
ab	15,630	544.5	1	544.5	60.5	.001	.08
w	15,666	36.0	4	9.0			
t		6,821.5	7				

Tests on main effects and interactions meet this requirement. Thus, as long as the sample sizes are equal or nearly equal, the F tests on main effects and interactions will be robust to the assumption of equal variances.

DATA TRANSFORMATION

In Chapter 2 we considered the possibility of equalizing variances by appropriately transforming the data values. For example, we saw that when the standard deviations of the scores were proportional to their means, the transformation $Y = \log(X)$ would make them approximately equal. However, we noted that the F ratio then tested the means of the transformed scores rather than the means of the original scores.

Data transformations may also be made in the two-way analysis of variance, but there is an additional complication; the transformation may have a large effect on the nature and size of the interaction. Consider the example in Tables 5.7 and 5.8. In Table 5.7 both main effects and the interaction are large and highly significant. Table 5.8 contains the same data as Table 5.7, but the values have been transformed by the function $Y = X^{1/2}$. The main effects are still large and significant, but the interaction has disappeared.

TABLE 5.8. Theoretical data illustrating effects of transformations on interactions (square root transformation on data).

		B_1	B_2	Row Sum
A_1		9.2	4.8	
		8.8	5.2	
		—	—	
		$\Sigma = 18.0$	$\Sigma = 10.0$	$\Sigma = 28.0$
		$\overline{X} = 9.0$	$\overline{X} = 5.0$	$\overline{X} = 7.0$
A_2		5.1	2.0	
		4.9	0.0	
		—	—	
		$\Sigma = 10.0$	$\Sigma = 2.0$	$\Sigma = 12.0$
		$\overline{X} = 5.0$	$\overline{X} = 1.0$	$\overline{X} = 3.0$
Column		$\Sigma = 28.0$	$\Sigma = 12.0$	$\Sigma = 40.0$
Sum.		$\overline{X} = 7.0$	$\overline{X} = 3.0$	$\overline{X} = 5.0$

	RS	SS	df	MS	F	p	$\hat{\omega}^2$
m		200.00	1				
a	232.00	32.00	1	32.00	58.7	.002	.47
b	232.00	32.00	1	32.00	58.7	.002	.47
ab	264.00	0.00	1	0.00	0.0	—	$(-.01)$
w	266.18	2.18	4	0.545			
t		66.18	7				

This effect of a transformation on the interaction is not necessarily bad: in fact, it may be good. If, by transforming the scores, we eliminate the interaction, we also eliminate the problems with interpretation of main effects, discussed earlier. However, be aware that transformations on the data do affect the interaction.

Pooling Sums of Squares

If there is good reason to believe that the interaction is zero or nearly zero for a particular set of data, it is possible to increase the power of tests on main effects by *pooling* SS_{ab} and SS_w. Under the null hypothesis $H_0(ab)$, that there is no interaction,

$$SS_{ab}/\sigma_e^2 \sim \chi^2_{(IJ-I-J+1)},$$

and SS_{ab} and SS_w are statistically independent, so we can define

$$SS_{\text{pooled}} = SS_{ab} + SS_w.$$

SS_{pooled} is the sum of two independent chi-square variables with degrees of freedom $(IJ-I-J+1)$ and $(N-IJ)$; therefore, SS_{pooled} is itself a chi-square variable,

$$SS_{\text{pooled}}/\sigma_e^2 \sim \chi^2_{(N-I-J+1)},$$

and

$$MS_{\text{pooled}} = SS_{\text{pooled}}/(N - I - J + 1)$$

can be used as the denominator of the F ratio, sometimes with a sizable increase in the degrees of freedom.

The same kind of argument can be used to prove the general principle that the sum of squares for any effect (main effect, interaction, or even the grand mean), if the true value is zero, can be pooled with other sums of squares, increasing the power of the remaining tests. However, usually we design our study so that significant main effects are expected. Consequently, as a general rule, only the interaction term is pooled in this way.

Be cautious about pooling any term with MS_w. Some experimenters (and, unfortunately, some statisticians) routinely test the interaction before testing the main effects. If the interaction is not significant at some previously specified significance level, it is automatically pooled with MS_w for the tests of main effects.

This practice is dangerous for two reasons. First, it unfairly biases the tests of main effects. In effect, one is waiting until the data have been obtained, then guessing which of two mean squares will produce a more significant result, and using that mean square. The bias introduced by such a procedure should be obvious. Second, the significance level specified is often fairly high; the .05 level, for example, is commonly specified. There is then some danger of making a Type II error. As a result, MS_{pooled} may be larger than MS_w, and the difference may be great enough to more than offset the increase in degrees of freedom. The net result of pooling is then a loss of power.

The decision to use MS_{pooled} should be based on considerations that are independent of the obtained data. Preferably, the decision should be made before the data are obtained. Furthermore, we should in theory be willing to abide by the decision to use MS_{pooled} even though the data strongly contradict the assumption of no interaction.

In practice, however, this last requirement is probably a little unrealistic, even if theoretically correct. In general, it is not reasonable for a person to abide by a decision after data have clearly shown that the decision was wrong. The theory is not seriously violated by the practice of using MS_w when the interaction is significant, even though the initial decision was to use MS_{pooled}. A good rule of thumb is to use MS_{pooled} only when (1) there are strong a priori reasons to assume that there is no interaction and (2) the obtained data give you no good reason to doubt that assumption. The basic difference between this procedure and that criticized in the paragraph above is that the recommended procedure correctly uses a significant interaction as reason to reject the null hypothesis of no interaction. The procedure criticized above involves the rather dubious practice of accepting the null hypothesis merely because the test was not significant (with an unknown probability of a Type II error and no a priori reasons for believing we haven't made one).

One Score Per Cell

In one commonly occurring case the assumption of no interaction must be made. In a two-way analysis of variance the total number of cells may be rather large. In a 10×12 analysis, for example, there are 120 cells. If we are limited in time and financing, we may be unable to obtain more than one score in each cell. There are then no degrees of freedom for MS_w. However, if we can assume that there is no interaction, we can use MS_{ab} as the denominator in our F ratio. The tests of main effects would then be

$$MS_a/MS_{ab} = F_{[I-1,(I-1)(J-1)]},$$
$$MS_b/MS_{ab} = F_{[J-1,(I-1)(J-1)]}.$$

This procedure, like that of pooling sums of squares, is technically invalid if there is an interaction. However, the procedure is somewhat excusable because no valid test can be made otherwise. Furthermore, even though an interaction exists, the test may be useful. When we use MS_{ab} in the denominator of F, a nonzero interaction reduces the probability that the test will be significant. Consequently, if our test is significant by this procedure, we can be confident that the results are in fact significant. If the test is not significant, there is no way of telling whether it might have been significant by a valid test.

Estimates of variances and proportions of variance accounted for are obtained somewhat differently when there is only one score per cell because there is no separate estimate of σ_e^2. However, if there is no interaction, MS_{ab} is an unbiased estimate of σ_e^2. Therefore, we can follow the estimation formulas given above, using MS_{ab} in place of MS_w and with $\hat{\sigma}_{ab}^2 = 0$.

Post Hoc and Multiple Tests on Main Effects

Post hoc and other multiple tests on individual cells can be conducted by simply regarding the experiment as a one-way instead of a two-way design. Similar tests on main effects can also be made with only a minor modification of the procedures described in Chapter 4. However, complications occur if tests are made on both main effects.

The modification consists of regarding each marginal mean of a main effect as if it were the mean of a single group of scores. If we are doing tests on the A main effect, we regard each $\overline{X}_{i..}$ as the mean of (nJ) scores; if we are doing tests on the B main effect, we regard each $\overline{X}_{.j.}$ as the mean of (nI) scores. The formulas in Chapter 4 can then be used, with n replaced by either (nJ) or (nI).

To illustrate, we will test

$$H_0 : \mu_{.3} - \mu_{.2} = 0$$

on the data in Table 5.2, using the Tukey method. We use the formulas in Chapter 4, substituting $(nI) = (3)(2) = 6$ for n in each case:

$$
\begin{aligned}
C &= 9 - 5 = 4, \\
C'' &= (6^{1/2})(4)/[(1/2)(1+1)] = 9.798, \\
T'_{(3,12)} &= 9.798/8.833^{1/2} = 3.30, \\
p &< .10.
\end{aligned}
$$

Notice that we use $MS_w^{1/2}$ in the denominator; MS_w is still our estimate of error. Notice also that the numerator parameter of t' is 3 because there are three marginal means. This works well if we are doing post hoc tests on only one main effect, but it does not control experimentwise errors if we are doing tests on both main effects. The only ways to control for experimentwise errors when doing post hoc tests on both main effects by the Tukey method are to treat the experiment as a one-way design, or to reduce the value of α for each set of tests.

However, we can control directly for experimentwise errors if we use the Scheffé method. Then we follow the procedures in Chapter 4, using either (nJ) or (nI) in place of n, and using $(I + J - 2)$, the sum of the main effects' degrees of freedom, in place of $(I - 1)$.

Applying the same principles, we can also conduct multiple pairwise tests by, e.g., Ryan's method, using the Studentized range version or the F version.

Unequal Sample Sizes

The above discussion assumed that every cell contained the same number of scores. If they do not, serious problems arise. To begin with, the formulas for SS_a, SS_b, and SS_{ab} become much more complicated. More important, however, these three sums of squares may not be independent; according to Equation 3.13, independence of two tests depends partly on relative sample sizes. If we wish to analyze the data as a two-way design, then it is advisable to have the same number of scores in each cell.

Occasionally, however, we cannot achieve that goal. In psychological or biological studies, for example, human subjects may quit before the experiment is completed or animals may become sick or die. Consequently, some cells may contain fewer scores than others. Moreover, occasionally there may be good experimental reasons for having unequal numbers of scores. For example, fewer subjects may be available to be run under some conditions. In such a case, three courses of action are possible.

If each cell contains approximately the same number of scores, approximate tests can be made. If the sample sizes are proportional (in a sense to be explained later), then exact tests are fairly simple. If the sample sizes are not proportional and they are highly unequal, then exact tests can still be made; however, the calculations and interpretations are both complicated.

APPROXIMATE TESTS

To make the approximate tests, we first calculate MS_w just as for a one-way design with unequal sample sizes. To calculate $SS_a, SS_b,$ and SS_{ab}, however, we first find the mean of each cell and then calculate the sums of squares using these means, as though every cell contained the same number of scores, $\bar{n}$. The value used for $\bar{n}$ is the *harmonic mean* of the sample sizes in the individual cells. If n_{ij} is the number of scores in cell AB_{ij}, then

$$
\begin{aligned}
\bar{n} &= IJ/[\Sigma_i\Sigma_j(1/n_{ij})], \\
\overline{X}_{i..} &= (1/J)\Sigma_j\overline{X}_{ij.}, \\
\overline{X}_{.j.} &= (1/I)\Sigma_i\overline{X}_{ij.}, \\
\overline{X}_{...} &= (1/I)\Sigma_i\overline{X}_{i..} = (1/J)\Sigma_j\overline{X}_{.j.} = (1/IJ)\Sigma_i\Sigma_j\overline{X}_{ij.}, \\
SS_m &= \bar{n}IJ\overline{X}_{...}^2, \\
RS_a &= \bar{n}J\Sigma_i\overline{X}_{i..}^2, \\
RS_b &= \bar{n}I\Sigma_j\overline{X}_{.j.}^2, \\
RS_{ab} &= \bar{n}\Sigma_i\Sigma_j\overline{X}_{ij.}^2.
\end{aligned}
$$

The rest of the values in the summary table, the SSs, dfs, MSs, Fs, and p values, are calculated exactly as for equal sample sizes. Table 5.9 shows the calculations when one score each in cells AB_{13} and AB_{21} of Table 5.2 is missing.

PROPORTIONAL SAMPLE SIZES

Sample sizes are said to be proportional when the ratio of the sample sizes of any two cells in the same column, but in different rows, is independent of the column that they are in. Table 5.10 shows an example of a three by four design in which the n_{ij} values are proportional. In each cell of the first row, the sample size is exactly 3/4 the sample size of the cell in the same column of the second row and exactly 3/2 the sample size of the cell in the third row. Incidentally, the same proportionality holds with respect to cells in the same row, but in different columns. For example, the sample size in each cell of the first column is exactly 2/3 the sample size of the cell in the same row of the second column. (The example in Table 5.10 is an extreme case of variations in sample size; it is presented to illustrate the generality of the procedure for analyzing such cases. Usually proportional sample sizes exist when all the values in a given row, say, are equal, but the sample sizes are different for different rows.)

The analysis with proportional sample sizes is analogous to the analysis for

TABLE 5.9. Data from Table 5.2 with one score each missing in cells AB_{12} and AB_{21} (approximate analysis).

	B_1	B_2	B_3	$\overline{X}_{i.}$
	8	8	4	
	4	10	8	
A_1	0	6		
	–	–	–	
	$\Sigma = 12$	$\Sigma = 24$	$\Sigma = 12$	
	$\overline{X} = 4$	$\overline{X} = 8$	$\overline{X} = 6$	6.000
	14	0	15	
	10	4	9	
A_2		2	12	
	–		–	
	$\Sigma = 14$	$\Sigma = 6$	$\Sigma = 36$	
	$\overline{X} = 12$	$\overline{X} = 2$	$\overline{X} = 12$	8.667
$\overline{X}_{.j}$	8	5	9	7.333

$$\overline{n} = (2)(3)/[4(1/3) + 2(1/2)] = 18/7$$

One-Way Analysis (MS_w)

	RS	SS	df	MS
bet	1,044			
w	1,126	82	10	8.20

Two-Way Analysis

SS_m = $(18/7)(2)(3)(7.333)^2 = 829.71$
RS_a = $(18/7)(3)(6^2 + 8.667^2) = 857.14$
RB_b = $(18/7)(4^2 + 8^2 + 6^2 + 12^2 + 2^2 + 12^2) = 1,049.14$

Summary

	RS	SS	df	MS	F	p	$\hat{\omega}^2$
m		829.71	1				
a	857.14	27.43	1	27.43	3.35	.10	.06
b	874.29	44.57	2	22.29	2.72	.11	.09
ab	1,049.14	147.43	2	73.72	8.99	.006	.43
w		82.00	10	8.200			
t		301.43	15				

TABLE 5.10. Sample two-way analysis with proportional sample sizes.

		B_1	B_2	B_3	B_4	Sum
	n	6	9	3	6	24
	ΣX	104	159	64	95	422
A_1	ΣX^2	1824	2909	1414	1539	7,686
	$\overline{X}$	17.33	17.67	21.33	15.83	17.58
	n	8	12	4	8	32
	ΣX	190	292	91	136	709
A_2	ΣX^2	4548	7270	2085	2380	16,283
	$\overline{X}$	23.75	24.33	22.75	17.00	22.16
	n	4	6	2	4	16
	ΣX	89	123	40	68	320
A_3	ΣX^2	2001	2621	832	1176	6,630
	$\overline{X}$	22.25	20.50	20.00	17.00	20.00
	n	18	27	9	18	72
	ΣX	383	574	195	299	1,451
Sum	ΣX^2	8,383	12,800	4,331	5,095	30,599
	$\overline{X}$	21.28	21.26	21.67	16.61	20.15

Calculations

SS_m = $1,451^2/72 = 29,241.68$
RS_a = $422^2/24 + 709^2/32 + 320^2/16 = 29,528.95$
RS_b = $383^2/18 + 574^2/27 + 195^2/9 + 299^2/18 = 29,543.93$
RS_{ab} = $104^2/6 + 159^2/9 + \cdots + 68^2/4 = 29,939.00$

Summary

	RS	SS	df	MS	F	p	$\hat{\omega}^2$
m		29,241.68					
a	29,528.95	287.27	2	143.6	13.1	$<.001$	.09
b	29,543.93	302.25	3	100.7	9.16	$<.001$	.07
ab	29,939.00	107.80	6	17.97	1.63	.15	.00
w	30,599.00	660.00	60	11.00			
t		1,357.32	71				

unequal sample sizes in the one-way design of Chapter 2. With equal n_{ij} we square each total, sum the squares, and then divide by the number of scores that were totalled. With unequal n_{ij} we square each total, divide the square by the number of scores that were summed to obtain that total, and then sum the resulting ratios. First we let

$$
\begin{aligned}
n_{i.} &= \Sigma_j n_{ij}, \\
n_{.j} &= \Sigma_i n_{ij}, \\
N &= \Sigma_i n_{i.} = \Sigma_j n_{.j} = \Sigma_i \Sigma_j n_{ij.}, \\
t_{ij} &= \Sigma_k X_{ijk}, \\
t_{i.} &= \Sigma_j t_{ij} = \Sigma_j \Sigma_k X_{ijk}, \\
t_{.j} &= \Sigma_i t_{ij} = \Sigma_i \Sigma_k X_{ijk}, \\
T &= \Sigma_i t_{i.} = \Sigma_j t_{.j} = \Sigma_i \Sigma_j t_{ij} = \Sigma_i \Sigma_j \Sigma_k X_{ijk}.
\end{aligned}
$$

Then the formulas are

$$
\begin{aligned}
SS_m &= NT^2, \\
RS_a &= \Sigma_i t_{i.}^2 / n_{i.}, \\
RS_b &= \Sigma_j t_{.j}^2 / n_{.j}, \\
RS_{ab} &= \Sigma_i \Sigma_j t_{ij}^2 / n_{ij}, \\
RS_w &= \Sigma_i \Sigma_j \Sigma_k X_{ijk}^2.
\end{aligned}
$$

SSs, dfs, MSs, and Fs are again calculated as for the design with equal sample sizes. The actual calculations are shown in Table 5.10.

EXACT TESTS, NONPROPORTIONAL SAMPLE SIZES

If the sample sizes are not proportional and the approximate analysis is not appropriate, we can still do an exact test. However, the exact test is much more complicated, both in the calculations that are required and in the interpretation. The problem is that the main effects are no longer orthogonal. If there is a B main effect, it can affect the test of the A main effect and vice versa. If, for a given level of A, small sample sizes are associated with small β_j (i.e., small values of the B main effect), while large sample sizes are associated with large values of β_j, then $\hat{\alpha}_i$ (the estimated value of the A main effect) for that level of A will be larger than it would otherwise be. This is due to the extra weight given to the large β_j when summing over the scores to obtain the $t_{i.}$ values. The opposite relationship would make $\hat{\alpha}_i$ too small. The A main effect can affect the $\hat{\beta}_j$ similarly.

Fortunately, the grand mean and the interaction are not affected, so they can be tested in a straightforward way. However, two different types of mean squares can be calculated for each main effect. The first type, which we will designate as MS_a and MS_b for the A and B main effects, respectively, is calculated exactly as for the main effects with proportional sample sizes. It can be used to test each main effect, *ignoring the other* (i.e., pretending that the other doesn't exist). The second type of mean square, which we will designate by $MS_{a:b}$ and $MS_{b:a}$, is used to test each main effect after all possible "contamination" by the other has been eliminated. In principle, this is found for the A main effect by first estimating the β_j, subtracting $\hat{\beta}_j$ from each X_{ijk}, and then calculating a mean square on the remainders; similarly, it is found for the B main effect by subtracting $\hat{\alpha}_i$ from each X_{ijk}.

In practice, these mean squares are found by first estimating the α_is and β_js and then calculating a new sum of squares, $SS_{a,b}$. The $\hat{\alpha}_i$s and $\hat{\beta}_j$s are calculated by solving the set of linear equations:

$$\Sigma_i n_{i.}\hat{\alpha}_i = \Sigma_j n_{.j}\hat{\beta}_j = 0,$$
$$n_{i.}\hat{\alpha}_i + \Sigma_j n_{ij}\hat{\beta}_j = t_{i.} - n_i\overline{X}_{...} \quad (i = 1 \ldots I),$$
$$n_{.j}\hat{\beta}_j + \Sigma_i n_{ij}\hat{\alpha}_i = t_{.j} - n_{.j}\overline{X}_{...} \quad (j = 1 \ldots J).$$

This is obviously a tedious process if there are more than two levels of each factor; it is certainly best done on a computer. (If you actually attempt to solve the equations, you will find that they are redundant; there are two more equations than there are unknowns. Nevertheless, the equations are consistent; they have a unique solution.)

Then

$$\widehat{\alpha\beta}_{ij} = \overline{X}_{ij.} - \hat{\alpha}_i - \hat{\beta}_j - \hat{\mu}.$$

To find sums of squares, we first calculate

$$SS_{a,b} = \Sigma_i \hat{\alpha}_i t_{i.} + \Sigma_j \hat{\beta}_j t_{.j}.$$

Then

$$
\begin{aligned}
RS_a &= \Sigma_i(t_{i.}^2/n_{i.}), \\
RS_b &= \Sigma_j(t_{.j}^2/n_{.j}), \\
RS_{ab} &= \Sigma_i\Sigma_j(t_{ij}^2/n_{ij}), \\
RS_w &= \Sigma_i\Sigma_j\Sigma_k X_{ijk}^2, \\
SS_m &= T^2/N, \\
SS_a &= RS_a - SS_m, \\
SS_b &= RS_b - SS_m, \\
SS_{a:b} &= SS_{a,b} - SS_b, \\
SS_{b:a} &= SS_{a,b} - SS_a, \\
SS_{ab} &= RS_{ab} - SS_{a,b} - SS_m, \\
SS_w &= RS_w - SS_{ab} - SS_{a,b} - SS_m, \\
SS_t &= RS_w - SS_m.
\end{aligned}
$$

All of these sums of squares, including both types of main effects, have the same number of degrees of freedom as for the case of equal or proportional sample sizes. Mean squares and F ratios are calculated in the usual way.

It is obviously not desirable to calculate and test both types of mean squares because that amounts to using the same degrees of freedom twice. Which type then should you test? Here is where the difficulties in interpretation occur. Some statisticians recommend testing $MS_{a:b}$ and $MS_{b:a}$ exclusively; they argue that one should never test one factor without first eliminating the effects of the other. However, this ignores the fact that each main effect may only partially "contaminate" the other. The test of $MS_{a:b}$ assumes that all of the contamination is from B to A, while the test of $MS_{b:a}$ assumes that all of the contamination is in the other direction. Thus both may overcompensate for the actual amount of contamination. As a consequence, $SS_{a:b}$ and $SS_{b:a}$ are not orthogonal, and $SS_{a:b} + SS_{b:a} + SS_{ab} + SS_w \neq SS_t$; i.e., these sums of squares do not analyze SS_t into orthogonal parts.

Of course, we do not fare any better if we use SS_a and SS_b because then we ignore the "contamination" entirely. The only way to analyze SS_t into orthogonal sums of squares is to test either SS_a and $SS_{b:a}$, or SS_b and $SS_{a:b}$. Then we find that

$$\begin{aligned} SS_t &= SS_a + SS_{b:a} + SS_{ab} + SS_w \\ &= SS_b + SS_{a:b} + SS_{ab} + SS_w. \end{aligned}$$

Now, however, we must choose between these two ways of dividing SS_t. If we are lucky, we will obtain essentially the same results no matter which tests we make. Alternatively, it may be clear that one main effect, say B, is small if it exists at all. We can then test for B using the adjusted mean square, $MS_{b:a}$, and for A using the unadjusted MS_a because presumably there is no B main effect to be adjusted for. At other times there may be nonstatistical reasons, related to the nature of the data, for dividing SS_t one way rather than the other. However, in the general case there seems to be no completely satisfactory solution to this problem.

The actual calculations for the data in Table 5.9 are shown in Table 5.11.

Note that the marginal means in Table 5.11 are different from the ones in Table 5.9. This is because the means in Table 5.9 are found by averaging the cell means, while the means in Table 5.11 are found by averaging the raw scores. The equations for estimating the main effect are:

$$\begin{aligned} 8\hat{\alpha}_1 + 8\hat{\alpha}_2 = 5\hat{\beta}_1 + 6\hat{\beta}_2 + 5\hat{\beta}_3 &= 0, \\ 8\hat{\alpha}_1 + 3\hat{\beta}_1 + 3\hat{\beta}_2 + 2\hat{\beta}_3 &= -9, \\ 8\hat{\alpha}_2 + 2\hat{\beta}_1 + 3\hat{\beta}_2 + 3\hat{\beta}_3 &= 9, \\ 5\hat{\beta}_1 + 3\hat{\alpha}_1 + 2\hat{\alpha}_2 &= .375, \\ 6\hat{\beta}_2 + 3\hat{\alpha}_1 + 3\hat{\alpha}_2 &= -12.75, \\ 5\hat{\beta}_3 + 2\hat{\alpha}_1 + 3\hat{\alpha}_2 &= 12.375. \end{aligned}$$

The results of this analysis are somewhat different from those of the approximate analysis in Table 5.9. The loss of a single score can have large consequences when the sample size is only three to begin with. With larger samples, the loss of one or two scores from a few cells would have little effect.

TABLE 5.11. Exact solution for data in Table 5.9. All main-effects tests are given here for illustrative purposes. Ordinarily one would calculate *either* SS_a and $SS_{b:a}$ or SS_b and $SS_{a:b}$, but one would not calculate all four main-effects sums of squares.

		B_1	B_2	B_3	Row Sum
	n	3	3	2	8
A_1	ΣX	12	24	12	48
	$\overline{X}$	4	8	6	6
	n	2	3	3	8
A_2	ΣX	24	6	36	66
	$\overline{X}$	12	2	12	8.25
	n	5	6	5	16
Column	ΣX	36	30	48	114
Sum	$\overline{X}$	7.2	5.0	9.6	7.125

Estimates of $\alpha\beta_{ij}$ (cells), α_i (row margins), β_j (column margins), and μ (lower right corner)

	B_1	B_2	B_3	$\hat{\alpha}_i$
A_1	-2.400	4.000	-2.400	-1.000
A_2	3.600	-4.000	1.600	1.000
$\hat{\beta}_j$	0.275	-2.125	2.275	7.125

Calculations

$$SS_{a,b} = (-1)(48) + (1)(66) + (.275)(36) + (-2.125)(30) +$$
$$(2.275)(48) = 73.35$$
$$RS_a = 48^2/8 + 66^2/8 = 832.5$$
$$RS_b = 36^2/5 + 30^2/6 + 48^2/5 = 870$$
$$RS_{ab} = 12^2/3 + 24^2/3 + 12^2/2 + 24^2/2 + 6^2/3 + 36^2/3 = 1044$$
$$RS_w = 1{,}126 \text{ (See Table 5.9)}$$
$$SS_m = 812.25$$
$$SS_a = 832.5 - 812.25 = 20.25$$
$$SS_b = 870 - 812.25 = 57.75$$
$$SS_{a:b} = 73.35 - 57.75 = 15.6$$
$$SS_{b:a} = 73.35 - 20.25 = 53.1$$
$$SS_{ab} = 1{,}044 - 73.35 - 812.25 = 158.4$$
$$SS_w = 1{,}126 - 158.4 - 73.35 - 812.25 = 82.00$$
$$SS_t = 1{,}126 - 812.25 = 313.75$$

Summary

	RS	SS	df	MS	F	p
m		812.25	1			
a	832.50	20.25	1	20.25	2.47	.15
b	870.00	57.75	2	28.88	3.52	.070
a, b		73.35				
$a : b$		15.60	1	15.60	1.90	.20
$b : a$		53.10	2	26.55	3.24	.082
ab	1,044.00	158.40	2	79.20	9.66	.005
w	1,126.00	82.00	10	8.200		
t		313.75	15			

Exercises

(1.) The following data are from an animal experiment. Three strains of rats were used: bright, mixed, and dull. Four rats from each strain were reared under "free" environmental conditions, and four were reared under "restricted" conditions. The data are the number of errors made by each rat in a maze.

	Bright		Mixed		Dull	
Free	26	41	41	26	36	39
	14	16	82	86	87	99
	51	96	39	104	42	92
Restricted	35	36	114	92	133	124

(a.) Do a two-way analysis of variance on the above data. Calculate the p values and the $\hat{\omega}^2$.

(b.) Graph the cell means and give a verbal interpretation of the nature of the interaction.

(c.) Test the following null hypotheses using planned comparison. Use either the regular or the approximate method, whichever you consider most appropriate. Calculate a single p value for each H_o.

$$H_0(1) : \mu_{11} = \mu_{23} \text{ and } \mu_{21} = \mu_{13},$$
$$H_0(2) : \mu_{12} = \mu_{22},$$
$$H_0(3) : \mu_{11} = \mu_{21}.$$

(2.) In the following table the subjects in A_1 are males and the subjects in A_2 are females. Level B_1 is a control group, and levels B_2 through B_5 are different experimental treatments.

	B_1	B_2	B_3	B_4	B_5
A_1	4	10	5	9	15
	4	14	4	15	8
	9	14	9	11	15
	7	18	5	13	10
A_2	12	9	11	14	15
	2	12	14	8	2
	0	8	15	17	7
	10	3	16	17	8

(a.) Do a two-way analysis of variance on the above data. Calculate the p values and the $\hat{\omega}^2$.

(b.) Graph the cell means and give a verbal interpretation of the nature of the interaction.

(c.) Do a simple-effects test, across levels of B, on each level of A.

(d.) Test the following two null hypotheses, using the results from part a to reduce the number of additional calculations as much as possible:

$H_0(1)$: There is no difference between the marginal mean for the control group and the average of the marginal means for the four experimental groups.

$H_0(2)$: There are no overall differences among the marginal means for the four experimental groups.

(e.) Suppose the experimenter intends to test some orthogonal contrasts among the marginal means for the four experimental groups. The null hypothesis will be rejected only if $p \leq .01$. Without actually making any tests, tell whether it would be possible for any of these contrasts to be significant.

(f.) Repeat the tests in part d without the assumption of equal variances.

(g.) From the data it appears that males have higher scores in B_2 and B_5 and females have higher scores in B_3 and B_4. Formulate a null hypothesis that would be appropriate to test for this difference with a *single* contrast, and test the contrast by both the Tukey method and the Scheffé method. Assume that the experimenter made no prior decision to limit the set of post hoc tests in the experiment.

(h.) Repeat part g assuming that the experimenter decided in advance to limit the post hoc tests to contrasts within the interaction.

(i.) Repeat part g assuming that the experimenter decided in advance to limit the post hoc tests to contrasts that were within the interaction and involved only the experimental groups B_2 through B_5.

(3.) Reanalyze the data in Problem 1, assuming that there are only three scores (26, 41, and 14) in cell AB_{11}, and that there are only two scores (92 and 124) in cell AB_{23}.

(a.) Use the approximate method of analysis.

(b.) Use the exact method of analysis.

(c.) Compare the results from the two methods.

(4.) Reanalyze the data in Problem 1, assuming that there are only three scores (26, 41, and 14) in cell AB_{11}, and that there are only three scores (51, 96, an 36) in cell AB_{21} (the lower left cell).

(a.) Use the approximate method,

(b.) Use the exact method for proportional cell frequencies.

(c.) Use the exact method without assuming proportional cell frequencies.

(d.) Compare the results from the three methods.

(5.) Twenty-four overweight people participated in four different weight-reduction programs. There were four people in each of six age groups. Each of the four people in an age group participated in a different program, making a 4×6 design with only one score per cell. The data are shown below (negative values indicate weight losses, positive values indicate weight gains). Analyze the data, calculating the p values and the $\hat{\omega}^2$, and interpret any significant effects. As part of your analysis, test $H_0 : \mu = 0$.

			Age			
Program	21–35	26–30	31–35	36–40	41–45	46–50
1	-59	22	-29	10	-23	72
2	-7	25	7	-8	40	64
3	-21	-54	-1	-44	-22	79
4	-27	-2	-77	-46	-11	93

(6.) Phyllis Psyche decided to do an experiment in which rats learned to run down a straight runway to receive either 3, 6, or 9 pellets of food. Thirty-six rats, twelve in each group, were trained for 50 trials, with each rat in a group receiving the same level of food reward on all 50 trials. After the training, she divided each group into three subgroups, giving four of the rats in each group three pellets on trial 51, giving another four rats six pellets, etc. Finally, she ran each rat one more time, measuring its running speed in the runway. The data were the running speeds of the rats on this last trial. The design is thus a conventional 3×3 design.

However, Phyl had some specific hypotheses to test.

H_1: Among rats receiving the same amount of food in all 51 trials, running speed is a function of the number of pellets received.

H_2: On the average, rats receiving a different number of pellets on the 51st trial from the number they had received on the first 50 trials will run more slowly than rats receiving the same number of pellets on the 51st trial.

H_3: The difference in running speed predicted by H_2 will be greater if the absolute value of the difference between the rewards obtained is six pellets rather than three.

H_4 : The difference predicted by H_2 will be greater if there is a decrease rather than an increase in the number of pellets received on the 51st trial.

Phyl has come to you for help. First, formulate the above experimental hypotheses as null hypotheses appropriate for testing; then tell *briefly* how you would recommend that she test each null hypothesis.

(7.) For a master's thesis, Phil Psyche ran a paired-associate learning study on subjects taking three different tranquilizers. He also used a fourth control group. Phil had four subjects in each group, and he analyzed the data as a simple fixed-effects one-way analysis of variance with four levels. However, his thesis advisor noted that half the subjects in each group were college freshmen and half were seniors. She suggested that Phil do a two-way analysis, with one factor being class ranking. When Phil reanalyzed the data as a two-way design, he obtained the following table:

	SS	df	MS	F	p	$\hat{\omega}^2$
Drugs	137.00	3	45.67	5.45	.025	.28
Class	30.25	1	30.25	3.61	.094	.05
Drug × Class	22.75	3	7.583	.91	—	(−.01)
Within	67	8	8.375			
Total	257.00	15				

(a.) Using only this information, find the values of F, p, and $\hat{\omega}^2$ that Phil got in his original one-way analysis.

(b.) When Phil discussed his problem with a statistician, she agreed that class-ranking should be studied as a factor, but suggested that Phil should have been interested in testing the following null hypotheses about the main effect due to drugs;

$H_0(1)$: The average performance under the three drug conditions is the same as the performance of the control group.

$H_0(2)$: There are no differences among the three drugs in their effects on performance.

Given that the following scores were obtained for the *sums* (not the means) of the scores of the subjects under each condition, test these two hypotheses (remember to perform the tests on the data represented as a two-way, not a one-way design):

Drug		Control	
A_1	A_2	A_3	A_4
52	60	42	74

(c.) Test for differences in performance under the three drugs, using the Studentized-range version of Ryan's method, with $\alpha = .05$.

(d.) Repeat part c, using the F version of Ryan's method.

6

Random Effects

In the previous chapters, we assumed that scores within a group were randomly selected. However, the groups from which scores were to be taken were assumed to have been the deliberate choice of the experimenter. For the data in Table 4.1, for example, the two drugs were assumed to have been chosen because we were particularly interested in those two drugs. They were not assumed to have been selected randomly from a large population of potential drugs. In some cases, however, the groups or "treatments" themselves may have been selected randomly from a large number of potential treatments. In this chapter we will consider methods for analyzing such data.

Note: When the groups are selected randomly, considerable complications arise with unequal sample sizes. Therefore, in this chapter we will consider only designs having equal sample sizes.

One-Way Model

Consider the following modification of the problem in Chapter 2. A large university is considering a comprehensive survey of the "intellectual maturity" of students in all courses in the university. However, the survey will be very expensive, requiring every student in every course to be tested. The university does not wish to undertake the study unless there is a good chance that large differences among the courses will be found.

The university decides to do a pretest on ten courses. To avoid biasing the pretest, the ten courses are selected randomly from among all the courses in the university. Five students from each course are given the test of "intellectual maturity." The basic data are shown in Table 6.1.

MODEL EQUATION

The model for this experiment, if X_{ij} is the score of the jth student in the ith class, is

$$X_{ij} = \mu + \alpha_i + \epsilon_{ij}, \tag{6.1}$$

where μ is the mean score of all students, α_i is the deviation of the ith course mean from the mean for all students (i.e., $\mu + \alpha_i$ is the mean score in the ith course), and ϵ_{ij} represents the deviation of the score of the jth student from the mean for the ith course.

Equation 6.1 is identical in form to Equation 2.8, but there is an important

TABLE 6.1. Hypothetical data on 10 randomly selected university courses ($n = 5$).

	A_1	A_2	A_3	A_4	A_5
ΣX	340	368	251	267	292
$\overline{X}_{i.}$	68.0	73.6	50.2	53.4	58.4
ΣX^2	23,451	27,943	13,200	14,609	17,182
$\hat{\sigma}$	9.1	14.6	12.2	9.4	5.7

	A_6	A_7	A_8	A_9	A_{10}	Sum
ΣX	255	310	271	300	333	2,987
$\overline{X}_{i.}$	51.0	62.0	54.2	60.0	66.6	
ΣX^2	13,411	19,825	14, 941	18,404	22,882	185,848
$\hat{\sigma}$	10.1	12.3	7.9	10.0	13.3	

	RS	SS	df	MS	F	p
m		178,433.4				
bet	181,206.6	2,763.2	9	307.0	2.65	.020
w	185,848.0	4,641.4	40	116.0		
t		7,404.6	49			

$$\hat{\sigma}^2_{bet} = (307.0 - 116.0)/5 = 38.2$$
$$\hat{\sigma}^2_t = 38.2 + 116.0 = 154.2$$
$$\hat{\omega}^2_{bet} = 39/154.2 = .25$$

difference. In Equation 2.8 the α_i were the effects of I specifically selected treatments; in Equation 6.1, the treatments (i.e., the classes) are selected randomly, so the α_i, like the ϵ_{ij}, are values of a random variable.

ASSUMPTIONS

In the random-effects model, like the fixed-effects model, we must make some assumptions. However, in the random-effects model, assumptions must be made about the distributions of both the ϵ_{ij} and the α_i. The assumptions are

(1) $\epsilon_{ij} \sim N_{(0,\sigma^2_e)}$,   (6.2)

(2) $\alpha_i \sim N_{(0,\sigma^2_{bet})}$,

(3) all α_I and ϵ_{ij} are jointly independent.

The assumptions that the ϵ_{ij} and the α_i have a mean of zero are inconsequential. The α_i have a mean of zero because they are deviations from their population mean (the mean of the population of potential courses), and the ϵ_{ij} have a mean of zero for a similar reason. The important assumptions about the α_i are that they are independent and normally distributed. The important assumptions about the ϵ_{ij} are that they are independent and normally distributed, and their variance does not depend on the particular course being considered. The assumptions about the ϵ_{ij} are basically the same as in Chapter 2. However, the assumptions about the α_i are a unique feature of the random-effects model; they have no counterpart in Chapter 2. The assumptions of independence are equivalent to the assumption that both the groups and the scores within the groups are sampled randomly.

These considerations, as well as the example given, may appear to be somewhat contrived. In fact, experiments employing a one-way analysis of variance with a random-effects model are relatively rare. Occasionally, however, such situations can arise; in addition, random effects are importantly involved in some two-way and higher-way models. The primary purpose of this section is to develop a general understanding of random-effects models in a simpler, though admittedly somewhat artificial, context.

CHARACTERISTICS OF RANDOM EFFECTS

The random-effects model differs from the fixed-effects model principally in three ways. First, the fixed-effects model applies when the particular groups being compared have been chosen because they are of interest. In the random-effects model the particular groups being compared have been selected randomly from a very large population of potential groups. In the experiment described above, for example, the "groups" are the courses, which were selected randomly

Second, because the groups being compared in a random-effects experiment have been selected randomly, we are not interested in the means of the groups that were actually tested. We are interested, rather, in the information that these group means can give us about the population of potential groups from which they were selected. In the above example, the university is no more interested in these ten courses than in any other courses in the university. Planned comparisons between the means of the groups tested are therefore uninteresting because we have no particular interest in specific comparisons between randomly selected courses. Instead, our interest lies in such population variables as μ, σ_{bet}^2, and σ_e^2.

The advantage of the random-effects experiment is that it enables us to generalize statistically beyond the I groups actually studied. In the experiment discussed in Chapter 2, we cannot generalize statistically beyond the three courses we actually tested. (Extrastatistical considerations, such as similarities between the courses tested and other courses, may allow us to make some tentative generalizations.) The random-effects design lets us generalize statistically to all courses in the university.

Finally, the above considerations imply a difference between the fixed-effects and random-effects models in any replication of the experiment. To replicate a fixed-effects experiment, we would obtain more scores from the same I groups used in the original experiment. To replicate a random-effects experiment, we would select a new random sample of I groups. If the university wished to replicate its experiment, it would not retest the same ten courses. Instead, it would test a different random sample of ten courses.

MEAN SQUARES

The values of μ, σ_{bet}^2, and σ_e^2 can be estimated from the data by first calculating $\overline{X}_{..}, SS_m, SS_{bet}, MS_{bet}, SS_w$, and MS_w, using the formulas developed for the fixed-effects model in Chapter 2.

Because MS_w is based only on deviations within groups, it has exactly the same distribution in both models. That is,

$$(N - I)MS_w/\sigma_e^2 \sim \chi_{(N-I)}^2, \tag{6.3}$$

$$\underline{E}(MS_w) = \sigma_e^2. \tag{6.4}$$

However, SS_m and MS_{bet} are distributed differently in the random-effects and fixed-effects models. To derive the expected values of SS_m and MS_{bet}, it is essential that all I groups contain the same number of scores (n). Then

$$\begin{aligned} \overline{X}_{i.} &= (1/n)\Sigma_j X_{ij} = (1/n)\Sigma_j(\mu + \alpha_i + \epsilon_{ij}) \\ &= \mu + \alpha_i + \overline{\epsilon}_{i.}. \end{aligned}$$

Because the X_{ij} are normally distributed, the $\overline{X}_{i.}$ are also normally distributed:

$$\overline{X}_{i.} \sim N_{(\mu, \sigma_{bet}^2 + \sigma_e^2/n)}.$$

This follows directly from Equation 6.2. To find an unbiased estimate of the variance of $\overline{X}_{i.}$, we calculate

$$S^2 = [\Sigma_i(\overline{X}_{i.} - \overline{X}_{..})^2]/(I - 1).$$

Now, S^2 is an estimate of a variance based on a sample from a normally distributed population, so

$$(I - 1)S^2/(\sigma_{bet}^2 + \sigma_e^2/n) \sim \chi_{(I-1)}^2,$$

$$\underline{E}(S^2) = \sigma_e^2/n + \sigma_{bet}^2.$$

$MS_{bet} = nS^2$, so

$$(I - 1)MS_{bet}/(\sigma_e^2 + n\sigma_{bet}^2) \sim \chi_{(I-1)}^2, \tag{6.5}$$

$$\underline{E}(MS_{bet}) = \sigma_e^2 + n\sigma_{bet}^2. \tag{6.6}$$

VARIANCE ESTIMATES

From these values we can obtain the unbiased estimates of σ_e^2 and σ_{bet}^2:

$$\hat{\sigma}_e^2 = MS_w, \tag{6.7}$$

$$\hat{\sigma}_{bet}^2 = (MS_{bet} - MS_w)/n. \tag{6.8}$$

Just as for the fixed-effects model, the total variance in the entire population of potential scores is

$$\sigma_t^2 = \sigma_e^2 + \sigma_{bet}^2,$$

so the proportion of variance accounted for is

$$\omega_{bet}^2 = \sigma_{bet}^2/\sigma_t^2.$$

An approximately unbiased estimate of ω_{bet}^2 is

$$
\begin{aligned}
\hat{\omega}_{bet}^2 &= \hat{\sigma}_{bet}^2/(\hat{\sigma}_e^2 + \hat{\sigma}_{bet}^2) \\
&= [SS_{bet} - (I-1)MS_w]/[SS_t - (n-1)MS_w] \\
&= (F-1)/(F-1+n).
\end{aligned}
\tag{6.9}
$$

COMPARISON WITH THE FIXED-EFFECTS MODEL

Comparisons between Equations 2.20 and 6.8 and Equations 2.22 and 6.9 show differences that, though not usually large in practice, are nevertheless important theoretically. The differences lie in the difference in meaning of σ_{bet}^2 in the fixed-effects and random-effects models. In the fixed-effects model the "population" of α_i (whose variance is σ_{bet}^2) is a more or less artificial population consisting only of the means of the groups actually used in the experiment. In the random-effects model the population of α_i is much larger than the number of groups actually tested. (In theory it is infinite.) In the fixed-effects model the α_i of the groups actually tested must sum to zero; in the random-effects model they need not do so. These differences result in a basic difference in the expected value of MS_{bet} in the two models, as a comparison of Equations 2.19 and 6.6 shows.

DEFINITION OF τ_{bet}^2

By defining the population parameter, τ_{bet}^2, Equations 2.19 and 6.6 can be modified so that they are identical in form. If we let N_A be the total number of groups in the population, we can define

$$\tau_{bet}^2 = [N_A/(N_A - 1)]\sigma_{bet}^2.$$

Then, for the random-effects model, for which N_A can be assumed to be infinite,

$$\tau_{bet}^2 = \sigma_{bet}^2 \quad \text{(random-effects model)}. \tag{6.10}$$

However, for the fixed-effects model the total population consists only of the I groups actually studied. In this case $N_A = I$ and

$$\tau_{bet}^2 = [I/(I-1)]\sigma_{bet}^2 \quad \text{(fixed-effects model)}. \tag{6.11}$$

Using Equation 6.11 to modify Equation 2.19 and Equation 6.10 to modify Equation 6.6, we find that for both models

$$E(MS_{bet}) = \sigma_e^2 + n\tau_{bet}^2. \tag{6.12}$$

The value of the redefinition in Equation 6.12 (along with many other concepts discussed in this chapter) cannot be fully appreciated until more complex analyses of variance are discussed in later chapters.

THE F TEST

An F ratio can be computed for the random-effects model just as for the fixed-effects model, and again it tests the null hypothesis that there are no differences between groups. This is equivalent to the hypothesis

$$H_0 : \sigma_{bet}^2 = 0. \tag{6.13}$$

Under this null hypothesis, the same theory can be applied as in Chapter 2 to show that

$$MS_{bet}/MS_w \sim F_{(I-1,N-I)}.$$

Therefore, the F test for the random-effects model is the same as for the fixed-effects model. (Table 6.1 shows the calculations for the university data.) However, the two tests differ in the distribution of the F ratio when the null hypothesis is false. For the fixed-effects model, the distribution is noncentral F (see Chapter 2); from Equations 6.3 and 6.4, we can show that for the random-effects model,

$$(MS_{bet}/MS_w)/[1 + n\omega_{bet}^2/(1 - \omega_{bet}^2)] \sim F_{(I-1,N-I)}.$$

This is a complicated function, but there are definite advantages in being able to use the central F instead of the noncentral F distribution.

Power. The above discussion shows that the power of the F test, for given degrees of freedom, depends entirely on the value of ω_{bet}^2. Furthermore, the power can be calculated using an ordinary F table without the more complicated methods for evaluating noncentral F required in the fixed-effects model. For a given significance level, α, the power is found by finding $F_{\alpha(I-1,N-I)}$, the value of F (one-tailed) needed to reach the α significance level, and calculating

$$F_{\alpha(I-1,N-I)}/[1 + n\omega_{bet}^2/(1 - \omega_{bet}^2)].$$

The p value for this ratio, with $(I-1)$ and $(N-I)$ degrees of freedom, is $1 - \beta$. (Normally, this ratio will be smaller than one; to find its p value we must find the p value for $1/F$ with $(N - I)$ and $(I - 1)$ degrees of freedom. The p value obtained in this way is β, and the power is $(1 - \beta)$.)

More General Null Hypotheses. Similar considerations lead to a more general null hypothesis test and a confidence interval for ω_{bet}^2. Equation 6.13 is equivalent to the null hypothesis

$$H_0 : \omega_{bet}^2 = 0.$$

Instead of this, a more general null hypothesis,

$$H_0 : \omega_{bet}^2 = \omega_{bet}^{*2},$$

for any arbitrary value of ω_{bet}^{*2}, can be tested. To do so, we need only note that under the null hypothesis,

$$(MS_{bet}/MS_w)/[1 + n\omega_{bet}^{*2}/(1 - \omega_{bet}^{*2})] \sim F_{(I-1,N-I)}.$$

If ω_{bet}^{*2} is larger than zero, either a one-tailed or a two-tailed test can be made. If a two-tailed test is made, the upper and lower limits of the $\alpha/2$ tails of the F distribution can be used for a test at the α level. (An improved test can be obtained by distributing the α probability unequally between the two tails, but the test is more complicated and the improvement is usually small.)

For example, suppose the university will pay to conduct the survey if ω_{bet}^2 is larger than .10. We can test $H_0 : \omega_{bet}^2 = .10$, with the ratio

$$2.65/[1 + (5)(.10)/(1 - .10)] = 1.70.$$

Under the null hypothesis, this is distributed as F with $\nu_1 = 9, \nu_2 = 40$, giving $p = .12$, one-tailed. For this null hypothesis a two-tailed test is also possible; to make it, we can just double the one-tailed p value, giving, for this example, $p = .24$.

Confidence Intervals. In the same way, we can find a $100(1 - \alpha)\%$ confidence interval for ω_{bet}^2 by solving for ω_{bet}^2 in the inequalities

$$F_{(1-\alpha/2)(I-1,N-I)} \leq (MS_{bet}/MS_w)/[1 + n\omega_{bet}^2/(1 - \omega_{bet})^2] \leq$$
$$F_{(\alpha/2)(I-1,N-I)}.$$

Solving for ω_{bet}^2, we obtain

$$(MS_{bet}/MS_w - G)/[MS_{bet}/MS_w + (n - 1)G] \leq \omega_{bet}^2 \leq$$
$$(MS_{bet}/MS_w - H)/[MS_{bet}/MS_w + (n - 1)H]$$

where

$$G = F_{(\alpha/2)(I-1,N-I)},$$
$$H = F_{(1-\alpha/2)(I-1,N-I)}.$$

For the data in Table 6.1, the 95% confidence interval is

$$.016 \leq \omega_{bet}^2 \leq .62.$$

A more detailed treatment of these problems, along with a way to find approximate confidence intervals for both σ_{bet}^2 and σ_e^2, can be found in Scheffé (1959, Chapter 7). In many cases, however, only the test of Equation 6.13 can be made with real data. The reasons why are given in the next section.

IMPORTANCE OF ASSUMPTIONS

The F test on the random-effects model, like that on the fixed-effects model, is robust with respect to many of the assumptions. The random-effects model shares with the fixed-effects model robustness with respect to nonnormality and inequality of variances of the ϵ_{ij}; the two models also share a lack of robustness with respect to nonindependence of the scores. In the random-effects model we also assume that the α_i are independent (i.e., randomly sampled) and normally distributed. As usual, the test is never robust to the assumption of random sampling. The assumption of normal α_i does not apply when the null hypothesis is $H_0 : \omega_{bet}^2 = 0$ because this hypothesizes that the α_i are all equal to zero. However, when testing other null hypotheses, estimating power, or obtaining a confidence interval for ω_{bet}^2, the assumption is important. As with the ϵ_{ij} (see Chapter 2), the most important effect here is the effect of nonzero kurtosis on the mean square. For the ϵ_{ij}, the effect is small if the sample size is relatively large. However, nonzero kurtosis of the α_i strongly affects MS_{bet}, even with large samples.

It should be pointed out that the point estimates (Equations 6.7 through 6.9) do not require the assumption of normality. They can therefore be made for any data. For example, suppose the university decided that it would pay to conduct the survey if ω_{bet}^2 were larger than .10. Even though the assumption of normality might not be valid, the estimate of .25 (see Table 6.1) suggests that the survey would be worthwhile.

HYPOTHESES ABOUT THE GRAND MEAN

As in the fixed-effects model, it is possible in the random-effects model to test the general hypothesis

$$H_0 : \mu = \mu^*. \tag{6.14}$$

However, the test differs radically from that of the fixed-effects model because SS_m is distributed very differently. In this case

$$\overline{X}_{..} = \mu + \overline{\alpha}_{.} + \overline{\epsilon}_{..}, \tag{6.15}$$

where

$$\overline{\alpha}_{.} = (1/I)\Sigma_i \alpha_i \tag{6.16}$$

and

$$\overline{\epsilon}_{..} = (1/N)\Sigma_i \Sigma_j \epsilon_{ij}. \tag{6.17}$$

as before. In the fixed-effects model $\overline{\alpha}_.$ was assumed to be zero, but this assumption cannot be made in the random-effects model. Although the population mean of the α_i is zero, that is not necessarily true for any finite sample.

From Equations 6.15, 6.16, and 6.17, we get

$$\overline{X}_{..} \sim N_{(\mu, \sigma_{bet}^2/I + \sigma_e^2/N)},$$

so

$$E(\overline{X}_{..} - \mu^*)^2 = \sigma_e^2/N + \sigma_{bet}^2/I + (\mu - \mu^*)^2,$$

and

$$E(SS_m^*) = E[N(\overline{X}_{..} - \mu^*)^2] = \sigma_e^2 + n\sigma_{bet}^2 + N(\mu - \mu^*)^2.$$

When the null hypothesis is true, $\mu = \mu^*$ and

$$E(SS_m^*) = \sigma_e^2 + n\sigma_{bet}^2.$$

For an appropriate test, the denominator of our F ratio must have the same expected value. Comparing this with Equations 6.4 and 6.6, we see that the appropriate denominator for the test is MS_{bet}, not MS_w. Under the null hypothesis, SS_m^*/MS_{bet} has an F distribution with 1 and $(I - 1)$ degrees of freedom.

The test of Equation 6.14 is valid only if the α_i are normally distributed or if I is large. However, it provides a good illustration of a test for which MS_w is not an appropriate error term. Later, more important cases will be studied in which the same problem arises. This example illustrates the general procedure of determining F ratios by comparing expected values of mean squares.

Two-Way Design

In the two-way design either or both factors can be selected randomly. If the levels of both factors are selected randomly, it is called a *random-effects design*; if the levels of only one factor are selected randomly, it is a *mixed-effects design*.

In the fixed-effects design, the two-way analysis of variance is basically a variation on the one-way analysis of variance. That is, the two-way analysis is simply an alternative to the overall one-way F, or a set of planned comparisons. The data from such a study can be analyzed as easily with a one-way as with a two-way analysis of variance (see Chapter 5). The same is not true of the random-effects and mixed-effects designs. For these the alternative of performing a simple one-way analysis of variance is not available.

EXAMPLES

We can explain the problem most easily with two examples, the first involving the mixed-effects model and the second involving the random-effects model.

Mixed-Effects Design. For an example of the mixed-effects design, suppose we obtain a list of all colleges and universities in the United States containing more than one thousand students; we randomly sample twenty schools from this list. We then sample thirty students from each of the twenty schools and divide the sampled students into three groups of ten each. Each group of ten students is given one of three forms of a proposed college entrance examination. The basic purpose of the study is to determine the extent to which the three forms are interchangeable. We choose to sample schools randomly rather than to sample students directly because we feel that by sampling schools we are more likely to obtain a representative cross section of students. A truly random sample of all college students in the United States would be very difficult to obtain, and even if one were obtained, it would be possible for almost all of the students (by pure chance) to come from only a few very large schools. By first randomly sampling schools and then randomly sampling students within each school, we make our random sampling problem simpler and increase the likelihood that we will obtain a more representative sample. In addition, as a sort of side benefit, this design enables us to obtain an estimate of the variability among schools in average student ability.

If the three different forms of the entrance examination are considered to be the three levels of Factor A and the twenty schools are considered to be the levels of factor B, then we have a 3×20 two-way analysis of variance with ten scores per cell. If we regarded this same study as a one-way analysis of variance in which each cell was a group, there would then be sixty (three times twenty) groups. But would it be a fixed-effects or a random-effects design? Clearly, the choice of cells was neither completely random nor completely fixed. Therefore, neither the fixed-effects model nor the random-effects model of the one-way analysis of variance would be appropriate for these data. We solve the problem by treating the study as a two-way design with Factor A fixed and Factor B random. Hypothetical data from such a study are shown in Table 6.2.

Random-Effects Design. For an example of a random-effects design, consider a modification of the one-way design in Table 6.1. Suppose that in addition to assessing differences among classes, we are also interested in studying different universities. Four universities (Factor A) and six courses (Factor B) are selected; although the universities and the courses are selected randomly, the same six courses are studied at all four universities. Table 6.3 shows hypothetical data from such a study, with scores from two subjects in each course in each university.

Suppose we again regard this as a one-way analysis of variance with 24 levels. Now clearly the groups have all been selected randomly. Or have they? From what population are they a random sample? The population must of course be regarded as the collection of all possible combinations of universities and courses. Suppose, for example, that there are 100 universities from which the four were selected, and that the six courses were selected from a total set of two hundred. Then the total population of cells would consist of the 20,000 possible ways that the 100 universities could be paired with the 200 courses.

TABLE 6.2. Cell means and data analysis for three tests (Factor A) and twenty randomly selected colleges (Factor B), with $n = 10$.

	A_1	A_2	A_3	$\overline{X}$
B_1	22	18	13	17.7
B_2	24	30	19	24.3
B_3	26	21	22	23.0
B_4	21	27	13	20.3
B_5	26	23	17	22.0
B_6	18	21	22	20.3
B_7	16	15	21	17.3
B_8	22	25	23	23.3
B_9	14	22	20	18.7
B_{10}	11	25	17	17.7
B_{11}	24	15	17	18.7
B_{12}	21	19	21	20.3
B_{13}	20	22	20	20.7
B_{14}	22	25	19	22.0
B_{15}	25	19	24	22.7
B_{16}	23	24	23	23.3
B_{17}	22	24	21	22.3
B_{18}	22	14	17	17.7
B_{19}	21	23	12	18.7
B_{20}	17	18	22	19.0
$\overline{X}$ 20.85	21.50	19.15		20.5

	RS	SS	df	MS	F	p
m		252,150	1	252,150	1,661	<.001
a	252,739	589	2	294.5	1.95	.16
b	255,033	2,883	19	151.8	1.77	.023
ab	261,360	5,738	38	151.0	1.76	.004
w	307,800	46,440	540	86.00		
t		56,650	599			

TABLE 6.3. Hypothetical data from four universities (Factor A) and six courses (Factor B), with $n = 2$.

	B_1	B_2	B_3	B_4	B_5	B_6	$\overline{X}$
A_1	20	8	14	15	19	26	
	14	10	16	13	11	21	
	—	—	—	—	—	—	
	17.0	9.0	15.0	14.0	15.0	23.5	15.58
A_2	16	15	6	2	10	23	
	8	6	2	10	11	14	
	—	—	—	—	—	—	
	12.0	10.5	4.0	6.0	10.5	18.5	10.25
A_3	4	7	8	7	14	18	
	20	5	14	2	10	17	
	—	—	—	—	—	—	
	12.0	6.0	11.0	4.5	12.0	17.5	10.50
A_4	8	5	10	10	15	15	
	16	14	6	17	21	22	
	—	—	—	—	—	—	
	12.0	9.5	8.0	13.5	18.0	18.5	13.25

| $\overline{X}$ | 13.25 | 8.75 | 9.50 | 9.50 | 13.88 | 19.50 | 12.40 |

	RS	SS	df	MS	F	p
m		7,375.5	1			
a	7,604.6	229.1	3	76.35	5.25	.011
b	8,043.1	667.6	5	133.5	9.18	<.001
ab	8,490.5	218.3	15	14.55	0.65	—
w	9,029.0	538.5	24	22.44		
t		1,653.5				

Selecting a random sample from this population would be the same as selecting 24 cells from a table with 100 rows and 200 columns. The probability of selecting the same course in two universities is the same as the probability of selecting more than one cell from the same column, i.e., less than one in a hundred. The probability of selecting the same course in four different universities is infinitesimal. Yet, in this design *every* course is selected in four universities—and the same four universities at that! Obviously, we cannot consider all of the cells to have been randomly sampled from the same population. Instead, we must consider it to be a two-way design in which the individual cells were determined by randomly (and independently) sampling the levels of each factor.

THE MODELS AND THE ASSUMPTIONS

The model equation for the two-way random- and mixed-effects models, like that for the fixed-effects model, is

$$X_{ijk} = \mu + \alpha_i + \beta_j + \alpha\beta_{ij} + \epsilon_{ijk}.$$

The assumptions about the ϵ_{ijk} are the same as for the fixed-effects model; i.e.,

$$\epsilon_{ijk} \sim N_{(0,\sigma_e^2)}.$$

However, different assumptions are made about the other terms in the model.

Mixed-Effects Model. In the mixed-effects model, when Factor A is fixed and Factor B is random, both the β_j and the $\alpha\beta_{ij}$ are sampled randomly. The assumptions are

$$\begin{aligned} \beta_j &\sim& N_{(0,\sigma_b^2)}, \\ \alpha\beta_{ij} &\sim& N_{(0,\sigma_{ab}^2)}. \end{aligned}$$

In addition, several assumptions of independence must be made. These requirements are usually expressed in abstract terms, but it is not difficult to state them in terms of the scores themselves. For the above requirements to be met, the scores must satisfy four conditions. First, all the scores within a cell must be sampled randomly and independently. Second, the scores must all be normally distributed with mean $(\mu + \alpha_i)$ and variance $(\sigma_b^2 + \sigma_{ab}^2 + \sigma_e^2)$. The important points here are that the scores are all normally distributed and, although scores in different cells may have different means, they must all have the same variance. Third, the levels of B must be sampled randomly and independently, and all cell means, $\overline{X}_{ij}$, within a given level of A must be normally distributed with the same variance.

The fourth condition is more difficult to explain. The cell means $(\overline{X}_{ij.})$ within the same level (j) of B are usually not independent. That is,

$$\underline{E}(\overline{X}_{ij.}\overline{X}_{i'j.}) \neq \underline{E}(\overline{X}_{ij})\underline{E}(\overline{X}_{i'j.}), (i' \neq 1).$$

TABLE 6.4. Estimates of variances, covariances, and correlations for data in Table 6.2. (Variances are on the diagonal, covariances above the diagonal, and correlations below the diagonal.)

	A_1	A_2	A_3	$\hat{\sigma}$
A_1	15.40	.71	.34	3.92
A_2	.04	17.63	-.97	4.20
A_3	.02	-.07	12.34	3.51

The expression

$$C(\overline{X}_{ij.}, \overline{X}_{i'j.}) = \underline{E}(\overline{X}_{ij.}\overline{X}_{i'j.}) - \underline{E}(\overline{X}_{ij.})\underline{E}(\overline{X}_{i'j.})$$

measures the degree to which the $\overline{X}_{ij.}$ are dependent. The term $C(\overline{X}_{ij.}, \overline{X}_{i'j.})$ is called the *covariance* of $\overline{X}_{ij.}$ and $\overline{X}_{i'j.}$. It can be estimated from the data by

$$\hat{C}(\overline{X}_{ij.}, \overline{X}_{i'j.}) = (\Sigma_j \overline{X}_{ij.}\overline{X}_{i'j.})/(J-1) - [J/(J-1)]\overline{X}_{i..}\overline{X}_{i'..}.$$

A different estimate, $\hat{C}(\overline{X}_{ij.}, \overline{X}_{i'j.})$, can be calculated for each pair of levels of Factor A. The fourth condition that must be met is that all $C(\overline{X}_{i'j.})$ must be equal. (Technically, this condition is not absolutely necessary; a somewhat weaker condition, which is more difficult to explain, will suffice. However, for most practical purposes the two conditions are equivalent. The weaker condition is described in Chapter 13.)

This condition will perhaps be more familiar to some readers if we point out that the Pearson product-moment correlation between any two random variables is defined to be their covariance divided by the product of their standard deviations:

$$r_{xy} = C(x, y)/(\sigma_x \sigma_y).$$

By the third condition, all of the $\overline{X}_{ij.}$ have the same variance, so the fourth condition is equivalent to the stipulation that all pairs of cell means in the same level of B must have the same correlation.

Estimates of the relevant variances, covariances, and correlations for the data in Table 6.2 are shown in Table 6.4; they are found by treating the cell means in Table 6.2 as though they were a random sample of 20 triplets of scores. The 20 values in columnA_1, for example, are the data from which the variance of A_1 is calculated; the 20 pairs of means in columns A_1 and A_2 are the data from which the covariance of A_1 and A_2 ($C(\overline{X}_{1j.}, \overline{X}_{2j.})$) is calculated, and so on.

There is no simple method of directly testing the assumption of equal variances and covariances, but the estimates can be examined to determine whether the assumption of approximately equal variances and covariances is reasonable. Usually the standard deviations and correlations are a better index than are the variances and covariances. In Table 6.4, for example, there appear to be large differences among the covariances. However, these differences are small compared to the size of the variances, so the correlations are very close to zero. Because the correlations are all approximately zero and the standard deviations are not very different, the assumption of approximately equal variances and covariances is probably reasonable for these data.

The assumption of equal covariances may be clearer if we give an example of its probable violation. In a study of nonsense syllable learning the experimenter gives each of 30 subjects a list of twenty nonsense syllables to learn by the anticipation method. The number of nonsense syllables correctly anticipated on each of ten trials is recorded. Factor A is trials, Factor B is subjects, and a score is defined to be the number of syllables correctly anticipated by a given subject on a given trial. It is unlikely, in this case, that the covariances will be equal; trials nine and ten, for example, are likely to be much more highly correlated than trials one and ten.

The above list of assumptions is long and complicated; however, we shall see in the section on robustness that only two conditions (in addition to the requirements of independent random sampling) are generally important. The important conditions are that the $\overline{X}_{ij.}$ have equal variances and equal covariances.

Random-Effects Model. All of the assumptions made for the mixed-effects model must also be made for the random-effects model. In addition, the assumptions made about the β_j are also made about the α_i; they are assumed to be randomly sampled from a normal distribution with mean zero and variance σ_a^2. In practice, this means that the conditions given above on cell means within different levels of A but within the same level of B must be extended to cell means within different levels of B but the same level of A; that is, the variances and the covariances, $C(\overline{X}_{ij.}, \overline{X}_{ij'.})$, must be constant.

Once again the relevant variances and covariances can be estimated from the data, but for the random-effects design two sets of variances and covariances must be estimated. To find the first set of estimates, we treat the cell means at different levels of Factor B as though they were randomly sampled single scores, just as we did for the mixed-effects model of Table 6.2. To find the other set of estimates, we reverse the roles of Factors A and B, treating the levels of Factor A as randomly sampled. Both sets of variances, covariances, and correlations are shown in Table 6.5 for the data in Table 6.3.

The assumption of equal variances and covariances among levels of factor A appears reasonable. However, among levels of Factor B, the assumptions are more doubtful; estimates of standard deviations range from 1.94 to 4.95, and correlation estimates range from -.50 to +.98. These estimates are not very accurate, because each is based on only four "scores" (the cell means at the four levels of Factor A) in the case of the standard deviations, or pairs of "scores" in the case of the correlations. Therefore, it is still possible

TABLE 6.5. Estimates of variances, covariances, and correlations for the data in Table 6.3. (Variances are on the diagonal, covariances above the diagonal, and correlations below the diagonal.)

	A_1	A_2	A_3	A_4	$\hat{\sigma}$
A_1	22.24	15.28	19.10	12.88	4.72
A_2	.64	25.68	16.40	14.63	5.07
A_3	.86	.69	22.00	10.95	4.69
A_4	.63	.67	.54	18.68	4.32

	B_1	B_2	B_3	B_4	B_5	B_6	$\hat{\sigma}$
B_1	6.25	.42	9.17	7.50	1.88	6.67	2.50
B_2	.09	3.75	-4.50	3.92	.88	1.33	1.94
B_3	.79	-.50	21.67	10.17	5.25	8.67	4.65
B_4	.61	.41	.44	24.50	14.25	9.17	4.95
B_5	.23	.14	.34	.87	11.06	2.50	3.33
B_6	.98	.25	.69	.68	.28	7.33	2.71

that the population variances and covariances are equal, but the data suggest otherwise.

EXPECTED MEAN SQUARES

In the random- and mixed-effects models the same sums of squares and mean squares are calculated as in the fixed-effects model; moreover, they are calculated in exactly the same way. (Table 5.5 gives the formulas, and Tables 6.2 and 6.3 show the calculations for the two examples in the tables.) However, the expected values of the mean squares in the random and mixed models are different from those in the fixed-effects model. The expected mean squares for the fixed-, mixed-, and random-effects models are given in Table 6.6, along with the appropriate F ratio for testing each effect.

The τ^2 values in Table 6.7 are related to the σ^2 by a general formula that is easily applied. The value of τ^2 for any given effect is found by multiplying σ^2 (the variance of the effect) by as many coefficients as there are factors in the effect being considered. The coefficient for each factor is the number of levels in the *total population* of that factor divided by the same number of levels minus one. Thus, if we let N_a be the total number of levels in the population of Factor A—i.e., $N_a = I$ if Factor A is fixed, and N_a is infinite if Factor A is random—and N_b be the total number of levels of Factor B, then

$$\tau_a^2 = [N_a/(N_a - 1)]\sigma_a^2,$$
$$\tau_b^2 = [N_b/(N_b - 1)]\sigma_b^2,$$
$$\tau_{ab}^2 = [N_a/(N_a - 1)][N_b/(N_b - 1)]\sigma_{ab}^2.$$

TABLE 6.6. Expected mean squares in the mixed- (A fixed, B random) and random-effects models of the two-way design.

	Expected MS	F Ratio	df
	Mixed effects		
m	$\sigma_e^2 + nI\tau_b^2 + N\mu^2$	SS_m/MS_b	$1, J-1$
a	$\sigma_e^2 + n\tau_{ab}^2 + nJ\tau_a^2$	MS_a/MS_{ab}	$I-1, (I-1)(J-1)$
b	$\sigma_e^2 + nI\tau_b^2$	MS_b/MS_w	$J-1, N-IJ$
ab	$\sigma_e^2 + n\tau_{ab}^2$	MS_{ab}/MS_w	$(I-1)(J-1), N-IJ$
w	σ_e^2		
	Random effects		
m	$\sigma_3^2 + n\tau_{ab}^2 + nJ\tau_a^2 + nI\tau_b^2 + n\mu^2$	*	*
a	$\sigma_e^2 + n\tau_{ab}^2 + nJ\tau_a^2$	MS_a/MS_{ab}	$I-1, (I-1)(J-1)$
b	$\sigma_e^2 + n\tau_{ab}^2 + nI\tau_b^2$	MS_b/MS_{ab}	$J-1, (I-1)(J-1)$
ab	$\sigma_e^2 + n\tau_{ab}^2$	MS_{ab}/MS_w	$(I-1)(J-1), N-IJ$
w	σ_e^2		

* No standard F ratio can be computed for this test (see text).

TABLE 6.7. Estimates of τ^2, σ^2, and ω^2 for data in Tables 6.2 and 6.3.

	Table 6.2			Table 6.3		
	$\hat{\tau}^2$	$\hat{\sigma}^2$	$\hat{\omega}^2$	$\hat{\tau}^2$	$\hat{\sigma}^2$	$\hat{\omega}^2$
a	0.72	0.48	.01	5.15	5.15	.13
b	2.19	2.19	.02	14.87	14.87	.30
ab	6.50	4.33	.05	(-3.95)	(-3.95)	(-.10)
w	86.00	86.00	.92	22.44	22.44	.58
t		93.01			38.51	

The exact relationships between the τ^2 and the σ^2 depend on the model being considered. In the random model both N_a and N_b are infinite, so the coefficients are all equal to one:

$$\tau_a^2 = \sigma_a^2, \quad \tau_b^2 = \sigma_b^2, \quad \tau_{ab}^2 = \sigma_{ab}^2 \quad \text{(random model)}. \tag{6.18}$$

In the mixed-effects model, $N_a = I$ and N_b is infinite:

$$\begin{aligned} \tau_a^2 &= [I/(I-1)]\sigma_a^2, \quad \tau_b^2 = \sigma_b^2, \\ \tau_{ab}^2 &= [I/(I-1)]\sigma_{ab}^2 \quad \text{(mixed model)}. \end{aligned} \tag{6.19}$$

The same principle applies to the fixed-effects model (Chapter 5), with $N_a = I$ an $N_b = J$.

The derivation of the expected mean squares is straightforward but too long and involved to present here. However, it may seem puzzling that when Factor B is changed from fixed to random, the expected mean square of Factor A changes but the expected mean square of Factor B remains the same. A partial derivation will give an idea why this is so. From the basic model (Eq. 5.12) for the two-way analysis of variance we can obtain the estimates.

$$\begin{aligned} \overline{X}_{i..} &= \Sigma_j \Sigma_k (\mu + \alpha_i + \beta_j + \alpha\beta_{ij} + \epsilon_{ijk})/(nJ) \\ &= \mu + \alpha_i + \overline{\beta}_. + \overline{\alpha\beta}_{i.} + \overline{\epsilon}_{i..}, \\ \overline{X}_{...} &= (1/I)\Sigma_i \overline{X}_{i..} = \mu + \overline{\beta}_. + \overline{\epsilon}_{...}, \end{aligned}$$

where, as before, a dot replacing a subscript indicates that an average has been taken over the missing subscript. Note that the α_i drop out of the equation for $\overline{X}_{...}$; because A is a fixed effect, they sum to zero. However, the β_j do not drop out; because B is a random effect, they do not necessarily sum to zero for any finite sample. Similarly, the $\alpha\beta_{ij}$ do not drop out of the equation for $\overline{X}_{i..}$, although they do drop out of the equation for $\overline{X}_{...}$, because then they are summed over the subscript of a fixed factor. We can now find

$$\hat{\alpha}_i = \overline{X}_{i..} - \overline{X}_{...} = \alpha_i + \overline{\alpha\beta}_{i.} + (\overline{\epsilon}_{i..} - \overline{\epsilon}_{...}).$$

Here, μ and $\overline{\beta}_.$ cancel out of the difference, but $\overline{\alpha\beta}_{i.}$ does not. Because MS_a is based on the sum of the squares of these estimates (Eq. 5.13), and the estimates depend partly on the interaction, it is reasonable that MS_a should also depend on the interaction.

F RATIOS

The logic behind the choices of F ratios in Table 6.6 is easily given. In general, the appropriate denominator term for testing a given effect is that term which would have the same expected value as the effect mean square if the effect did not exist, that is, if τ^2 for the effect were zero. In the mixed-effects model, for example, if there were no A main effect, τ_a^2 would be zero, and the expected value of MS_a would be exactly the same as that of MS_{ab}. Accordingly, MS_{ab} is

TABLE 6.8. Hypothetical population means for a population of 500
colleges, illustrating the effect of an
interaction on the test of the fixed
effect.

College	A_1	A_2	A_3	$\mu_{.j}$
1–250	28	12	20	20
252–500	12	28	20	20
μ_i	20	20	20	20

the appropriate term to use in the denominator of the F ratio. The difference between these models and the fixed-effects model should be obvious. In the fixed-effects model the use of MS_{ab} in the denominator of the F ratio was recommended only as a last resort when there was only one score per cell. In the mixed and random models MS_{ab} is the appropriate denominator for testing some effects no matter how many scores there are per cell. Of course, if there are *good* reasons to assume that there is no interaction, S_{ab} and SS_w may be pooled, just as in the fixed-effects model; then the pooled mean square is the appropriate denominator for testing both main effects.

A practical illustration of the effect of an interaction on MS_a, and the reason for testing MS_a against MS_{ab} instead of MS_w, can be found in the highly idealized example in Table 6.8. This table shows true mean scores for a hypothetical population of 500 colleges from which the data in Table 6.2 might have been drawn. Note that for half the colleges students score more highly on test A_1, and for the other half they score more highly on test A_2; i.e., there is a large interaction, but there are no main effects.

When drawing a sample of 20 colleges, it could happen by pure chance that, say, 15 are from the first half of the population and only 5 are from the second half. Our obtained means would probably then be approximately

$$\overline{X}_{1..} \simeq [(15)(28) + (5)(12)]/20 = 24,$$
$$\overline{X}_{2..} \simeq [(15)(12) + (5)(28)]/20 = 16,$$
$$\overline{X}_{3..} \simeq [(15)(20) + (5)(20)]/20 = 20.$$

We would obtain a large MS_a and therefore be inclined to conclude (erroneously) that there was a large main effect. However, we would also have a large MS_{ab}, so the F ratio would probably be close to one. If we had used MS_w as our denominator, we would not have had that protection against falsely rejecting the null hypothesis. An example with no interaction can easily be constructed to show that the problem arises only when an interaction is present.

GRAND MEAN

Testing for the grand mean in the random model presents a special problem. To test for it, we should use a denominator mean square whose expected value is $\sigma_e^2 + nJ\tau_a^2 + nI\tau_{ab}^2$. However, there is no mean square with this expected value, and no method of pooling sums of squares will solve the problem. The linear combination of mean squares, $MS_a + MS_b - MS_{ab}$, has the right expected value, but it does not have the right distribution for use in the F ratio. (This problem seldom arises in practice with the two-way design because the grand mean is seldom tested. However, as we shall see, it can also arise in more complicated designs. It arises regularly when there are two or more random factors in the experiment.) One solution to this problem is an approximate F ratio, F^*, sometimes called a *quasi-F*, which has approximately an F distribution. It can be shown that $F^* = SS_m/(MS_a + MS_b - MS_{ab})$ has approximately an F distribution (under the null hypothesis $\mu = 0$), with numerator and denominator degrees of freedom, respectively,

$$\nu_1 = 1,$$

$$\nu_2 = \frac{\underline{E}(MS_a + MS_b - MS_{ab})^2}{\underline{E}(MS_a)^2/(I-1) + \underline{E}(MS_b)^2/(J-1) + \underline{E}(MS_{ab})^2/(IJ - I - J + 1)}. \tag{6.20}$$

The denominator degrees of freedom cannot be known exactly because the expected mean squares cannot be known exactly. They can be estimated from the obtained mean squares:

$$\hat{\nu}_2 = \frac{(MS_a + MS_b - MS_{ab})^2}{MS_a^2/(I-1) + MS_b^2/(J-1) + MS_{ab}^2/(IJ - I - J + 1)}.$$

(The numerator and denominator terms in this formula are only approximately unbiased; unbiased estimates can be found but they are more complicated and have some undesirable properties.)

These conclusions can be generalized in a way that will be of use later. In general, any combination of sums and differences of mean squares,

$$MS_{combined} = \sigma_{eff}k_{eff}MS_{eff},$$

is itself distributed approximately as though it were an ordinary mean square with degrees of freedom,

$$\nu = [\underline{E}(MS_{combined})]^2/[\Sigma_{eff}k_{eff}^2\underline{E}(MS_{eff})^2/\nu_{eff}]. \tag{6.21}$$

The meaning of Equation 6.21 can be clarified by a comparison with Equation 6.20. The degrees of freedom can then be approximated by

$$\hat{\nu} = (MS_{combined})^2/[\Sigma_{eff}k_{eff}^2MS_{eff}^2/\nu_{eff}].$$

For the data in Table 6.3 we have

$$\hat{\nu} = 6.9 \simeq 7, \quad F^* = 37.76, \quad p < .001.$$

Unfortunately, F^* may sometimes be a poor approximation; in fact, $MS_a + MS_b - MS_{ab}$ may sometimes be less than zero. Although that is unlikely, it is a possibility, and F^* would then be negative and uninterpretable. We can avoid this problem by noting (from Table 6.6) that under the null hypothesis, $SS_m + MS_{ab}$ and $MS_a + MS_b$ both have the same expected value. This suggests using the quasi-F,

$$F^* = (SS_m + MS_{ab})/(MS_a + MS_b).$$

Both the numerator and the denominator of this quasi-F are sums of positive values, so the ratio can never be negative. For this ratio both the numerator and the denominator degrees of freedom must be estimated. The estimates can be derived easily from Equation 6.21:

$$\begin{aligned}
\hat{\nu}_1 &= (SS_m + MS_{ab})^2/[SS_m^2 + MS_{ab}^2/(IJ - I - J + 1)], \\
\hat{\nu}_2 &= (MS_a + MS_b)^2/[MS_a^2/(I - 1) + MS_b^2/(J - 1)].
\end{aligned}$$

For the data in Table 6.3 we obtain

$$\hat{\nu}_1 \simeq 1, \quad \hat{\nu}_2 \simeq 8, \quad F^* = 35.21, \quad p < .001.$$

These results are essentially the same (as they usually will be) as those obtained with the other quasi-F above. The F^* is slightly smaller, but the degrees of freedom are slightly larger.

NOTE ON TERMINOLOGY

The most commonly used mixed-effects design is the *repeated measures design*, in which each subject receives every treatment. In this design the treatments are the levels of the fixed factor, and the subjects are the levels of the random factor. Some textbooks treat the repeated measures design as a special design, referring to MS_b as the "between subjects" mean square and to MS_{ab} as the "within subjects" mean square. It is more parsimonious to treat such a design as a special case of the mixed-effects model. However, you should be familiar with the alternate terminology because it is used frequently.

ROBUSTNESS

The problem of robustness in the random and mixed models is more serious than in the fixed-effects model, simply because there are more assumptions to be met. Concerning the assumptions shared by all three models, about the same things can be said as were said about the fixed model. In general, the random and mixed models are robust to the assumptions of normality and equal variances of the errors so long as each cell contains the same number of scores. They are not robust to the assumption of random sampling. The random and mixed models require the additional assumption that the random effects themselves are normally distributed, and about the same things can be said about these assumptions as were said about the one-way random model. In general, when testing the null hypothesis that the variance of a random effect is some specified value *different from zero*, the test is not robust to the normality assumption. For all other tests (including any null hypothesis that the variance of a random effect is zero), the normality assumption is not critical.

In the fixed-effects models we assumed that the scores in different cells were statistically independent. In random and mixed models this assumption is replaced by the assumption that the cell means have equal variances and are equally correlated. Just as the fixed-effects model is not robust to the assumption of independence, the random and mixed models are not robust to the assumption of equal variances and correlations. Violation of this assumption generally results in too small a p value. This tendency to underestimate the p value will be present in any F ratio for which MS_{ab} is the denominator. Thus, it is present when testing the main effect of the fixed factor in the mixed-effects model.

Fortunately, corrections for this exist. There is a test that does not assume equal variances and covariances; it is a generalization of the matched-sample t test to more than two matched groups, treating the data as coming from a multivariate distribution. Its use has been limited because of its complexity and frequent lack of power relative to F; however, its use is increasing with availability of computer programs. It is discussed in Chapter 13.

APPROXIMATE TESTS

Approximate tests, obtained by modifying the F test, are also possible. Box (1953) showed that an approximate test can be obtained by multiplying both the numerator and the denominator degrees of freedom of the F ratio by the same constant, K_B. The correct value of K_B is a rather complicated function of the population variances and covariances. However, even though K_B cannot be known exactly, we can still base approximate tests on it. It is always true that $1/\nu_1 \leq K_B \leq 1$. If we test each hypothesis twice, letting K_B be as large as possible for one test and as small as possible for the other, we can find approximate lower and upper limits for the p value. If K_B is set to 1, its upper limit, we have the standard F test with ν_1 and ν_2 degrees of freedom. If K_B is set to $1/\nu_1$, its lower limit, we have an F whose degrees of freedom are 1 and ν_2/ν_1. The p values obtained from these two tests are the lower and upper limits on the actual p value.

For example, when testing the A main effect in Table 6.3, the lower limit on p, as determined by the conventional F test with 3 and 15 degrees of freedom, is .012. The upper limit is .071, the p value of the same F when its degrees of freedom are 1 and 5. We can thus conclude that the p value for the A main effect, even though the assumptions of equal variances and correlations may be violated, is between .012 and .071. If the violations are small, the p value is closer to .012; if they are large, it is closer to .071.

If the upper limit on p is less than α, then the test is certainly significant;

TABLE 6.9. Estimated variances and covariances,
and their marginal sums, for the A main effect in
Table 6.5 (data from Table 6.3).

	A_1	A_2	A_3	A_4	Sum
A_1	22.24	15.28	19.10	12.88	69.50
A_2	15.28	25.68	16.40	14.63	71.99
A_3	19.10	16.40	22.00	10.95	68.45
A_4	12.88	14.63	10.95	18.68	57.14
					267.08

if the lower limit on p is greater than α, then it is certainly not significant. If
we are not satisfied with the limits, we can estimate K_B from the data. The
actual calculations are simplified if we add some new notation. Let $v_{ii'}$ be
the estimated variance or covariance in row i and column i' of the covariance
table (i.e., Table 6.4 or the tables in Table 6.5; however, each cell must contain
a variance or covariance, unlike these tables which also contain correlation
coefficients). Then let

$$V_1 = \Sigma_i v_{ii} \ (= \text{ the sum of the diagonal elements}),$$
$$V_2 = \Sigma_i \Sigma_{i'} v_{ii'} \ (= \text{ the sum of all the elements}),$$
$$V_3 = \Sigma_i \Sigma_{i'} v_{ii}^2 \ (= \text{ the sum of the squared elements}),$$
$$V_4 = \Sigma_i (\Sigma_{i'} v_{ii'})^2 \ (= \text{ the sum of the squared marginal totals}),$$

and if the A main effect is being tested:

$$\hat{K}_B = (IV_1 = -V_2)^2 / [(I-1)(I^2 V_3 - 2IV_4 + V_2^2)].$$

If the B main effect is being tested, I in the above equation is changed to J.

Example. We will illustrate again with the A main effect in Table 6.3. The
variances and covariances are found in Table 6.5. However, first the correlation
coefficients in the lower half of that table have to be replaced by covariances
so the table will be symmetric about its main diagonal. The result is Table
6.9.
     We then have

$$V_1 = 88.60,$$
$$V_2 = 267.08,$$
$$V_3 = 4{,}721.2,$$
$$V_4 = 17{,}963,$$
$$\hat{K}_B = .80.$$

We use the original F ratio of 5.25, but the numerator degrees of freedom are
now $(.80)(3) = 2.4$, and the denominator degrees of freedom are $(.80)(15) =$
12. Interpolating between the tabled degrees of freedom, we find that $p = .019$.

The same test on the B main effect yields

$$
\begin{aligned}
V_1 &= 74.56, \\
V_2 &= 229.12, \\
V_3 &= 2{,}751.5, \\
V_4 &= 1{,}0981, \\
\hat{K}_B &= .48.
\end{aligned}
$$

The degrees of freedom are $(.48)(5) = 2.4$ and $(.48)(15) = 7.2$, giving $p = .009$. The minimum p value, using 5 and 15 degrees of freedom, is $< .001$ (see Table 6.3), and the maximum, using 1 and 3 degrees of freedom, is .056.

VARIANCE ESTIMATES

Unbiased estimates of the σ^2 are found most easily by estimating the τ^2 from appropriate linear combinations—determined by Table 6.6—of the mean squares, and using Equations 6.18 and 6.19 to solve for the σ^2. To calculate each $\hat{\tau}^2$, we first find the difference between the associated mean square and the mean square used in the denominator for testing that effect. We then divide this difference by the multiplier of the τ^2 for the effect in the expected mean square. To find each $\hat{\sigma}^2$, we multiply the associated $\hat{\tau}^2$ by $(N_a - 1)/N_a$, for each *fixed* factor in the effect, where N_a is the number of levels of that factor.

As an example, we can see from Table 6.6 that in both the random and mixed models

$$
\hat{\tau}_a^2 = (1/nJ)(MS_a - MS_{ab})
$$

because MS_{ab} is the denominator for testing the A main effect, and nJ is the multiplier of τ_a^2 in $\underline{E}(MS_a)$. In the random-effects model, $\hat{\sigma}_a^2 = \hat{\tau}_a^2$. In the mixed-effects model, with Factor A fixed,

$$
\hat{\sigma}_a^2 = [(I - 1)/I]\hat{\tau}_a^2.
$$

The other estimates are found in the same way.

The total variance of the X_{ijk} can be estimated as the sum of the variances of the effects and the error:

$$
\hat{\sigma}_t^2 = \hat{\sigma}_a^2 + \hat{\sigma}_b^2 + \hat{\sigma}_{ab}^2 + \hat{\sigma}_e^2.
$$

If there is no MS_w, it is customary to let $\sigma_e^2 = 0$ when obtaining these estimates.

The proportion of variance accounted for by any effect is the variance of that effect divided by the total variance. For example, the proportion of variance accounted for by the AB interaction is $\hat{\omega}_{ab}^2 = \hat{\sigma}_{ab}^2/\hat{\sigma}_t^2$.

The complete set of $\hat{\tau}^2$, $\hat{\sigma}^2$ and $\hat{\omega}^2$ is given in Table 6.6 for the data in Tables 6.2 and 6.3.

Note that a special problem arises when there is only one score per cell. In that case, MS_w cannot be calculated, so it cannot be subtracted from the other mean squares. There is no alternative to proceeding as if $\sigma_e^2 = 0$. In practice this is equivalent to setting $MS_w = 0$. This is not an entirely satisfactory solution, but it cannot be avoided. In effect, the within cells error is absorbed into the random effects, and it cannot be estimated separately. Note, however, that these assumptions are made *only* when estimating variances; they should not be made when performing any other procedures.

POWER CALCULATIONS

Power formulas are equally easy once some general principles are understood. If the effect being tested is a random effect, then

$$(MS_{num}/MS_{den})/[\underline{E}(MS_{num}/\underline{E}(MS_{den})] \sim F_{(\nu_1, \nu_2)},$$

where MS_{num} is the mean square in the numerator and MS_{den} is the mean square in the denominator of the F ratio. For example, for the B main effect in the random-effects model,

$$\frac{MS_b/MS_{ab}}{(\sigma_e^2 + n\tau_{ab}^2 + nI\tau_b^2)/(\sigma_e^2 + n\tau_{ab}^2)} \sim F_{(J-1, IJ-I-J+1)}.$$

Power calculations are then made using tables of the ordinary central F distribution, just as for the one-way random-effects model.

If the effect being tested is a fixed effect, the F ratio is distributed as noncentral F, with ϕ^2 found by substituting $\underline{E}(MS_{den})$ for σ_e^2 in Equation 5.17 to obtain

$$\phi^2 = \nu_1[\underline{E}(MS_{num})/\underline{E}(MS_{den}) - 1].$$

For example, for the A main effect in the mixed model,

$$\begin{aligned} \phi_a^2 &= (I-1)[\sigma_e^2 + n\tau_{ab}^2 + nJ\tau_a^2)/(\sigma_e^2 + n\tau_{ab}^2) - 1] \\ &= (I-1)nJ\tau_a^2/(\sigma_e^2 + n\tau_{ab}^2). \end{aligned}$$

The final principle is that if an effect varies over one or more random factors, it should be regarded as a random effect. Thus, MS_{ab}, because it varies over the random Factor B, should be treated as a random effect in both the random and the mixed model.

POOLING SUMS OF SQUARES

As in the fixed-effects model, sums of squares in random and mixed models can be pooled to increase the power of statistical tests. In Table 6.3, for example, the AB interaction is not significant. If there is no AB interaction, then, according to Table 6.6, the expected values of MS_{ab} and MS_w are the same. Then SS_w and SS_{ab} can be added to make a pooled sum of squares with

$39 \,(= 15 + 24)$ degrees of freedom. The new denominator mean square would then be $MS_{\text{pooled}} = (218.3 + 538.5)/39 = 19.41$.

Its expected value would just be σ_e^2. The expected values of MS_a and MS_b, if we assume no interaction, would be

$$
\begin{aligned}
\underline{E}(MS_a) &= \sigma_e^2 + nJ\tau_a^2, \\
\underline{E}(MS_b) &= \sigma_e^2 + nI\tau_b^2,
\end{aligned}
$$

so that MS_{pooled} would be an appropriate denominator. The tests on the A and B main effects would be

$$
\begin{aligned}
F_{(3,39)} &= 76.35/19.41 = 3.93, \quad p = .015 \quad A \text{ main effect)}, \\
F_{(5,39)} &= 133.5/19.41 = 6.88, \quad p < .001 \quad B \text{ main effect)}.
\end{aligned}
$$

In this example, the tests are not materially improved by pooling the sums of squares. If MS_{ab} had had fewer degrees of freedom or MS_w had had more degrees of freedom, there might have been some improvement.

The general rule for pooling sums of squares is as follows: If, when the τ^2 for a given effect is assumed to be zero, two mean squares have the same expected value, those mean squares can be pooled. They are pooled by adding their sums of squares and dividing by the sum of the degrees of freedom. In the mixed-effects design, if τ_{ab}^2 is assumed to be zero, it can be pooled with MS_w just as in the random-effects design. The pooled mean square would then be used to test both the A and B main effects. Remember again, however, that there should be good a priori reasons for assuming that τ_{ab}^2 is zero.

PLANNED AND POST HOC COMPARISONS

Planned and post hoc comparisons are meaningful only when fixed effects are involved. They are therefore not meaningful in the random-effects model, and in the mixed-effects model they are meaningful only when testing the A main effect. That is, any planned comparison must be of the form

$$ H_0 : \Sigma_i c_i \mu_{i.} = 0. \tag{6.22} $$

In theory it is possible to test any planned comparison of this form, but special methods are required if the test is not a contrast. For that reason we shall limit our discussion first to tests of contrasts.

Contrasts. It can be shown (although the derivations are somewhat complicated because the $\overline{X}_{i..}$ are not independent) that the theory given in Chapter 3 can be applied to tests of contrasts in the mixed model. If we let

$$
\begin{aligned}
\psi_k &= \Sigma_i c_{ik} \mu_{i.}, \\
C_k &= \Sigma_i c_{ik} \overline{X}_{i..}, \\
C_k' &= C_k (nJ/\Sigma_i c_{ik}^2)^{1/2}, \\
SS_k &= (C_k')^2 = nJC_k^2/(\Sigma_i c_{ik}^2).
\end{aligned}
$$

then Equation 6.22 can be tested by

$$F_{(1,IJ-I-J+1)} = SS_k/MS_{ab},$$
$$t_{(Ij-I-J+1)} = C_k'/(MS_{ab})^{1/2}.$$

Basically, the only differences between these equations and those in Chapter 3 are that nJ replaces n wherever it appears and MS_{ab} replaces MS_w. All of the theory relating to contrasts and based on F ratios in Chapters 3 and 4, including post hoc tests (by the S method), confidence intervals (for both planned and post hoc tests), power calculations, proportions of variance accounted for, and combinations of planned and post hoc tests, can be applied to contrasts of the form of Equation 6.22. It only is necessary to replace n by nJ and MS_w by MS_{ab} in the appropriate formula in Chapter 3 or 4. Post hoc tests by the T method, and other tests based on the Studentized range distribution, cannot be made because the means being tested are not independent.

Considerable caution must be exercised when testing contrasts by this method; its robustness is not well known, but it is probably not very robust with respect to violations of the assumption that the $\overline{X}_{ij.}$ have equal variances and covariances.

More General Method. There is another more general method for making planned comparisons. It can be used for any planned comparisons, including those that are not contrasts, and it does not require the assumption of equal variances and covariances. It thus has very general applicability. It pays for this generality, however, by a decrease in power and an increase in the amount of work required to calculate the statistic. The method depends on the fact that any linear combination of normal random variables is also normal.

With a little algebra on the summations, we can show that

$$C_k = \Sigma_i c_{ik} \overline{X}_{i..} = (1/J)\Sigma_j (\Sigma_i c_{ik} \overline{X}_{ij.}).$$

Therefore, if we let

$$Y_{jk} = \Sigma_i c_{ik} \overline{X}_{ij.},$$

then C_k is the mean of the Y_{jk}. Furthermore, the Y_{ik}, being a linear combination of normal random variables, must also be normally distributed. Therefore, we simply make the planned comparison by calculating Y_{jk} for each level (B_j) of Factor B and doing a t test of the null hypothesis that the mean of the Y_{jk} is zero. The t test will have $(J-1)$ degrees of freedom because the sample contains J values of Y_{jk}.

It is not necessary to actually calculate the Y_{jk} if we have available the estimated means, variances, and covariances among the levels of A (e.g., the values in Table 6.4). The t test requires only that we estimate the mean and variance of the Y_{jk}, and these can be found from the formulas:

$$
\begin{aligned}
\overline{Y}_{.k} &= \Sigma_i c_{ik} \overline{X}_{i..}, \\
\hat{V}(Y_{jk}) &= \Sigma_i c_{ik}^2 \hat{V}(\overline{X}_{ij.}) + 2\Sigma_i \Sigma_{i'} c_{ik} c_{i'k} \hat{C}(\overline{X}_{i'j.}, \overline{X}_{i'j.}).
\end{aligned}
\tag{6.23}
$$

Here, $\overline{X}_{.k}$ and $\hat{V}(Y_{jk})$ are the estimated mean and variance of the Y_{jk}, and $\hat{V}(\overline{X}_{ij.})$ and $C(\overline{X}_{ij.}, \overline{X}_{i'j.})$ are the estimates of variances and covariances found as in Table 6.4.

We will illustrate this method by testing the null hypothesis $H_0(1) : \mu_1 + \mu_2 - 2\mu_3 = 0$. In principle, we will test it by finding $Y_{j1} = \overline{X}_{1j.} + \overline{X}_{2j.} - 2\overline{X}_{3j.}$ for each level of Factor B and then doing a t test on the null hypothesis that the mean of the Y_{j1} is zero. However, the only values needed for the t test are the mean and variance of the Y_{j1}, and we can find them from Equation 6.23:

$$
\begin{aligned}
\overline{Y}_{.1} &= \overline{X}_{1..} + \overline{X}_{2..} - 2\overline{X}_{3..} = 20.8 + 21.5 - (2)(19.2) \\
&= 3.9, \\
\hat{V}(Y_{j1}) &= [(1)(15.40) + (1)(17.63) + (4)(12.34)] \\
&\quad + 2[(1)(1)(.71) + (1)(-2)(.34) + (1)(-2)(-97)] \\
&= 86.33.
\end{aligned}
$$

The total number of scores for this test is considered to be $J = 20$, so the standard error of the mean is $(86.33/19)^{1/2} = 2.13$, and

$$
\begin{aligned}
t_{(10)} &= 3.9/2.13 = 1.83, \\
F_{(1,19)} &= 1.83^2 = 3.35, \\
p &= .083.
\end{aligned}
$$

If we had made this test using the method described in the previous section, the F would still have been 3.35, but the degrees of freedom would have been 57, and the p value would have been .072. (It is mostly coincidental that both tests give the same F value in this example; the values will be similar when the estimated variances and covariances are approximately equal, but they will seldom be identical.) Of course, the test described in the previous section depends on the assumption of equal variances and covariances, whereas the test described here does not. In a sense, the added assumptions "buy" us 38 more degrees of freedom. However, the assumptions are seldom known to be satisfied, so the test described here is probably better for routine use. Moreover, this method can be extended to planned comparisons that are not contrasts, while the method described in the previous section cannot.

The method described here cannot be used when one wishes to combine two planned comparisons into a single test with two degrees of freedom in the numerator. The problem is that no two tests of planned comparisons have the same denominator term when this method is used. The method described in the previous section can be used if the assumptions are met; if the assumptions are not met, a multivariate test (Chapter 13) can be used.

Exercises

(1.) In a study of IQs of first-grade school children from slum areas in a major city, the following procedure was used. Seven schools were chosen at random from among all of the schools located in what were commonly regarded as slum areas. A list of all children whose parents finished high school but did not attend college was obtained for each school. (Subjects were limited to this subpopulation to control for economic factors.) From among these children in each school six were chosen at random $(n = 6)$. The following table gives the mean measured IQs of the sampled first graders from each school $(MS_w = 200)$:

	\multicolumn{7}{c}{School}
	A_1	A_2	A_3	A_4	A_5	A_6	A_7
Mean IQ	96	99	89	115	97	99	101

(a.) Test the null hypothesis that there are no differences in average IQ among the seven schools.

(b.) Estimate σ^2_{bet} and ω^2_{bet}.

(c.) How much power does the test in question (a) have against the alterative $\sigma^2_{bet} = 2\sigma^2_e$, if $\alpha = .01$?

(d.) Test the null hypothesis that $\sigma^2_{bet} = 2\sigma^2_e$, using a two-tailed test.

(e.) Test the null hypothesis that the average IQ of all these children is not different from the average for the nation; i.e., $H_0 : \mu = 100$.

(f.) How much power does the test in question (d) have against the alternative $\sigma^2_{bet} = 0$ if $\alpha = .01$?

(General hint: The calculations are much simpler if, instead of operating on the means given above, you first subtract 100 from each mean and then perform the analysis on the differences.)

(2.) An experimenter has done an experiment with four groups of subjects and obtained the following results:

A_1	A_2	A_3	A_4
8	-1	4	-13
12	-2	-4	-1
1	5	7	-3
4	-5	1	-12
6	3	5	3

Her results were significant at the .05 level.

A second experimenter replicated the study using subjects from exactly the sample population. His results were

A_1	A_2	A_3	A_4
-6	2	3	9
-6	-8	-1	2
5	-4	1	7
-3	3	10	-1
4	-3	-2	1

His results were not significant.

Both experimenters drew their subjects from the same population, so the results from the two experiments can be combined. Analyze the combined data, assuming that the second experimenter replicated the experiment in the most appropriate manner and the original experiment was (a) fixed-effects, (b) random-effects. Explain the difference in the results of parts (a) and (b). In your opinion, is it more probable that the design was actually fixed-effects or random-effects?

(3.) The following data are correct responses, out of a possible 25, on four difficult concept-learning tasks, involving a total of 20, 40, 60, and 80 trials, respectively, for six subjects.

Subjects	Total Trials			
	20	40	60	80
S_1	11	10	15	23
S_2	2	5	9	13
S_3	6	3	12	13
S_4	11	12	9	19
S_5	7	5	12	20
S_6	8	6	13	14

(a.) Analyze these data as a two-way, mixed-effects design.

(b.) Do post hoc tests on all possible pairs of means.

(4.) An experimenter was interested in comparing the effects of repeated versus unique words in a verbal learning situation. The subjects were five randomly selected college students. Each S saw 12 different 15-word lists during the hour, and, after seeing each list, was to recall as many words as possible. There were three lists each of four types: Type 1 contained an item followed by an immediate repetition of the same item in positions 8 and 9; Type 2 contained a starred item in position 8; Type 3 contained an item printed in red in position 8; and Type 4 was a control list, with no unique items. The data for lists of Types 1 through 3 were a one if the subject recalled the unique word, and a zero if not. The data for type 4 lists were the proportion of the middle three words (7, 8, and 9) recalled. The data were

| | List Type | | | |
	A_1	A_2	A_3	A_4
	1	0	1	.000
S_1	1	1	0	.333
	1	1	0	.000
	1	1	0	.667
S_2	0	1	0	.333
	1	0	1	.000
	1	0	1	.000
S_3	1	1	1	.000
	1	1	1	.333
	1	1	1	.000
S_4	0	1	1	.000
	0	1	1	.000
	0	0	1	.667
S_5	1	0	0	.000
	0	0	1	.333

(a.) By summing over certain sets of values in the above table, this can be turned into a simple two-way mixed-effects design with subjects as one factor. Sum the values and test the null hypothesis of no differences among types of lists. Calculate the p value and the $\hat{\omega}^2$.

(b.) Find the variances and covariances of the data analyzed in part (a) and comment on the validity of the assumptions for that analysis.

(c.) Analyze the data in the original table as a two-way, mixed-effects design, in which Factor A (fixed) is the type of list, with four levels, and Factor B (random) has 15 levels, corresponding to the 15 rows in the table. Compare this analysis with the one in part (a) and comment on the validity of analyzing the data in this way.

(d.) Using the analysis in part (a), test the following two null hypotheses; find a single p value and $\hat{\omega}^2$ for each hypothesis.

$$H_0(1): \quad \mu_{1.} + \mu_{2.} + \mu_{3.} = 3\mu_{4.},$$
$$H_0(2): \quad \mu_{1.} = \mu_{2.} = \mu_{3.}.$$

(5.) Reanalyze the data in Problem 2, Chapter 5, assuming that

(a.) Factor B is random and Factor A is fixed;

(b.) Factor A is random and Factor B is fixed.

(c.) Comment on the relationship of the power of the test on the fixed factor to the number of levels of the random factor.

(6.) Suppose an experimenter had good reasons to believe that there was no A main effect for a two-way design, and that the data supported this belief. He wants to test the grand mean, the B main effect, and the AB interaction. How might he use the absence of an A main effect to increase his power by pooling sums of squares, if we assume that it is

(a.) a fixed-effects model,

(b.) a random-effects model,

(c.) a mixed-effects model with Factor A random, or

(d.) a mixed-effects model with Factor B random.

(7.) Consider a one-way fixed-effects design with repeated measures (actually, a two-way design where the second factor is subjects), with $I = 2$. Suppose a third level of factor A (the fixed factor) is added to the analysis. Tell whether each condition listed below is likely to increase the F ratio for three levels over that obtained with only two, to decrease it, or to leave it the same.

(a.) σ_e^2 is larger for level A_3 than for levels A_1 and A_2.

(b.) $\mu_3 = (\mu + \mu_2)/2$.

(c.) The correlation between the scores for levels A_3 and A_2 is higher than that between the scores for A_1 and A_2.

7
Higher-Way Designs

The principles discussed so far with respect to one-way and two-way designs can be extended to designs having any arbitrary number of factors. We need only extend the basic model to include the additional factors; however, the extension involves additional interaction effects. For example, consider a three-way analysis of variance ($2 \times 2 \times 3$), such as the one in Table 7.1.

The cells in this design vary on three dimensions (factors). We might obtain such a design if we did an experiment like the one on mental patients, but we divided our subjects by sex. The sex of the subject would then be a third factor with two levels.

We can label the factors A, B, and C, and let ABC_{ijk} be the cell at the ith level of Factor A, the jth level of Factor B, and the kth level of Factor C. In keeping with our previous notation, we will let X_{ijkl} be the lth score in cell ABC_{ijk}. We will let μ_{ijk} and $\overline{X}_{ijk.}$ be the population and sample means, respectively, of cell ABC_{ijk}, and we will use dots to replace subscripts of factors over which we have summed or averaged.

As in previous chapters, X_{ijkl} can be expressed as the sum of a number of fixed and random effects plus an error term. However, a new set of effects must be introduced. The grand mean, μ, is included in the equation, as are the main effects (α_i, β_j, and γ_k) of Factors A, B, and C. The main effects are defined as

$$
\begin{aligned}
\alpha_i &= \mu_{i..} - \mu, \\
\beta_j &= \mu_{.j.} - \mu, \\
\gamma_k &= \mu_{..k} - \mu.
\end{aligned}
$$

In the three-way model there are three two-way interactions, one for each pair of factors, defined as

$$
\begin{aligned}
\alpha\beta_{ij} &= (\mu_{ij.} - \mu) - \alpha_i - \beta_j \\
&= \mu_{ij.} - \mu_{i..} - \mu_{.j.} + \mu, \\
\alpha\gamma_{ik} &= (\mu_{i.k} - \mu) - \alpha_i - \gamma_k \\
&= \mu_{i.k} - \mu_{i..} - \mu_{..k} + \mu, \\
\beta\gamma_{jk} &= (\mu_{.jk} - \mu) - \beta_j - \gamma_k \\
&= \mu_{.jk} - \mu_{.j.} - \mu_{..k} + \mu.
\end{aligned}
$$

All of these terms are still not sufficient to express X_{ijkl}. The complete expression is

$$
X_{ijkl} = \mu_{ijk} + \epsilon_{ijkl}
$$

TABLE 7.1. Hypothetical three-way analysis of variance.

	A_1			A_2		
	C_1	C_2	C_3	C_1	C_2	C_3
B_1	2	8	14	10	8	4
	3	5	11	5	9	7
	0	5	9	3	14	3
	2	11	10	7	12	6
	1.75	7.25	11.00	6.25	10.75	5.00
B_2	9	4	3	3	5	5
	3	11	9	5	5	7
	12	9	10	9	7	5
	8	8	8	5	6	6
	8.00	8.00	7.50	5.00	5.75	5.75

$$\begin{aligned} = \ & \mu + (\alpha_i + \beta_j + \gamma_k) + (\alpha\beta_{ij} + \alpha\gamma_{ik} + \beta\gamma_{jk}) \\ & + \alpha\beta\gamma_{ijk} + \epsilon_{ijkl}, \end{aligned}$$

where $\alpha\beta\gamma_{ijk}$ is the *three-way interaction*:

$$\begin{aligned} \alpha\beta\gamma_{ijk} = \ & (\mu_{ijk} - \mu) - (\alpha_i + \beta_j + \gamma_k) - (\alpha\beta_{ij} + \alpha\gamma_{ik} + \beta\gamma_{jk}) \\ = \ & \mu_{ijk} - (\mu_{ij.} + \mu_{i.k} + \mu_{.jk}) + (\mu_{i..} + \mu_{.j.} + \mu_{..k}) \\ & - \mu. \end{aligned}$$

(The parentheses in the above equations are used to mark off groups of similar terms.) The three-way interaction is the extent to which the deviation of each cell mean from the grand mean cannot be expressed as the sum of the main effects and two-way interactions.

Similarly, if we extend the design to a four-way analysis of variance, the model becomes

$$\begin{aligned} X_{ijklm} = \ & \mu + (\alpha_i + \beta_j + \gamma_k + \delta_1) \\ & + (\alpha\beta_{ij} + \alpha\gamma_{ik} + \alpha\delta_{il} + \beta\gamma_{jk} + \beta\delta_{jl} + \gamma\delta_{kl}) \\ & + (\alpha\beta\gamma_{ijk} + \alpha\beta\delta_{ijl} + \alpha\gamma\delta_{ikl} + \beta\gamma\delta_{jkl}) \\ & + \alpha\beta\gamma\delta_{ijkl} + \epsilon_{ijklm}. \end{aligned}$$

In the four-way design there are four main effects, six two-way interactions, four three-way interactions, and one four-way interaction.

In an m-way analysis of variance there are m main effects, $\binom{m}{2}$ two-way interactions, $\binom{m}{3}$ three-way interactions, and in general $\binom{m}{r}$ r-way interactions, for each value of r from two through m. An r-way interaction is defined

TABLE 7.2. Cell totals (upper values) and means (lower values) for data in Table 7.1.

Three-Way Table

		A_1			A_2	
	C_1	C_2	C_3	C_1	C_2	C_3
B_1	7	29	44	25	43	20
	1.75	7.25	11.00	6.25	10.75	5.00
B_2	32	32	30	22	23	23
	8.00	8.00	7.50	5.50	5.75	5.75

Two-Way Tables

	A_1	A_2
B_1	80	88
	6.67	7.33
B_2	94	68
	7.83	5.67

	A_1	A_2
C_1	39	47
	4.88	5.88
C_2	61	66
	7.63	8.25
C_3	74	43
	9.25	5.38

	C_1	C_2	C_3
B_1	32	72	64
	4.00	9.00	8.00
B_2	54	55	53
	6.75	6.88	6.63

One-Way Tables

A_1	A_2
174	156
7.25	6.50

B_1	B_2
168	162
7.00	6.75

C_1	C_2	C_3
86	127	117
5.38	7.94	7.31

$\overline{X}$
330
6.88

as follows. Suppose we average over all of the scores in each cell and then average over all factors *except* the r factors involved in the interaction being considered. The r-way interaction is the extent to which the expected values of the cell means of the resulting r-way table cannot be expressed as the sums of the main effects and lower-order interactions of the r factors. (The r-way tables produced by this procedure, for Table 7.1, are shown in Table 7.2.)

The following selected examples of two-, three-, and four-way interactions taken from the four-way analysis of variance illustrate this definition:

$$
\begin{aligned}
\beta\delta_{jl} &= (\mu_{.j.l} - \mu) - (\beta_j + \delta_l), \\
\alpha\gamma\delta_{ikl} &= (\mu_{i.kl} - \mu) - (\alpha_i + \gamma_k + \delta_l) \\
&\quad - (\alpha\gamma_{ik} + \alpha\delta_{il} + \gamma\delta_{kl}), \\
\alpha\beta\gamma\delta_{ijkl} &= (\mu_{ikl} - \mu) - (\alpha_i + \beta_j + \gamma_k + \delta_l) \\
&\quad - (\alpha\beta_{ij} + \alpha\gamma_{ik} + \alpha\delta_{il} + \beta\gamma_{jk} + \beta\delta_{jl} + \gamma\delta_{kl})
\end{aligned}
$$

$$-(\alpha\beta\gamma_{ijk} + \alpha\beta\delta_{ijl} + \alpha\gamma\delta_{ikl} + \beta\gamma\delta_{jkl}).$$

Advantages of Higher-Way Designs

The disadvantages of using higher-way designs are obvious; the more factors we add to the design, the more difficult will be both the analysis and the interpretation. However, there may also be advantages in adding factors to our design.

One obvious advantage is that it enables us to study the effects of more variables within a single experiment. This saves us the effort and expense of setting up additional experiments. Another is that it enables us to learn how a number of different variables interact with each other to produce the effect.

Another advantage is less obvious. Even if we are not interested in the additional factors, they may have an effect on the dependent variable. If the factors are ignored, their effect will be included in the error term, inflating it and reducing the power of our tests. For example, suppose we are studying the effects of two different teaching techniques on children of high, medium, and low IQ. Our subjects are children in the 4th, 5th, and 6th grades. Even though we may not be interested in differences among grades, they will almost surely exist. If we ignore them, using a two-way, teaching technique × IQ design, the differences in grade levels will inflate our error terms. If we include them, using a three-way, teaching technique × IQ × grade design, we will have smaller error terms and more power for testing the effects we are interested in.

Kinds of Models

In the one-way analysis of variance we saw that we could have either a fixed or random model. In the two-way analysis of variance three distinct models—fixed, random, and mixed—were possible. In higher-way analyses of variance there are even more possibilities; each factor in the design may be either fixed or random. Thus, in the three-way design we may have zero, one, two, or three random factors, with the remaining factors fixed; in the four-way design we add the possibility of four random factors; and so on. Usually, the number of random factors in a design is either zero or one (the single random factor is usually subjects), but sometimes designs are constructed with two or more random factors. The general model we will consider in this chapter applies to designs with arbitrary numbers of fixed and random factors.

TABLE 7.3. Assumptions in three-way analysis of variance with factor C random.

(1.) All X_{ijkl} *within a cell* are randomly sampled and normally distributed with the same mean and variance.

(2.) The levels of C are randomly sampled.

(3.) The $\overline{X}_{ijk.}$ are normally distributed over Factor C (within each combination of levels of A and B), with equal variances.

(4.) For all $i, j, k, i' \neq 1$, and $j' \neq j$:

   (a) $C(\overline{X}_{ijk.}, \overline{X}_{i'jk.}) = $ a constant;

   (b) $C(\overline{X}_{ijk.}, \overline{X}_{ij'k.}) = $ a constant;

   (c) $C(\overline{X}_{ijk.}, \overline{X}_{i'j'k.}) = $ a constant.

Assumptions

The assumptions made in a higher-way analysis of variance are easily stated in terms of the individual effects, but translating these assumptions into statements about the scores themselves is not easy.

Some assumptions are obvious. We must assume that the scores are sampled randomly from a normal distribution with constant variance. We must also assume that the random factors are in fact randomly and independently sampled. The assumption of equal covariances is similar to that for the mixed-model two-way analysis of variance but not exactly the same. A comprehensive statement of the conditions for the m-way design would be too complicated to give here. However, the general approach should be clear from Tables 7.3 and 7.4, which list the assumptions in the three-way design with one and two random factors, respectively. For practical purposes, we need only note that the conditions are probably not satisfied in most experiments. A later section will give corrections on the F test for data for which the variances and covariances cannot be assumed equal. In addition, methods described in Chapter 13 can be used.

However, two further points must be made because they are important later. The first involves models in which all of the factors are fixed. We saw previously that the assumptions in a fixed-effects model were essentially the same whether there were one or two factors. This is true for any fixed-effects design, no matter how many factors there are. The necessary assumptions for a fixed-effects model are that the scores within the cells are independently sampled from normal populations with the same variance. The only important assumption, if the sample sizes are equal and not too small, is that the sampling is independent. All of the tests for main effects and interactions in

TABLE 7.4. Assumptions in three-way analysis of variance with factors B and C random.

(1.) All X_{ijkl} *within a cell* are sampled randomly and normally distributed with the same mean and variance.

(2.) (a) The levels of B are sampled randomly.

   (b) The levels of C are sampled randomly and independently of the levels of B.

(3.) (a) The $\overline{X}_{ijk.}$ are normally distributed over Factor B (within each combination of levels of A and C).

   (b) The $\overline{X}_{ijk.}$ are normally distributed over Factor C (within each combination of levels of A and B).

(4.) For all $i, j, k, i' \neq i, j' \neq j, k' \neq k$:
   (a) $C(\overline{X}_{ijk.}, \overline{X}_{i'jk.}) =$ a constant;
   (b) $C(\overline{X}_{ijk.}, \overline{X}_{ij'k.}) =$ a constant;
   (c) $C(\overline{X}_{ijk.}, \overline{X}_{ijk'.}) =$ a constant;
   (d) $C(\overline{X}_{ijk.}, \overline{X}_{i'j'k.}) =$ a constant;
   (e) $C(\overline{X}_{ijk.}, \overline{X}_{i'jk'.}) =$ a constant;

---

the m-way fixed-effects model can be expressed as combinations of planned comparisons on a one-way design.

The other point is that this same principle does not extend to designs with one or more random factors. For example, a three-way design with two factors fixed and one random cannot simply be thought of as a two-way design with one factor fixed and one random. For the two fixed factors to be legitimately combined into one, more restrictive additional assumptions would have to be made.

Mean Squares

As in the two-way analysis of variance, the calculation of sums of squares and mean squares in the m-way analysis of variance does not depend on the particular model. In theory, when calculating the sum of squares for an effect, estimates of the values of the effect are first obtained. These estimates are then squared and summed, and the sum is multiplied by the total number of scores over which sums were taken to estimate the effects. In the three-way

TABLE 7.5. Summary table for data in Tables 7.1 and 7.2 (all factors fixed).

	RS	SS	T	df	MS	F	p	$\hat{\omega}^2$
m		2,268.750		1				
a	2,275.500	6.750	2	1	6.750	1.08	.31	.00
b	2,269.500	0.750	2	1	.7500	0.12	—	(-.01)
c	2,325.875	57.125	3	2	28.56	4.58	.017	.09
ab	2,300.333	24.083	4	1	24.08	3.86	.057	.04
ac	2,391.500	58.875	6	2	29.44	4.72	.015	.09
bc	2,381.750	55.125	6	2	27.56	4.42	.019	.08
abc	2,547.500	76.042	12	2	38.02	6.10	.005	.12
w	2,772.000	224.500	48	36	6.236			
t		503.250		47				

design, for example,

$$SS_a = nJK\Sigma_i\hat{\alpha}_i^2,$$
$$SS_{ab} = nK\Sigma_i\Sigma_j\widehat{\alpha\beta}_{ij}^2,$$
$$SS_{abc} = n\Sigma_i\Sigma_j\Sigma_k\widehat{\alpha\beta\gamma}_{ijk}^2,$$

where I, J, and K are the numbers of levels of Factors A, B, and C, respectively. The other sums of squares are found similarly.

An easy way to determine the multiplier for each sum of squares is to note that it is equal to the total number of scores in the experiment divided by the total number of estimates of that effect. Thus, the total number of scores in the experiment is $N = nIJK$, and there are I estimated values of α_i, so the multiplier for SS_a is $nIJK/I = nJK$. For SS_{ab}, IJ values of $\alpha\beta_{ij}$ are estimated, so the multiplier is $nIJK/IJ = nK$. The same principle holds for SS_{abc} and all of the other sums of squares.

As in the one- and two-way analyses of variance, SS_t is the sum of all of the squared deviations of the scores from the grand mean, and SS_w is the sum of the squared deviations of the scores from their cell means.

It is tedious to estimate all of the effects and then square and sum them. Special calculating formulas, like those in Tables 2.4 and 5.5, could be given to simplify each higher-way analysis of variance. However, those formulas are generated by a small number of rules that apply to any m-way design. Instead of the individual formulas, we will give the general rules for calculating sums of squares, mean squares, and degrees of freedom in the m-way design. The application of the rules to the data in Table 7.1 is illustrated in Tables 7.2 and 7.5.

Rules for Calculating Raw Sums of Squares

(1) Find the sum of the squared values of all the scores; this is RS_w, the raw sum of squares within. For the data in Table 7.1, $RS_w = 2772$.

(2) Find the sum of the scores in each cell (if there are more than one). Sum the squares of these values and divide by n, the number of values over which each sum is taken. This is the raw sum of squares for the m-way interaction. For Table 7.1, this is the three-way table in Table 7.2, and $RS_{abc} = (1/4)(7^2 + 29^2 + 44^2 + \ldots) = 2,547.5$.

(3) Sum the cell totals over each factor in turn to obtain m different tables, each of which has $(m-1)$ dimensions. For the example of Table 7.1, these are the three two-way tables in Table 7.2.

(4) For each table separately, find the sum of the squared cell totals, and divide this by the *total* number of scores summed over to obtain each cell total. This divisor will be N divided by the number of cell totals that were squared and summed. The results are the m raw sums of squares for the $(m-1)$-way interactions. For the AB interaction in Table 7.2, for example, $RS_{ab} = (1/12)(80^2 + 88^2 + 94^2 + 68^2) = 2,300.333$.

(5) Sum the cell totals of each table obtained in Step (3) separately over each factor in the table to obtain $(m^2 - m)/2$ *different* $(m-2)$-dimensional tables. (Actually, $m^2 - m$ tables will be obtained, but one-half of these will be duplicates of the others.) These are the one-way tables in Table 7.2. Then follow the procedure in Step (4) to obtain the $(m^2 - m)/2$ raw sums of squares for the $(m-2)$-way interactions. For the data in Tables 7.1 and 7.2, these are actually the raw sums of squares for the main effects. As an example, $RS_a = (1/24)(174^2 + 156^2) = 2,275.5$.

(6) Continue as in Steps 3, 4, and 5 until the total of all of the scores is found by summing over any of the m one-dimensional tables eventually obtained. Square this grand total and divide the squared total by N to obtain SS_m. For Tables 7.1 and 7.2, $SS_m = (1/48)(330)^2 = 2,268.75$.

Rules for Calculating Sums of Squares

(7) List the raw sums of squares in a table such as Table 7.5 as an aid in calculating the sums of squares.

(8) The sum of squares for each main effect is the RS for that main effect minus SS_m. For example, $SS_a = RS_a - SS_m$. The sum of squares for each effect should be listed in the second column of the table.

(9) The sum of squares for any interactions is the raw sum of squares minus SS_m and minus the sums of squares for *all* main effects and lower-order interactions involving *only* the factors involved in the interaction whose

sum of squares is being found, i.e., involving no *other* factors. For example, for the three-way table,

$$SS_{ab} = RS_{ab} - SS_m - SS_a - SS_b,$$
$$SS_{abc} = RS_{abc} - SS_m - SS_a - SS_b - SS_c - SS_{ab} - SS_{ac} - SS_{bc}.$$

The value SS_{ac} is not subtracted when finding SS_{ab} because it involves a factor (C) that is not involved in the AB interaction. As another example, for the four-way analysis of variance,

$$SS_{acd} = RS_{acd} - SS_m - SS_a - SS_c - SS_d - SS_{ac} - SS_{ad} - SS_{cd}.$$

The terms SS_b, SS_{ab}, and SS_{bd} are not subtracted because they involve Factor B, and the ACE interaction does not involve Factor B.

(10) To find SS_w, subtract the RS for the m-way interaction from RS_w. For Table 7.5, $SS_w = RS_w - RS_{abc} = 2772 - 2547.5 - 224.5$.

(11) $SS_t = RS_w - SS_m$.

The tabular method of summarizing shown in Table 7.5 is especially useful for calculating sums of squares because you need only scan the list of row headings to determine which sums of squares should be subtracted. In the table, main effects should precede interactions, and lower-order interactions should precede higher-order interactions; SS_m should be at the top, and SS_w and SS_t should be at the bottom.

THE HIERARCHY

The rules for calculating sums of squares can be expressed in terms of a hierarchy as in previous chapters. The hierarchy for the three-way design is presented in Figure 7.1. Note that, as in Chapter 5, it is a partial ordering; some effects are neither above nor below others. One effect is above another if and only if its table can be obtained by summing over values in the other's table. In Figure 7.1, one effect is above another if and only if it can be reached from the other by moving upward along connecting lines. A comparison of Rule (3) above with Figure 7.1 will show that the SS for any effect is obtained by subtracting the sums of squares for all effects that are above it in the hierarchy. This fact helps to clarify the reason for the subtractions. The RS for any effect will be large if the values in its table are large. These values will be large not only if the effect is large but if any of the effects above it in the hierarchy are large (i.e., if the values in any table derived from its table are large). We eliminate the influence of these other effects by subtracting their sums of squares.

FIGURE 7.1. Hierarchy of factors in three-way analysis of variance.

RULES FOR CALCULATING DEGREES OF FREEDOM AND MEAN SQUARES

(1) For each raw sum of squares, RS_{eff} (where eff refers to any effect), let T_{eff} be the total number of values that were squared and summed to obtain the RS_{eff}. For example, in the three-way design, $T_a = I$ (because I values were squared and summed in finding RS_a), $T_{ab} = IJ, T_{abc} = IJK$, and $T_w = N$. For the data of Table 7.1, these values are listed in the third column of Table 7.5.

(2) From each T_{eff}, subtract the number of degrees of freedom (df) of each SS (including SS_m) that was subtracted from the RS in finding the SS. Again, for the three-way design (note that $df_m = 1$),

$$df_a = T_a - 1,$$
$$df_{ab} = T_{ab} - 1 - df_a - df_b,$$
$$df_{abc} = T_{abc} - 1 - df_a - df_b - df_c - df_{ab} - df_{ac} - df_{bc}.$$

Similarly, in the four-way design,

$$df_{acd} = T_{acd} - 1 - df_a - df_c - df_d - df_{ac} - df_{ad} - df_{cd}.$$

(3) The degrees of freedom for SS_w are the total number of scores (N) minus the total number of cells in the original table. In the three-way design, $df_w = N - IJK = T_w - T_{abc}$.

(4) The degrees of freedom for SS_t are the total number of scores minus one: $df_t = N - 1 = T_w - 1$. The degrees of freedom are shown in the fourth column of Table 7.5.

(5) Each mean square is then found by dividing the sum of squares by its degrees of freedom.

Expected Mean Squares

As we saw in the previous chapters, estimation of variances and proportions of variance accounted for, F tests, and power calculations all depend on finding the expected mean squares of the effects. The following rules can be used to find the expected mean squares:

(1) Write the appropriate model equation for the design. For the three-way analysis of variance, for example,

$$X_{ijkl} = \mu + \alpha_i + \beta_j + \gamma_k + \alpha\beta_{ij} + \alpha\gamma_{ik} + \beta\gamma_{jk}$$
$$+ \alpha\beta\gamma_{ijk} + \epsilon_{ijkl}.$$

(2) Each expected mean square is the sum of a set of terms. Each term is a constant times the τ^2 for an effect. The constant multiplier of each τ^2 is the divisor used in finding the RS for that effect. That is, it is the total number of scores divided by the number of cells in the table of sums for that effect. For the A main effect, for example, there are I cells; in the three-way design, there are $nIJK$ scores. Therefore, the multiplier of τ_a^2 is $nIJK/I = nJK$, the number of scores summed over in finding the table for the A main effect. The complete set of terms in the three-way model is

$$N\mu^2, \; nJK\tau_a^2, \; nIK\tau_b^2, \; nIJ\tau_c^2, \; nK\tau_{ab}^2, \; nJ\tau_{ac}^2, \; nI\tau_{bc}^2, \; n\tau_{abc}^2, \; \sigma_e^2.$$

These terms are found by referring to the model equation. They should be written out completely for the experiment being analyzed to simplify the calculation of the expected mean squares.

(3) Not all of the above enter into the sum for a given expected mean square. The τ^2 values in the expected mean square for a given effect must satisfy two criteria:

(a) Their subscripts must include all of the subscripts in the effect whose expected mean square is being calculated and

(b) they can have no other subscripts of *fixed* factors (they may have subscripts of random factors).

Finally, σ_e^2 is always a term in the sum.

As an example, consider the three-way design with Factors A and B fixed and Factor C random. We will first find $\underline{E}(MS_a)$, the expected mean square of the A main effect. We want the sum of all τ^2 that have an a in their subscript and that do not have subscripts of any other *fixed* factors. Since B is the only other fixed factor, we must find all τ^2 that have an a but not b in their subscript. There are just two that meet this criterion, so the expected mean square of the A main effect is

$$\underline{E}(MS_a) = \sigma_e^2 + nJ\tau_{ac}^2 + nJK\tau_a^2.$$

It is a common (and useful) practice when writing an expected mean square to write σ_e^2 first. The τ^2 terms then appear, with the higher-order interactions preceding lower-order interactions in the sum. The order of the terms in the expected mean square is thus the reverse of their order in the list in Step 2. If this practice is followed, the τ^2 for the effect whose expected mean square is being written will always be the last term in the sum, and F ratios will be easier to find.

If B and C were both random factors, τ_{ab}^2 and τ_{abc}^2 would also satisfy the criterion because b would not then be the subscript of a fixed factor. The expected mean square would then be

$$\underline{E}(MS_a) = \sigma_e^2 + n\tau_{abc}^2 + nJ\tau_{ac}^2 + nK\tau_{ab}^2 + nJK\tau_a^2.$$

To find the expected mean square of the AB interaction, we must find terms whose subscripts contain both a and b, but no other subscripts of fixed factors. If C is fixed, τ_{ab}^2 is the only term that satisfies this criterion, so

$$\underline{E}(MS_{ab}) = \sigma_e^2 + nK\tau_{ab}^2.$$

However, if C is random, τ_{abc}^2 also satisfies the criterion, and

$$\underline{E}(MS_{ab}) = \sigma_e^2 + n\tau_{abc}^2 + nK\tau_{ab}^2.$$

(4) When finding expected mean squares, the grand mean is regarded as an effect without subscripts. Therefore, the terms that enter into its expected mean square are those that have no subscripts of fixed factors, i.e., all of whose subscripts are of random factors; the term $N\mu^2$ is added, of course. Thus, in the three-way design, if there are no random factors (all factors are fixed),

$$\underline{E}(SS_m) = \sigma_e^2 + N\mu^2;$$

if Factor C is the only random factor,

$$\underline{E}(SS_m) = \sigma_e^2 = nIJ\tau_c^2 + N\mu^2;$$

if B and C are both random,

$$\underline{E}(SS_m) = \sigma_e^2 + nI\tau_{bc}^2 + nIJ\tau_c^2 + nIK\tau_b^2 + N\mu^2,$$

and so on.

(5) The expected value of MS_w is always σ_e^2.

A careful study of Tables 7.6 and 7.7 will further help to clarify the application of these rules to specific designs.

TABLE 7.6. Expected mean squares
in the three-way design with Factor
C random.

Effect	$\underline{E}(MS)$
m	$\sigma_e^2 + nIJ\tau_c^2 + N\mu^2$
a	$\sigma_e^2 + nJ\tau_{ac}^2 + nJK\tau_a^2$
b	$\sigma_e^2 + nI\tau_{bc}^2 + nIk\tau_b^2$
c	$\sigma_e^2 + nIJ\tau_c^2$
ab	$\sigma_e^2 + n\tau_{abc}^2 + nK\tau_{ab}^2$
ac	$\sigma_e^2 + nJ\tau_{ac}^2$
bc	$\sigma_e^2 + nI\tau_{bc}^2$
abc	$\sigma_e^2 + n\tau_{abc}^2$
w	σ_e^2

F Ratios and Variance Estimates

The appropriate denominator terms for calculating F ratios are found as in Chapter 6. The denominator mean square for testing each effect is the mean square whose expected value would be the same as that of the effect tested if the null hypothesis were true.

Estimates of variances of effects are obtained from expected mean squares, as in Chapter 6. Estimates of the τ^2 are found by first subtracting the denominator mean square from the numerator mean square, and then dividing this difference by the multiplier of the τ^2. If there is no denominator mean square, we don't subtract anything. (Ignore σ_e^2 if there is only one score per cell.) To estimate the σ^2, we multiply each τ^2 by $(N_k - 1)/N_k$, where N_k is the number of levels of Factor K, for each *fixed* factor. (For a review of this procedure, see Chapter 6.)

To estimate proportions of variance accounted for, we first calculate $\hat{\sigma}_t^2$ as the sum of the variance estimates for all of the effects. We then divide $\hat{\sigma}_t^2$ into each variance estimate to obtain $\hat{\omega}^2$.

Power calculations in three-way and higher designs follow the general formulas given in Chapter 6.

Pooling Sums of Squares

Sums of squares can be pooled in much the same way as described in Chapter 6. They can be pooled whenever a main effect or interaction can legitimately be assumed to have a true value of zero, i.e., whenever a τ^2 or, equivalently, a σ^2 can be assumed to be zero. To determine which sums of squares can be pooled, we once again refer to the expected mean squares. We first cross out from each expected mean square the τ^2 of the effect assumed to have a

TABLE 7.7. Expected mean squares in the three-way design with Factors B and C random.

Effect	$\underline{E}(MS)$
m	$\sigma_e^2 + nI\tau_{bc}^2 + nIJ\tau_c^2 + nIK\tau_b^2 + N\mu^2$
a	$\sigma_e^2 + n\tau_{abc}^2 + nJ\tau_{ac}^2 + nK\tau_{ab}^2 + nJK\tau_a^2$
b	$\sigma_e^2 + nI\tau_{bc}^2 + nIK\tau_b^2$
c	$\sigma_e^2 + nI\tau_{bc}^2 + nIJ\tau_c^2$
ab	$\sigma_e^2 + n\tau_{abc}^2 + nK\tau_{ab}^2$
ac	$\sigma_e^2 + n\tau_{abc}^2 + nJ\tau_{ac}^2$
bc	$\sigma_e^2 + nI\tau_{bc}^2$
abc	$\sigma_e^2 + n\tau_{abc}^2$
w	σ_e^2

true value of zero. We can then pool the sums of squares of any effects whose expected mean squares are equal. The pooling is accomplished by adding the associated sums of squares together to obtain SS_{pooled}. The degrees of freedom of SS_{pooled} are the sum of the degrees of freedom of the sums of squares that were added.

For example, suppose that the true value of τ_{abc}^2 in the design illustrated in Table 7.6 is assumed to be zero. Two expected mean squares, $\underline{E}(MS_{ab})$ and $\underline{E}(MS_{abc})$, are affected by this assumption. Under the assumption, their expected mean squares become

$$\underline{E}(MS_{ab}) = \sigma_e^2 + nK\tau_{ab}^2,$$
$$\underline{E}(MS_{abc}) = \sigma_e^2.$$

In this case SS_{abc} and SS_w can be pooled because their expected mean squares are equal; MS_{pooled} is then the denominator for testing the AB interaction as well as the C main effect and the AC and BC interactions.

Theoretically, numerator sums of squares can also be pooled, although this rarely occurs in practice because the necessary assumptions are seldom reasonable. Suppose we can assume there is no A main effect. Then

$$\underline{E}(MS_a) = \underline{E}(MS_{ac}) = \sigma_e^2 + nJ\tau_{ac}^2.$$

In this case, $SS_{\text{pooled}} = SS_a + SS_{ac}$ can be used to increase numerator degrees of freedom when testing for the AC interaction.

ONE SCORE PER CELL

As in the two-way design, special assumptions may be necessary when there is only one score per cell. To find an error term for testing some or all of the effects, we may have to assume that at least one of the effects has a true value of zero. Usually, the effect chosen is the highest interaction in the design, i.e., the three-way interaction in the three-way design, the four-way interaction in the four-way design, and so on. There is no mathematical reason for choosing the highest interaction, but it is usually the most difficult effect to interpret, and experimenters usually feel (wishfully?) that it is least likely to be very large. The selected error term becomes the denominator mean square for testing other effects for which MS_w would normally be appropriate.

If the effect chosen for the denominator does not have a true value of zero (i.e., if the assumptions necessary for using the effect as an error estimate are not met), the expected value of its mean square will tend to be too large. The corresponding F ratios will tend to be too small, so the obtained p values will be larger than they should be. Whenever a null hypothesis is rejected with this type of test, we can be confident that there is, in fact, an effect. However, when designing an experiment with only one score per cell, we run the risk that otherwise significant differences will be masked by too large a denominator mean square.

QUASI-F RATIOS

If no single mean square can be used, mean squares can be combined as in Chapter 6 to obtain quasi-F ratios. The same principles for constructing quasi-F ratios and degrees of freedom can be applied. As in Chapter 6, either the denominator alone or both the numerator and the denominator may be linear combinations of mean squares.

For example, note that when Factors B and C are both random (Table 7.7), the A main effect has to be tested with a quasi-F. We might use

$$F^* = MS_a/(MS_{ab} + MS_{ac} - MS_{abc}).$$

The numerator degrees of freedom for this F^* are $(I-1)$, and the denominator degrees of freedom are estimated to be

$$\hat{\nu}_2 = \frac{(MS_{ab} + MS_{ac} - MS_{abc})^2}{\dfrac{MS_{ab}^2}{(I-1)(J-1)} + \dfrac{MS_{ac}^2}{(I-1)(K-1)} + \dfrac{MS_{abc}^2}{(I-1)(J-1)(K-1)}}.$$

For the data in Table 7.1 (see Table 7.5 for the mean squares), we have $\hat{\nu}_2 < .2$. For these data, a better quasi-F would be $F^* = (MS_a + MS_{abc})/(MS_{ab} + MS_{ac})$. For the data in Table 7.1, this quasi-F would be $F^*_{(2.6, 2.8)} = .84$. For this example, the latter quasi-F produces more reasonable degrees of freedom.

Interpreting Interactions

We said in Chapter 5 that a two-way interaction was due to a lack of parallelism when the data were plotted against the levels of one factor. These interactions can be found by studying the cell means in the table corresponding to the interaction. Figure 7.2 plots the cell means for the BC table. From it we can see that the means at level B_1 are very different, across levels of C, while the means at level B_2 are nearly equal; this difference in parallelism accounts for the BC interaction.

Interpretation of higher-order interactions is not easy. A three-way interaction is a difference in the degree of parallelism when the curves for two factors are plotted against the third. This difference is most easily visualized in a graph like Figure 7.3. In Figure 7.3 the data for the four combinations of A and B are plotted against the levels of C. The curves for levels AB_{12} and AB_{22} are nearly parallel; if the experiment had included only level B_2, there would have been no AC interaction. However, the curves for levels AB_{11} and AB_{21} are definitely not parallel. This difference in parallelism accounts for the three-way interaction in these data.

With a four-way interaction, the problem is even more complex. Basically, a four-way interaction is—hold your breath—the difference, over one factor, in the differences in parallelism for the other three factors. Higher interactions become increasingly difficult to interpret.

Fortunately, many higher-way interactions are due to a single commonly occurring pattern of results. A higher-way interaction can often be interpreted by looking in the associated r-way table for one cell (or perhaps two or three scattered cells) that has a mean greatly different (either smaller or larger) from the means in the other cells. This phenomenon is likely to result not only in an r-way interaction but in main effects and lower-order interactions as well. However, its greatest effect will tend to be in the r-way table itself. It is often the main cause of a large interaction. For the data in Table 7.1, $\overline{X}_{111.}$ is considerably smaller than the other $\overline{X}_{ijk.}$, whereas $\overline{X}_{113}$ and $\overline{X}_{212.}$ are somewhat larger.

Estimating Effects

Interpretation can also be facilitated with estimates of the effects. To estimate the values for any effect, we begin with tables containing cell means as in Table 7.2. We then subtract, from each cell mean, the estimated values for effects above it in the hierarchy (i.e., for which we subtracted sums of squares when obtaining the SS for the effect in question).

Using Table 7.2, for example, we get $\hat{\mu} = \overline{X}_{....} = 6.875$. We then obtain estimates for the A main effect by subtracting this from the means in the A table:

$$\hat{\alpha}_1 = \overline{X}_{1...} - \hat{\mu} = 7.25 - 6.875 = 0.375,$$
$$\hat{\alpha}_2 = \overline{X}_{2...} - \hat{\mu} = 6.50 - 6.875 = -0.375.$$

Similarly, the estimates for the C main effects are

$$\hat{\gamma}_1 = 5.375 - 6.875 = -1.50,$$

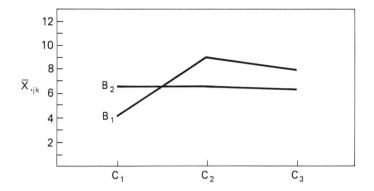

FIGURE 7.2. Graph of *BC* interacting in Table 7.2.

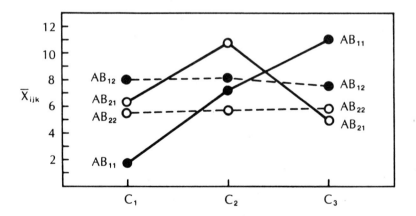

FIGURE 7.3. Data from Table 7.1 plotted against factor *C*.

TABLE 7.8. Estimated effects from data in Table 7.1.

	M			A_1	A_2			B_1	B_2
	6.88			.38	-.38			.12	-.12

$$\hat{\mu} = \overline{X}_{...} \qquad\qquad \hat{\alpha}_i = \overline{X}_{i...} - \hat{\mu} \qquad\qquad \hat{\beta}_j = \overline{X}_{.j..} - \hat{\mu}$$

	C_1	C_2	C_3
	-1.50	1.06	.44

	A_1	A_2
B_1	-.71	.71
B_2	.71	-.71

$$\hat{\gamma}_K = \overline{X}_{..k.} - \hat{\mu} \qquad\qquad \hat{\alpha\beta}_{ij} = \overline{X}_{ij..} - \hat{\mu} - \hat{\alpha}_i - \hat{\beta}_j$$

	A_1	A_2
C_1	-.88	.88
C_2	-.69	.69
C_3	1.56	-1.56

	C_1	C_2	C_3
B_1	-1.50	.94	.56
B_2	1.50	-.94	-.56

$$\hat{\alpha\gamma}_{ik} = \overline{X}_{i.k.} - \hat{\mu} - \hat{\alpha}_i - \hat{\gamma}_k \qquad\qquad \hat{\beta\gamma}_{jk} = \overline{X}_{.jk.} - \hat{\mu} - \hat{\beta}_j - \hat{\gamma}_k$$

	A_1			A_2		
	C_1	C_2	C_3	C_1	C_2	C_3
B_1	-1.04	-.73	1.77	1.04	.73	-1.77
B_2	1.04	.73	-1.77	-1.04	-.73	1.77

$$\hat{\alpha\beta\gamma}_{ijk} = \overline{X}_{ijk.} - \hat{\mu} - \hat{\alpha}_i - \hat{\beta}_j - \hat{\gamma}_k - \hat{\alpha\beta}_{ij} - \hat{\alpha\gamma}_{ik} - \hat{\beta\gamma}_{jk}$$

$$\hat{\gamma}_2 = 7.9375 - 6.875 = 1.0625,$$
$$\hat{\gamma}_3 = 7.3125 - 6.875 = 0.4375.$$

To estimate the values for the AC interaction, we begin with the cell means in the AC table. From these we subtract the corresponding estimates of the A main effect, the C main effect, and the grand mean:

$$\hat{\alpha\gamma}_{11} = \overline{X}_{11..} - \hat{\alpha}_1 - \hat{\gamma}_1 - \hat{\mu}$$
$$= 4.875 - 0.375 - (-1.50) - 6.875 = -0.875,$$
$$\hat{\alpha\gamma}_{12} = \overline{X}_{12..} - \hat{\alpha}_1 - \hat{\gamma}_2 - \hat{\mu}$$
$$= 7.625 - 0.375 - 1.0625 - 6.875 = -0.6875,$$

and so on.

The complete set of formulas and estimates is in Table 7.8. From the ABC table we see that C_3 is markedly different from C_1 and C_2 under all combinations of levels of A and B. Comparing this again with Figure 7.3, we see that the difference in parallelism is at least partly due to the fact that the solid lines cross between C_2 and C_3. A reanalysis of the data, using only levels C_1 and C_2, gives $F_{(1,24)} = 0.11$ for the ABC interaction, showing that level C_3 is indeed accountable for the three-way interaction in these data.

(Note that three interpretations have now been given for the three-way interaction. All of them can contribute to our understanding of the effect. It is not necessary to, say, choose one as the "correct" interpretation to the exclusion of the others.)

Comparisons

Standard planned and post hoc comparisons can be carried out in higher-way analyses of variance just as in the one-way and two-way designs. However, unless all of the factors are fixed, some complications may arise. First, if there are one or more random factors, such comparisons are not robust to the assumptions of equal variances and covariances.

Second, if there are more than two fixed factors, not all planned comparisons are easily tested. The problem is that different denominator mean squares may be needed for testing different effects. For some planned comparisons there is no appropriate denominator mean square. The following three examples of orthogonal contrasts will make the problem clearer. Appropriate denominator mean squares can be found for the first two, but not for the third. All three examples are from the $2 \times 3 \times 5$ design with Factor C random and $n = 1$ shown in Table 7.9.

The first example is a test of the null hypothesis

$$H_0(1) : \mu_{11.} + \mu_{21.} - \mu_{12.} - \mu_{22.} = 0. \tag{7.1}$$

This hypothesis can be expressed more simply by noting that

$$\mu_{11.} + \mu_{21.} = 2\mu_{.1.},$$
$$\mu_{12.} + \mu_{22.} = 2\mu_{.2.}.$$

Remembering that multiplication or division of the coefficients by a constant does not change the null hypothesis itself, $H_0(1)$ can be rewritten

$$H_0(1) : \mu_{.1.} - \mu_{.2.} = 0. \tag{7.2}$$

Equation 7.2 expresses $H_0(1)$ entirely in terms of the B main effect. In fact, $H_0(1)$ is one of two orthogonal contrasts making up the B main effect, the other contrast being

$$H_0(1') : \mu_{.1.} + \mu_{.2.} - 2\mu_{.3.} = 0.$$

The appropriate denominator term for testing the B main effect is MS_{bc}; it is also the appropriate denominator for testing $H_0(1)$.

The test of $H_0(1)$ can be performed equally well on the $\overline{X}_{ijk.}$, the $\overline{X}_{ij..}$, or the $\overline{X}_{.k..}$. It is necessary only to modify the multiplier when finding SS_1. In general, the multiplier is the number of scores over which the means used in computing C were taken. This is best explained by an illustration. We will

TABLE 7.9. Three-way design with factor C random, illustrating tests of comparisons.

	A_1					A_2				
	C_1	C_2	C_3	C_4	C_5	C_1	C_2	C_3	C_4	C_5
B_1	11	10	10	14	4	1	7	2	-1	0
B_2	6	-2	3	8	3	-1	1	4	6	-7
B_3	-7	-7	-7	-3	-1	-3	-1	2	4	1

	A_1	A_2			C_1	C_2	C_3	C_4	C_5
B_1	49	9	A_1		10	1	6	19	6
	9.8	1.8			3.33	0.33	2.00	6.33	2.00
B_2	18	3	A_2		-3	7	8	9	-6
	3.6	0.6			-1.00	2.33	2.67	3.00	-2.00
B_3	-25	3							
	-5.0	0.6							

	C_1	C_2	C_3	C_4	C_5		A_1	A_2
B_1	12	17	12	13	4		42	15
	6.0	8.5	6.0	6.5	2.0		2.8	1.0
B_2	5	-1	7	14	-4			
	2.5	-0.5	3.5	7.0	-2.0	B_1	B_2	B_3
B_3	-10	-8	-5	1	0	58	21	-22
	-5.0	-4.0	-2.5	0.5	0.0	5.8	2.1	-2.2

	C_1	C_2	C_3	C_4	C_5	$\overline{X}$
	7	8	14	28	0	57
	1.17	1.33	2.33	4.67	0.00	1.90

	RS	SS	T	df	MS	Den	F	p	$\hat{\omega}^2$
m		108.30		1	108.3	c	5.86	.073	
a	132.60	24.30	2	1	24.30	ac	1.90	.24	.01
b	428.90	320.60	3	2	160.3	bc	11.0	.005	.32
c	182.17	73.87	5	4	18.47				.10
ab	689.80	236.60	6	2	118.3	abc	13.6	.003	.24
ac	257.67	51.20	10	4	12.80				.07
bc	619.50	116.73	15	8	14.59				.16
abc	1,001.00	69.40	30	8	8.675				.10
t		892.70		29					

test $H_0(1)$ first by Equation 7.1 (using the $\overline{X}_{ij..}$) and then by Equation 7.2 (using the $\overline{X}_{.j..}$). Using Equation 7.1, we calculate

$$
\begin{aligned}
C_1 &= 9.8 + 1.8 - 3.6 - 0.6 = 7.4, \\
SS_1 &= 5(7.4)^2/4 = 68.45, \\
F_{(1,8)} &= 68.45/14.59 = 4.69, \\
p &= .062.
\end{aligned}
$$

In the equation for SS_1, the multiplier was 5 because it was necessary to average over 5 scores to obtain each $\overline{X}_{ij..}$. If there had been n scores per cell (n larger than one), the multiplier would have been $5n$.

With the null hypothesis as in Equation 7.2, the test would be

$$
\begin{aligned}
C_1 &= 5.8 - 2.1 = 3.7, \\
SS_1 &= 10(3.7)^2/2 = 68.45.
\end{aligned}
$$

The multiplier is now 10 because C_1 is a linear combination of the $\overline{X}_{.j..}$, each of which is the average of 10 scores. This example shows that C depends on the way in which the null hypothesis is formulated and tested, but the SS (and therefore the test itself) does not, so long as the correct multiplier is used when calculating the SS. If the null hypothesis had been calculated from the entire ABC table, as

$$
\begin{aligned}
H_0(1): \quad &(\mu_{111} + \mu_{112} + \mu_{113} + \mu_{114} + \mu_{115}) + \\
&(\mu_{211} + \mu_{212} + \mu_{213} + \mu_{214} + \mu_{215}) - \\
&(\mu_{121} + \mu_{122} + \mu_{123} + \mu_{124} + \mu_{125}) - \\
&(\mu_{221} + \mu_{222} + \mu_{223} + \mu_{224} + \mu_{225}) = 0,
\end{aligned}
$$

then C_1 would have been calculated as a linear combination of the $\overline{X}_{ijk.}$, but SS_1 would have had the same value. (Verify this for yourself.)

In general, if the comparison can be expressed entirely in terms of a main effect, the appropriate denominator is the mean square for testing that main effect.

The second example is the null hypothesis

$$
H_0(2): \mu_{11.} + \mu_{22.} - \mu_{12.} - \mu_{21.} = 0. \tag{7.3}
$$

This is one of two orthogonal contrasts making up the AB interaction. The other is

$$
H_0(2'): \mu_{11.} + \mu_{12.} + 2\mu_{23.} - \mu_{21.} - \mu_{22.} - 2\mu_{13.} = 0.
$$

To see this, we add the SSs for these two null hypotheses:

$$
\begin{aligned}
C_2 &= 9.8 + 0.6 - 3.6 - 1.8 = 5.0, \\
SS_2 &= 5(5.0)^2/4 = 31.25, \\
C_{2'} &= 9.8 + 3.6 + 2(0.6) - 1.8 - 0.6 - 2(-5.0) = 22.2, \\
SS_{2'} &= 5(22.2)^2/12 = 205.35, \\
SS_2 + SS_{2'} &= 31.25 + 205.35 = 236.60 = SS_{ab}.
\end{aligned}
$$

Because $H_0(2)$ is part of the AB interaction, the correct denominator for testing it is MS_{abc}, the denominator used in testing the AB interaction. The result is

$$F_{(1,8)} = 31.25/8.675 = 3.60; \quad p = .094.$$

The third example is the contrast

$$H_0(3) : \mu_{13.} - \mu_{23.} = 0. \tag{7.4}$$

This test is orthogonal to the B main effect because it involves a difference only between two levels of A. However, the null hypothesis may be false if there is *either* an A main effect *or* an AB interaction. Because the test is on the difference between two levels of Factor A, the relevance of the A main effect should be clear. However, even if there were no A main effect, there could still be a difference between these two means, due to an AB interaction. Consequently, both the A main effect and the AB interaction are involved in this contrast. Which denominator should we use to test it, MS_{ac} or MS_{abc}? The answer is neither. The null hypothesis $H_0(3)$ cannot be tested with a single mean square. In general, to be testable with a single mean square, a planned comparison must be independent of all but one effect. The denominator for testing it is then the denominator that would be used in testing that effect. Methods to be described later must be used to test planned comparisons that cannot be tested with a single mean square.

The generalization of the above discussion to S-method post hoc comparisons should be clear. If a planned comparison can be tested by the standard method, it is transformed to a post hoc comparison by merely treating SS as if it has more than one degree of freedom. The appropriate degrees of freedom are determined by the considerations discussed in Chapter 4. In most cases they are the sum of the degrees of freedom of all fixed effects in the design (post hoc comparisons are normally independent of all random effects). For example, in the three-way design with only A fixed, it would be $(I-1)$; in the same design with both A and B fixed it would be $(I-1)+(J-1)+(I-1)(J-1) = IJ-1$. However, if the actual post hoc comparisons are chosen from a more limited set of potential comparisons (as discussed in Chapter 4), fewer degrees of freedom might be appropriate.

Power calculations for planned comparisons follow the formulas given in Chapter 3.

ANALYZING COMPARISONS

We saw in the last section that a comparison may be part of a single effect only or it may involve two or more effects. To test the comparison appropriately, we first have to know which effects are involved. With some comparisons, this is a relatively simple problem; with others, the following method for analyzing comparisons can be used.

To determine the effects involved in a comparison and the extent to which each effect is involved, we need only do an analysis of variance on the coefficients of the comparison, testing them as data with one score per cell. Any

TABLE 7.10. Approximate test of $H_0(3)$ (Eq. 7.4). Data from Table 7.9.

	B_1	B_2	B_3	Sum		rs	ss	$ss*$	Den
A_1	0	0	1	1	m		0	0	c
A_2	0	0	-1	-1	a	2/3	2/3	1/3	ac
	0	0	0	0	b	0	0	0	bc
					ab	2	4/3	2/3	abc

$$F^* = 78.40/[(1/3)(12.80) + (2/3)(8.675)] = 78.40/10.05 = 7.80$$

$$\hat{\nu} = \frac{(10.05)^2}{(1/9)(12.80^2/4) + (4/9)(8.675^2/8)} \simeq 12, \quad p = .016$$

effect with a nonzero sum of squares in this ana'ysis is involved in the comparison. Any effect with a zero sum of squares is independent of the comparison.

We will demonstrate this on the contrast $H_0(3)$, Equation 7.4. The first step is to put the coefficients into the cells of an AB table:

	B_1	B_2	B_3
A_1	0	0	1
A_2	0	0	-1

The analysis of these coefficients is shown in Table 7.10. The coefficients are treated just as though they were the data in a two-way analysis of variance with one score per cell. The analysis follows the procedure outlined earlier. However, this analysis need not be done on all three factors. Instead, it need be done only on those factors that may be involved in the comparison. In this example Factor C is clearly not involved in the comparison, so it is omitted for the analysis in Table 7.10. By omitting factors clearly not involved in the comparison, we simplify the analysis.

In Table 7.10, the analysis proceeds in the usual way up to the calculation of sums of squares. The symbols rs and ss, in lower case, indicate that our "data" are the coefficients of the comparison. Because ss_a and ss_b are the only nonzero sums of squares, we conclude that this particular comparison involves both the A main effect and the AB interaction, but no other effect. Unfortunately, in this design, the A main effect and the AB interaction are tested with different denominator mean squares. Consequently, the ordinary methods of testing comparisons cannot be used. However, the next section tells how to find a denominator term for any comparison.

APPROXIMATE TESTS OF COMPARISONS

When there is no appropriate denominator mean square for a comparison, an approximate denominator can be created. The approximate denominator is a linear combination of mean squares, following the procedures for a quasi-F. Both the mean squares and their coefficients in the linear combination are found from the analysis described in the last section. The mean squares are those that would be used to test each of the effects involved in the comparison. To find the coefficients, we normalize the ss values found in the above analysis, dividing them by their sum, so that the normalized ss values add up to one.

The step-by-step procedure is described below. Each step is illustrated using the planned comparison in Table 7.10 on the data in Table 7.9.

(1) Normalize the ss values. First, sum the ss values, then divide each ss by the sum. The resulting values (labeled ss^* in Table 7.10) will sum to one. In Table 7.10 the ss values sum to 2; consequently, we normalize these by dividing each ss by 2.

(2) The denominator for the quasi-F is then

$$\Sigma_{eff}ss^*_{eff}MS_{den(eff)},$$

where $MS_{den(eff)}$ refers to the mean square that would appear in the denominator when testing the effect. For the example, these denominator terms (taken from Table 7.9) are in the last column of Table 7.10. Basically, the above sum is the sum of the cross-products of the values in the last two columns of Table 7.10; for the example, the sum would be $(1/3)MS_{ac} + (2/3)MS_{abc}$, and the quasi-$F$ would be

$$F^* = SS_3/[(1/3)MS_{ac} + (2/3)MS_{abc}].$$

From Table 7.9 we see that the denominator is $(1/2)(12.80+(2/3)(8.675) = 10.05$. The value of SS_3 is 78.40, so the F ratio is 7.80.

(3) As usual, the numerator has one degree of freedom. The denominator degrees of freedom are estimated from Equation 6.21, letting $k_{eff} = ss^*_{eff}$:

$$\hat{\nu} = (MS_{combined})^2/(\Sigma_{eff}ss^{*2}_{eff}MS^2_{eff}/\nu_{eff}).$$

For the current problem,

$$\hat{\nu} = (10.05)^2/[(1/3)^2(12.80)^2/4 + (2/3)^2(8.675)^2/8] \simeq 12.$$

With 1 and 12 degrees of freedom, an F^* of 7.80 has a p value of .016.

If more terms had been involved in the comparison, more terms would have entered into the denominator, but the principle would have been the same.

This approach can be used even when the effects involved are themselves testable only by quasi-F ratios. For example, suppose that in the design in Table 7.9 we wanted to test $H_0 : \mu_{1..} = 0$. This is not a contrast, and the relative weights of the grand mean and the A main effect are in Table 7.11. Suppose also that B and C are both random. The denominator for testing

TABLE 7.11. Test of $H_0 : \mu_1.. = 0$, data in
Table 7.9.

Coefficients				Analysis		
A_1	A_2	Sum		rs	ss	ss^*
1	0	1	m		1/2	1/2
			a	1	1/2	1/2

the grand mean would then be $(MS_b + MS_c - MS_{bc})$, and the denominator
for testing the A main effect would be $(MS_{ab} + MS_{ac} - MS_{abc})$. The F ratio
would then be

$$
\begin{aligned}
F^* &= SS_4/[(1/2)(MS_b + MS_c - MS_{bc}) + (1/2)(MS_{ab} + MS_{ac} - MS_{abc})] \\
&= SS_4/[(1/2)(MS_b + MS_c - MS_{bc} + MS_{ab} + MS_{ac} - MS_{abc})] \\
&= 117.6/143.3 = .82.
\end{aligned}
$$

The denominator degrees of freedom would be

$$
\begin{aligned}
\hat{\nu} &= \frac{143.3^2}{\dfrac{1}{4}\left(\dfrac{160.3^2}{2} + \dfrac{18.47^2}{4} + \dfrac{14.59^2}{8} + \dfrac{118.3^2}{2} + \dfrac{12.8^2}{4} + \dfrac{8.68^2}{8}\right)} \\
&= 20{,}535/5{,}002 \simeq 4.
\end{aligned}
$$

The F is not significant.

As in our previous examples of quasi-F ratios, we could eliminate the negative terms from the denominator by adding them to the numerator. We would
then have to use Equation 6.21 to estimate both the numerator and the denominator degrees of freedom.

EXACT PLANNED COMPARISONS: SPECIAL CASES

The method described next for testing planned comparisons applies to any
comparison made on fixed factors, and the only necessary assumption is that
the random factor is normally distributed. However, the method is limited
to mixed-effects models with only one random factor. Fortunately, most commonly used mixed-model designs have only one random factor.

Also, this method is usually less powerful than the method described in
the last section; as a general rule, the fewer assumptions we make, the less
power the test has. However, the fact that it is an exact test under minimal
assumptions is a strong argument for its use whenever appropriate.

The method is simplicity itself, although some of the calculations are a little
tedious. Instead of averaging over the levels of the random factor and then
calculating C from these averages, we calculate C separately for each level of
the random factor. This generates as many different values of C as there are

levels of the random factor. The average of these separate C values is equal to the C that would normally be calculated for the planned comparison. More to the point, however, is that under the null hypothesis the separate C values are a random sample from a normally distributed population with a mean of zero. To test the planned comparison, we just do a one-sample t test on the C values. We will illustrate this method with the three null hypotheses that we tested before on the data in Table 7.9.

To test $H_0(1)$ (Equation 7.1), on the B main effect, we use the cell means in the BC table, calculating $\overline{X}_{.1k} - \overline{X}_{.2k}$ for each level of the random factor (Factor C) to obtain the five values, 3.5, 9.0, 2.5, -0.5, 4.0. An ordinary t test on the null hypothesis that the mean of these five values is zero results in a t of 2.41, with $\nu = 4$ and $p = .074$.

To test $H_0(2)$ (Equation 7.3) and $H_0(3)$ (Equation 7.4), we must use the complete ABC table. For $H_0(2)$ we calculate $X_{11k} + X_{22k} - X_{12k} - X_{21k}$ for each level of Factor C, giving the five values, 3, 6, 9, 13, -6. The t value is 1.56, $\nu = 4$, and $p = .19$.

Finally, for $H_0(3)$, we calculate the five values of $X_{13k} - X_{23k}$. The values are -4, -6, -9, -7, -2. The value of t is -4.63, $\nu = 4$, and $p = .010$.

The loss of power is not evident from a comparison of these results with the previous tests. However, due to the reduction in degrees of freedom, the obtained p value usually will not be as small with this exact test as with either the standard or approximate method of testing planned comparisons.

Box's Correction

If the assumptions of equal variances and covariances are not satisfied, Box's correction can be used for tests of main effects and interactions. It is appropriate whenever the denominator of the F ratio is an interaction mean square. (The one exception to this rule occurs when the appearance of an interaction term in the denominator is due to pooling; then Box's correction cannot be used.) The p value found for the ordinary F ratio is the lower limit on p; for the upper limit, we find the p value of the same F ratio with one degree of freedom in the numerator and ν_2/ν_1 degrees of freedom in the denominator. The "true" p value is between these limits.

Consider the data in Table 7.9. The test on the B main effect has a p value of .006, with 2 and 8 degrees of freedom. This is the minimum possible value of p. The maximum value, .003, is found using the same F ratio with one and $8/2 = 4$ degrees of freedom.

To estimate K_B we use the formulas in Chapter 6, using the covariance table calculated from the BC table of Table 7.9. The covariance table (with correlations below the diagonal) appears in Table 7.12.

Applying the formulas of Chapter 6 to that table, we find that $K_B = .95, \hat{\nu}_1 = 1.9, \hat{\nu}_2 = 7.6, p = .006$. K_B is nearly one, so the estimated p value is very near the minimum. (Note that for this example the assumption of equal variances and covariances is violated, yet the ordinary F test seems

TABLE 7.12. Estimated variances (on the diagonal), covariances (above the diagonal), and correlations (below the diagonal) for the B main effect on the data in Table 7.9.

	B_1	B_2	B_3
B_1	5.575	3.0875	-2.9875
B_2	.37	12.425	1.8375
B_3	-.52	.22	5.825

basically valid. Technically, the assumption is not strictly necessary, but for most practical purposes it can be regarded as necessary.)

For the AB interaction, the minimum p value is .003, and the maximum, with one and four degrees of freedom, is .021. The estimation of K_B is more complicated. First, we must find two orthogonal contrasts that divide the two degrees of freedom in the interaction. (The finding of these contrasts will be discussed later.) Moreover, we must express these contrasts so that the squares of the coefficients will sum to one. (To do this, we first find the coefficients and the sum of their squares; we then divide each coefficient by the square root of that sum.) The coefficients for two possible contrasts are at the top of Table 7.13.

Next, we apply these coefficients to the AB table at each level of the random factor, C, just as we did for the exact test of a contrast, described in the previous section. (We use the values in the full ABC table for this purpose.) The C values for each comparison at each level of Factor C are shown in the center of Table 7.13. Now we construct a table of the estimated variances and covariances of these quantities (the bottom table in Table 7.13). Letting $v_{ii'}$ be the element in the ith row and i'th column of this table, and letting I_V be the number of rows and columns in the table,

$$K_B = (\Sigma_i v_{ii})^2 / (I_V \Sigma_i \Sigma_{i'} v_{ii'}^2).$$

The numerator is the square of the sum of the diagonal elements; the denominator is the sum of the squares of all the elements, multiplied by the number of rows in the variance-covariance table. For the AB interaction, $I_V = 2$ and

$$
\begin{aligned}
K_B &= (12.875 + 4.475)^2 / [(2)(12.875^2 + 4.042^2 + 4.042^2 + 4.475^2)] \\
&= .69.
\end{aligned}
$$

Thus,

$$\hat{\nu}_1 = (.69)(2) = 1.4, \quad \hat{\nu}_2 = (.69)(8) = 5.5, \quad \text{and} \quad p = .009.$$

This general approach can be used on any main effect or interaction. For example, the B main effect could have been expressed as a combination of

TABLE 7.13. Estimating K_B for the AB interaction, data from Table 7.9.

	A_1	A_2	A_1	A_2
B_1	.5	-.5	.2887	-.2887
B_2	-.5	.5	.2887	-.2887
B_3	0	0	-.5774	.5774

C Values

C_1	1.5	7.217
C_2	3.0	3.464
C_3	4.5	7.217
C_4	6.5	8.949
C_5	-3.0	5.196

Estimated Variances (diagonal), Covariances
(above diagonal), and Correlations (below diagonal)

	Cont. 1	Cont. 2
Cont. 1	12.875	4.042
Cont. 2	.53	4.475

$$K_B = .69$$

two orthogonal contrasts, and the same procedure could have been followed. The ordinary F test is valid if and only if (1) the variances (i.e., the diagonal values) in the resulting table are all equal to each other and (2) the covariances are all zero. The formula given in Chapter 6 is more complicated but does not require that we find orthogonal contrasts. The formula given here is simpler once the contrasts have been found and their C values have been calculated. It also gives a necessary and sufficient set of conditions for the ordinary F test to be valid.

FINDING ORTHOGONAL CONTRASTS

Although no simple rule can be given for finding orthogonal contrasts that divide the degrees of freedom for a particular effect, the task is usually not difficult in practice. The search can be simplified by noting that the coefficients can always be fitted into the cells of the table for that effect. Moreover, it is easy to test whether or not a given contrast is strictly within a given effect; if it is, the analysis described previously will show that only that effect is involved. As a general rule, when the coefficients are analyzed, all tables found by collapsing over the initial one should contain zeroes. Finally, recall that a contrast testing differences between cells is orthogonal to one that compares the average of those cells with other cells. Tests for orthogonality were described previously.

Exercises

(1.) As part of a program testing athletic ability, six randomly selected students were asked to take a basketball-shooting test. Each subject was asked to take two shots from each of three distances—5, 10, and 20 feet—and then to repeat the set of shots. Each student then took 100 practice shots, after which the tests were repeated. The following table shows the number of baskets made by each student under each condition:

5 Feet

		S_1	S_2	S_3	S_4	S_5	S_6
Practice	Set 1	2	1	2	1	1	2
	Set 2	2	2	0	2	2	1
No	Set 1	0	2	2	1	1	2
practice	Set 2	0	1	2	2	1	1

10 Feet

S_1	S_2	S_3	S_4	S_5	S_6
2	0	1	2	2	1
2	1	0	1	2	0
0	1	1	0	0	0
0	1	2	0	1	0

20 Feet

S_1	S_2	S_3	S_4	S_5	S_6
0	0	0	1	0	1
1	1	0	0	1	0
0	1	0	0	1	0
0	1	1	0	0	0

(a.) Analyze these data as a four-way, distance × practice × repetition × subjects design, finding cell means, the p value, and $\hat{\omega}^2$ for each effect. When appropriate, find upper and lower limits on the p values.

(b.) Estimate K_B for each effect for which that is appropriate, and find the associated p value.

(c.) Reanalyze the data, assuming that the scores on the two repetitions are independent of each other, i.e., that they are the two scores in a single cell.

(d.) Comparing the results of parts a and c, comment on the validity of the assumption in part c. If it is not valid, how might it affect conclusions about the other three factors?

(e.) Average the scores for Set 1 and Set 2 for each subject under each condition. Then analyze the resulting data as a three-way, distance × practice × subjects design. Compare these results with the results in part a. Comment on the differences, if any.

(f.) Reanalyze the data, keeping all four factors as in part a, but assume that the three distances used were selected randomly from the set of all distances between 1/2 foot and 49 1/2 feet.

(2.) Reanalyze the data in Problem 4, Chapter 6, treating the three lists of each type as the levels of a third factor. Assume that (1) this third factor is fixed, and (2) it is random. Find limits on the p value and estimate K_B and its associated p value, when appropriate.

8

Nested Designs

In each of the designs discussed so far, every cell contained scores. However, sometimes it is impossible, impractical, or otherwise undesirable to obtain scores in every cell. This chapter and the next discuss some designs in which not all cells contain scores. These *incomplete designs* are convenient for some purposes, but they also have disadvantages. When choosing a design, the advantages and disadvantages must be weighed against each other.

The designs in this chapter are probably the most commonly used type of incomplete design. Characteristically each level of some factor is paired with one and only one level of some other. For example, consider the following experiment. We are interested in the extent to which high-school students' knowledge of algebra depends on the schools and the teacher. We get the cooperation of the three high schools in town, each of which has two algebra teachers. Three randomly selected students are assigned to each teacher's class, making a total of eighteen students in all (three students in each of two classes in each of three schools). At the end of the term each student is given a standardized test in algebra; the scores are shown in Table 8.1.

If we treat the schools as the levels of Factor A and the teachers as the levels of Factor B, we have a 3×6 design with three scores in each cell that contains scores. However, not all cells contain scores. This is illustrated in Table 8.2. (Perhaps we wanted to have each teacher teach one class in each of the three schools, but the teachers felt that that would be stretching cooperation a little too far.)

The important point in Table 8.2 is that each level of Factor B is paired with one and only one level of Factor A, whereas each level of Factor A is paired with two levels of Factor B. Whenever each level of one factor is paired with one and only one level of another factor, the former is said to be *nested* in the latter. In our example, Factor B is nested in Factor A. Notice that nesting is strictly a "one-way" relationship; Factor A is not nested in Factor B, because each level of Factor A is paired with two levels of Factor B. When some factors are nested in others, the design is called a *nested design*. When each level of one factor is paired with each level of the other, the factors are said to be *crossed*, and if all factors are crossed with each other, the design is called *completely crossed*. All of the designs up until now have been completely crossed.

TABLE 8.1. Hypothetical data comparing teachers (B) and schools (A).

	A_1		A_2		A_3	
	B_1	B_2	B_3	B_4	B_5	B_6
	20	19	14	12	13	9
Scores	18	20	18	12	16	4
	14	20	14	9	13	4
	52	59	46	33	42	17
B	17.33	19.67	15.33	11.00	14.00	5.67
	111		79		59	
A	18.50		13.17		9.83	
			249			
$\overline{X}$			13.83			

Two-Way Models

Consider the design of Tables 8.1 and 8.2 in terms of the model developed previously. For two-way designs this model is

$$X_{ijk} = \mu + \alpha_i + \beta_j + \alpha\beta_{ij} + \epsilon_{ijk}. \tag{8.1}$$

The average of the scores in each cell is

$$\overline{X}_{ij.} = (1/n)\Sigma_k X_{ijk} = \mu + \alpha_i + \beta_j + \alpha\beta_{ij} + \overline{\epsilon}_{ij.}.$$

To get additional averages, we must have some new definitions. First, instead of letting J equal the total number of levels of Factor B, we will let it be the number of levels of B that are paired with each level of A. In the example of Tables 8.1 and 8.2, J equals 2 (instead of 6, as it would if it were the total number of levels of B).

Second, we have to define a new summation index. The average over the levels of B, within a level of A, can be taken over only the levels of B that are paired with that level of A. We will use the symbol $j(i)$ to indicate that level j of Factor B is paired with level i of Factor A. The average over the levels of B, for a given level of A, is then

$$\begin{aligned} \overline{X}_{i..} &= (1/J)\Sigma_{j(i)}\overline{X}_{ij.} \\ &= \mu + \alpha_i + (1/J)\Sigma_{j(i)}(\beta_j + \alpha\beta_{ij}) + \overline{\epsilon}_{i..}. \end{aligned} \tag{8.2}$$

The term $\overline{\epsilon}_{i..}$ is of course an average over only the levels of Factor B that are paired with level A_i.

To estimate the α_i and test for the A main effect, we must make the assumption

$$\Sigma_{j(i)}(\beta_j + \alpha\beta_{ij}) = 0. \tag{8.3}$$

TABLE 8.2. Data from Table 8.1 arranged as
a 3×6 design.

	A_1	A_2	A_3	Sum mean
B_1	20 18 14			52 17.33
B_2	19 20 20			59 19.67
B_3		14 18 14		46 15.33
B_4		12 12 9		33 11.00
B_5			13 16 13	42 14.00
B_6			9 4 4	17 5.67
Sum	111	79	59	249
Mean	18.50	13.17	9.83	13.83

Notice that this assumption is different from the assumptions made in the completely crossed two-way design. In the completely crossed design the β_j and $\alpha\beta_{ij}$ summed to zero separately, and the summation in each case was over all of the levels of B. In the nested design the terms need not sum to zero separately, but their combined sum must be zero *when taken only over those levels of B that are paired with A_i*. Notice also that this assumption, like the similar assumptions with the completely crossed designs, is one that is *imposed* on the model; that is, by defining the terms of the model appropriately, we can *require* that Equation 8.3 hold. The assumption does not depend on the nature of the data themselves. (We will have more to say on this later.) We are assuming that Equation 8.3 is true. Equation 8.2 becomes

$$\overline{X}_{i..} = \mu + \alpha_i + \overline{\epsilon}_{i..}. \tag{8.4}$$

Finally, we can average over the levels of A to obtain

$$\overline{X}_{...} = \mu + \overline{\epsilon}_{...}. \tag{8.5}$$

The α_i cancel out of this equation by the assumption (identical to that in the completely crossed designs) that they sum to zero. The $\overline{X}_{ij.}$ for the above examples are shown in the row labelled B in Table 8.1 and in the last column of Table 8.2; the $\overline{X}_{i..}$ are shown in the row labelled A in Table 8.1 and in the last row of Table 8.2.

From Equations 8.4 and 8.5, we can easily derive the estimates

$$
\begin{aligned}
\hat{\mu} &= \overline{X}_{...}, \\
\hat{\alpha}_i &= \overline{X}_{i..} - \overline{X}_{...}.
\end{aligned}
$$

Estimating β_j and $\alpha\beta_{ij}$ is a different problem. Ordinarily, we would use the row means in Table 8.2 to estimate the β_j and then use the means of the individual cells to estimate the $\alpha\beta_{ij}$. However, the means of the rows and the means of the individual cells are identical. Each row mean is the mean of the single cell in that row. Consequently, we cannot obtain separate estimates of β_j and $\alpha\beta_{ij}$; the best we can do is estimate their sum for the cells containing data. If we let

$$
\beta_{j(i)} = \beta_j + \alpha\beta_{ij}, \tag{8.6}
$$

then

$$
\hat{\beta}_{j(i)} = \overline{X}_{ij.} - \overline{X}_{i..},
$$

for each cell (AB_{ij}) that contains scores. The parentheses around the subscript i indicate that B is nested in A rather than vice versa.

Because the β_j and $\alpha\beta_{ij}$ cannot be estimated separately, we modify the model by inserting Equation 8.6 into Equation 8.1 to obtain

$$
X_{ijk} = \mu + \alpha_i + \beta_{j(i)} + \epsilon_{ijk}. \tag{8.7}
$$

Equation 8.3 now becomes

$$
\Sigma_{j(i)}\beta_{j(i)} = 0.
$$

The estimates of μ, α_i, and $\beta_{j(i)}$, for the example of Table 8.1, are shown in Table 8.3. There appear to be large differences among schools, and the difference between teachers in the same school appears to vary from school to school.

Note that this design can compare teachers in the same school only. Because the $\hat{\beta}_{j(i)}$ sum to zero within each school, we cannot directly compare $\hat{\beta}_{j(i)}$ in different schools. We cannot say, for example, that teacher B_5 is a better teacher than teacher B_2 even though $\hat{\beta}_{5(3)} = 4.17$ and $\hat{\beta}_{2(1)} = 1.17$. In fact, the *average* score for teacher B_2, shown in Table 8.1, is *higher* than the average for teacher B_5. Of course, the difference in *average* scores may be due to the difference between schools rather than teachers, but we cannot tell from this experiment. We will discuss this problem in more detail later.

SUMS OF SQUARES AND DEGREES OF FREEDOM

Statistical tests follow the same pattern as in the crossed model. A statistical test exists for each effect in Equation 8.7. The sum of squares for each effect is found by squaring and summing the estimates of the effect and multiplying

TABLE 8.3. Estimates on effects for data in Table 8.1.

	A_1		A_2		A_3	
	B_1	B_2	B_3	B_4	B_5	B_6
$\hat{\beta}_{j(i)}$	-1.17	1.17	2.17	-2.17	4.17	-4.17
$\hat{\alpha}_i$		4.67		-0.67		-4.00
$\hat{\mu}$				13.83		

$$\hat{\mu} = \overline{X}_{...}$$
$$\hat{\alpha}_i = \overline{X}_{i..} - \hat{\mu}$$
$$\hat{\beta}_{j(i)} = \overline{X}_{ij.} - \hat{\mu} - \hat{\alpha}_i$$

that sum by the number of scores averaged over to obtain each estimate. (A simpler computational method will be given later.)

$$SS_a = nJ\Sigma_i\hat{\alpha}_i^2,$$
$$SS_{b(a)} = n\Sigma_{j(i)}\hat{\beta}_{j(i)}^2,$$
$$SS_w = \Sigma_i\Sigma_{j(i)}\Sigma_k\hat{\epsilon}_{ijk}^2,$$
$$SS_m = N\overline{X}_{...}^2.$$

As usual, the grand mean has one degree of freedom and the A main effect has $(I-1)$ degrees of freedom. The degrees of freedom for error are equal to $(n-1)$ times the total number of cells containing scores, or $(n-1)IJ = N-IJ$. The number of degrees of freedom remaining to test the $B(A)$ "main effect" are $N-1-(I-1)-(N-IJ) = I(J-1)$. This can also be seen by noting that the J estimates of $\beta_{j(i)}$ within a given level of A must sum to zero, so only $(J-1)$ of them are free to vary. Because there are $(J-1)$ "free" estimates in each of the I levels of A, the total degrees of freedom are $I(J-1)$. We might also note that in a completely crossed $I \times J$ design the sum of the degrees of freedom for the B main effect and the AB interaction is $(J-1)+(I-1)(J-1) = I(J-1)$. We saw earlier that $\beta_{j(i)}$ was the sum of the B main effect and the AB interaction; correspondingly, the test of the B main effect is really a combined test on both the B main effect and the AB interaction. The B main effect and the AB interaction are *confounded*. This confounding is one price that must be paid for the practical advantages of a nested design.

The sums of squares and mean squares for the data in Table 8.1 are given in Table 8.4.

CONFOUNDED EFFECTS

There is a sense in which a similar problem occurs with respect to the A main effect, although the problem is then in the interpretation rather than in the statistical test itself. One is tempted, in a model such as this, to attribute

TABLE 8.4. Sums of squares and
mean squares from Table 8.1.

	SS	df	MS
m	3,444.50	1	3,444.5
a	229.33	2	114.7
$b(a)$	140.50	3	46.83
w	58.67	12	4.889
t	428.50	17	

the A main effect to factors independent of the levels of B. Thus, in the schools example a significant effect due to schools is likely to be interpreted as due to differences in physical facilities, administration, and other factors that are independent of the teaching abilities of the teachers themselves. The temptation to interpret the A main effect in that way rests on the argument that the test on differences between teachers is independent of the test on differences between schools. In fact, however, the test on the B main effect is only a test on differences between teachers in the *same school*. Differences between teachers in *different* schools are part of the A main effect, and the observed differences between schools could be due entirely to the fact that some schools have better teachers. They could also be due to the fact that some schools have smarter children attending them. Both of these possibilities could have been ruled out by randomly assigning both the teachers and the students to schools as well as classrooms; however, in the described design this was impractical.

The preceding considerations can also help to clarify the nature of the confounding between the B main effect and the AB interaction. The B main effect in this design is the extent to which each teacher is an *inherently* better (or worse) teacher than a colleague regardless of the school. The AB interaction is the extent to which the teaching abilities of the teachers are differentially affected by the school in which they teach. Thus, teacher B_5 appears to be much better than B_6 in the example of Tables 8.1 and 8.3. However, it may be that B_6 is having difficulties with the administration of school A_3 and that his teaching would greatly improve if he were transferred to school A_1. The teaching of B_5, on the other hand, might deteriorate if she were transferred to A_1. Thus, it is entirely possible that in school A_1, B_6 might turn out to be a better teacher than B_5. The existence of a significant $B(A)$ effect does not tell us whether the cell mean at one level of B is generally higher than another or whether it is higher only when paired with that particular level of A. It tells us only that there are differences among cells that lie within the same level of A. It tells us that the variance of the true means of the J cells *within each level of A* is not zero.

TABLE 8.5. Expected mean squares in a nested two-way design.

Model	Effect	$\underline{X}(MS)$
	μ	$\sigma_e^2 + N\mu^2$
A and B fixed	α_i	$\sigma_e^2 + nJ\tau_a^2$
	$\beta_{j(i)}$	$\sigma_e^2 + n\tau_{b(a)}^2$
	μ	$\sigma_e^2 + n\tau_{b(a)}^2 + N\mu^2$
A fixed, B random	α_i	$\sigma_e^2 + n\tau_{b(a)}^2 + nJ\tau_a^2$
	$\beta_{j(i)}$	$\sigma_e^2 + n\tau_{b(a)}^2$
	μ	$\sigma_e^2 + nJ\tau_a^2 + N\mu^2$
A random, B fixed	α_i	$\sigma_e^2 + nJ\tau_a^2$
	$\beta_{j(i)}$	$\sigma_e^2 + n\tau_{b(a)}^2$
	μ	$\sigma_e^2 + n\tau_{b(a)}^2 + nJ\tau_a^2 + N\mu^2$
A and B random	α_i	$\sigma_e^2 + n\tau_{b(a)}^2 + nJ\tau_a^2$
	$\beta_{j(i)}$	$\sigma_e^2 + n\tau_{b(a)}^2$

MODELS

We saw that three distinct models were possible in the completely crossed two-way analysis of variance. The models corresponded to zero, one, or two random factors. In the nested design there are four different models in all. The four models are listed in Table 8.5 with the expected values of the mean squares for each model (the derivation of the expected mean squares is straightforward but tedious). In Table 8.5, $\tau_a^2 = \sigma_a^2$ when A is random and, similarly, $\tau_{b(a)}^2 = \sigma_{b(a)}^2$ when B is random. The corresponding equations when the factors are fixed are

$$\tau_a^2 = [I/(I-1)]\sigma_a^2,$$
$$\tau_{b(a)}^2 = [J/(J-1)]\sigma_{b(a)}^2.$$

Whether Factor B is considered to be random or fixed depends on how the levels of B within each level of A are chosen. Accordingly, whether B is random or fixed does not depend on the nature of Factor A. For example, suppose that the schools were chosen randomly from a large population of potential schools, each school having only two algebra teachers. Factor B would then be fixed, even though Factor A was random. Conversely, if each school had a large number of teachers from which two had been chosen randomly, Factor B would be random no matter how the schools had been chosen.

The information in Table 8.5, together with that given above, can be used in exactly the same way as outlined in previous chapters to estimate the variances of the effects and the proportions of variance accounted for, as well as to make the usual statistical tests.

ASSUMPTIONS

As with previous models, the necessary assumptions can be given most easily in terms of the effects, but they can be *applied* most easily in terms of the data themselves. Accordingly, we will give the assumptions here in terms of the data.

First, for all models we must assume that the scores within each cell are randomly sampled from a normal distribution with constant variance. Once again, the assumptions of normality and constant variance are generally unimportant, but the assumption of independent random sampling is crucial.

When Factor A is random, the levels of A must be selected randomly. The same must be true of B if it is random. We must also assume that any random factor has a normal distribution. Tests using the random factor in an error term probably are robust to this assumption, but tests on effects involving the factor probably are not.

Higher-Way Models

We can generalize directly to higher-way models. The number and complexity of the possible models increases considerably as a result, but any model, no matter how complex, can be broken down into relatively simple components to which specific rules apply in the calculation of mean squares and expected mean squares.

The complexity arises because, in the general m-way design, each factor may be either fixed or random, and each pair of factors may be either crossed or nested. Whether a given pair of factors is crossed or nested can be determined most easily by a consideration of the two-way table containing the results of summing or averaging over all other factors. If every cell in this table contains a sum or average, the factors are crossed. If each level of one factor is paired with only one level of the other, the former is nested in the latter. The only exception to this rule is the case in which two factors are both nested in a third. An example of this is given below, and the nature of the exception is explained there.

EXAMPLES

The following are a number of examples to help clarify the points made here.

$A \times B \times C(B)$ *Design.* Suppose subjects are required to learn four lists of nonsense syllables under three different sets of instructions. Each subject learns all four lists of nonsense syllables but is given only one set of instructions.

TABLE 8.6. $A \times B \times C(B)$ design.

	B_1			B_2			B_3		
	C_1	C_2	C_3	C_4	C_5	C_6	C_7	C_8	C_9
A_1	8	20	14	21	23	26	15	6	9
A_2	15	24	20	18	21	29	12	11	18
A_3	12	16	19	17	17	26	13	10	9
A_4	17	20	20	28	31	30	18	12	23

	A_1	A_2	A_3	A_4
B_1	42	59	47	57
	14.0	19.7	15.7	19.0
B_2	70	68	60	89
	23.3	22.7	20.0	29.7
B_3	30	41	32	53
	10.0	13.7	10.7	17.7

	A_c	A_2	A_3	A_4
	142	168	139	199
	15.8	18.7	15.4	22.1

	B_1	B_2	B_3
	205	287	156
	17.1	23.9	13.0

	B_1			B_2			B_3		
	C_1	C_2	C_3	C_4	C_5	C_6	C_7	C_8	C_9
	52	80	73	84	92	111	58	39	59
	13.0	20.0	18.3	21.0	23.0	27.8	14.5	9.8	14.8

$\overline{X}$
648
18.0

The levels of Factor A are the lists of nonsense syllables, the levels of Factor B are the sets of instructions, and the levels of Factor C are the subjects. Each level of Factor C (each individual subject) is paired with only one level of Factor B, so C is nested in B. However, all four lists of nonsense syllables are learned under each condition, so A and B are crossed. Similarly, A and C are crossed because every subject learns all four lists of nonsense syllables. Hypothetical data from this design are shown in Table 8.6. The data from this table will be used later to illustrate the general rules for analyzing data from nested designs.

This type of design is sometimes referred to as a two-way design with repeated measures on Factor A. It may also be referred to as a "partially replicated" design. Still another term that may be used is "split-plot" design, the name being derived from its use in agricultural experiments. A number of plots of ground are assigned to groups, with a different level of Factor B applied to the plots in each group. Each plot is then divided ("split") into a number of subplots, with a different level of Factor A applied to each subplot. For this design the plots are regarded as Factor C, and because each plot is given only one level of B, C is nested in B. The special name *split plot* implies that designs of this type are somehow basically different from other designs that might have nested factors. As a result, special terms such as "within subjects" and "between subjects" variance have been invented. Both the terminology and the analysis are simplified (as will be seen later) by merely treating it as a three-way design with a nested factor.

$A \times B \times C(AB)$ *Design.* Two different textbooks are to be tested. Three teachers are recruited to do the teaching, and each teaches four classes. Two of the classes are taught using textbook A_1 and the other two using textbook A_2. Thus, each teacher teaches two classes from each textbook for a total of twelve classes in all. The levels of Factor A are the textbooks, the levels of Factor B are the teachers, and the levels of Factor C are the classes. Factors A and B are crossed because each teacher uses both textbooks. However, each class has only one teacher and one textbook, so C is nested in both A and B. If the classes were randomly assigned to both teachers and textbooks, they would be regarded as a random factor. This is true even though there may have been a total of only twelve potential classes to begin with. (Scheffé (1959) points out that a "permutation" model is actually more appropriate when students are randomly assigned to classes in this way; however, he argues that a good approximation to the permutation model can be had by regarding Factor C as random.)

$A \times B(A) \times C(A)$ *Design.* Two kinds of intelligence tests are to be tested on first-grade children. Both kinds are paper-and-pencil tests designed for group administration; one kind requires minimal reading ability while the other does not. Two forms of each type of test are to be administered, making a total of four tests. The tests are given to twenty randomly selected children in each of six randomly selected first-grade classes. All the children in a single class receive the same *type* of test, but half (10) of the children in the class receive one *form* of the test and the other half receive the other form. In this design the levels of Factor A are the two types of tests and the levels of Factor B are the two tests of each type. Each form can be of only one type, so Factor B is nested in Factor A. The levels of Factor C are the classes, and because each class is given only one type of test, C is nested in A. The scores of the individual children are the scores within the cells ($n = 10$).

This design is diagrammed in Table 8.7, where an X in a cell means that that cell contains data. Notice that neither B nor C is nested in the other but, at the same time, the levels of B are not paired with all levels of C; e.g., B_1 is not paired with C_4. However, each level of B nested in level A_1 is paired with every level of C that is also nested in level A_1, i.e., B_1 and B_2 are paired with C_1, C_2, and C_3; similarly, each level of B nested in A_2 is paired with every level of C that is also nested in A_2. Here, B and C are considered to be crossed. In general, when two factors are both nested in a third, they are considered to be crossed if they are crossed within each level of the third factor. Hypothetical data from this design are shown in Table 8.8.

TABLE 8.7. $A \times B(A) \times C(A)$ design.

	A_1							A_2					
	C_1	C_2	C_3	C_4	C_5	C_6		C_1	C_2	C_3	C_4	C_5	C_6
B_1	X	X	X										
B_2	X	X	X										
B_3											X	X	X
B_4											X	X	X

They will be used later as another illustration of the procedures for finding mean squares and expected mean squares.

$A \times B(A) \times C(AB)$ *Design.* An experiment is conducted to assess the effects of personality variables on teacher performance. Three teachers with high scores and three with low scores on an "extroversion scale" each teach two different sections of an introductory psychology class (making twelve sections in all). The two levels of Factor A are the high and low extroversion scores, the levels of Factor B are the teachers, and the levels of Factor C are the classes. No teacher will have both a high and a low score, so Factor B is nested in Factor A. Also, because each class is taught by only one teacher, Factor C is nested in Factor B. However, this means that no class is taught by teachers having both high and low scores. That is, C is also nested in A. Nesting is always transitive; whenever C is nested in B and B is nested in A, C is nested in A.

THE MODEL EQUATIONS

As we have seen, writing out the model equation for a given design can help to specify the effects that are involved. In nested designs the model equation shows which main effects and interactions are confounded. This is helpful for finding expected mean squares. For the $A \times B \times C(B)$ design described above, for example, the equation is

$$X_{ijkl} = \mu + \alpha_i + \beta_j + \gamma_{k(j)} + \alpha\beta_{ij} + \alpha\gamma_{ik(j)} + \epsilon_{ijkl}. \tag{8.8}$$

This equation tells us that six different effects (in addition to the grand mean) can be estimated and tested, that the C main effect is confounded with the BC interaction, and that the AB and AC interactions are confounded with the ABC interaction. To find the confounded effects in each term, we combine the letters outside the parentheses with each letter or set of letters (including the "empty" set) inside the parentheses. Thus, the term $\alpha\gamma_{ik(j)}$ involves the AC (i and k) and ABC (i, j, and k) interactions. If there were a fourth factor, D, the term $\alpha\gamma_{ik(jl)}$ would involve the AC, ABC, ACD, and $ABCD$ interactions.

TABLE 8.8. Hypothetical cell totals (upper values) and means (lower values) from $A \times B(A) \times C(A)$ design, assuming 10 scores per cell.

		BC(A) table						
		A_1					A_2	
	C_1	C_2	C_3			C_4	C_5	C_6
B_1	147	199	197		B_3	144	108	126
	14.7	19.9	19.7			14.4	10.8	12.6
B_2	192	156	149		B_4	156	101	154
	19.2	15.6	14.9			15.6	10.1	15.4

B(A) table			
A_1		B_4	
543	497	378	411
18.1	16.6	12.6	13.7

C(A) table					
A_1			C_4	C_5	C_6
339	355	346	300	209	280
17.0	17.8	17.3	15.0	10.5	14.0

A table			
A_1	A_2	$\overline{X}$	
1,040	789	1,829	
17.3	13.2	15.2	

Once written, the model equation appears complicated, but it is really very easy to write. We will illustrate the procedure by deriving Equation 8.8. The first step is to write the equation that we would obtain if all of the factors were completely crossed. For the three-way design this is

$$X_{ijkl} = \mu + \alpha_i + \beta_j + \gamma_k + \alpha\beta_{ij} + \alpha\gamma_{ik} + \beta\gamma_{jk} + \alpha\beta\gamma_{ijk}$$
$$+ \epsilon_{ijkl}.$$

Next, the subscripts of all effects (except error) involving nested factors are modified by adding parentheses containing the subscripts of the factors in which they are nested. In the $A \times B \times C(B)$ design, for example, any effect containing k as a subscript is modified by appending (j). The equation then becomes

$$X_{jkl} = \mu + \alpha_i + \beta_j + \gamma_{k(j)} + \alpha\beta_{ij} + \alpha\gamma_{ik(j)} + \beta\gamma_{jk(j)}$$
$$+ \alpha\beta\gamma_{ijk(j)} + \epsilon_{ijkl}.$$

Finally, each term in the equation that contains the same subscript both inside and outside parentheses is struck out. In our example there are two such terms: $\beta\gamma_{jk(j)}$ and $\alpha\beta\gamma_{ijk(j)}$. Eliminating these gives Equation 8.8, the final form of the model equation. In the same way, the equations for other designs described can be shown to be

$$A \times B \times C(AB): \quad X_{ijkl} = \mu + \alpha_i + \beta_j + \gamma_{k(ij)} + \alpha\beta_{ij} + \epsilon_{ijkl},$$
$$A \times B(A) \times C(A): \quad X_{ijkl} = \mu + \alpha_i + \beta_{j(i)} + \gamma_{k(i)} + \beta\gamma_{jk(i)} + \epsilon_{ijk},$$
$$A \times B(A) \times C(AB): \quad X_{ijkl} = \mu + \alpha_i + \beta_{j(i)} + \gamma_{k(ij)} + \epsilon_{ijkl}.$$

You should derive each of these equations, using the rules given above, to test your understanding of the procedures.

Note that no letter ever appears more than once in any subscript. For example, in the term $\beta\gamma_{jk(i)}$, in the $A \times B(A) \times C(A)$ design, the i appears only once even though j and k both represent factors nested in A. Note also that there is only one set of parentheses for each term (e.g., we write $\gamma_{k(ij)}$, not $\gamma_{k(i)(j)}$), and that the parentheses always enclose the last subscripts to be listed (we write $\beta_{j(i)}$, not $\beta_{(i)j}$). Confusion can arise if these practices are not followed.

SUMS OF SQUARES AND DEGREES OF FREEDOM

The rules given in Chapter 7 for finding sums of squares and degrees of freedom in completely crossed designs can be extended to nested designs with little difficulty. The rules are extended by making the following changes.

The number of different tables obtained by summing over factors in a nested design is smaller than in a completely crossed design. This is because any attempt to sum over the levels of a factor before first summing over the levels of all factors nested in it would result in "summing" over only one cell— the result would of course be simply the original table. This can be seen most clearly by referring again to Table 8.2, in which any attempt to sum over Factor A before summing over Factor B would simply reproduce the six original cell totals. Tables 8.7 and 8.8 also show that in the $A \times B(A) \times C(A)$ design we cannot sum over the levels of A without first summing over the levels of B and C. The total number of different tables obtained will be exactly equal to the number of terms in the model equation, and each table will correspond to one of the terms. For the $A \times B \times C(B)$ design, for example, Equation 8.8 shows that six tables, one for the grand mean and one for each of the five effects, will be obtained. If each cell contains more than one score, a seventh table (containing the raw data) will correspond to the error term.

An RS is obtained for each term in the model equation, i.e., for each table, by the usual procedure of squaring each total in the table, summing the squares, and dividing the result by the number of scores over which the totals were taken. The number of scores over which they were taken will always be the total number of scores (N) divided by the number of cells in the table. For the design of Table 8.6, for example, the divisor for $RS_{c(b)}$ is $36/9 = 4$. The

TABLE 8.9. Summary table and expected mean squares for $A \times B \times C(B)$ design in Table 8.6.

	RS	SS	T	df	MS	F	p	$\hat{\omega}^2$
m		11,664.00	1	1				
a	11,923.33	259.33	4	3	86.44	10.1	<.001	.15
b	12,394.17	730.17	3	2	365.1	8.24	.020	.42
$c(b)$	12,660.00	265.83	9	6	44.31			.26
ab	12,720.67	67.17	12	6	11.19	1.31	.31	.01
$ac(b)$	13,140.00	153.50	36	18	8.528			.15
t		1,476.00	35					

Effect	$E(MS)$
m	$\sigma_e^2 + nI\tau_{c(b)}^2 + N\mu^2$
a	$\sigma_e^2 + n\tau_{ac(b)}^2 + nJK\tau_a^2$
b	$\sigma_e^2 + nI\tau_{c(b)}^2 + nIK\tau_b^2$
$c(b)$	$\sigma_e^2 + nI\tau_{c(b)}^2$
ab	$\sigma_e^2 + n\tau_{ac(b)}^2 + nK\tau_{ab}^2$
$ac(b)$	$\sigma_e^2 + n\tau_{ac(b)}^2$

complete equation is

$$RS_{c(b)} = (52^2 + 80^2 + 73^2 + 84^2 + 92^2 + 111^2 + 58^2 + 39^2 + 59^2)/4.$$

The raw sums of squares are then placed in a summary table with the same form as Tables 8.9 and 8.10, which show the summaries for the data in Tables 8.6 and 8.8, respectively. It is important, in the summary table, to indicate nesting with appropriate parentheses as was done in Tables 8.9 and 8.10. Although the parentheses are ignored in these calculations, they make it clear that the effect is nested, and this information is used in later calculations.

The sums of squares are then calculated from the raw sums of squares in exactly the same way as with the completely crossed design. That is, from each RS we subtract SS_m and the SS for each effect involving only factors in the effect whose mean square is being found. For this purpose, parentheses are ignored. In Table 8.9, for example, the SS for the C main effect is

$$SS_{c(b)} = RS_{c(b)} - SS_m - SS_b,$$

subtracting the SSs for all effects that do not involve Factor A. Similarly, in Table 8.10 the $BC(A)$ interaction is treated as the ABC interaction for purposes of calculating $SS_{bc(a)}$:

$$SS_{bc(a)} = RS_{bc(a)} - SS_m - SS_a - SS_{b(a)} - SS_{c(a)}.$$

If the design has only one score per cell, SS_w cannot be found. Otherwise, SS_w is found by subtracting all of the other SSs from RS_w. Alternatively, SS_w

TABLE 8.10. Summary table and expected mean squares for $A \times B(A) \times C(A)$ design in Table 8.8 (assuming $n = 10$, with $RS_w = 36,000$).

	RS	SS	T	df	MS	F	p	$\hat{\omega}^2$
m		27,877.01	1	1	27,877	474		
a	28,402.02	525.01	2	1	525.0	8.93	.041	.05
$b(a)$	28,455.43	53.42	4	2	26.71	0.35	—	(-.01)
$c(a)$	28,637.15	235.13	6	4	58.78	0.91	—	.00
$bc(a)$	28,994.90	304.33	12	4	76.08	1.17	.33	.01
w	36,000.00	7,005.10	120	108	64.86			.95
t		8,122.99		119				

	$E(MS)$
m	$\sigma_e^2 + nJ\tau_{c(a)}^2 + N\mu^2$
a	$\sigma_e^2 + nJ\tau_{c(a)}^2 + nJK\tau_a^2$
$b(a)$	$\sigma_e^2 + n\tau_{bc(a)}^2 + nK\tau_{b(a)}^2$
$c(a)$	$\sigma_e^2 + nJ\tau_{c(a)}^2$
$bc(a)$	$\sigma_e^2 + n\tau_{bc(a)}^2$

can be found as follows. In every design there will be one effect that has the subscripts (either in or out of parentheses) of all of the factors in the design. The value of SS_w is found by subtracting the RS for this highest interaction from RS_w. The value of SS_t is found as usual, by subtracting SS_m from RS_w. The sums of squares for the data in Tables 8.6 and 8.8 are shown in the columns labeled SS of Tables 8.9 and 8.10.

Degrees of freedom are found by basically the same procedure as that given in Chapter 7. First, in the column headed T, we list the number of cells in each table. Then, to find the degrees of freedom for a particular effect, we subtract from that T value the degrees of freedom for each effect (including SS_m) that was subtracted from RS to find the SS for that effect. In Table 8.9, for example,

$$df_{c(b)} = T_{c(b)} - 1 - df_b.$$

(Note, again, that $df_m = 1$.) In Table 8.10

$$df_{bc(a)} = T_{bc(a)} - 1 - df_a - df_{b(a)} - df_{c(a)}.$$

Mean squares are then found by dividing each sum of squares by its degrees of freedom.

HIERARCHIES

The concept of a hierarchy of effects can be employed here just as in previous chapters. In fact, the hierarchy is constructed in the same way. An effect is above another if and only if its associated table can be derived by summing

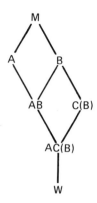

FIGURE 8.1. Hierarchy of effects for design in Table 8.6.

over the table associated with the other effect. Figures 8.1 and 8.2 show the hierarchies for the designs in Tables 8.6 and 8.8, respectively.

The rule for finding sums of squares can be stated as follows: from the RS for the effect, subtract those sums of squares that are above the effect in the hierarchy. The rule for finding degrees of freedom can be similarly restated.

EXPECTED MEAN SQUARES

The rest of the information necessary for estimating and testing effects is found, as usual, from the expected values of the mean squares. The rules for finding expected mean squares in nested designs can be obtained from some relatively simple modifications of the rules given in Chapter 7 for completely crossed designs. We begin, as in Chapter 7, by listing all of the terms that can enter into the expected mean squares. As in Chapter 7, the appropriate terms are found from the model equation; there is a term for each effect, and each term is a τ^2 multiplied by the divisor used in finding the RS for the associated effect. For the data in Table 8.6 the terms are

$$N\mu^2, nJK\tau_a^2, nIK\tau_b^2, nI\tau_{c(b)}^2, nK\tau_{ab}^2, n\tau_{ac(b)}^2.$$

(Remember that for this purpose the number of levels of a nested factor is regarded as the number of levels over which we must sum when summing over that factor. Factor C in Table 8.6 has nine different levels, but we sum over only three levels of C to obtain each cell of the $A \times B$ table, so $K = 3$.) For the data in Table 8.8, the terms are

$$N\mu^2, \; nJK\tau_a^2, \; nK\tau_{b(a)}^2, \; nJ\tau_{c(a)}^2, \; n\tau_{bc(a)}^2.$$

In Chapter 7 the terms included in the expected mean squares for a given effect had to meet two criteria: They had to have subscripts of all of the factors

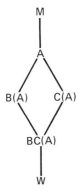

FIGURE 8.2. Hierarchy of effects for design in Tables 8.7 and 8.8.

involved in the effect whose expected mean square was being considered, and they could have no other fixed factors as subscripts. For nested designs the first criterion remains the same; each term entering into an expected mean square must contain subscripts of all of the factors involved in the effect whose expected mean square is being considered. It does not matter whether the subscripts are inside or outside of the parentheses, so long as they are present in the subscript of the term. For the data in Table 8.9, for example, $\tau^2_{c(b)}$ is part of the expected mean square of the B main effect even though the b subscript is found inside the parentheses. Similarly, every term entering into $\underline{E}(MS_{c(b)})$ must have both b and c as subscripts.

The second criterion in Chapter 7 is modified slightly for nested designs. The subscript of each term in an expected mean square must contain no other symbols for fixed factors *outside of the parentheses*. Fixed factors unrelated to the effect whose expected mean square is found may appear inside but not outside the parentheses. This point is best illustrated by finding $\underline{E}(MS_a)$ for the $A \times B \times C(AB)$ design with Factor C random and Factors A and B fixed. The appropriate model for this design is

$$X_{ijkl} = \mu + \alpha_i + \beta_j + \alpha\beta_{ij} + \gamma_{k(ij)} + \epsilon_{ijkl}.$$

The terms entering into the expected mean squares are, accordingly,

$$N\mu^2, \ nJK\tau^2_a, \ nIK\tau^2_b, \ nK\tau^2_{ab}, \ n\tau^2_{c(ab)}.$$

When finding the expected mean square for the A main effect, the terms $N\mu^2$ and $nIK\tau^2_b$ are eliminated because they do not contain an a in their subscripts. The term $nK\tau^2_{ab}$ is also eliminated because it contains b (a fixed factor) in its subscript. However, the term $n\tau^2_{c(ab)}$ is not because, unlike in τ^2_{ab}, the b appears inside the parentheses. (The presence of a c in the subscript does not disqualify the term because c refers to a random rather than to a fixed factor.) The expected mean square for the A main effect is therefore

$$\underline{E}(MS_a) = \sigma_e^2 + n\tau_{c(ab)}^2 + nJK\tau_a^2.$$

In summary, the two qualifications that must be met for a term to be included in an expected mean square are (1) that it include all of the subscripts (including any in parentheses) of the effect whose expected mean square is being found, and (2) that it include no *other* subscripts of *fixed* factors *outside* the parentheses. Additional examples of expected mean squares are found in Tables 8.5, 8.9, and 8.10. These tables can be used most profitably by first finding for yourself the expected mean squares and then checking the results against those in the tables.

F Ratios

F ratios are formed in the usual way. To test for a given effect we find another effect whose expected mean square would be exactly equal to that of the effect tested if the null hypothesis were true. If no such mean square exists, a quasi-F must be used.

The following rules are usually not useful for *finding* the correct F ratios, but they can serve as a check on the F ratios after they have been found. If these rules are violated by any of the F ratios, there is an error somewhere in the original calculations.

(1) The subscript of the denominator mean square (if it is not MS_w) must contain symbols (inside or outside of the parentheses) for all of the factors in the numerator effect.

(2) The subscript of the denominator mean square (if it is not MS_w) must contain the subscript of at least one *random* factor, *outside of the parentheses*, that is not also in the subscript of the numerator effect.

(3) The subscript of the denominator mean square must contain no symbols for fixed factors, *outside of the parentheses*, that are not also in the numerator effect.

(4) An effect must be tested with a quasi-F *if and only if* it is crossed with *at least two random* factors. Otherwise, it can always be tested with an ordinary F ratio.

Variance Estimates

Estimates of τ^2 and σ^2 for each effect are found in the same way as with completely crossed designs. They are found most easily by examining the F ratio. The τ^2 for an effect can be estimated by first subtracting the denominator from the numerator of the F ratio used to test for that effect. The resulting difference is then divided by the multiplier of the τ^2 for that effect in the expected mean square. Earlier it was stated that the multiplier of τ^2 was the divisor used in finding the associated RS.

As an example, we will derive the equation for the B main effect in the $A \times B \times C(B)$ design of Table 8.6. According to Table 8.9, $MS_{c(b)}$ is the appropriate denominator for testing the B main effect, and

$$
\begin{aligned}
\underline{E}(MS_b) &= \sigma_e^2 + nI\tau_{c(b)}^2 + nIK\tau_b^2, \\
\underline{E}(MS_{c(b)}) &= \sigma_e^2 + nI\tau_{c(b)}^2, \\
\underline{E}(MS_b - MS_{c(b)}) &= \underline{E}(MS_b) - \underline{E}(MS_{c(b)}) = nIK\tau_b^2, \\
\hat{\tau}_b^2 &= (MS_b - MS_{c(b)})/(nIK).
\end{aligned}
$$

In studies with only one score per cell, σ_e^2 cannot be estimated, so it is assumed to be zero when estimating the τ^2. Basically, the within variance is "absorbed" in the other terms. Thus, in the same $A \times B \times C(B)$ design, if $n = 1$ so σ_e^2 cannot be estimated,

$$
\begin{aligned}
\hat{\tau}_{c(b)}^2 &= MS_{c(b)}/(nI) = MS_{c(b)}/I, \\
\hat{\tau}_{ac(b)}^2 &= MS_{ac(b)}/n = MS_{ac(b)}.
\end{aligned}
$$

To find the variance (σ^2) for any effect, the τ^2 for the effect must be multiplied by as many constants as there are *fixed factors outside the parentheses* in the subscript of the τ^2. Subscripts in parentheses are ignored, as are subscripts of random factors. The appropriate multiplier of each factor is the number of levels of that factor divided into the number of levels minus one. (Compare with the discussion in Chapter 7.) For this purpose the number of levels of each nested factor is defined as the number of levels over which one must sum when summing over that factor. In Table 8.6, for example, Factor C is considered to have only three levels instead of nine, and in Table 8.8 Factor C is considered to have three levels instead of six. Table 8.11 illustrates these principles for the data in Tables 8.6 and 8.8, and Table 8.12 shows the estimates for the same data.

The estimate of ω^2 for each effect is the same as in previous chapters; the estimated variances are summed to obtain an estimate of σ_t^2, and this is divided into each variance estimate to obtain $\hat{\omega}^2$.

POWER

The rules for calculating the power of a test are also easily adapted to nested designs. When calculating power, an effect is considered to be fixed if all of the letters outside parentheses in its subscript represent fixed factors; the effect is considered to be random if one or more of the letters correspond to random factors.

The F distribution for a fixed effect when the null hypothesis is false is noncentral F, with parameters ϕ^2 and ϕ'^2 defined in the same way as in Chapter 6; the F distribution for a random effect is a constant times an ordinary central F, as in Chapter 6.

TABLE 8.11. Formulas relating σ^2 to τ^2 in Tables 8.9 and 8.10.

Table 8.9		
σ_a^2	$=$	$[(I-1)/I]\tau_a^2 = (3/4)\tau_a^2$
σ_b^2	$=$	$[(J-1)/J]\tau_b^2 = (2/3)\tau_b^2$
$\sigma_{c(b)}^2$	$=$	$\tau_{c(b)}^2$
σ_{ab}^2	$=$	$[(I-1)/I][(J-1)/J]\tau_{ab}^2 = (1/2)\tau_{ab}^2$
$\sigma_{ac(b)}^2$	$=$	$[(I-1)/I]\tau_{ac(b)}^2 = (3/4)\tau_{ac(b)}^2$

Table 8.10		
σ_a^2	$=$	$[(I-1)/I]\tau_a^2 = (1/2)\tau_a^2$
$\sigma_{b(a)}^2$	$=$	$[(J-1)/J]\tau_{b(a)}^2 = (1/2)\tau_{b(a)}^2$
$\sigma_{c(a)}^2$	$=$	$\tau_{c(a)}^2$
$\sigma_{bc(a)}^2$	$=$	$[(J-1)/J]\tau_{bc(a)}^2 = (1/2)\tau_{bc(a)}^2$

TABLE 8.12. Variance estimates for data in Tables 8.9 and 8.10.

	Table 8.9				Table 8.10		
	$\hat{\tau}^2$	$\hat{\sigma}^2$	$\hat{\omega}^2$		$\hat{\tau}^2$	$\hat{\sigma}^2$	$\hat{\omega}^2$
a	8.66	6.49	.15	a	7.77	3.89	.06
b	26.73	17.82	.42	$b(a)$	(-1.65)	(-0.82)	(-.01)
$c(b)$	11.08	11.08	.26	$c(a)$	(-0.30)	(-0.30)	.00
ab	0.89	0.44	.01	$bc(a)$	1.12	0.56	.01
$ac(b)$	8.53	6.40	.15	w	64.86	64.86	.95
t		42.23		t		68.19	

ASSUMPTIONS AND BOX'S CORRECTION

The necessary assumptions are even more difficult to state in detail for multi-factor nested designs than for multifactor crossed designs. Some assumptions are basically the same as in previous chapters. The scores within a cell are assumed to be normally distributed with constant variance, but as usual, this assumption is not critical if each cell contains the same number of scores. In addition, the cell means for random factors must be normally distributed, and covariances must be assumed to be equal.

However, it is not necessary to spell out the assumptions in detail in order to tell when Box's correction (or, alternatively, a more exact multivariate test; see Chapter 13) may be appropriate. For each effect to be tested, compare the subscripts *outside the parentheses* on both the numerator and the denominator mean square. If the subscripts outside the parentheses on the denominator mean square include *any* of those on the numerator mean square, Box's correction is applicable. If not, the test is robust, and Box's correction does not apply. In Table 8.9, for example, the denominator for testing the A main effect is $MS_{ac(b)}$. The letter a is the only subscript of MS_a, and that letter also appears outside the parentheses in the subscript of $MS_{ac(b)}$. The letter a also appears in both the numerator and the denominator when testing the AB interaction. Therefore, Box's correction may be appropriate for both of these tests. However, the appropriate denominator for the B main effect is $MS_{c(b)}$. The numerator MS_b contains only the subscript b; $MS_{c(b)}$ also contains b, but the b is within the parentheses. Therefore, Box's correction does not apply.

In general, a factor may be thought of as *directly* involved in an effect if its symbol appears as a subscript *outside of the parentheses* on the mean square for that effect. It is *indirectly* involved in the effect if its symbol appears *inside the parentheses*. With these terms, the principle above can be restated as follows. Box's correction applies only when one or more factors *directly* involved in the numerator mean square are also *directly* involved in the denominator mean square.

Two special cases should also be mentioned; effects tested by quasi-F ratios, and effects tested by a pooled mean square. Both of these are, in at least some sense, only approximate tests anyway, and the appropriateness of Box's correction in general has not been determined for them. For these cases it is best just to make the usual tests but to interpret the results with some caution.

If Box's correction applies, estimates of K_B are found as in Chapter 7. (Later, we will tell how to determine whether a given contrast is part of a particular effect.) Upper and lower limits on the p value are also found as in Chapter 7, with one modification. The lower limit is that of the ordinary F ratio, as before, but the upper limit depends on the effect being tested. If the effect being tested is nested in one or more other factors, then the numerator degrees of freedom are the number of levels, or combinations of levels, of those factors. The denominator degrees of freedom are ν_2/ν_1 times the numerator degrees of freedom.

For example, in Table 8.10, $b(a)$ is tested with $bc(a)$. The letter b is outside the parentheses in both effects, so Box's correction appears to be applicable. However, B is nested in A, so the numerator degrees of freedom for finding the upper limit on p are two (the number of levels of A). The denominator degrees of freedom are ν_2/ν_1 times 2, i.e., $(4/2)2 = 4$. In this example, the upper and lower limits on p are found using the same degrees of freedom in both the numerator and denominator. In effect, Box's correction needn't be used with these data. If J had been three instead of two, Box's correction would have applied. Then the degrees of freedom for the ordinary F ratio would have been $\nu_1 = 4$, $\nu_2 = 8$, while the degrees of freedom for finding the maximum p value would have been $\nu_1 = 2$, $\nu_2 = 4$.

For another example, consider an $A \times B \times C(AB) \times D(A)$ design with D random. To test $c(ab)$, we use $cd(ab)$. (You should check this for yourself.) The degrees of freedom for the ordinary F ratio are $\nu_1 = IJ(K-1)$, $\nu_2 = IJ(K-1)(L-1)$. Because C is nested in both A and B, and their combined

TABLE 8.13. Estimates of effects for data in Table 8.6.

$$M$$

$$\boxed{18.0}$$

$$\hat{\mu} = \overline{X}...$$

A_1	A_2	A_3	A_4		B_1	B_2	B_3
-2.2	0.7	-2.6	4.1		-0.9	5.9	-5.0

$$\hat{\alpha}_i = \overline{X}_{i..} - \hat{\mu} \qquad \hat{\beta}_j = \overline{X}_{.j.} - \hat{\mu}$$

	B_1			B_2			B_3	
C_1	C_2	C_3	C_4	C_5	C_6	C_7	C_8	C_9
-4.1	2.9	1.2	-2.9	-0.9	3.8	1.5	-3.3	1.8

$$\hat{\gamma}_{j(i)} = \overline{X}_{ij.} - \hat{\mu} - \hat{\beta}_j$$

	A_1	A_2	A_3	A_4
B_1	-0.8	1.9	1.1	-2.2
B_2	1.6	-1.9	-1.4	1.6
B_3	-0.8	0.0	0.2	0.6

$$\hat{\alpha\beta}_{ij} = \overline{X}_{ij.} - \hat{\mu} - \hat{\alpha}_i - \hat{\beta}_j$$

	B_1			B_2			B_3		
	C_1	C_2	C_3	C_4	C_5	C_6	C_7	C_8	C_9
A_1	-1.9	3.1	-1.2	0.6	0.6	-1.2	3.5	-0.8	-2.8
A_2	-0.6	1.4	-0.8	-1.8	-0.8	2.5	-3.2	0.6	2.6
A_3	0.4	-2.6	2.2	-0.1	-2.1	2.2	0.8	2.6	-3.4
A_4	2.1	-1.9	-0.2	1.3	2.3	-3.5	-1.2	-2.4	3.6

$$\hat{\beta\gamma}_{jk(i)} = X_{ijk} - \hat{\mu} - \hat{\alpha}_i - \hat{\beta}_j - \hat{\gamma}_{k(j)} - \hat{\alpha\beta}_{ij}$$

number of levels is IJ, the maximum p value is found by using $\nu_1 = IJ$, $\nu_2 = IJ(L-1)$.

ESTIMATING EFFECTS

Estimates of effects are found very much as in previous chapters. We begin with the cell means for the table associated with a given effect, and we subtract the estimates of all effects above it in the hierarchy (i.e., all effects whose sums of squares were subtracted when finding the SS for the effect in question). Estimates of effects, with formulas, for the data in Tables 8.6 and 8.8, appear in Tables 8.13 and 8.14.

TABLE 8.14. Estimates of effects for data in Table 8.8.

A table

M	A_1	A_2
15.2	2.1	-2.1

$$\hat{\mu} = \overline{X}_{....} \qquad \hat{\alpha}_i = \overline{X}_{i...} - \hat{\mu}$$

B(A) table

A_1		A_2	
B_1	B_2	B_3	B_4
0.8	-0.8	-0.6	0.6

$$\hat{\beta}_{j(i)} = \overline{X}_{ij..} - \hat{\mu} - \hat{\alpha}_i$$

C(A) table

A_1			A_2		
C_1	C_2	C_3	C_4	C_5	C_6
-0.4	0.4	0.0	1.9	-2.7	0.9

$$\hat{\gamma}_{k(i)} = \overline{X}_{i.k.} - \hat{\mu} - \hat{\alpha}_i$$

BC(A) BC(A) table

	A_1				A_2		
	C_1	C_2	C_3		C_4	C_5	C_6
B_1	-3.0	1.4	1.6	B_3	-0.1	0.9	-0.9
B_2	3.0	-1.4	-1.6	B_4	0.1	-0.9	0.9

$$\hat{\beta\gamma}_{jk(i)} = \overline{X}_{ijk.} - \hat{\mu} - \hat{\alpha}_i - \hat{\beta}_{j(i)} - \hat{\gamma}_{k(i)}$$

COMPARISONS

The basic principles for testing comparisons are the same as those in Chapter 7; however, there are some differences in their application. First, the exact method described in Chapter 7 is more complicated and more difficult to apply; it will not be discussed here.

Second, there are times when a planned comparison may involve more than one effect and still be testable by the standard method. We will illustrate by testing μ_{43} against the average of the other three values in the same row of the AB table in Table 8.6. Our null hypothesis is

$$H_0(1) : 3\mu_{43.} - \mu_{13.} - \mu_{23.} - \mu_{33.} = 0. \tag{8.9}$$

Just as in Chapter 7, we first put the coefficients into the cells of the AB table and then do an analysis of variance on them to determine the effects involved.

TABLE 8.15. Analysis of coefficients in Equation 8.9.

	\multicolumn{5}{c}{Coefficients}				\multicolumn{3}{c}{Analysis}
	A_1	A_2	A_3	A_4	Σ		rs	ss	ss^*
B_1	0	0	0	0	0	m	0	0	0
B_2	0	0	0	0	0	a	4	4	1/3
B_3	-1	-1	-1	3	0	b	0	0	0
Σ	-1	-1	-1	3	0	ab	12	8	2/3

TABLE 8.16. Analysis of coefficients in Equation 8.10.

	\multicolumn{5}{c}{Coefficients}				\multicolumn{3}{c}{Analysis}
	A_1	A_2	A_3	A_4	Σ		rs	ss	ss^*
B_1	0	0	0	-1	-1	m	0	0	0
B_2	0	0	0	0	0	a	2/3	2/3	1/3
B_3	-1	0	0	0	1	b	1/2	1/2	1/4
Σ	1	0	0	-1	0	ab	2	5/6	5/12

This analysis, shown in Table 8.15, reveals that the A main effect and the AB interaction are involved. However, Table 8.9 shows that these effects are both tested against $MS_{ac(b)}$.

Consequently, $MS_{ac(b)}$ is the appropriate denominator for testing Equation 8.9:

$$F_{(1,18)} = SS_1/MS_{c(ab)} = 87.11/8.528 = 10.2, \; p = .006.$$

Note that the analysis of variance on the coefficients must take exactly the same form as on the original data. That is, any nesting relationships in the original data must be preserved in the analysis on the coefficients.

A third difference is that care is sometimes needed when estimating degrees of freedom for a quasi-F. We will illustrate this with the contrast

$$H_0(2) : \mu_{13.} - \mu_{41.} = 0 \tag{8.10}$$

on the data in Table 8.6. From the analysis of these coefficients (Table 8.16), both main effects and the interaction are involved. Comparing these results with the analysis in Table 8.9, we see that the appropriate F ratio is

$$F^* = SS_2/[(1/4)MS_{ac(b)} + (1/3)MS_{c(b)} + (5/12)MS_{ac(b)}].$$

Our estimate of ν_2 may be in error, however, unless we combine the two terms involving $MS_{ac(b)}$ to obtain

$$F^* = SS_2/[(1/4 + 5/12)MS_{ac(b)} + (1/3)MS_{c(b)}]$$

$$= SS_2/[(2/3)MS_{ac(b)} + (1/3)MS_{c(b)}]$$
$$= 121.5/20.45 = 5.94.$$

The estimated degrees of freedom are then

$$\hat{\nu}_2 = 20.45^2/[(4/9)(8.528^2/18) + (1/9)(44.30^2/6)] \simeq 11.$$

In general, before using Equation 7.1 to estimate degrees of freedom, all terms involving the same mean square in the F ratio must be combined.

TREATING ERROR AS A RANDOM EFFECT

It is possible to treat random error itself as a random effect. Like the levels of a random factor, the random errors in a design are always due to a specific random-selection procedure. In a typical design, where only one score is obtained from each subject, differences between scores of subjects in the same cell are treated as due to error. If we modify the design so that two or more scores are taken from each subject, subjects will be treated as levels of a random factor in the experiment. (Whenever more than one score is taken on each subject, subjects should be treated as levels of a random factor.) Thus, differences between subjects are treated as random error in the former design and as the subjects main effect in the second.

Because random error is always the result of a specific random sampling of the levels of some variable (e.g., subjects), it is equally logical to treat it either as random error or as the levels of a random factor. The advantage of the latter concept of error is parsimony. Instead of having a design with factors plus error, we have only a design with different factors. Some computer programs take advantage of this parsimony to simplify their operation. To use such programs effectively, we must understand how random error can be treated as a specific factor in the experiment. Moreover, an understanding of this point can lead to a better understanding of the general nature of error in all of the designs that can occur in the analysis of variance.

To see just exactly how random error can be treated as a factor, consider a two-way fixed-effects design with n subjects per cell. Ordinarily, differences between subjects in the same cell are regarded as due to random error. If we had obtained more than one score from each subject, however, we would have had to treat subjects as a random effect. For example, if each subject had been given two trials, we would have had an $A \times B \times C(AB) \times D$ design, where D would have been the trials factor (two levels) and $C(AB)$ the subjects factor (n levels per cell).

Because there is only one trial per subject in the simple two-way design, it would seem logical to think of it as the same kind of design, but with the trials factor omitted, leaving an $A \times B \times C(AB)$ design. Factor C (subjects) is now treated as a random factor, nested in both A and B, whereas in the simple two-way design it would be treated as random error. The following two model equations should then be equivalent:

$$X_{ijk} = \mu + \alpha_i + \beta_j + \alpha\beta_{ij} + \epsilon_{ijk},$$

$$X_{ijk} \;=\; \mu + \alpha_i + \beta_j + \alpha\beta_{ij} + \gamma_{k(ij)}.$$

The first equation is the standard equation for the two-way design; the second is the equation for the same design with error (subjects) treated as a random factor. Notice that no additional error term appears in the second equation; Factor C is the "error."

In general, the error term can be treated either as random error or as a random effect that is nested in all of the other effects (both random and fixed) in a design. MS_w is then equivalent to the mean square for that random effect.

The point can be clarified if we review the rules for finding sums of squares and expected mean squares, treating error as a factor. We will again use the $A \times B \times C(AB)$ design, with C random, as our example. Because the subscript of $SS_{c(ab)}$ contains symbols for all of the factors in the design, the RS table for the $C(AB)$ main effect must be the table of original scores. Consequently, $RS_{c(ab)} = RS_w$. In addition $SS_{c(ab)}$ is found (following the method described previously) by subtracting out the sums of squares for all of the other effects in the design (including SS_m). $SS_{c(ab)}$ is therefore a residual sum of squares, consisting of the remainder when the sums of squares of all other effects have been subtracted from $RS_{c(ab)}$. By definition, however, SS_w is the sum of squares that remains after subtracting from RS_w the sums of squares for all systematic effects. Therefore, $SS_{c(ab)}$ in the $A \times B \times C(AB)$ design is equivalent to SS_w in the simple two-way design. Similarly, $MS_{c(ab)}$ is equivalent to MS_w, and both have the same number of degrees of freedom. The simple two-way design can be treated either as a two-way design with error or as an $A \times B \times C(AB)$ design with only one score per cell. The resulting mean squares are identical.

Identical expected mean squares are also obtained. We can see this by reviewing the rules given previously, but without adding a separate error term to the expected mean squares. For the simple two-way design with error treated as a random factor, the terms that can enter into the expected mean square are

$$N\mu^2, \; nJ\tau_a^2, \; nI\tau_b^2, \; n\tau_{ab}^2, \; \tau_{c(ab)}^2.$$

Here we regard n as the number of levels of Factor C. Because $\tau_{c(ab)}^2$ contains both of the other factors in its subscript, and the only symbol outside of the parentheses is for a random factor, $\tau_{c(ab)}^2$ enters into the expected mean squares of all of the other effects. The complete list of expected mean squares (calculated according to the rules given previously) is given in Table 8.17. If we substitute σ_e^2 for $\tau_{c(ab)}^2$, these equations are identical to those in Chapter 5.

These results can be generalized to any design. The error always can be treated as a random factor nested in all other factors. For the one-way design, the model can be written

$$X_{ij} = \mu + \alpha_i + \epsilon_{j(i)};$$

TABLE 8.17. Expected mean squares for two-way design regarded as an $A \times B \times C(AB)$ design with Factor C random.

	$E(MS)$
m	$\tau_{c(ab)}^2 + N\mu^2$
a	$\tau_{c(ab)}^2 + nJK\tau_a^2$
b	$\tau_{c(ab)}^2 + nIK\tau_b^2$
ab	$\tau_{c(ab)}^2 + nK\tau_{ab}^2$
$c(ab)$	$\tau_{c(ab)}^2$

for the two-way, completely crossed design, it can be written

$$X_{ijk} = \mu + \alpha_i + \beta_j + \alpha\beta_{ij} + \epsilon_{k(ij)},$$

and so on.

When the error is so treated, MS_w is the mean square and σ_e^2 the τ^2 for the main effect of the error random factor. (There cannot be any interactions involving the error factor because it is nested in all other factors.) Although this point is important primarily in computer programs for the analysis of variance, it should help to clarify the nature of the error terms in all of the designs we have discussed.

Exercises

(1.) *Life* Magazine once reported that the California climate was so healthy the women in California were frequently stronger than the average man in the East. A certain researcher decided to test whether California women were stronger than Michigan women. She chose three California towns and three Michigan towns at random, and she chose two female high-school students at random from each town. She then tested each of these twelve subjects for the number of push-ups they could do (Test 1) and the number of times they could chin themselves (Test 2). She obtained the following data:

	California						Michigan				
Town	A		B		C		D		E		F	
Test	1	2	1	2	1	2	1	2	1	2	1	2
Subjects	11	7	13	5	10	6	20	15	11	13	18	16
	12	9	12	3	10	3	19	13	13	14	12	16

(a.) Analyze the data and interpret the results.

(b.) Can you draw any conclusions about the main effect of subjects or the interaction between subjects and tests? What assumptions would such conclusions require?

(2.) Specify the correct design in Problem 6, Chapter 2, and analyze the data. Assume that Groups A_1 and A_3 were given the same form, and that Groups A_2 and A_4 were given the same form.

(3.) Suppose that in Problem 7, Chapter 2, each odd answer was the answer of the husband, and each even answer was that of the wife in the same family. Specify the correct design for this experiment and analyze the data.

(4.) Reanalyze the data in Problem 4 of Chapter 6. Assume that the lists are not comparable across list types (i.e., assume that there is no natural way to pair the three lists of one type with those of another).

(5.) Experimenters were puzzled by an apparent difference in average running times of rats under seemingly identical conditions in different laboratories. One investigator speculated that rats run in southern climates were healthier and thus faster. The investigator chose four laboratories at random from each of three ranges of latitude: 30° to 35° N, 35° to 40° N, and 40° to 45° N. Ten randomly selected rats were run in each laboratory. The running times for each rat in each laboratory were as follows:

Laboratory

Lat. 30–35°				Lat. 30–40°				Lat. 40–45°			
1	2	3	4	5	6	7	8	9	10	11	12
11	5	3	2	13	11	9	10	13	15	12	12
15	1	2	4	1	1	7	6	14	21	16	14
2	3	19	12	5	34	15	4	16	18	6	6
9	1	10	10	17	5	1	5	17	16	19	18
2	13	20	5	16	6	22	14	12	10	20	12
9	2	7	11	10	13	14	5	23	14	13	27
2	4	8	4	15	3	11	8	8	18	12	16
13	14	11	14	5	9	18	3	13	18	4	17
8	10	12	4	2	6	16	11	20	17	16	8
10	20	18	11	4	1	10	12	20	7	15	13

(a.) Analyze these data and interpret the results.

(b.) Compare the three latitudes using multiple comparisons.

(c.) If the only test of interest is the one comparing different latitudes, how might the data be combined to simplify the analysis?

(6.) A study was made of the effect of ward conditions in mental hospitals on patients with mild catatonia. The experimenter suspected that subtle differences among conditions in different wards of the same hospital could affect the patients. Eight patients were randomly selected in each of two hospitals; each patient was placed first in one ward and then in another, with each patient spending three months in each ward. (Each patient spent time in two different wards, with half the patients in one ward first and the other half in the other ward first.) The data (scores on an objective test of mental health) were as follows:

	Hospital 1		Hospital 2	
	Ward 1A	Ward 1B	Ward 2A	Ward 2B
Patient	3	2	3	3
in Ward	3	3	2	3
A first	1	2	0	1
	3	1	3	2
Patient	1	0	3	1
in Ward	2	2	1	2
B first	1	0	2	0
	0	0	1	1

Analyze these data and interpret the results; do not assume that wards in different hospitals are comparable (i.e., do not assume that the wards in one hospital can be paired with those in the other).

(7.) Twelve randomly selected male college students who had spent two years in the army were compared with twelve who had not. Six students from each group were in engineering school, the other six were in liberal arts. Each student was evaluated in two courses. In each course the student received a score of one if his grade was B or better, and a score of zero if it was not. For the engineering students, the courses were freshman math and freshman chemistry; for the liberal arts students the courses were freshman English and American history. The data were as follows:

		Army					
		S_1	S_2	S_3	S_4	S_5	S_6
L.A.	Engl.	0	0	0	0	0	0
	Hist.	0	0	1	1	0	0
Engin.	Math	1	1	1	1	1	0
	Chem.	0	0	0	0	0	0

No Army					
S_1	S_2	S_3	S_4	S_5	S_6
1	0	1	1	1	1
0	0	0	0	1	1
0	0	0	0	0	0
0	1	0	0	0	0

Analyze these data

(a.) Analyze these data and interpret the results.

(b.) Test the following null hypotheses by planned comparisons, using the modified method whenever no existing mean square can be used, and using both methods whenever an existing mean square can be used.

$H_0(1)$: There is no difference between army veterans and others in performance in English.

$H_0(2)$: There is no difference between army veterans and others in overall performance in engineering courses.

$H_0(3)$: There is no overall difference in performance between mathematics and chemistry.

(c.) How might you have extracted more information from the data if the engineering courses had been mathematics and English instead of mathematics and chemistry?

(8.) An experimenter studied the effects of early environment on the problem-solving ability of rats raised in "normal," "enriched," and "restricted" environments. Each rat, of course, was raised in only one environment. The experimenter measured the number of trials required to learn a visual discrimination task (Task A) and a tactile discrimination task (Task B). Half the rats were given Task A first and the other half were given Task B first. The data, in number of trials required to learn, are as follows:

		Task A first			Task B first	
	Subj.	A	B	Subj.	A	B
Enriched	E1	45	62	E5	50	68
	E2	20	39	E6	43	63
	E3	20	39	E7	66	84
	E4	33	55	E8	59	82
Normal	N1	28	44	N5	50	71
	N2	53	69	N6	43	59
	N3	36	53	N7	42	66
	N4	52	75	N8	50	62
Restricted	R1	69	86	R5	63	78
	R2	59	77	R6	48	60
	R3	81	107	R7	61	82
	R4	83	101	R8	37	63

(a.) Analyze the data and interpret the results.

(b.) Test, by a planned comparison, the null hypothesis that the difference between enriched and restricted groups is the same for Task A as for Task B.

(c.) Do multiple comparison tests on the three environments.

(9.) In an $A \times B \times C(A)$ design, with C random, how should each of the following planned comparisons be tested when (1) all assumptions are met, (2) assumptions of equality of variances and covariances are not met?

(a.) $\mu_{1..} = \mu_{2..}$

(b.) $\mu_{11.} = \mu_{12.}$

(c.) $\mu_{11.} = \mu_{21.}$

(When answering, consider each hypothesis by itself; i.e., do not worry about whether or not the tests are orthogonal.)

(10.) For each of the following designs (1) tell what type of design it is (e.g., $A \times B \times C(A)$); (2) write the model equation; (3) write the expected mean square for each effect; and (4) tell what denominator mean square would be used to test each effect.

(a.) Thirty randomly selected subjects were tested in an experiment on the effects of sunglasses on visual acuity. The tests were conducted using all possible combinations of two types of sunglasses and five different lighting conditions. In addition, each subject was tested under three different levels of initial light adaptation. A single visual acuity score was obtained from each subject under each level of adaptation. (The total N is 90.)

(b.) A memory experiment used three different lists of 10 words each. List N contained 10 nouns, List V contained 10 verbs, and List A contained 10 adjectives. The words in each list were selected independently of (and thus were not directly comparable to) the words in each of the other two lists (but the words were not selected randomly). In addition, the words in each list were presented in each of four different random orders, so that there were a grand total of 12 lists in all. Each subject received one of the three types of lists in one of the four random orders. After hearing the entire list, the subject was asked to recall as many of the words as he could. If he recalled a word correctly, he received a score of 1 for that word; if not, he received a score of 0. Each subject's scores, then, were a series of 10 ones and zeroes. A total of 144 randomly selected subjects were run: 12 under each of the 12 combinations of type of list and word order ($N = 1,440$).

(c.) The design is identical to that in part b except that the words within each list were selected randomly.

(d.) Six randomly selected first-grade teachers were each assigned 12 randomly selected pupils. Each teacher taught the pupils two subjects, reading and arithmetic. At the end of the year all six classes were given the same standard tests in these two subjects. The data were the test scores of the individual pupils in each of the two subjects.

(e.) A large corporation was interested in the efficiency of its workers. The corporation manufactured four different products, each in a different plant. Three different assembly jobs were selected in each plant, and ten workers were selected randomly in each job. The workers were each given efficiency ratings twice: once in the winter and again in the summer. The specific jobs were not randomly chosen, nor were the jobs in one plant comparable with those in another (because the plants did not produce similar products).

(f.) Three rather expensive cancer treatments were being compared. Forty hospitals participated in the experiment, with ten hospitals using each treatment, and ten using traditional treatments (as a control). In each hospital, twenty patients were treated, each having one of four types of cancer (i.e., there were five patients with each type of cancer). The health of each patient was rated on a ten-point scale after three days, seven days, and two weeks of treatment. The patients were randomly selected. The hospitals were not randomly selected, but they were randomly assigned to treatments; this is properly called a *permutation design*, and it is most easily handled by assuming that hospitals are random (Scheffé, 1956).

(g.) (This is a description of an actual experiment; the author was asked to advise on the data analysis.) A study of dichotic listening had subjects attempt to identify tunes played in one ear while white noise was played in the other. Each subject heard nine tunes, differing in tonal difficulty (three levels) and rhythmic difficulty (three levels). Each subject heard the complete set of nine tunes twice; once in the left ear and once in the right. Thus 18 measures of ease of recognition were obtained from each subject.

There were 40 normal subjects and 40 diagnosed as learning-delayed. Each group of 40 was divided by age, using a median split.

9

Other Incomplete Designs

If Factor B is nested in Factor A, it is impossible to separate the B main effect from the AB interaction. When effects cannot be separated, they are *confounded*. Whenever the factors in a design are not completely crossed, some effects will be confounded. However, in some designs the factors are neither completely crossed nor completed nested. The 4×6 design in Table 9.1 is an example; in this design, the factors are not completely crossed, yet neither is nested within the other. The analysis of most such *partially nested* designs is very difficult, but for certain designs it is relatively simple.

Advantages of Partially Nested Designs

Partially nested designs like the one shown in Table 9.1 offer a number of advantages, the most obvious being a savings in experimental labor. In a completely crossed 4×6 design there would be 24 cells, requiring at least 24 scores. In the design of Table 9.1, scores are obtained for only 12 of the 24 cells. For designs with three or more factors the savings may be even greater. Ordinarily, for example, a $4 \times 4 \times 4$ design requires at least 64 scores, but a $4 \times 4 \times 4$ *latin square* (discussed later in this chapter) requires only 16.

In addition, partially nested designs sometimes make it possible to improve precision by adding "control factors" to the design. For example, suppose we want to study the effects of four different drugs on the behavior of mice. We have a number of mice on which to experiment, but we are concerned that the heredities of the mice might affect their susceptibilities to the drugs. We are

TABLE 9.1. Partially nested (balanced incomplete blocks) 4×6 design. The Xs indicate cells containing scores.

	B_1	B_2	B_3	B_4	B_5	B_6
A_1	X		X		X	
A_2	X			X		X
A_3		X	X			X
A_4		X		X	X	

not especially interested in hereditary differences, but if we do not control for them, they might increase the error mean square and reduce our chances of obtaining significant results. One way to control heredity would be to separate the mice according to the litters they came from and choose four animals for each litter. The litter an animal came from could then be treated as a second factor. However, although we have six litters, some litters contain only two animals each. Fortunately, we may be able to solve the problem by using the design of Table 9.1, with Factor A being drugs and Factor B being litters; as can be seen in Table 9.1, this design requires no more than two animals from each litter.

Disadvantages of Partially Nested Designs

Balancing the advantages, however, there are a number of disadvantages; in fact, the disadvantages usually outweigh the advantages. To begin with, the number and variety of relatively simple partially nested designs available for a given research problem are usually limited. If, say, we had had only five litters, no simple design would have been available. In general, partially nested designs are feasible only when the numbers of levels of the factors meet certain rather rigid requirements.

Partially nested designs have a more serious problem; in most of them all interactions are confounded not only with each other but with main effects as well. Moreover, they are confounded in complicated ways. In a completely nested design, the B main effect and the AB interaction are confounded in a simple additive way: The $B(A)$ main effect is the sum of the B main effect and AB interaction. In a partially nested design the problem is much more complicated. In a latin square design, for example, the A main effect is confounded with part of the BC interaction, and the confounding is not simple; because only part of the BC interaction is confounded with the A main effect, a large BC interaction may either increase or decrease SS_a. Consequently, the results of most nested experiments are uninterpretable unless all interactions can be assumed to be negligible. In our example, we must assume that the mice do not inherit differential susceptibilities to specific drugs. If that assumption is not plausible, the design of Table 9.1 is a poor choice. Unless interactions are specifically discussed, all of the designs considered in this chapter require the assumption that there are no interactions.

Finally, partially nested designs are basically multifactor designs with unequal sample sizes. (Some sample sizes are zero.) As such, many of them share the problem, described in Chapter 5, that simple main effects are not orthogonal. This problem will be more clear when we discuss the analysis of incomplete blocks designs below.

As stated above, many kinds of partially nested designs are possible. Most of them have been developed to meet specific needs. It would not be fruitful to discuss them all in this text. Instead, we will discuss a small number of relatively simple designs that are in common use and are related to each

TABLE 9.2. The design of Table 9.1
diagrammed in the conventional form.

B_1	B_2	B_3	B_4	B_5	B_6
A_1	A_3	A_1	A_2	A_1	A_2
A_2	A_4	A_3	A_4	A_4	A_3

other. Then we will show how these designs can be modified and expanded to produce other, more complicated designs. Still other designs can be found in various books and journal articles.

Two-Factor Designs: Incomplete Blocks

An *incomplete blocks* design is a two-factor design in which each level of Factor A is paired with only a small number of levels of Factor B. In the drug experiment described above and in Table 9.1, each drug is administered to animals from only three of the six litters. Letting the drugs be the levels of Factor A and the litters be the levels of Factor B, we have an incomplete blocks design in which each level of Factor A is paired with only three of the six levels of Factor B.

Suppose there are I levels of Factor A, and each is paired with g levels of B. (In Table 9.1, $g = 3$.) Then the total number of cells containing data must be gI. We can reason about Factor B in the same way; each level of B is paired with the same number (h) of different levels of A. Therefore, the total number of cells containing scores must be hJ. Consequently, $gI = hJ$. In the design of Table 9.1, $I = 4, J = 6, g = 3, h = 2$, and $gI = hJ = 12$.

Incomplete blocks designs are frequently diagrammed in a table with J columns representing the J levels of B; each column contains the h levels of A that are paired with that level of B. This way of diagramming is illustrated in Table 9.2 for the design of Table 9.1.

The term *incomplete blocks* derives from the fact that the incomplete blocks design is often thought of as an incomplete version of the *randomized blocks design*. Basically, the randomized blocks design is one in which we are primarily interested in only one factor, but we perform the experiment as a crossed two-factor design in order to reduce MS_w. In the design of Tables 9.1 and 9.2 our primary interest is in the drugs, but if we had used a one-way design, differences among litters would have contributed to error and increased MS_w. In this case, the introduction of Factor B is equivalent to dividing the scores within the levels of Factor A into distinct groups, or *blocks*, and treating the blocks as levels of a second factor. The entities referred to as blocks in some statistics texts are thus identical to what we here call the *levels* of B.

TABLE 9.3. Example of a 7×7 balanced incomplete blocks design.

B_1	B_2	B_3	B_4	B_5	B_6	B_7
A_1	A_1	A_1	A_2	A_2	A_1	A_3
A_2	A_3	A_2	A_3	A_4	A_4	A_5
A_3	A_4	A_5	A_4	A_5	A_6	A_6
A_6	A_5	A_7	A_7	A_6	A_7	A_7

BALANCED INCOMPLETE BLOCKS

The incomplete blocks design can be regarded as a two-way analysis of variance with unequal sample sizes (some samples are of size zero). As such, it can be analyzed using the methods described in Chapter 5. However, the analysis is complicated, involving the solution of several simultaneous linear equations. For one special design, the *balanced incomplete blocks design,* the task is much simpler.

If an incomplete blocks design is represented as in Table 9.2, we can count the number of times that any two levels of Factor A, say A_i and $A_{i'}$, appear together in the same column (i.e., within the same level of B). We can let this number be $\lambda_{ii'}$. An incomplete blocks design is said to be *balanced* if $\lambda_{ii'} = \lambda$ (i.e., a constant) for all $i, i' \neq i$. That is, a balanced incomplete blocks design is one in which each level of A appears together with each other level of A within the same level of B, the same number of times. In the design in Table 9.2, levels A_1 and A_2 appear together under B_1 but under no other level of B; similarly, every other pair of levels of A_i occurs under one and only one level of B. The design of Tables 9.1 and 9.2 is thus a balanced incomplete blocks design with $\lambda = 1$.

Another example of a balanced incomplete blocks design is in Table 9.3. This table shows a 7×7 design with $g = h = 4$. In Table 9.3 each level of A appears with each other level in two different columns. For example, levels A_2 and A_6 appear together in columns B_1 and B_5 but in no other columns. This is a balanced incomplete blocks design with $\lambda = 2$.

The balanced incomplete blocks design places additional restrictions on the number of levels of each factor. With h levels of A paired with each level of B, level A_i will appear with $(h-1)$ other levels of A within each level of B with which it is paired. If A_i is paired with g different levels of B, then the number of times A_i appears with all other levels of A combined is $g(h-1)$. On the other hand, because each level of A is paired with each other level of A λ times, the number of times A_i is paired with all of the other $I-1$ levels of A combined must be $\lambda(I-1)$. Therefore, it must be true that $g(h-1) = \lambda(I-1)$. These requirements are highly restrictive because g, h, I, J, and λ must all be integers. However, an additional restriction can also be shown to hold: $I \leq J$.

A list of possible designs can be found in Table B.10 of Appendix B.
   The following list summarizes the definitions of symbols introduced so far:

$I =$ number of levels of Factor A,

$J =$ number of levels of Factor B,

$g =$ number of levels of B with which each A_i is paired,

$h =$ number of levels of A with which each B_j is paired, and

$\lambda =$ number of times any two levels of A are paired with the same level of B.

Randomization Condition. All of the theory to follow is heavily dependent on the particular pairings of AB_{ij} that are chosen. To avoid systematic biases it is essential that these pairings be chosen randomly. The pairings often can be made most readily by choosing a particular design and then randomly assigning the labels B_1, B_2, etc., to the levels of Factor B. An equally good alternative is to randomly assign the labels to the levels of Factor A.

Estimating Effects. The analyses to be given here assume that all cells containing scores have the same number of scores, n. Because we assume that there is no interaction, n may equal one.
   With balanced incomplete blocks designs, $t_{i.}$ and $t_{.j}$ are found by summing over the B and A factors, respectively, as in previous designs. As with nested designs, we sum only over those cells containing scores. This summation is illustrated in Table 9.4 for hypothetical data, for the design in Tables 9.1 and 9.2, with two scores per cell.
   If n_{ij} is the number of scores in cell AB_{ij}, then $n_{ij} = n$ if cell AB_{ij} contains scores, and $n_{ij} = 0$ if it does not. That is, $n_{ij} = n$ if and only if level A_i is paired with level B_j. To simplify later formulas, we will define Q_{ij} to be one if cell AB_{ij} contains data, and to be zero if cell AB_{ij} does not contain data. That is, $Q_{ij} = 1$ if and only if A_i is paired with B_j, and $n_{ij} = nQ_{ij}$. We also define t_{ij} as the sum of the scores in cell AB_{ij}, as usual. With these definitions, the formulas for $t_{i.}$, $t_{.j}$, and T can be written

$$\begin{aligned} t_{i.} &= \Sigma_j Q_{ij} t_{ij}, \\ t_{.j} &= \Sigma_i Q_{ij} t_{ij}, \\ T &= \Sigma_i \Sigma_j Q_{ij} t_{ij}. \end{aligned}$$

   The model for the balanced incomplete blocks design, if we assume there are no interactions, is

$$X_{ijk} = \mu + \alpha_i + \beta_j + \epsilon_{ijk}.$$

By simple algebra,

$$\begin{aligned} \underline{E}(T) &= E\Sigma_i \Sigma_j n_{ij} \overline{X}_{ij.} = \Sigma_i \Sigma_j n_{ij} \underline{E}(\mu + \alpha_i + \beta_j + \epsilon_{ijk}) \\ &= N\mu + \Sigma_i \alpha_i \Sigma_j n_{ij} + \Sigma_j \beta_j \Sigma_i n_{ij}. \end{aligned}$$

TABLE 9.4. Sample analysis of 4×6 balanced incomplete blocks design with $n = 2$.

	B_1	B_2	B_3	B_4	B_5	B_6	$t_{i.}$	t_i^*	$\hat{\alpha}_i$
A_1	19		22		25				
	17		16		21		120	104	4.0
A_2	20			24		35			
	20			18		29	146	124	5.5
A_3		6	18			23			
		8	12			17	84	100	-4.0
A_4		3		10	11				
		11		16	7		58	80	-5.5
$t_{.j}$	76	28	68	68	64	104			
$\overline{X}_{.j}$	19	7	17	17	16	26	$T = 408$		
$\hat{\beta}_j$	-2.75	-5.25	0.00	0.00	-0.25	8.25	$\overline{X}_{...} = 17$		

The last equality is found by distributing both the summations and the expected value over the terms in the parentheses, with N being the total number of scores in the table. Now, a reference to Table 9.1 makes it clear that

$$\Sigma_j n_{ij} = ng, \quad \Sigma_i n_{ij} = nh.$$

Consequently,

$$\underline{E}(T) = N\mu + ng\Sigma_i \alpha_i + nh\Sigma_j \beta_j = N\mu.$$

The last equality derives from the usual restriction that the α_i and β_j sum to zero. An unbiased estimate of μ is therefore

$$\hat{\mu} = \overline{X}_{...} = T/N.$$

To obtain unbiased estimates of the α_i, we must first define

$$t_i^* = \Sigma_j Q_{ij} t_{.j}/h.$$

It can then be shown, by a derivation too involved to present here, that the least-squares estimate of α_i is

$$\hat{\alpha}_i = (t_{i.} - t_i^*)/(ngH),$$

where

$$H = [I(h - 1)]/[h(I - 1)].$$

The error variance of each estimate $(\hat{\alpha}_i)$ is $\sigma_e^2/(ngH)$. If each level of A were replicated g times in a complete design, the error variance would be $\sigma_e^2/(ng)$, so the balanced incomplete blocks design results in an error variance $1/H$ times as large as an equivalent complete design. For this reason, H (which is always

TABLE 9.5. Calculating formulas for balanced incomplete blocks design with n scores per cell. (MS_{rem} is the denominator for every F ratio. See test for calculation of $\hat{\alpha}_i$.)

	RS	SS	df
m		T^2/N	1
a	$\Sigma_i t_{i.}^2/(ng)$	$RS_a - SS_m$	$I-1$
$a:b$		$ngH\Sigma_i\hat{\alpha}_i^2$	$I-1$
b	$\Sigma_j t_{.j}^2/(nh)$	$RS_b - SS_m$	$J-1$
$b:a$		$RS_b + SS_{a:b} - RS_a$	$J-1$
rem	$\Sigma_{ij}Q_{ij}\Sigma_k X_{ijk}^2$	$RS_{rem} - SS_{a:b} - RS_b =$	$N-I-J+1$
		$RS_{rem} - RS_a - SS_{b:a}$	
t		$RS_{rem} - SS_m$	$N-1$

smaller than one) is called the *efficiency factor* of the balanced incomplete blocks design. Put differently, $1 - H$ is the proportional loss in accuracy due to pairing only h of the I levels of A with each level of B. In general, the smaller the value of h, the less efficient the experiment. The efficiency factors are 0.67 for the design in Table 9.2 and 0.88 for the design in Table 9.3.

If we let

$$\overline{X}_{.j.} = t_{.j}/(nh),$$

the estimates of the β_j are

$$\hat{\beta} = \overline{X}_{.j.} - \overline{X}_{...} - (\Sigma_i Q_{ij}\hat{\alpha}_i)/h.$$

These calculations are illustrated in Table 9.4.

Significance Tests. The sum of squares for testing the grand mean is, as usual, $SS_m = T^2/N$, and it has one degree of freedom.

As in the case of unequal sample sizes discussed in Chapter 5, simple sums of squares for the A and B main effects are not independent. Therefore, we can calculate four different sums of squares for main effects, namely, SS_a (A ignoring B), $SS_{a:b}$ (A accounting for B), SS_b (B ignoring A), and $SS_{b:a}$ (B accounting for A). In many applications of the balanced incomplete blocks design, the levels of B are blocks which are included to remove their effects from the analysis rather than because they are interesting in themselves. In that case, it is natural to calculate SS_b and $SS_{a:b}$, taking B into account when testing A. However, there is no mathematical reason for such a choice.

Calculating formulas for sums of squares and degrees of freedom are in Table 9.5 Each mean square is the associated sum of squares divided by its degrees of freedom, as usual. MS_{rem} is the denominator term for all F ratios. Table 9.6 shows the summary for the data of Table 9.4.

TABLE 9.6. Summary table for data of Table 9.4.

	RS	SS	df	MS	RF	p
m		6,936.00	1			
a	7,689.33	753.33	3	251.1	17.6	<.001
$a:b$		370.00	3	123.3	8.64	.002
b	7,680.00	744.00	5	148.8	10.4	<.001
$b:a$		360.67	5	72.13	5.06	.007
rem	8,264.00	214.00	15	14.27		
t		1,328.00	23			

General planned comparisons can be difficult to make, but planned comparisons on the A main effect, with the effects of B taken into account, are simple. The quantity

$$SS_k/MS_{rem} = ngH(\Sigma_i c_{ik}\hat{\alpha}_i)^2/[MS_{rem}(\Sigma_i c_{ik}^2)]$$

is distributed as $F_{(1,N-I-J+1)}$. To make post hoc comparisons by the Scheffé method, this F ratio is divided by $(I-1)$ and treated as $F_{(I-1,N-I-J+1)}$. Tests on the B main effect, ignoring A, are also simple. They are made directly on the marginal means. Multiple pairwise comparisons can be made similarly, using F tests. Tests based on the Studentized range distribution cannot be made.

All of the above theory applies whether Factors A and B are fixed or random. However, occasionally it is possible to improve the estimates of the α_i, and thus to improve the test for the A main effect, when Factor B is random. The method for doing this was developed by Yates (1940); it is also presented in Scheffé (1959, p. 170). The method is useful only if the number of levels of B is fairly large (about 15 or more) and the B main effect is not very large.

Three-Factor Designs

The balanced incomplete blocks design discussed in the previous section can be extended in a number of ways to designs with three factors. Incomplete three-factor designs are more commonly used than incomplete two-factor designs because many completely crossed three-factor designs require a prohibitive number of subjects. However, the requirement that there be no interactions is much more restrictive in three-factor than in two-factor designs. In most three-factor designs all three of the two-factor interactions as well as the three-factor interaction are assumed to be zero. Thus, the model for these designs is usually

$$X_{ijkl} = \mu + \alpha_i + \beta_j + \gamma_k + \epsilon_{ijkl}.$$

TABLE 9.7. Example of 4×4
Youden square with $g = h = 3$,
$\lambda = 2$.

	B_1	B_2	B_3	B_4
C_1	A_1	A_2	A_3	A_4
C_2	A_2	A_3	A_4	A_1
C_3	A_3	A_4	A_1	A_2

YOUDEN SQUARES

When a balanced incomplete blocks design is represented as in Table 9.2, it may be possible to balance the positions of the levels of A so that each appears the same number of times in each row of the table. It is impossible to do this with the designs in Tables 9.2 or 9.3, but it has been done with the 4×4 design shown in Table 9.7.

In this design each level of A appears once in each row. In general, if each level of A appears q times in each row, the number of columns (levels of B) must be q times the number of levels of A; i.e., $J = qI$. In Table 9.7 $q = 1$, but other examples exist for which q is larger than one.

Any balanced incomplete blocks design for which $J = qI$, with q an integer, can be arranged so that each level of A appears q times in each row of the table. Such a design is called a *Youden square*. In Table A.10, the designs that can be extended to Youden squares are marked with asterisks.

One common reason for arranging a balanced incomplete blocks design in such a way is to balance out the effects of the position of a treatment in a sequence. Suppose the experiment involves tasting four different sucrose solutions, with a number of subjects each tasting three of the four. In such an experiment, the order in which the solutions are tasted is likely to be important. With Factor A being sucrose solutions and Factor B being subjects, the three levels of Factor A can be given to each subject in the order in which they appear in that subject's column in Table 9.7. The order in which the levels of A are presented will then be balanced out across subjects, so that estimates and sums of squares for Factors A and B can be calculated as for the usual balanced incomplete blocks design. In addition, the position effects can be estimated and tested for.

To test for position effects, we regard the rows of Table 9.7 as the levels of a third factor, C. Then we define $t_{..k}$ as the sum of the scores in row k,

$$t_{..k} = \Sigma_i \Sigma_j Q_{ijk} t_{ijk},$$

where Q_{ijk} is equal to one if treatment AB_{ij} appears in row C_k and is equal to zero if it does not. (In other words, $Q_{ijk} = 1$ if and only if cell ABC_{ijk} contains scores.) In summing across a row of the table, each level of B (each column) will appear once in the summation, and because A is balanced with respect to C, each level of A will appear exactly q times. The expected value of $t_{..k}$ is thus easily shown to be

$$
\begin{aligned}
\underline{E}(t_{..k}) &= n(J\mu + q\Sigma_i\alpha_i + \Sigma_j\beta_j + J\gamma_k) \\
&= nJ(\mu + \gamma_k),
\end{aligned}
$$

because the α_i and β_j both sum to zero. The estimate of γ_k is then

$$
\begin{aligned}
\hat{\gamma}_k &= t_{..k}/(nJ) - \overline{X}_{...} \\
&= \overline{X}_{..k.} - \overline{X}_{....},
\end{aligned}
$$

and

$$
SS_c = nJ\Sigma_k\hat{\gamma}_k^2,
$$

which can be shown to be

$$
SS_c = \Sigma_k t_{..k}^2/(nJ) = T^2/N = RS_c - SS_m,
$$

where $RS_c = \Sigma_k t_{..k}^2/(nJ)$, and $SS_m = T^2/N$ as usual. SS_c has $(K-1) = (h-1)$ degrees of freedom because the number of levels of C must be equal to h, the number of rows in the table.

The values of SS_m, SS_a, SS_b, $SS_{a:b}$, and $SS_{b:a}$ are found just as in the ordinary balanced incomplete blocks design because the effects of C are balanced out across the levels of both A and B. SS_{rem} is the SS_{rem} of the balanced incomplete blocks design (see Tables 9.5 and 9.7) with SS_c subtracted out:

$$
\begin{aligned}
SS_{rem} &= RS_{rem} - SS_{a:b} - RS_b - SS_c \\
&= RS_{rem} - RS_a - SS_{b:a} - SS_c,
\end{aligned}
$$

where RS_{rem} is the sum of all of the squared scores:

$$
RS_{rem} = \Sigma_i\Sigma_j\Sigma_k Q_{ijk}\Sigma_l X_{ijkl}^2.
$$

The degrees of freedom of SS_{rem} are those of Table 9.5 minus the degrees of freedom of SS_c, or $(N - I - J - K + 2)$. Table 9.8 summarizes the calculations for the Youden square design, and Table 9.9 gives a numerical example for the 4×4 design of Table 9.7.

The levels of Factor C do not have to be the positions in a sequence of presentations. They may be the levels of any third factor. Once again, however, it is important that levels be paired in a random manner. In addition to pairing the levels of A randomly with the levels of B, as in the balanced incomplete blocks design, it is also important to assign the levels of C randomly to the rows of the table.

TABLE 9.8. Calculating formulas for the Youden square design. (See text for calculation of $\hat{\alpha}_i$.)

	RS	SS	df
m		T^2/N	1
a	$(\Sigma_i t_{i..}^2)/(ng)$	$RS_a - SS_m$	$I - 1$
$a:b$		$ngH\Sigma_i\hat{\alpha}_i^2$	$I - 1$
b	$(\Sigma_j t_{.j.}^2)/(nh)$	$RS_b - SS_m$	$J - 1$
$b:a$		$RS_b + SS_a - RS_a$	$J - 1$
c	$(\Sigma_k t_{..k}^2)/(nJ)$	$RS_c - SS_m$	$K - 1$
rem	$\Sigma_{ijk}Q_{ijk}\Sigma_1 X_{ijkl}^2$	$RS_{rem} - SS_{a:b} - RS_b - SS_c$	
		$= RS_{rem} - RS_a - SS_{b:a} - SS_c$	$N - I - J - K + 2$
t		$RS_{rem} - SS_m$	$N - 1$

LATIN SQUARES

The incomplete blocks design can be extended to a complete blocks design by making t equal to I so every level of Factor A is paired with every level of Factor B. The result is then simply a completely crossed two-factor design. When such a design is represented in the same way as the incomplete blocks designs in Tables 9.2, 9.3, and 9.7, it may be possible to add a third factor by balancing the positions of the levels of A so each appears the same number of times in each row of the table. This is possible if and only if $J = qI$ and $K = I$. The special case in which $q = 1$, i.e., $I = J = K$, is called a *latin square design*, but the general method of analysis for such a design applies to any positive integral value of q.

The analysis of a latin square design, or a generalized latin square with $q > 1$, can be undertaken in exactly the same manner as for a Youden square. However, because every level of A is paired with every level of B, the calculations are much simpler. For latin square designs,

$$\hat{\alpha}_i = \overline{X}_{i...} - \overline{X}_{....},$$
$$\hat{\beta}_j = \overline{X}_{.j..} - \overline{X}_{....},$$
$$\hat{\gamma}_k = \overline{X}_{..k.} - \overline{X}_{....}.$$

Calculations of sums of squares are similarly simplified. SS_a, SS_b, and SS_c are all orthogonal, so there is no need to calculate quantities such as $S_{a:b}$. The calculational formulas are given in Table 9.10; a numerical example of a latin square is given in Table 9.11, and a numerical example of a generalized latin square is given in Table 9.12.

TABLE 9.9. Numerical example of 4×4 Youden square design with $g = h = 3, \lambda = 2, n = 1$, and $H = 8/9$.

	B_1	B_2	B_3	B_4	$t_{..k}$	$\overline{X}_{..k}$	$\hat{\gamma}_k$
C_1	A_1 15	A_2 7	A_3 14	A_4 8	44	11.0	-1.0
C_2	A_2 17	A_3 6	A_4 8	A_1 7	38	9.5	-2.5
C_3	A_3 22	A_4 14	A_1 11	A_2 15	62	15.5	3.5
$t_{.j.}$	54	27	33	30			
$\overline{X}_{.j.}$	18	9	11	10	$T = 144$		
$\hat{\beta}_j$	6.00	-3.75	-0.75	-1.50	$\overline{X}_{...} = 12$		

	A_1	A_2	A_3	A_4
$t_{i..}$	33	39	42	30
t_i^*	39	37	38	30
$\hat{\alpha}_i$	-2.25	0.75	1.50	0.00

	RS	SS	df	MS	F	p
m		1,728	1			
a	1,758	30	3	10	1.43	.39
$a:b$		21	3	7	1.00	.50
b	1,878	150	3	50	7.14	.070
$b:a$		141	3	47	6.71	.077
c	1,806	78	2	39	5.57	.098
rem	1,998	21	3	7		
t		270	11			

TABLE 9.10. Calculating formulas for the generalized latin square design $(I = K, J = qI)$.

	RS	SS	df
m		T^2/N	1
a	$(\Sigma_i t_{i..}^2)/(nI)$	$RS_a - SS_m$	$I - 1$
b	$(\Sigma_j t_{.j.}^2)/(nJ)$	$RS_b - SS_m$	$J - 1$
c	$(\Sigma_k t_{..k}^2)/(nK)$	$RS_c - SS_m$	$K - 1$
rem	$\Sigma_{ijk} Q_{ijk} \Sigma_l X_{ijkl}^2$	$RS_{rem} - SS_m - SS_a$ $- SS_b - SS_c$	$N - I - J - K + 2$
t		$RS_{rem} - SS_m$	$N - 1$

TABLE 9.11. Numerical example of 4×4 latin square design.

	B_1	B_2	B_3	B_4	$t_{..k}$	$\overline{X}_{..k}$	$\hat{\gamma}_k$
	A_1	A_2	A_3	A_4			
C_1	10	8	5	4	27	6.75	-3.25
	A_2	A_4	A_1	A_3			
C_2	11	13	16	12	52	13.00	3.00
	A_3	A_1	A_4	A_2			
C_3	10	14	9	10	43	10.75	0.75
	A_4	A_3	A_2	A_1			
C_4	8	6	11	13	38	9.50	-0.50
$t_{.j.}$	39	41	41	39			
$\overline{X}_{.j.}$	9.75	10.25	10.25	9.75	$T = 160$		
$\hat{\beta}_j$	-0.25	0.25	0.25	-0.25	$\overline{X}_{...} = 10$		

	A_1	A_2	A_3	A_4
$t_{i..}$	53	40	33	34
$\overline{X}_{i..}$	13.25	10.00	8.25	8.50
$\hat{\alpha}_i$	3.25	0.00	-1.75	-1.50

	RS	SS	df	MS	F	p
m		1,600.0	1			
a	1,663.5	63.5	3	21.17	7.94	.017
b	1,601.0	1.0	3	0.333	0.12	—
c	1,681.5	81.5	3	27.17	10.2	.010
rem	1,762.0	16.0	6	2.667		
t		162.0	15			

Once again, it is important that levels of different factors be paired randomly. Cochran and Cox (1957, pp. 145–147) give a number of plans for latin squares in "standard" form. A latin square is put into standard form by permuting both the rows and columns so that the levels of A appear in numerical order in both the first row and the first column. The latin square in Table 9.11, for example, is in standard form. The requirement of random pairings is met by first randomly selecting a design in standard form and then randomly permuting first the rows and then the columns of the matrix.

REPEATED BALANCED INCOMPLETE BLOCKS

The latin square and the Youden square are the most commonly used special three-way designs. However, other variations on the balanced incomplete

TABLE 9.12. Numerical example of 3×6 generalized latin square design.

	B_1	B_2	B_3	B_4	B_5	B_6	$t_{..k}$	$\overline{X}_{..k}$	$\hat{\gamma}_k$
	A_1	A_3	A_3	A_1	A_2	A_2			
C_1	-1	-2	1	-1	-3	-8	-13	-1.08	-1.50
	4	2	1	1	1	-8			
	—	—	—	—	—	—			
	3	0	2	0	-2	-16			
	A_3	A_2	A_2	A_3	A_1	A_1			
C_2	-1	-2	2	2	0	-2	4	0.33	-0.08
	2	-2	1	4	1	-1			
	—	—	—	—	—	—			
	1	-4	3	6	1	-3			
	A_2	A_1	A_1	A_2	A_3	A_3			
C_3	1	6	-3	-4	6	-1	24	2.00	1.58
	4	4	5	-1	5	2			
	—	—	—	—	—	—			
	5	10	2	-5	11	1			
$T_{.j.}$	9	6	7	1	10	-18			
$\overline{X}_{.j..}$	1.50	1.00	1.17	0.17	1.67	-3.00	$T = 15$		
$\hat{\beta}_j$	1.08	0.58	0.75	-0.25	1.25	-3.42	$\overline{X}_{...} = 42$		

	A_1	A_2	A_3
$t_{i...}$	13	-19	21
$\overline{X}_{i...}$	1.08	-1.58	1.75
$\hat{\alpha}_i$	0.67	-2.00	1.33

	RS	SS	df	MS	F	p
m		6.25	1			
a	80.92	74.67	2	37.33	5.82	.009
b	98.50	92.25	5	18.45	2.88	.034
c	63.42	57.17	2	28.58	4.46	.022
rem	397.00	166.67	26	6.410		
t		390.75	36			

blocks design are also possible. The design to be described here is not commonly used, but it can be useful in special situations. It is a relatively simple, logical extension of the balanced incomplete blocks design; it is an example of how any incomplete design can be extended to include additional factors.

Suppose, in the example of drugs and mice given at the beginning of the chapter, we had had three different strains of mice, with six litters from each strain, and at least two animals per litter. We could then have done a separate balanced incomplete blocks experiment, in the manner of Tables 9.1 and 9.2, on each strain of mice. By doing a separate statistical analysis on each strain, however, we lose information on differences between strains, as well as on overall A and B main effects, averaged over strains. If we can reasonably assume that there are no interactions between drugs, litters, or strains, we can treat strains as a third factor, C, and do a three-way analysis. This will enable us to recover the information that we would lose if we did three separate analyses.

The three-way analysis is simple if the balanced incomplete blocks design is understood. Table 9.13 gives a worked example of an experiment in which the design of Table 9.4 is repeated twice.

Because we assume that the same balanced incomplete blocks design has been repeated K times over the K levels of Factor C, the estimates of α_i and β_j are averages of the estimates from the separate balanced incomplete blocks designs:

$$\hat{\alpha}_i = (1/K)\Sigma_k\hat{\alpha}_{i(k)},$$
$$\hat{\beta}_j = (1/K)\Sigma_k\hat{\beta}_{j(k)},$$

where $\hat{\alpha}_{i(k)}$ and $\hat{\beta}_{j(k)}$ are the estimates obtained from the balanced incomplete blocks design under the kth level of Factor C. Then

$$RS_a = \Sigma_i t_{i..}^2/(ngK),$$
$$RS_b = \Sigma_j t_{.j.}^2/(nhK).$$

(In the first equation $t_{i..}$ is a sum over all cells in the entire design that are paired with level A_i; in the second equation $t_{.j.}$ is similarly defined with respect to B_j.) Then

$$SS_m = T^2/N,$$
$$SS_a = RS_a - SS_m,$$
$$SS_b = RS_b - SS_m,$$
$$SS_{a:b} = ngHK\Sigma_i\hat{\alpha}_i^2,$$
$$SS_{b:a} = RS_b + SS_a - RS_a.$$

The C main effect is found much as it would be in a three-way design. Each $\hat{\gamma}_k$ is found by averaging the values in all cells paired with level C_k and subtracting $\hat{\mu} = \overline{X}_{....}$ from the average:

$$\hat{\gamma}_k = t_{..k}/(nhJ) - \overline{X}_{....}.$$

TABLE 9.13. Repeated 4 × 6 balanced incomplete blocks design with $k = 2$, $n = 2$.

	B_1	B_2	B_3	B_4	B_5	B_6	$t_{i.(1)}$	$t^*_{i(1)}$	$\hat{\alpha}_{i(1)}$
					C_1				
A_1	19 17		22 16		25 21		120	104	4.0
A_2	20 20			24 18		35 29	146	124	5.5
A_3		6 8	18 12			23 17	84	100	-4.0
A_4		3 11		10 16	11 7		58	80	-5.5
$t_{.j(1)}$	76	28	68	68	64	104	$t_{..1} = 408$		
$X_{.j(1)}$	19	7	17	17	16	26	$\overline{X}_{..1} = 17$		
$\hat{\beta}_{j(1)}$	-2.75	-5.25	0.00	0.00	-0.25	8.25	$\hat{\gamma}_1 = 0.25$		

	B_1	B_2	B_3	B_4	B_5	B_6	$t_{i.(2)}$	$t^*_{i(2)}$	$\hat{\alpha}_{i(2)}$
					C_2				
A_1	22 16		26 13		29 20		126	103	6.0
A_2	21 13			23 13		36 29	135	117	4.50
A_3		7 8	16 13			20 17	81	100	-4.75
A_4		4 11		13 11	8 7		54	77	-5.75
$t_{.j(2)}$	72	30	68	60	64	102	$t_{..2} = 396$		
$X_{.j(2)}$	18.0	7.5	17.0	15.0	16.0	25.5	$\overline{X}_{..2} = 16.5$		
$t_{.j.}$	148	58	136	128	128	206	$T = 804$		
$\hat{\beta}_{j(2)}$	-3.75	-3.75	-.125	-.875	-.625	9.125	$\overline{X}_{...} = 16.75$		
$\hat{\beta}_{j}$	-3.25	-4.50	-0.06	-0.44	-0.44	8.69	$\hat{\gamma}_2 = -0.25$		

	A_1	A_2	A_3	A_4
$t_{i..}$	246	281	165	112
$\hat{\alpha}_i$	5.000	5.000	-4.375	-5.625

	RS	SS	df	MS	F	p
m		13,467.00	1			
a	14,937.17	1,470.17	3	490.1	30.0	<.001
$a:b$		806.25	3	268.8	16.4	<.001
b	14,871.00	1,404.00	5	280.8	17.2	<.001
$b:a$		740.08	5	148.0	9.05	<.001
c	13,470.00	3.00	1	3.000	0.18	—
rem	16,302.00	621.75	38	16.36		
t		2,835.00				

SS_c, with $(K-1)$ degrees of freedom, is found by

$$
\begin{aligned}
SS_c &= nhJ\Sigma_k\hat{\gamma}_k^2 = (1/nhJ)\Sigma_k t_{..k}^2 - T^2/N \\
&= RS_c - SS_m.
\end{aligned}
$$

The formula for SS_{rem} is

$$
\begin{aligned}
SS_{rem} &= RS_{rem} - RS_a - SS_{b:a} - SS_c \\
&= RS_{rem} - RS_b - SS_{a:b} - SS_c.
\end{aligned}
$$

The degrees of freedom for SS_{rem} are $(N - I - J - K + 2)$, as can be seen by subtracting the degrees of freedom for the various effects from N.

Note: It is important that exactly the same design be repeated under each level of C; if levels A_i and B_j are paired (i.e., if cell AB_{ij} contains data) under one level of C, then they must be paired under all levels of C. The pairings must be determined randomly, as in the simple balanced incomplete blocks design, but once determined, the same pairings must be repeated under all levels of C.

Higher-Way Designs

The approach used in the repeated balanced incomplete blocks design can be applied to any of the designs discussed in this chapter. Either the Youden square or the latin square can be expanded to a four-way design by repeating the basic three-way design over the levels of a fourth factor. For the latin square, the analysis of such a repeated design is especially simple. To find the sum of squares for, say, the A main effect, find RS_a by first finding the total of the scores $(t_{i...})$ for all cells under each level of A_i. Then square each sum, $t_{i...}$, add the squares, and divide the final total by the number of scores over which you summed to get each $t_{i...}$. In other words, RS_a is found by basically the same formulas as those in Table 9.10, except that the sum is taken over the levels of Factor D as well as of B and C. Because of the extra summation, the divisor is changed from (nJ) to (nJL), where L is the number of levels of the fourth factor. You then find SS_a, as usual, by $SS_a = RS_a - SS_m = RS_a - T^2/N$. The other sums of squares, SS_b, SS_c, and SS_d, are found the same way.

In the repeated Youden square design, Factors A and B are treated just as in the repeated balanced incomplete blocks design. Factors C and D are treated in the manner described in the preceding paragraph.

The balanced incomplete blocks design can also be extended to a four-factor design. Suppose, for example, we want to add two factors, C with three levels and D with two levels, to a basic $A \times B$ balanced incomplete blocks design. The three levels of C and the two levels of D combine to form a six-celled table; thus, six repetitions of the balanced incomplete blocks design (one for each combination of levels of C and D) are needed. To analyze the A and B main effects in such a design, we treat the six combinations of levels of Factors

C and D as though they were six levels of a single factor, and we follow the procedure outlined previously. The C and D main effects are analyzed by finding the RS for each and then subtracting SS_m as described above.

When the design is repeated over two factors, calculations for the C and D main effects can be simplified by first listing the total of all scores for each CD_{kl} combination in the cells of a C by D table (as though they were the cell totals in an ordinary two-way analysis of variance). Sums of squares for the C and D main effects are then found just as they would be for such a two-way table (as described in Chapter 6) if there were no A or B factors. In addition, a sum of squares for the CD interaction can be calculated, in the usual way, from the $C \times D$ table. This, then, is an exception to the usual rule that no sums of squares for interactions can be calculated in specialized incomplete designs.

The same approach can, of course, be used to extend the balanced incomplete blocks design to five or more factors, by repeating it over three or more additional factors. In addition, it can be used to extend the Youden square and latin square to designs of five or more factors. In every case, the same denominator mean square is used for testing all of the effects. The calculations are a straightforward extension of those described above.

Other Designs

The designs described in this chapter are only a few of the possible specialized designs available. To describe all of the available designs would require a rather large volume in itself. In fact, the larger portion of one entire textbook (Cochran and Cox, 1957) is devoted to such specialized designs, and many new designs have been devised since that text was published. Most of these designs are so specialized and require such strict assumptions (regarding interactions, etc.) that they are very rarely used. The reader who is interested in such designs or who has very specialized needs should refer to the relevant literature.

Exercises

(1.) Surgical lesions were made in ten areas of the brain (the ten levels of Factor A) in 30 cats. The cats were then tested for visual (B_1), auditory (B_2), tactile (B_3), gustatory (B_4), and olfactory (B_5) discrimination. The data are total correct discriminations in 20 learning trials. Each value is the score of a single cat.

A_1	A_2	A_3	A_4	A_5	A_6	A_7	A_8	A_9	A_{10}
B_1	B_1	B_1	B_2	B_1	B_1	B_1	B_2	B_2	B_3
9	7	5	8	1	10	6	2	13	2
B_2	B_2	B_2	B_3	B_3	B_3	B_4	B_3	B_4	B_4
10	2	4	10	6	1	4	8	4	2
B_3	B_4	B_5	B_4	B_5	B_4	B_5	B_5	B_5	B_5
8	2	1	11	0	2	0	0	10	0

(a.) What is λ?

(b.) What is the relative efficiency of this design?

(c.) Analyze and interpret the data.

(d.) What assumptions are needed for this analysis? Comment on the probable validity of the assumptions for this particular study.

(2.) In a study on judgments of TV sets, each of four different screen sizes of seven different brands of sets was tested. There were seven pairs of judges in all, but an incomplete design was used to save time. In the following table Factor A is the four screen sizes, Factor B is the seven brands, and Factor C is the seven pairs of judges (subjects). The scores are ratings of picture quality on a ten-point scale.

	C_1	C_2	C_3	C_4	C_5	C_6	C_7
	B_7	B_2	B_6	B_5	B_4	B_1	B_3
A_1	3	5	4	3	3	5	5
	6	1	4	1	3	2	5
	B_1	B_3	B_7	B_6	B_5	B_2	B_4
A_2	5	2	5	3	3	3	5
	6	1	3	4	4	5	4
	B_5	B_7	B_4	B_3	B_2	B_6	B_1
A_3	5	5	5	3	2	4	4
	6	3	5	2	3	6	6
	B_4	B_4	B_1	B_7	B_6	B_3	B_5
A_4	4	3	5	5	2	5	7
	5	4	7	5	5	7	6

(a.) Analyze these data, assuming that the two scores from each pair of subjects are independent.

(b.) Interpret these results. What can be said, on the basis of your analysis, about differences between sets and screen sizes?

(3.) Reanalyze and reinterpret the data in Problem 2. Assume the pairs of subjects are husbands and wives, the upper score being that of the wife and the lower score that of the husband. Treat the husband–wife dichotomy as the two levels of a fourth factor, D. How do the results of this analysis differ from those in Problem 2, and why?

(4.) Reanalyze the data in Problem 2, assuming that the two scores in each cell are independent. However, in this analysis ignore Factor A, treating the design as a 7×7 balanced incomplete blocks design. What is the value of λ for this design? What is the relative efficiency? How do the results of this analysis differ from those in Problem 2, and why?

(5.) Analyze the data in Problem 2 once more. Ignore Factor A, as in Problem 4, but treat the two scores within each cell as coming from husband–wife pairs, as in Problem 3.

(6.) The first latin square below is a study of the effects of different kinds of music on college students' studies. Factor A represents five different textbooks that the students were asked to study, Factor B represents five different kinds of music played during study, and Factor C is the five students. The data are scores on standardized tests. The second latin square is a replication of the first study using five different students.

	B_1	B_2	B_3	B_4	B_5		B_1	B_2	B_3	B_4	B_5
C_1	A_3 83	A_1 77	A_4 80	A_5 83	A_2 85	C_1	A_2 75	A_4 77	A_3 78	A_1 74	A_5 75
C_2	A_1 80	A_5 85	A_2 85	A_4 81	A_3 79	C_2	A_3 90	A_2 91	A_1 82	A_5 82	A_4 79
C_3	A_2 82	A_3 97	A_5 76	A_1 84	A_4 76	C_3	A_1 72	A_5 80	A_4 75	A_3 86	A_2 79
C_4	A_4 81	A_2 93	A_1 81	A_3 91	A_5 83	C_4	A_4 81	A_3 93	A_5 76	A_2 88	A_1 73
C_5	A_5 74	A_4 87	A_3 89	A_2 88	A_1 72	C_5	A_5 77	A_1 80	A_2 79	A_4 79	A_3 82

(a.) Analyze each set of data separately.

(b.) Analyze the data as a single experiment, treating the two replications as levels of a fourth factor. Assume the subjects are paired, i.e., C_1 in the first experiment is paired with C_1 in the second, etc.

(c.) Combine the two sets of data into a single generalized latin square with ten levels of C, and analyze the result.

(d.) These data, while artificial, were chosen so that the second set of data do in fact represent a replication of the first set. Comment, accordingly, on (1) the value of replicating small studies, and (2) the value of larger rather than smaller studies when larger studies are feasible.

(7.) You are interested in problem-solving activities of groups. Your method consists of bringing a number of children of about the same age together in a room, giving them a problem, and measuring the length of time required to solve it. You want to use four different group sizes (3, 5, 7, and 9 subjects), four different age levels (grades 1, 2, 3, and 4, each grade representing an age level), and three different problems. Because learning effects would produce confounding, you cannot give more than one problem to any group. A complete $4 \times 4 \times 3$ design would require 48 different groups, and a total of 288 subjects, or 72 at each grade level. The school where you are doing the study can provide only 60 subjects at each grade level, or 240 subjects in all. If you divide these equally among the four group sizes, you can get only 40 groups, or ten at each level.

(a.) Briefly describe two basically different designs that you might use to perform the study using all four group sizes, all four grades, and all three problems, without using more than 40 groups in all. (Note: you may use fewer than 40 groups.)

(b.) List the assumptions you must make to use each of the designs you proposed and discuss their possible importance for this particular problem.

(c.) Tell which of the two designs you would prefer, and why.

(8.) You want to study differences in performance of students on four supposedly equivalent tests. You have been given the use of four different sixth-grade classes for about 50 minutes each. You want to give each student each test but, unfortunately, each test takes 15 minutes so that each student can be given only three tests. What kind of design could you use? What assumptions would you have to make to be able to use the design you have chosen?

(9.) You want to test five different drugs for their effects on visual acuity, using subjects with a certain rare eye disease. You can find only 12 subjects who have this disease. You can test each subject under all five drugs (one at a time, of course), but while you expect no interactions among the effects of the drugs, you do expect a learning effect. Consequently, the sequence in which the drugs are given is important. What experimental design could you use?

10

One-Way Designs with Quantitative Factors

In some designs, meaningful numerical values can be assigned to the factor levels. An example of this might be a study of extinction (i.e., "unlearning") of a learned response after 10, 20, 30, 40, 50, and 60 learning trials. The six numbers of learning trials are the six levels of the factor being studied; the data are the numbers of trials to extinction. The labels on the factor levels in this experiment are meaningful numerical values: Thirty trials are 20 more than 10, 60 are 40 more than 20, and so on. If the cell means from such an experiment were plotted in a graph, they might look like those in Figure 10.1 (taken from the data in Table 10.1); in this graph the numerical values of the factor levels dictate both their order and their spacing along the X axis. By contrast, for the data plotted in Figure 3.1, both the ordering and the spacing of the factor levels were arbitrary.

Other examples of experiments with numerically valued factors would be studies of the effects of a person's age or the amount of some drug on the performance of a task. The *data* in both cases would be the values of some index of performance on that task. The *factor levels* in the former example would be the ages of the people being tested, e.g., people might be tested at ages 20, 30, 40, and 50 years, and in the latter example would be the amounts of the drug they had been given. In either experiment the levels of the factor would be readily characterized by specific numerical values.

Trend Analysis on One-Way Fixed-Effects Model

In any such design we can, of course, ignore the numerical values of the factors and do a conventional one-way analysis of variance, perhaps testing for some specific differences with planned comparisons. However, when the data from such an experiment are graphed as in Figure 10.1, they suggest that the cell means might be related to the numerical values of the factors by some specific continuous function. Trend analysis is one way to obtain evidence about the function relating numerical factor levels to cell means.

THE MODEL EQUATION

Trend analysis uses a polynomial function to describe the relationship between the cell means and the numerical values of the factor levels. If we let V_i

TABLE 10.1. Hypothetical extinction data illustrating trend analysis.

	Number of Learning Trials						
	A_1	A_2	A_3	A_4	A_5	A_6	
V_i	10	20	30	40	50	60	
Number of	13	12	13	14	18	17	
trials to	5	3	10	13	18	19	
extinction	9	9	17	21	18	18	
(X_{ij})	7	10	11	18	14	12	
	12	9	11	14	26	21	
	3	16	11	12	19	20	
	10	10	13	17	18	15	
	10	15	14	17	21	19	
	9	12	13	15	21	20	
	10	11	14	22	22	17	
t_i	88	107	127	163	195	178	858
$\overline{X}_{i.}$	8.8	10.7	12.7	16.3	19.5	17.8	14.3

	RS	SS	df	MS
m		12,269.4	1	
bet	13,160.0	890.6	5	178.1
w	13,656.0	496.0	54	9.185
t		1,386.6	59	

be the numerical value assigned to the ith group (e.g., in Table 10.1, $V_1 = 10$, $V_2 = 20$, etc.), and we let $\overline{V}_.$ be the average of the V_i (in Table 10.1, $\overline{V}_. = (10 + 20 + 30 + 40 + 50 + 60)/6 = 35$), then the polynomial function can be expressed as

$$\mu_i = \mu + a_1(V_i - \overline{V}_.) + a_2(V_i - \overline{V}_.)^2 + a_3(V_i - \overline{V}_.)^3 + \cdots. \qquad (10.1)$$

Recall that, by the linear model in Chapter 2, $\mu_i = \mu + \alpha_i$. The model in Equation 10.1 is basically the same model, but we assume that the α_i are a polynomial function of the V_i:

$$\alpha_i = a_1(V_i - \overline{V}_.) + a_2(V_i - \overline{V}_.)^2 + a_3(V_i - \overline{V}_.)^3 + \cdots. \qquad (10.2)$$

Inserting Equation 10.2 into Equation 2.8 gives us the fundamental model for trend analysis:

$$X_{ij} = \mu + a_1(V_i - \overline{V}_.) + a_2(V_i - \overline{V}_.)^2 + \cdots + \epsilon_{ij}. \qquad (10.3)$$

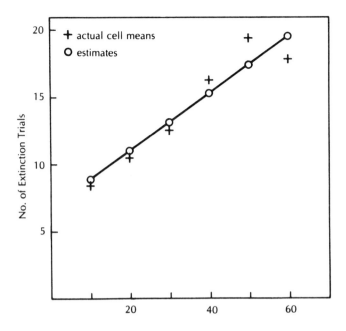

FIGURE 10.1. Data from Table 10.1, with best linear fit.

Trend analysis is a technique for testing, for each a_k in the equation, the null hypothesis that $a_k = 0$.

Caution: A polynomial function is one function that might describe the obtained data; however, it might not be the best function for this purpose. In a particular experiment, for example, a logarithmic function, or some other function, might also provide a good fit to the data. As we will show, any set of points can be fit perfectly by a polynomial function containing enough terms. In particular, the data from an experiment with I groups can always be fitted perfectly by a polynomial having $(I - 1)$ values of a_k. However, fitting a function by $(I - 1)$ different parameters can be an awkward procedure at best. Sometimes a function other than the polynomial will provide a more "natural" fit with fewer fitted parameters. By automatically applying trend analysis to data, you may miss this more natural fit.

Nevertheless, finding a more natural fit is often difficult, requiring considerable familiarity with both the research problem and the possible functions. Trend analysis has the virtue of being a relatively straightforward method of fitting the data to a relatively easily understood function.

However, if you find that a very large number of terms have to be incorporated in the function, it might be advisable to look for a different function that will provide a better fit. Fitting it would, of course, require more sophisticated techniques than are described here.

LINEAR TREND

Trend analysis is most easily illustrated by limiting the model in Equation 10.3 to the first term in the sum. The model then is

$$X_{ij} = \mu + a_1(V_i - \overline{V}.) + \epsilon_{ij}.$$

This model is applicable, of course, only if we have good a priori reasons to assume that the μ_i are related to the V_i by a linear function, i.e., a straight line. For example, the points in Figure 10.1 fit a straight line only rather crudely, but the crudeness of the fit could be due to errors in the data. If we had good a priori reasons for believing that the population means were related to the V_i by a straight line, Equation 10.3 would be justified.

Our problem, given this model, is to estimate μ and a_1 and to test the null hypothesis

$$H_0(1) : a_1 = 0. \tag{10.4}$$

Because the test depends directly on the estimates, we will find them first.

Parameter Estimates: Equal Sample Sizes. The best estimates of μ and a_1, for the purposes of trend analysis, are defined (as usual) to be those that minimize the variance of the estimated errors. The best estimates, for equal samples, can be shown to be

$$\begin{aligned} \hat{\mu} &= \overline{X}_{..}, \\ \hat{a}_1 &= \Sigma_i[(V_i - \overline{V}.)/\Sigma_{i'}(V_{i'} - \overline{V}.)^2]\overline{X}_{i.}. \end{aligned} \tag{10.5}$$

For the data in Table 10.1, $\hat{\mu} = \overline{X}_{..} = 14.3$, and

$$\begin{aligned} \overline{V}. &= 35, \\ \Sigma_{i'}(V_{i'} - \overline{V}.)^2 &= (10 - 35)^2 + (20 - 35)^2 + \cdots + (60 - 35)^2 = 1,750, \end{aligned}$$

so

$$\begin{aligned} \hat{a}_1 &= [(10 - 35)/1,750]8.8 + \cdots + [(60 - 35)/1,750]17.8 \\ &= .214. \end{aligned}$$

The line representing the function $\hat{X}_{ij} = 14.3 + .214(V_i - \overline{V}.)$ is shown in Figure 10.1.

Significance Test: Equal Sample Sizes. Because a_1 is the slope of the best-fitting straight line, $H_0(1)$ (Eq. 10.4) states that the best-fitting straight line has slope zero. In general, we will reject the null hypothesis if $\hat{a}_1$ is very different from zero, and we will fail to reject it if $\hat{a}_1$ is close to zero. The method of testing the null hypothesis becomes obvious if we note that Equation 10.5 is a linear contrast in the $\overline{X}_{i.}$ (see Chapter 3). If we let

$$c_{i1} = (V_i - \overline{V}.)/\Sigma_{i'}(V_{i'} - \overline{V}.)^2, \tag{10.6}$$

TABLE 10.2. Test for linear trend in extinction data, Table 10.1.

	A_1	A_2	A_3	A_4	A_5	A_6
V_i	10	20	30	40	50	60
$\overline{X}_{i.}$	8.8	10.7	12.7	16.3	19.5	17.8
c_{i1}	-5	-3	-1	1	3	5

$$c_1 = (-5)(8.8) + (-3)(10.7) + \cdots + (5)(17.8) = 75.0$$
$$SS_1 = (10)(75)^2/(25 + 9 + \cdots + 25) = 803.6$$
$$F = 803.6/9.185 = 87.5, \quad p < .001$$

then

$$\hat{a}_1 = \Sigma_i c_{i1} \overline{X}_{i.},$$

the equation of a linear contrast. The null hypothesis of no linear trend (Eq. 10.4) is thus tested as a simple contrast for which the c_{i1} values are defined in Equation 10.6.

The theory of planned comparisons was developed in Chapter 3; we will not repeat that theory here. However, the analysis is simplified by the fact that the test of a planned comparison is not changed when we multiply each c_{ik} by a constant. Usually a multiplier can be found that will make the c_{ik} integers. For this experiment, by multiplying every c_{i1} in Equation 10.6 by 350, we can transform them to the more easily used integers shown in Table 10.2. (Table 10.2 illustrates the test for a linear trend on the data in Table 10.1.)

Estimates and Tests: Unequal Sample Sizes. The theory of estimation and testing for linear trends is the same for unequal as for equal sample sizes; however, the specific equations must take into account the different numbers of scores in the cells. For unequal sample sizes, $\overline{V}_.$ is a weighted average of the V_i:

$$\overline{V}_. = (1/N)\Sigma_i n_i V_i.$$

As before, $\hat{\mu} = \overline{X}_{...}$ but

$$\hat{a}_1 = \Sigma_i n_i (V_i - \overline{V}_.)/[\Sigma_{i'} (V_{i'} - \overline{V}_.)^2 \overline{X}_i.$$

The test of $H_0(1)$ (Eq. 10.4) is a planned comparison with unequal sample sizes and with

$$c_{i1} = n_i (V_i - \overline{V}_.)/[\Sigma_{i'} n_{i'} (V_{i'} - \overline{V}_.)^2].$$

As with equal sample sizes, it may be possible to transform the c_{ik} into more easily handled integer values. Table 10.3 illustrates a test for linear trend on a variation, with some data missing, of the design in Table 10.1. In this case, for the linear trend test, we chose to let

$$c_{i1} = n_i (V_i - \overline{V}_.).$$

TABLE 10.3. Linear trend analysis: Unequal sample sizes.

	A_1	A_2	A_3	A_4	A_5	A_6	Sum
V_I	10	20	30	40	50	60	
X_{ij}	13	12	13	14	18	17	
	5	3	10	13	18	19	
	9	9	17	21	18	18	
	7	10	11	18	14	12	
	12	9	11	14	26	21	
	3	16	11	12		20	
	10		13	17			
	10		14				
	9		13				
			14				
n_i	9	6	10	7	5	6	43
t_i	78	59	127	109	94	107	574
$\overline{X}_{i.}$	8.67	9.83	12.70	15.57	18.80	17.83	13.35
c_{i1}	-203.0	-75.3	-25.6	52.1	87.2	164.7	$c_1 = 2,562$

$$SS_1 = 2,562^2/[(-203)^2/9 + (-75.3)^2/6 + \cdots + (164.7^2/6] = 546.1$$

	RS	SS	df	MS
m		7,662.2	1	
bet	8,241.7	579.5	5	115.9
w	8,642.0	400.3	37	10.82
t		979.8	42	

$$F_{(1,37)} = 546.1/10.82 = 50.5, \quad p < .001$$
$$SS_{rem} = 579.5 - 546.1 = 33.4, MS_{rem} = 33.4/4 = 8.35$$
$$F_{(4,37)} = 8.35/10.82 = .77$$

HIGHER-ORDER TRENDS

The test for a linear trend merely tells whether there is an overall tendency for the μ_i to increase or decrease with increases in V_i. Put another way, *if* the function is a straight line, the test for linear trend tests whether or not that straight line has a slope different from zero. The test for a linear trend *does not* test the appropriateness of using a straight line to fit the data in the first place.

We can test whether the relationship is truly linear by using the theory relating to MS_{rem} in Chapter 3. If the relationship is in fact linear, then Equation 10.1 should describe the cell means almost perfectly, and all of the significant differences in the data should be accounted for by the test for a linear trend. A general test of the null hypothesis that all differences are accounted for by the linear trend can be made by calculating (see Chapter 3)

$$
\begin{aligned}
SS_{rem} &= SS_{bet} - SS_1 = 890.6 - 803.6 = 87.0, \\
MS_{rem} &= SS_{rem}/(I-2) = 87.0/4 = 21.75, \\
F_{(4,54)} &= MS_{rem}/MS_w = 21.75/9.185 = 2.37, \\
p &= .064.
\end{aligned}
$$

A p value of .064 is suggestive but not compelling. However, we saw in Chapter 3 that, even though an overall test like this is not highly significant, some specific additional contrasts may be significant. Just as the linear trend was tested by a contrast, nonlinear trends can also be tested by contrasts. In particular, referring again to Equation 10.3, for each a_k in that equation we can test the null hypothesis

$$
H_0(k) : a_k = 0.
$$

Because a_2 is associated with the squares of the V_i, a test of $H_0(2)$ is called a test of *quadratic trend*. Similarly, a test of $H_0(3)$ is a test of *cubic trend*, a test of $H_0(4)$ is a test of *quartic trend*, and so on.

The test of quadratic trend ($H_0(2)$) will tend to be significant if there is a large curvature in a single direction. Figure 10.2 shows three examples of quadratic curvature. Notice that for each curve all of the curvature is in the same direction; none of the curves have *inflection points* (places where the direction of curvature changes).

If there is a cubic trend, there will usually be some point (called an inflection point) at which the direction of curvature changes. Figure 10.3 shows some examples of cubic curvature.

Of course, linear, quadratic, and cubic effects may all be present in the same data. In Functions B and C of Figure 10.2, for example, because the overall slope of each function is different from zero, a linear trend is present in addition to the quadratic trend. However, Function A has no linear trend. In Figure 10.3, even though Functions C and D have inflection points, most of the curvature is in only one direction; this indicates the presence of a quadratic as

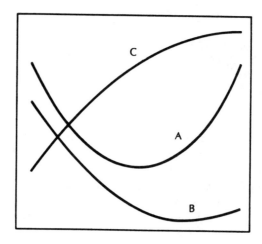

FIGURE 10.2. Examples of quadratic trends. Curve A is a pure quadratic trend; curves B and C have linear components.

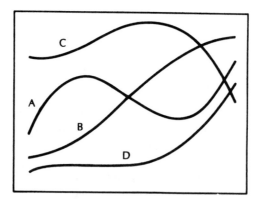

FIGURE 10.3. Examples of cubic trends. Curve A is a pure cubic trend; curve B has a linear component; curve C has a quadratic component; curve D has both linear and quadratic components.

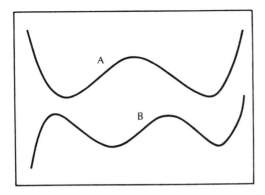

FIGURE 10.4. Examples of fourth powr (function A) and fifth power (function B) trends.

well as cubic trend. By contrast, Functions A and B have no quadratic trend. (But Functions B and D have linear trends.)

Similar considerations apply to higher-order trends. If there is a quartic (fourth power) trend, there will usually be two inflection points, i.e., the curve will usually change direction twice; if there is a fifth-power trend, there will usually be three inflection points, and so on. Figure 10.4 illustrates some of these trends.

Trend Tests: Equal Sample Sizes, Equal Intervals. Each trend is tested by a planned comparison. However, the c_{ik} for higher-order trends are not as easily derived as those for linear trends. For one special case the values have been derived and tabled; this is the common case in which the n_i are all equal and the intervals between the adjacent V_i are also equal. In the example in Table 10.1, each cell contains the same number of scores and the V_i are all exactly 10 units apart. Thus, they are at equal intervals. In the example of Table 10.3 these requirements are not met because the n_i are not all equal. Similarly, suppose the numbers of learning trials were 2, 4, 8, 16, 32, and 64 instead of 10, 20, 30, etc. Then, even though each cell contained the same number of scores, the intervals between the V_i would not be equal. For the method described here to apply, both the n_i and the intervals between the factor values V_i must be equal.

Note: The intervals between the V_i depend on the scale on which they are measured. In the example just given, with the numbers of trials equal to 2, 4, 8, 16, 32, and 64, the intervals were not equal. However, if we chose to let the V_i be the logarithms of the numbers of trials, the V_i would be $V_1 = \log 2 = .301, V_2 = \log 4 = .602, V_3 = \log 8 = .903$, etc. The intervals between the V_i would then be equal (each interval equals .301), and the method described here could be used. However, the interpretation of the results would then be very different; trends that do not exist for the untransformed factor values may exist for the transformed values, and vice versa. A test on the logarithms of

TABLE 10.4. Trend analysis on data from Table 10.1.

	A_1	A_2	A_3	A_4	A_5	A_6	Sum
V_i	10	20	30	40	50	60	
$\overline{X}_i$	8.8	10.7	12.7	16.3	19.5	17.8	

c_{ik}	1	-5	-3	-1	1	3	5	75.0
	2	5	-1	-4	-4	-1	5	-13.2
	3	-5	7	4	-4	-7	5	-31.0
	4	1	-3	2	2	-3	1	-6.0
	5	-1	5	-10	10	-5	1	1.0

Test	C_k	SS	F	p	$\hat{\omega}^2$	$\hat{a}^*$
1	75.0	803.57	87.5	<.001	.57	1.071
2	-13.2	20.74	2.26	.14	.01	
3	-31.0	53.39	5.81	.019	.03	-.172
4	-6.0	12.86	1.40	.24	.00	
5	1.0	0.04	0.00	—	(-.00)	

	\multicolumn{6}{c}{Estimates}
	A_1	A_2	A_3	A_4	A_5	A_6
V_i	10	20	30	40	50	60

	A_1	A_2	A_3	A_4	A_5	A_6
$\overline{X}_{i.}$	8.8	10.7	12.7	16.3	19.5	17.8
$\hat{\mu}_1$	9.8	9.9	12.5	16.1	18.7	18.8

the factor values may show a significant quadratic trend, for example, while a test on the untransformed values may not; the reverse may also be true. This is not an argument against transforming the factor values; the choice of factor values is always somewhat arbitrary, and you may change a difficult task into a relatively simple one by a transformation such as the logarithmic transformation used above. You might even get a better fit with fewer terms to the transformed factor values. However, a linear trend, for example, then implies that the data are a linear function of the *transformed* values, *not* of the original values.

Calculating Formulas. Table B.7 in Appendix B gives c_{ik} values for trend tests. If there are I groups, only $I - 1$ orthogonal contrasts can be tested, so the highest power at which a trend can be tested is $(I - 1)$. Each of the $(I - 1)$ trends is tested as a planned comparison using the c_{ik} values in Table A2.7. Table 10.4 shows the trend analysis on the data in Table 10.1.

It is clear in this example that most of the systematic variance is accounted for by the linear trend. The cubic trend is also significant, although it does not account for an especially high proportion of the variance. The quadratic trend approaches the 10% level of significance but accounts for only a very small proportion of the variance; it can probably be ignored for practical purposes. Both the quartic and quintic trends are nonsignificant.

Estimation. All of the theory in Chapter 3, including calculation of power, estimation of ω^2, and so on, applies to trend analysis. You can also apply the methods in Chapter 4.

However, we can use a special formula to estimate the cell means. This formula gives estimates that can be connected by a smooth continuous function. For the example in Table 10.1, for instance, we concluded from the tests that the data could be almost completely accounted for by a linear and a cubic trend. That is, they could be accounted for by a function of the form

$$X_{ij} = \mu + a_1(V_i - \overline{V}.) + a_3(V_i - \overline{V}.)^3 + \epsilon_{ij}$$

so

$$\mu_i = \mu + a_1(V_i - \overline{V}.) + a_3(V_i - \overline{V}.)^3. \tag{10.7}$$

If we could estimate a_1 and a_3, we could estimate the μ_i from this function. Unfortunately, the a_k are difficult to estimate directly. However, the μ_i can be estimated easily if we change the form of Equation 10.7 to

$$\mu_i = \mu + a_1^* c_{i1} + a_3^* c_{i3}. \tag{10.8}$$

This change is possible because the c_{i1} values are directly related to the linear trend and the c_{i3} values are directly related to the cubic trend. More generally, an equation of the form

$$\mu_i = \mu + a_1(V_i - \overline{V}.) + a_2(V_i - \overline{V}.)^2 + a_3(V_i - \overline{V}.)^3 + \cdots \tag{10.9}$$

can always be changed to an equivalent equation of the form

$$\mu_i = \mu + a_1^* c_{i1} + a_2^* c_{i2} + a_3^* c_{i3} + \cdots. \tag{10.10}$$

When we say that Equations 10.9 and 10.10 are equivalent, we mean that if we insert appropriate values for the a_k^* into Equation 10.10, we will obtain the same estimates of μ_i that we would have obtained by inserting estimates of the a_k into Equation 10.9.

Of course, in practice only those components that contribute meaningfully to the overall trend, i.e., those that are significant and account for a reasonably large proportion of the variance are included. In our example, only the linear and cubic trends are included.

The estimate of a_k^* is relatively simple. The formula is

$$\hat{a}_k^* = C_k/(\Sigma_i c_{ik}^2), \tag{10.11}$$

where C_k and c_{ik} are the values used in testing for the kth trend component. To illustrate, for the data in Table 10.1 we must estimate a_1^* and a_3^* (see

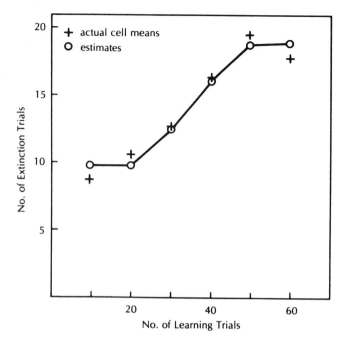

FIGURE 10.5. Data from Table 10.1, with best fit using linear and cubic trend components.

Eq. 10.8). The estimates use the c_{ik} and C_k values in Table 10.4:

$$\hat{a}_1^* = 75/[(-5)^2 + (-3)^2 + \cdots + (5)^2] = 75/70 = 1.071,$$
$$\hat{a}_3^* = -31/[(-5)^2 + (7)^2 + \cdots + (5)^2] = -31/180 = -.172.$$

The estimate of μ (from Table 10.1) is $\overline{X}_{..} = 14.3$, and the estimation formula is

$$\hat{\mu}_i = 14.3 + 1.071c_{i1} - .172c_{i3}.$$

From this equation, we estimate μ_1 by inserting $c_{11} = -5$ and $c_{13} = -5$ to obtain

$$\hat{\mu}_1 = 14.3 + 1.071(-5) - .172(-5) = 9.8.$$

Similarly, we estimate μ_2 by inserting $c_{21} = -3$ and $c_{23} = 7$ into the same equation to obtain

$$\hat{\mu}_2 = 14.3 + 1.071(-3) - .172(7) = 9.9.$$

The remaining four $\hat{\mu}_i$, found the same way, are shown in Table 10.4. These predicted values are plotted, along with the obtained means, in Figure 10.5. It can be seen that the cubic function provides a somewhat better fit than the linear function in Figure 10.1; it also gives a smoother function than that provided by the $\overline{X}_{..}$.

Trend Test: Unequal Sample Sizes or Intervals. Table A2.7 cannot be used if either the sample sizes or the intervals are unequal, or if we want to analyze trends that are not tabled. We must then calculate the c_{ik} values. The method is tedious but not difficult. We first calculate the c_{i1} for the linear trends as described above. We then calculate the c_{ik} for higher-degree trends, using the c_{ik} already calculated for lower-degree trends to simplify the calculations. The general formula for the kth trend is

$$c_{ik} = V_i^k - \Sigma_{k'=0}^{k-1} b_{k'} c_{ik'},$$

where

$$b_{k'} = (\Sigma_i c_{ik'} V_i^k)/(\Sigma_i c_{ik'}^2).$$

Of course, as usual, the c_{ik} may be multiplied by any convenient constant. For example, it may be possible to find a constant that makes them all integers. Alternatively, the formula for $b_{k'}$ may be simplified by choosing a constant that makes

$$\Sigma_i c_{ik}^2 = 1.$$

Some care is needed when using these formulas. Small rounding errors in the c_{ik} for lower-order trends can cause large errors in the c_{ik} for higher-order trends. Therefore, the c_{ik} should be calculated to several more places than you expect to use.

Note on Interpretation of Polynomial Functions. The use of straight lines to represent the function in Figure 10.5 may seem strange, since the purpose of the estimation equation is to generate a smooth curve. Nevertheless, my use of straight lines to connect the predicted values of the μ_i helps to make an important point.

The use of a polynomial function to describe the data implies that that function would also describe the data at values of V_i other than those actually sampled. Consider Figure 10.6, in which the smooth curve actually represented by the obtained cubic function is drawn; from it we would predict that with 45 learning trials the expected number of extinction trials would be 17.6. However, there is nothing in either the data or the statistical analysis to justify such a prediction on *statistical* grounds. Intuitively, we would expect such a prediction to be approximately correct, but there is no logical or statistical basis for using the obtained function to predict data other than those actually studied in the experiment.

In fact, the danger of extending the conclusions to conditions that were not studied is clearly shown in Figure 10.6, which shows that the predicted number of trials to extinction is at a minimum at about 15 learning trials, and that extinction would take about 14 trials if there were no learning at all!

Finally, consider Figure 10.7. The function in this figure fits the predicted points just as well as the function in Figure 10.6. It would lead, however, to considerably different predictions for points other than those that were actually studied. In short, there are usually good intuitive and theoretical reasons for assuming that a relatively simple, smooth function would describe potential data from points that were not actually studied. Nevertheless, there are no strictly logical or statistical reasons for preferring one curve over another. In practice, a smooth curve like that in Figure 10.6 will usually seem most

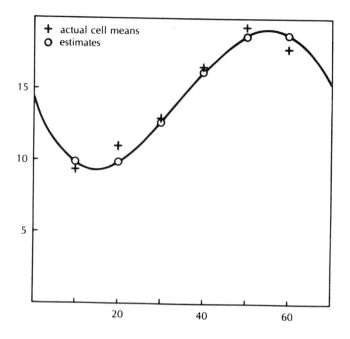

FIGURE 10.6. Data from Table 10.1, with complete function of best linear and cubic fit.

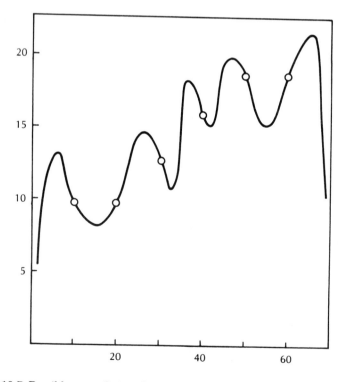

FIGURE 10.7. Possible curve fitting the same estimated data points as that in Figure 10.6.

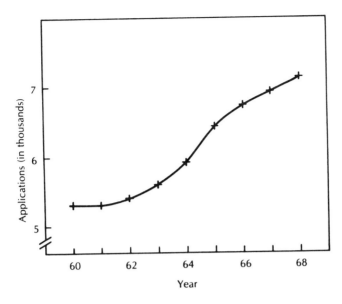

FIGURE 10.8. Hypothetical data: applications for admission to Knutsta University, 1960 to 1968.

reasonable, but the exact shape of that curve should not be taken too seriously. Sometimes a "curve" consisting simply of line segments like those in Figure 10.5 will fit the function about as well as a smooth curve. At other times a smooth curve might not be appropriate at all.

One Score per Cell

To calculate MS_w we must have more than one score in each cell of a one-way design. Consequently, in an ordinary one-way analysis of variance no test can be made with only one score per cell. However, when the factor levels have numerical values, there is a trick that allows us (provided our assumptions are reasonable) to analyze data with only one score per cell.

The use of a special method to deal with only one score per cell is best demonstrated by an example. Figure 10.8 shows the number of applications for admission to the University of Knutsta (affectionately known as "Knutsta U") during the years 1960 to 1968. The officials noticed what appeared to be a sharp rise in applications during the early 1960s and wondered if it might be due to the "baby boom" of the late 1940s. They had two questions to ask of the data. (1) Was there a significantly greater rise in applications during the years after 1962 or 1963 than previously? (2) Was there a significantly smaller rise in applications during the years after 1966 than between 1963 and 1965? If the answer to the first question is yes, the increase could probably be attributed to the baby boom; if the answer to the second question is yes, the officials could

TABLE 10.5. Trend analysis on data in Figure 10.8. The X_i are numbers of applications divided by 100.

	A_1	A_2	A_3	A_4	A_5	A_6	A_7	A_8	A_9	C_k
V_i	60	61	62	63	64	65	66	67	68	
X_i	53	53	54	56	59	64	67	69	71	
c_{ik} 1	-4	-3	-2	-1	0	1	2	3	4	154
2	28	7	-8	-17	-20	-17	-8	7	28	138
3	-14	7	13	9	0	-9	-13	-7	14	-101

$$SS_t = SS_{bet} = 414, \ df = 8$$

Test	C	SS	F	p	$\hat{\omega}^2$
1	154	395.267	1,268	<.001	.95
2	138	6.870	22.0	.005	.02
3	-101	10.304	33.0	.002	.02

$$df_{rem} = 5, MS_{rem} = 1.5592/5 = .3118$$

expect that the baby boom had about ended and that a relatively steady but smaller growth rate in applications would occur in the future.

One way to answer these questions would be to use a trend analysis. We would expect a significant linear trend, of course, because applications increased steadily throughout the period. The answer to the first question is yes if there is a significant nonlinear trend—most likely quadratic or cubic. The second question can be answered by testing specifically for a cubic trend. If the baby boom was in fact ending, the initial increase followed by a tendency to taper off would show up as a significant cubic trend (compare Figure 10.8 with curve B of Figure 10.3). If it is not tapering off, there should be a significant quadratic trend but no cubic trend.

TREND TESTS

One simple and straightforward way to fit these data into the framework of the analysis of variance would be to consider the numbers of applications as the scores in a one-way analysis of variance for which the years are the levels. This arrangement of the data in an ordinary one-way design is shown in Table 10.5. With such a design, however, there is only one score per cell. Thus, special methods are needed to find a denominator for the F tests on trends.

Even with only one score per cell, SS_k values for tests on trends can be calculated. Table 10.5 shows them for the first three trend components. Notice that for these data nearly all of the variance (in fact, 99.7%) is accounted for by the first three components. Furthermore, there are no a priori reasons to expect large trends higher than the third order. We thus have reason to believe, and the data support our belief, that the first three trend components account for all, or nearly all, of the true variation in the data. Any remaining variance must then be due largely to chance; that is, it must be mostly error variance. The remainder, after subtracting SS_1, SS_2, and SS_3 from SS_{bet}, is 1.559 (see Table 10.5), and the above argument makes it reasonable to believe that this variance represents error. We can thus treat it as a sum of squares for error when developing an F ratio to test for the three trends in which we are interested.

Because there were originally eight degrees of freedom, and the SS_k for three orthogonal tests have been subtracted out, SS_{rem} has five degrees of freedom. The appropriate denominator for the F ratio is therefore $MS_{rem} = 1.559/5 = .3118$. Table 10.5 shows the results of the F tests, with MS_{rem} in the denominator. Although the linear trend is dominant, as we expected, both the quadratic and the cubic trends are significant. We can conclude that applications were affected by the baby boom and that the increase in applications began to taper off near the end of the period studied. However, the positive quadratic trend indicates that the tapering off was not complete; the growth rate was still higher at the end of the period than it was at the beginning.

To devise a denominator for these tests we made use of a principle often repeated in this text: a variance that is not due to systematic effects is error variance and can be treated as such in forming F ratios. The same principle could have been used on the data of Tables 10.1 and 10.3. If we had had good a priori reasons for believing that there were no true fourth or fifth degree trend components, SS_4 and SS_5 could have been pooled with SS_w to increase the degrees of freedom in the denominator of each F ratio. (The same would hold for the quadratic effect if we had a priori reasons for expecting that there were no true quadratic effect, but it is extremely unlikely that we would have such reasons and still expect a cubic effect.) However, the increase in degrees of freedom would only be two (from 54 to 56), and MS_{pooled} would be 9.088, down only very slightly from $MS_w = 9.185$. The very small gain in power in this case must be balanced against all of the risks and cautions discussed in Chapter 5.

Pooling is therefore useful primarily when there is no other denominator available, i.e., when the only alternative to the risks discussed in Chapter 5 is to make no tests at all. In any case we can be reasonably certain whenever we use this approach that the error will tend to make rejection of the null hypothesis less, rather than more, probable.

The biggest problem in applying this approach to real data lies in determining which trend components are to be tested and which will comprise the error. In the example above, this problem was easily solved. The nature of the questions asked required the testing of linear, quadratic, and cubic trends, and the nature of the data made it reasonable to assume that higher-order trends would be small. In other cases, however, such assumptions are not easily made before gathering the data. When they can be made (and the data bear them out), it is usually best to proceed as though the assumptions were true.

When the assumptions cannot be made, a special procedure can be used to give approximate significance levels. Although the procedure is not completely satisfactory, it is sometimes useful when no other procedure is available.

With this procedure, we test each trend as if no higher-order trends existed. Thus, when testing the linear trend, all of the variance remaining after subtracting the linear trend from SS_{bet} is treated as error. For the example in Table 10.5 the error would be $SS_{rem} = SS_{bet} - SS_1 = 414 - 395.267 = 18.73$. This SS_{rem} would have seven degrees of freedom, so $MS_{rem} = 18.73/7 = 2.676$, and the linear effect is tested by $F_{(1,7)} = 395.267/2.676 = 148 \ (p < .001)$.

When testing the quadratic trend, we subtract the variances for the linear and quadratic trends from SS_{bet} and treat the remainder as error:

$$
\begin{aligned}
SS_{rem} &= SS_{bet} - SS_1 - SS_2 \\
&= 414 - 395.267 - 6.870 = 11.86, \\
MS_{rem} &= 11.86/6 = 1.977, \\
F_{(1,6)} &= 6.870/1.977 = 3.47, \\
p &= .11.
\end{aligned}
$$

Notice that by this method the quadratic effect is not significant, although in the analysis in Table 10.5 the quadratic effect is significant beyond the .01 level. The analysis usually proceeds until two or three nonsignificant trends in a row have been obtained. A complete analysis of the same data in Table 10.5 is shown in Table 10.6. The analysis was discontinued after finding that the fourth, fifth, and sixth degree trends were nonsignificant.

A comparison of Table 10.6 with Table 10.5 illustrates the relative loss of power resulting from an inability to make valid assumptions about the trends. In Table 10.5 all three trends are significant beyond the .01 level. In Table 10.6 the quadratic trend is not even significant at the .10 level. The F ratio for the linear trend is also much smaller in Table 10.6 than in Table 10.5, although it is highly significant in both tables. The loss of power derives from the fact that the "error" estimates used for testing the linear and quadratic trends are inflated by inclusion of the cubic effect. Thus, they are not pure error at all, but error plus a systematic effect. It would be possible, of course, to redo the analysis, once we had found the highest significant trend, using the final error term to retest all lower trends. It would be possible, but it would not be legitimate, because such a procedure involves choosing one's error term on the basis of the data.

ESTIMATING TRENDS

The formulas for estimating the trends are the ones for ordinary trend analysis. For the example in Table 10.5, we are fitting the data to the equation

$$ \hat{X}_i = \hat{\mu} + \hat{a}_1^* c_{i1} + \hat{a}_2^* c_{i2} + \hat{a}_3^* c_{i3}. $$

Table 10.7 illustrates the calculation of the $\hat{a}_k^*$, using Equation 10.11; it also compares the actual and estimated scores. In this example the actual and estimated scores are very close.

TABLE 10.6. Trend analysis of data in Figure 10.8, cubic function not assumed. The X_i are numbers of applications divided by 100.

		A_1	A_2	A_3	A_4	A_5	A_6	A_7	A_8	A_9	C_k
V_i		60	61	62	63	64	65	66	67	68	
X_i		53	53	54	56	59	64	67	69	71	
c_{ik}	1	-4	-3	-2	-1	0	1	2	3	4	154
	2	28	7	-8	-17	-20	-17	-8	7	28	138
	3	-14	7	13	9	0	-9	-13	-7	14	-101
	4	14	-21	-11	9	18	9	-11	-21	14	-15
	5	-4	11	-4	-9	0	9	4	-11	4	20
	6	4	-17	22	1	-20	1	22	-17	4	24

$$SS_t = SS_{bet} = 414, \quad df = 8$$

Test	C	SS	SS_{rem}	df_{rem}	MS_{rem}	F	p
1	154	395.267	18.733	7	2.676	148	<.001
2	138	6.870	11.863	6	1.977	3.47	.11
3	-101	10.304	1.559	5	0.3118	33.0	.001
4	-15	0.112	1.447	4	0.3618	0.31	—
5	20	0.855	0.592	3	0.1973	4.33	.13
6	24	0.291	0.301	2	0.1505	1.93	.30

TABLE 10.7. Calculation of estimated applications (in hundreds), data from Table 10.5.

$$\hat{\mu} = 546/9 = 60.67$$
$$\hat{a}_1^* = 154/[(-4)^2 + (-3)^2 + \cdots + (4)^2] = 2.567$$
$$\hat{a}_2^* = 138/[(28)^2 + (7)^2 + \cdots + (28)^2] = .0498$$
$$\hat{a}_3^* = -101/[(-14)^2 + (7)^2 + \cdots + (14)^2] = -.1020$$

	A_1	A_2	A_3	A_4	A_5	A_6	A_7	A_8	A_9
V_i	60	61	62	63	64	65	66	67	68
X_i	53	53	54	56	59	64	67	69	71
$\hat{X}_i$	53.2	52.6	53.8	56.3	59.7	63.3	66.7	69.4	70.9

Exercises

(1.) Do trend analyses on the data in (a) Problem 1, Chapter 2; (b) Problem 2, Chapter 2; and (c) Problem 6, Chapter 2. For each problem assume that there are equally spaced V_i.

(2.) For the data in Problem 1, Chapter 2, assume the following V_i values:

	A_1	A_2	A_3	A_4	A_5	A_6
V_i	0	1	2	4	8	10

(a.) Test for a linear trend; then do a general test for nonlinear trends.

(b.) Test for linear, quadratic, and cubic trends; then do a general test for higher-order trends.

(3.) Test for (a) linear and (b) nonlinear trends in the data in Problem 9, Chapter 2. Assume that the V_i are equally spaced.

(4.) A theorist believes that for the data of a certain one-way experiment with five equally spaced levels (i.e., the single factor is quantitative) the appropriate model is

$$X_{ij} = a_1(V_i - \overline{V}) + e_{ij}.$$

That is, she believes that the only effect is linear, and $\mu = 0$. You suspect that the data are nonlinear and the true mean is not zero, so you repeat the experiment. Financial considerations restrict you to only one score at each of the five levels.

(a.) Describe what you consider to be the best available way to test the theory, using the obtained data. Tell what assumptions your test(s) require(s), and the effects of violation of the assumptions.

(b.) The data are as follows:

	A_1	A_2	A_3	A_4	A_5
V_i	-10	-5	0	5	10
X_i	-3	0	-1	2	5

Perform the analysis (analyses) you described in part (a) and draw whatever conclusions you can. Justify your conclusions (or lack of them).

(c.) Suppose you had two scores in each cell. How would you then answer part (a)?

11

Trend Analyses in Multifactor Designs

Frequently, a multifactor design has one or more numerically valued factors. When this occurs, trend analyses can be performed on main effects and interactions involving these factors. Just as in the one-way design, the theory of trend analysis differs for fixed and random numerical factors. For multifactor designs with random factors, however, the problem can be complicated; random factors with numerical values exert important influences on tests involving only fixed effects. In this chapter we will discuss designs in which all of the numerical factors are fixed; in Chapter 15 we will discuss designs that can be regarded as having numerically valued random factors.

Main Effects

Because any trend analysis is basically a contrast, the theory for contrasts holds for trend analyses on main effects. The c_{ik} are the same as in Chapter 10, and the formulas are the same as those for contrasts on main effects.

The data in Table 11.1 illustrate a trend analysis in a two-way design. The hypothetical data are scores on standard arithmetic tests in three different schools in a medium-sized town. The tests were given to students whose measured IQs fell into the ranges 80 to 90, 90 to 100, 100 to 110, and 110 to 120. The IQ ranges are the levels of Factor A, and the schools are the levels of Factor B. Ten pupils in each IQ range were tested in each school, for a total of 120 pupils. The values in Table 11.1 are the average scores for the pupils in each cell. The results of a standard analysis of variance are shown below the table. Although all three effects are significant, we are not interested in the overall tests on the A main effect and the interaction. Instead, we are interested first in knowing what trends are significant in the A main effect, and, second, in knowing whether or not the trends differ among schools. The first question involves the A main effect; the second involves the AB interaction, as we shall see.

The trend analysis on the A main effect is just a set of planned comparisons, using the coefficients for the linear, quadratic, and cubic trends given in Table B.7. The tests can be made directly on the $\overline{X}_{i..}$ values, as in Chapter 5, remembering that because each $\overline{X}_{i..}$ is the mean of nJ scores (n scores at

TABLE 11.1. Hypothetical mean arithmetic scores for students in four IQ ranges (Factor A) at three schools (Factor B), $n = 10$.

	A_1	A_2	A_3	A_4	Mean
B_1	31	43	48	58	45.00
B_2	33	50	62	66	52.75
B_3	37	59	66	69	57.75
Mean	33.67	50.67	58.67	64.33	51.83

	SS	df	MS	F	p	$\hat{\omega}^2$
m	322,403.3	1				
a	16,030.0	3	5,343	134	<.001	.65
b	3,301.7	2	1,651	41.3	<.001	.13
ab	605.0	6	100.8	2.52	.025	.02
w	4,320.0	108	40.00			
t	24,256.7	119				

each of the J levels of B),

$$SS_k = nJ(\Sigma_i c_{ik} \overline{X}_{i..})^2 / (\Sigma_i c_{ik}^2).$$

Table 11.2 gives these calculations for the linear, quadratic, and cubic trends on A. The linear and quadratic trends are significant, but the cubic trend is not.

Variance estimates for trend components are the same as for contrasts (Chapter 5). Estimates of ω_k^2 are also the same.

Note that the calculations shown are those for a design in which Factor B is fixed. If Factor B were random, the appropriate denominator for the trend test would be MS_{ab} instead of MS_w. This is because the trends are part of the A main effect, and the appropriate denominator for the A main effect would be MS_{ab} in a mixed-effects design. The F ratio for the linear trend would then be $15,000/100.8 = 148.8$, with six degrees of freedom in the denominator. Tests of the other two trends would use the same denominator mean square.

Two-Way Interactions with One Numerical Factor

SIGNIFICANCE TESTS

The second question in the introduction to this chapter asked whether scores in all three schools could be described by a single function. To answer that question, we first separately calculate SS_k values for each trend at each level of B. To make the calculations, we assume that each level of B constitutes a complete experiment in itself (much as we did when testing simple effects

TABLE 11.2. Trend analysis on Factor A, data from Table
11.1.

		A_1 80–90	A_2 90–100	A_3 100-110	A_4 110-120
	Mean	33.67	50.67	58.67	64.33
	Linear	-3	-1	1	3
c_{ik}	Quad	1	-1	-1	1
	Cubic	-1	3	-3	1

	C	SS	F	p	$\hat{\omega}^2$
Linear	100.0	15,000	375	<.001	.62
Quad	-11.33	963.3	24.1	<.001	.04
Cubic	6.667	66.67	1.67	.20	.00

in Chapter 5). Then, because each $\overline{X}_{ij}$ is the average of only n rather than nJ scores, the multiplier for the separate levels is n rather than nJ. These calculations, for each trend and each level of B, are shown in Table 11.3.

Because each SS_k, calculated this way, is mathematically equivalent to that for a contrast in a simple one-way analysis of variance, each SS_k is proportional to chi-square with one degree of freedom; therefore, their sum is also distributed as chi-square.

To test the null hypothesis that the *linear* trend is the same for all three schools, we first sum the SS_1 (the values for the linear trend components) for the three separate levels of B. The resulting sum is proportional to chi-square with three degrees of freedom. We then subtract SS_1 from the A main effect to obtain a chi-square statistic with two degrees of freedom. The actual calculations, for these data, are

$$SS_a \text{ linear } (b) = 3,698.0 + 6,160.5 + 5,304.5 - 15,000 = 163.0.$$

We divide this by its degrees of freedom to obtain

$$MS_a \text{ linear } (b) = 163/2 = 81.5.$$

Finally, we divide this by MS_w to obtain the F ratio

$$F_{(2,108)} = 81.5/40 = 2.04.$$

Table 11.3 shows these calculations as well as calculations for the quadratic and cubic effects.

The general method for testing trend interactions should now be clear. First, the SS_k for the trend in question are calculated separately for each level of Factor B (if we assume that Factor B is the nonnumerical factor). These are summed, and the SS_k for the same trend on the A main effect is

TABLE 11.3. Dividing interaction into trend components, data from Table 11.1.

	Linear		Quadratic		Cubic	
	C	SS	C	SS	C	SS
B_1	86	3,698.0	-2	10.0	12	72.0
B_2	111	6,160.5	-13	422.5	-3	4.5
B_3	103	5,304.5	-19	902.5	11	60.5
Sum		15,163.0		1,335.0		137.0
A Main effect[a]		15,000.0		963.3		66.7
SS		163.0		371.7		70.3
df		2		2		2
MS		81.5		185.8		35.2
F		2.04		4.65		0.88
p		.14		.012		—
$\hat{\omega}^2$		.01		.01		.00

[a]From Table 11.2.

subtracted from their total. The result is treated as a sum of squares with $(J-1)$ degrees of freedom. The appropriate denominator is MS_w whether B is fixed or random because MS_w is always the denominator for the AB interaction in a two-way design.

The relationship between these tests and the test of the AB interaction can be seen from Table 11.3. Note, first, that for I levels of A there will be $(I-1)$ trend components to test, and each test has $(J-1)$ degrees of freedom. The total degrees of freedom for all of the tests are thus $(I-1)(J-1)$, which is exactly equal to the degrees of freedom of the AB interaction. Furthermore, if we add the sums of squares in Table 11.3, we get $163.0 + 371.7 + 70.3 = 605.0 = SS_{ab}$. We have just described a method for dividing the AB interaction into $(I-1)$ orthogonal tests, each with $(J-1)$ degrees of freedom.

Variance estimates are obtained from the mean squares in Table 11.3 by subtracting MS_w from each mean square, multiplying each difference by the degrees of freedom of the mean square, and dividing that result by N. For example, the estimated variance of the quadratic effect is $(186-40)(2)/120 = 2.4$. To estimate the proportion of variance accounted for, we divide this value by $\hat{\sigma}_t^2 = 202$ to get .012. These estimates, like the significance tests, are the same whether Factor B is fixed or random. Table 11.4 summarizes the trend analysis on the data in Table 11.1.

The interpretation of interaction trends is straightforward. If there is a linear trend in the interaction, the size of the linear trend, i.e., the overall slope of the best-fitting straight line, varies across levels of the nonnumerical factor. The presence of a significant quadratic trend in the data of Figure 11.1 confirms our suspicion that the overall curvature of the best-fitting function varies across levels of the nonnumerical factor (in this case, schools). Basically,

TABLE 11.4. Summary of trend analysis on data in Table 11.1
(bold numbers are totals).

	SS	df	MS	F	p	$\hat{\omega}^2$
m	332,403.3	1				
a	**16,030.0**	**3**				.65
Linear	15,000.0	1	15,000	375	<.001	.62
Quad	963.3	1	963.3	24.1	<.001	.04
Cubic	66.7	1	66.7	1.67	.20	.00
b	3,301.7	2	1,651	41.3	<.001	.13
ab	**605.0**	**6**				.02
Linear	163.0	2	81.5	2.04	.14	.01
Quad	371.7	2	185.8	4.65	.012	.01
Cubic	70.3	2	35.2	0.88	—	.00
w	4,320.0	108	40.0			
t	24, 256.7	119				

this means that a different estimate of a_2^* should be used for each level of B.
Similar considerations would apply to higher-order trends in the interaction.

ESTIMATION

Because both the B main effect and the quadratic trend component of the AB
interaction are significant, we cannot estimate cell means from the trends on
the A main effect alone. One solution to this problem might lie in a separate
trend analysis, across A, at each level of B. Such an approach would be in the
spirit of the simple-effects tests described in Chapter 6. However, it would not
be the most parsimonious approach because it would not take into account the
fact that the *linear trend* is approximately the same for all schools; i.e., there
is no linear trend component in the AB interaction. A more parsimonious
approach is to make our estimates from the model equation:

$$\hat{\mu}_{ij} = \hat{\mu} + \hat{a}_1(V_i - \overline{V}_.) + \hat{a}(b_j)_2(V_i - \overline{V}_.)^2 + \hat{\beta}_j.$$

In this equation, $\hat{a}(b_j)_2$ indicates that, because the quadratic component of
the interaction is significant, a different estimate of a_2 must be used for each
level of B. However, only one estimate of a_1 is needed.

There is no need to include $\hat{a}_2$ in this equation because it is implicitly

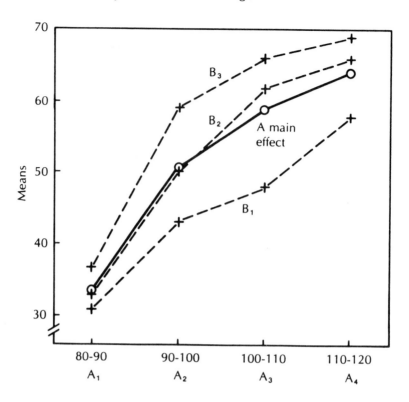

FIGURE 11.1. Obtained means for A main effect and for each level of B; data from Table 11.1.

included in $\hat{a}(b_j)_2$. Whenever a trend component is significant in both the main effect and the interaction, only the interaction term appears in the equation. To see how the quadratic interaction includes the quadratic component of the A main effect, consider the average of the C values in Table 11.3 for the quadratic interaction term $(-2 - 13 - 19)/3 = -11.33$, which, as we can see from Table 11.2, is the C value for the quadratic component of the A main effect. Because the estimates of the effects will be based on these C values, the trend on the A main effect will automatically be included in the average of the estimates for the interaction.

The equation can be shortened slightly by noting that

$$\overline{X}_{.j.} = \hat{\mu} + \hat{\beta}_j,$$

so

$$\hat{\mu}_{ij} = \overline{X}_{.j.} + \hat{a}_1(V_i - \overline{V}_.) + \hat{a}(b_j)_2(V_i - \overline{V}_.)^2.$$

As in Chapter 10, some of these values are difficult to estimate directly. However, estimates can be made easily if we change the equation to

$$\hat{\mu}_{ij} = \overline{X}_{.j.} + \hat{a}_1^* c_{i1} + \hat{a}^*(b_j)_2 c_{i2}.$$

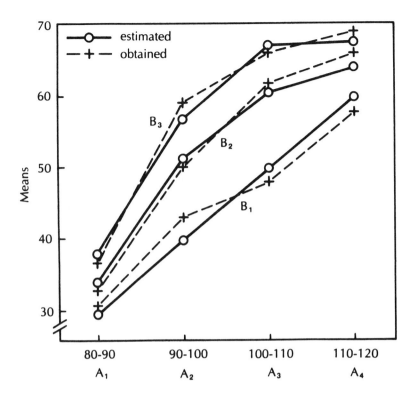

FIGURE 11.2. Estimated and obtained means, Factor A numerical; data from Table 11.1.

In this equation c_{i1} is the coefficient of $\overline{X}_{i..}$ when testing for a linear trend, and c_{i2} is the coefficient for the quadratic trend.

The value of a_1^* is estimated from C_1 for the A main effect just as in Chapter 10:

$$\hat{a}_1^* = C_1/\Sigma_i c_{i1}^2 = 100/20 = 5.$$

To estimate $a^*(b_j)_2$, we use the same formula (substituting C_2 for C_1 and C_{i2} for C_{i1}), but we use a different C_2 for each level of B. The values we use are in the quadratic column in Table 11.3:

$$
\begin{aligned}
\hat{a}^*(b_1)_2 &= -2/\Sigma_i c_{i2}^2 = -2/4 = -.50, \\
\hat{a}^*(b_2)_2 &= -13/\Sigma_i c_{i2}^2 = -13/4 = -3.25, \\
\hat{a}^*(b_3)_2 &= -19/\Sigma_i c_{i2}^2 = -19/4 = -4.75.
\end{aligned}
$$

The estimate of μ_{11} would then be $\hat{\mu}_{11} = 45.00 + 5(-3) - .50(1) = 29.50$. The estimate of μ_{12} would be $\mu_{12} = 52.75 + 54(-3) - 3.25(1) = 34.50$. The other estimates shown in Table 11.5 and plotted in Figure 11.2 are obtained similarly.

TABLE 11.5. Estimated effects and cell means for data in Table 11.1, Factor A numerical.

			Estimates of effects				
μ	β_1	β_2	β_3	a_1	$a^*(b_1)_2$	$a^*(b_2)_2$	$a^*(b_3)_2$
51.83	-6.83	0.92	5.92	5.00	-0.50	-3.25	-4.75

Obtained (upper value) and estimated (lower value) cell means

	A_1	A_2	A_3	A_4
	31	43	48	58
B_1	29.5	40.5	50.5	59.5
	33	50	62	66
B_2	34.5	51.0	61.0	64.5
	37	59	66	69
B_3	38.0	57.5	67.5	68.0

Higher-Way Interactions with One Numerical Factor

The principles described here are easily extended from two-way to higher-way designs. In a multiway design, tests for trends on main effects of numerical factors are carried out just as planned comparisons are performed on main effects. In addition, any two-way interaction involving one numerical and one nonnumerical factor can be divided into trend components as described above, by doing a separate trend test on each level of the nonnumerical factor, summing the SS_k values, and subtracting the SS_k for the same trend on the main effect. The number of numerator degrees of freedom for each such test will be one less than the number of levels of the nonnumerical factor.

These calculations are illustrated in Tables 11.6, 11.7, and 11.8 for an imaginary study on American buying habits. For the study, the nation was divided into three areas: west, midwest, and east; 75 stores in each area were selected for testing. Twenty-five stores each were located in low-income, middle-income, and high-income neighborhoods. A new brand of detergent was offered in each of these stores (the same brand in every store). The stores in each group of 25 were divided into subgroups of 5 each, in which the detergent sold for different prices. The five prices were 29 cents, 39 cents, 49 cents, 59 cents, and 69 cents.

In this design, Factor A, with three levels, is the area of the country; Factor B, with three levels, is the income level; and Factor C, the numerical factor with five levels, is the price. The five stores offering the detergent at the same price provide within-cell variance, so $n = 5$. The data are mean numbers of boxes purchased in each type of store at each price level in a single day. The data are plotted in Figure 11.3.

An overall analysis of variance on these data is shown in Table 11.7; for the most part, though, it will serve only as the basis for a more detailed trend analysis. Table 11.8 summarizes the trend analyses on the C main effect and on the AC and BC interactions. The significant B main effect is expected;

TABLE 11.6. Hypothetical data, cell means from a study of buying habits, $n = 5$.

		C_1	C_2	C_3	C_4	C_5		A_1	A_2	A_3
	B_1	24.0	28.8	22.8	13.2	4.8		21.8	20.3	19.1
A_1	B_2	31.8	20.4	15.6	19.2	30.6				
	B_3	2.4	13.2	32.4	33.6	34.8				
	B_1	26.4	26.4	16.8	10.8	3.6			b	
A_2	B_2	7.2	24.0	30.0	27.6	9.6		B_1	B_2	B_3
	B_3	10.8	21.6	25.2	32.4	32.4		16.3	21.0	24.0
	B_1	37.2	18.0	6.0	2.4	3.6				
A_3	B_2	31.2	12.0	10.8	15.6	28.8				
	B_3	9.6	9.6	18.0	31.2	52.8				

ac

	C_1	C_2	C_3	C_4	C_5
A_1	19.4	20.8	23.6	22.0	23.4
A_2	14.8	24.0	24.0	23.6	15.2
A_3	26.0	13.2	11.6	16.4	28.4

c

C_1	20.1
C_2	19.3
C_3	19.7
C_4	20.7
C_5	22.3

bc

	C_1	C_2	C_3	C_4	C_5
B_1	29.2	24.4	15.2	8.8	4.0
B_2	23.4	18.8	18.8	20.8	23.0
B_3	7.6	14.8	25.2	32.4	40.0

ab

	B_1	B_2	B_3
A_1	18.7	23.5	23.3
A_2	16.8	19.7	24.5
A_3	13.4	19.7	24.2

$\overline{X}$

20.4

TABLE 11.7. Overall analysis of data in Table 11.6.

	SS	df	MS	F	p	$\hat{\omega}^2$
m	93,881.0	1				
a	278.7	2	139.4	0.63	—	.00
b	2,243.8	2	1122	5.10	.007	.03
c	247.4	4	61.8	0.28	—	(-.01)
ab	344.3	4	86.1	0.39	—	(-.01)
ac	4,846.1	8	605.8	2.75	.007	.05
bc	16,880.2	8	2110	9.59	<.001	.22
abc	3,840.5	16	240.0	1.09	.37	.00
w	39,600.0	180	220.0			
t	68,281.0	224				

TABLE 11.8. Trend analysis on C main effect and AC and BC interactions; data from Table 11.6.

		C_1	C_2	C_3	C_4	C_5
			C Main Effect			
	V	29	39	49	59	69
	Mean	20.07	19.33	19.73	20.67	22.33
	Linear	-2	-1	0	1	2
	Quad	2	-1	-2	-1	2
	Cubic	-1	2	0	-2	1
	Quartic	1	-4	6	-4	1

	C	SS	F	p	$\hat{\omega}^2$
Linear	5.87	154.9	0.70	—	.00
Quad	5.33	91.4	0.42	—	.00
Cubic	-0.40	0.7	0.00	—	.00
Quartic	0.80	0.4	0.00	—	.00

		AC Interaction						
	Linear		Quadratic		Cubic		Quartic	
	C	SS	C	SS	C	SS	C	SS
A_1	9.2	127.0	-4.4	20.7	1.6	3.8	13.2	37.3
A_2	0.4	0.2	-35.6	1,357.9	1.2	2.2	-16.4	57.6
A_3	8.0	96.0	56.0	3,360	-4.0	24.0	5.6	6.7
Sum		223.2		4,738.6		30.0		101.6
C Main effect	154.9		91.4		0.7		0.4	
SS	68.3		4647.2		29.3		101.2	
df	2		2		2		2	
MS	34.2		2323.6		14.6		50.6	
F	0.16		10.56		0.07		0.23	
p	—		<.001		—		—	
$\hat{\omega}^2$	(-.01)		.06		(-.01)		.00	

TABLE 11.8. *Continued*

					BC Interaction			
	Linear		Quadratic		Cubic		Quartic	
	C	*SS*	*C*	*SS*	*C*	*SS*	*C*	*SS*
B_1	-66.0	6,534.0	2.8	8.4	-6.0	54.0	-8.4	15.1
B_2	1.2	2.2	15.6	260.7	-4.4	29.0	0.8	0.1
B_3	82.4	10,184.6	-2.4	6.2	-2.8	11.8	10.0	21.4
Sum		16,720.8		275.3		94.8		36.6
C Main Effect		154.9		91.4		0.7		0.4
SS		16,565.9		183.9		94.1		36.2
df		2		2		2		2
MS		8,283		92.0		47.0		18.1
F		37.7		0.42		0.21		0.08
p		<.001		—		—		—
$\hat{\omega}^2$		.24		.00		(-.01)		(-.01)

it simply means that, on the whole, rich people buy more detergent than poor people. The linear *BC* interaction is not surprising either. From Figure 11.3 it is clear that rich people tend to buy the detergent most readily when it is most expensive (a positive linear trend), perhaps believing that more expensive detergents must be better. Poor people, on the other hand, buy most readily when it is least expensive (a negative linear trend).

The significant quadratic trend in the *AC* interaction most probably comes from the curves for the middle-income group, which show a pronounced tendency for midwestern buyers to prefer the moderately priced product, whereas easterners and westerners buy most readily when the price is either very high or very low.

As usual, both the *F* tests and the approximately unbiased estimates of ω^2 depend on whether Factors *A* and *B* are random or fixed. The analyses made above assume that they are fixed; if one or both were random, the denominators for the *F* tests would be those appropriate for that type of design. In addition, the unbiased estimate of the variance of each effect would be different. Instead of subtracting MS_w from the mean square for the effect, we would subtract whatever mean square was used in the denominator for testing that effect. If *A* were random, for example, MS_{ac} would be subtracted from the *SS* for the trend components of the *C* main effect, and MS_{abc} would be subtracted from the mean squares for the *BC* interaction. The procedure after that would be the same as in the fixed-effects model; the difference would be multiplied by the degrees of freedom and divided by *n* to estimate the variance of the effect. Dividing that variance by $\hat{\sigma}_t^2$ would give $\hat{\omega}^2$.

Estimates of variances and proportions of variance accounted for are found from the sums of squares in the usual way.

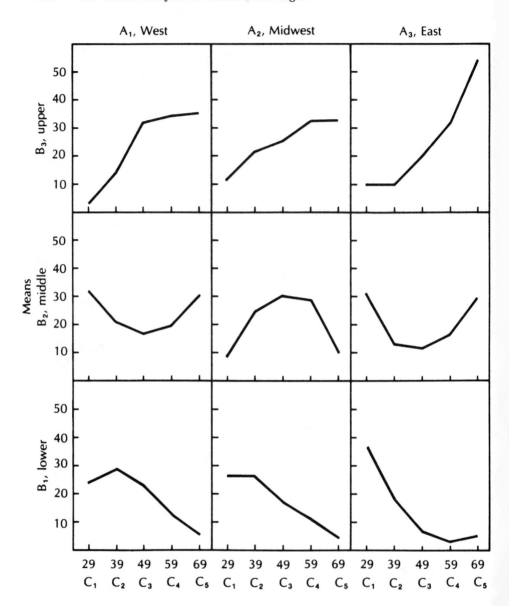

FIGURE 11.3. Obtained means; data from Table 11.6.

SIGNIFICANCE TESTS

The procedure for analyzing the two-way interaction into trend components is extended to three-way and higher interactions with only minor modifications. The procedure for the three-way interaction, for the data in Table 11.6, is illustrated in Table 11.9. First, values of C_k and SS_k are calculated for each trend component and for every combination of levels of Factors A and B (there are nine such combinations), making 36 values in all (nine for each of the four trend components).

To calculate the linear trend component, SS_1 is calculated separately for each of the nine combinations of levels of A and B. These nine values are then summed to give the value 17,720.6. From this sum are subtracted the sums of squares for the linear components of the C main effect, the AC interaction, and the BC interaction. The total of these three sums of squares is 16,789.1, and the difference, when this total is subtracted from 17,720.6, is 931.5. The total number of SS_1 that were summed is nine, and the total of the degrees of freedom of the terms subtracted out is five, leaving four degrees of freedom for the linear component of the ABC interaction. This gives a mean square of 232.9 and an F of 1.06, which is not significant.

The general rule for deciding which sums of squares to subtract is to find all lower-order interactions and main effects involving all of the *numerical* factors in the interaction in question. In our example Factor C alone is numerical, so we find all lower-order interactions involving C. These, as stated above, are the C main effect and the AC and BC interaction. The sums of squares for the trend components of these effects are subtracted from the totals of the SS_k for the corresponding trends on the ABC interaction.

The quadratic, cubic, and quartic components are calculated in exactly the same way. The result of each set of calculations is a sum of squares with four degrees of freedom, and there are four such sums of squares, so their total degrees of freedom are 16—the degrees of freedom of the ABC interaction. Moreover, the total of the sums of squares is $931.5 + 2{,}742.4 + 79.7 + 86.9 = 3{,}840.5 = SS_{abc}$.

Variances and proportions of variance accounted for are estimated in the usual way. The complete trend analysis on the data in Table 11.6 is summarized in Table 11.10.

ESTIMATION

The significant effects are the B main effect, the quadratic component of the AC interaction, the linear component of the BC interaction, and the quadratic component of the ABC interaction. In the three-way design, however, the estimates for the quadratic components of the ABC interaction will include those for the AC interaction. Thus, only the B main effect, the linear BC interaction, and the quadratic ABC interaction need to be included in the estimation equation.

To see that the quadratic components of the ABC interaction include those

TABLE 11.9. Trend analysis on three-way interaction; data from Table 11.6.

		Linear		Quadratic		Cubic		Quartic	
		C	SS	C	SS	C	SS	C	SS
A_1	B_1	-54.0	1,458.0	-30.0	321.4	12.0	72.0	-2.4	0.4
	B_2	-3.6	6.5	54.0	1,041.4	1.2	0.7	-2.4	0.4
	B_3	85.2	3,629.5	-37.2	494.2	-8.4	35.3	44.4	140.8
A_2	B_1	-61.2	1,872.7	-10.8	41.7	8.4	35.3	-18.0	23.1
	B_2	8.4	35.3	-78.0	2,172.9	-4.8	11.5	-9.6	6.6
	B_3	54.0	1,458.0	-18.0	115.7	0.0	0.0	-21.6	33.3
A_3	B_1	-82.8	3,427.9	49.2	864.5	-2.4	2.9	-4.8	1.6
	B_2	-1.2	0.7	70.8	1,790.2	-9.6	46.1	14.4	14.8
	B_3	108.0	5,832.0	48.0	822.9	0.0	0.0	7.2	3.7
Sum			17,720.6		7,664.9		203.8		224.7
C	Main effect		154.9		91.4		0.7		0.4
AC	Interaction		68.3		4,647.2		29.3		101.2
BC	Iinteraction		16,565.9		183.9		94.1		36.2
Sum			16,789.1		4,922.5		124.1		137.8
SS			931.5		2,742.4		79.7		86.9
df			4		4		4		4
MS			232.9		685.6		19.9		21.7
F			1.06		3.12		0.09		0.10
p			.38		.017		—		—
$\hat{\omega}^2$			.00		.03		(-.01)		(-.01)

TABLE 11.10. Summary of trend analyses on data in Table 11.6 (bold numbers are totals).

	SS	df	MS	F	p	$\hat{\omega}^2$
m	93,881.0	1				
a	278.7	2	139.4	0.63	—	.00
b	2,243.8	2	1,122	5.10	.007	.03
c	**247.4**	**4**			—	**(-.01)**
Linear	154.9	1	154.9	0.70	—	.00
Quad	91.4	1	91.4	0.42	—	.00
Cubic	0.7	1	0.7	0.00	—	.00
Quartic	0.4	1	0.4	0.00	—	.00
ab	344.3	4	86.1	0.39	—	(-.01)
ac	**4,846.1**	**8**				.05
Linear	68.3	2	34.2	0.16	—	(-.01)
Quad	4,647.2	2	2,324	10.6	<.001	.06
Cubic	29.3	2	14.6	0.07	—	(-.01)
Quartic	101.2	2	50.6	0.23	—	.00
bc	**16,880.2**	**8**				.22
Linear	16,565.9	2	8,283	37.7	<.001	.24
Quad	183.9	2	92.0	0.42	—	.00
Cubic	94.1	2	47.0	0.21	—	(-.01)
Quartic	36.2	2	18.1	0.08	—	(-.01)
abc	**3,840.5**	**16**				.00
Linear	931.5	4	232.9	1.06	.38	.00
Quad	2,742.4	4	685.6	3.12	.017	.03
Cubic	79.7	4	19.9	0.09	—	(-.01)
Quartic	86.9	4	21.7	0.10	—	(-.01)
w	39,600.0	180	220.0			
t	68,281.0	224				

for the AC interaction, take the average, over the levels of B, of the C values for the quadratic ABC interaction. The average for level A_1, for example, is $(-30.0 + 54.0 - 37.2)/3 = -4.4$. Table 11.8 shows that -4.4 is the C value of A_1 for the quadratic component of the AC interaction. Consequently, the averages of the C values for the quadratic ABC interaction take into account those for the AC interaction. In general, the terms for any trend component of an interaction will include, implicitly, the terms for the same trend on all main effects and lower-order interactions involving all of the *numerical* factors in the higher interaction (Factor C in this case). These are the same terms as those that are subtracted when finding the sum of squares of the higher interaction. For the ABC interaction in this design, it inclues the AC and BC interactions and the C main effect. Of these the only significant quadratic term is in the AC interaction.

The estimation equation, using the B main effect, the linear BC interaction, and the quadratic ABC interaction, is

$$\hat{\mu}_{ijk} = \hat{\mu} + \hat{\beta}_j + \hat{c}^*(b_j)_1 c_{k1} + \hat{c}^*(ab_{ij})_2 c_{k2}$$
$$= \overline{X}_{.j..} + \hat{c}^*(b_j)_1 c_{k1} + \hat{c}^*(ab_{ij})_2 c_{k2}.$$

Our symbols are now even more complex, and the lower-case c has taken on two meanings. In general, when the term is starred, it refers to the C_k factor; when it is not, it refers to a coefficient for calculating a trend. Thus, c_{k1} is the kth coefficient for testing the linear trend component ($c_{11} = -2, c_{21} = -1, \ldots, c_{51} = 2$), and c_{k2} is the kth coefficient for testing the quadratic trend component. The term $\hat{c}^*(b_1)_j$ refers to the estimated linear effect over Factor C at level B_j, and $\hat{c}^*(ab_{ij})_2$ refers to the estimated quadratic trend over Factor C at level AB_{ij}.

Each of the latter estimates is calculated in the usual way; i.e., the corresponding C_k value (the value from which SS_k is calculated) is divided by the sum of the squared coefficients from which it was derived. Thus, the $\hat{c}^*(b_j)_1$ are calculated from the column labeled C_1 under BC interaction in Table 11.8 as

$$\hat{c}^*(b_1)_1 = -66.00/10 = -6.6,$$
$$\hat{c}^*(b_2)_1 = 1.20/10 = 0.12,$$
$$\hat{c}^*(b_3)_1 = 82.40/10 = 8.24.$$

(The divisor, 10, is the sum of the c^2 for the linear trend on the C main effect; this can be seen at the top of Table 11.8.)

The nine values of $\hat{c}^*(ab_{ij})_2$ are calculated by the same formula from the values in the column labeled C_2 in Table 11.9:

$$\hat{c}^*(ab_{11})_2 = -30.0/14 = -2.14,$$
$$\hat{c}^*(ab_{12})_2 = 54.0/14 = 3.86,$$
$$\cdots$$
$$\hat{c}^*(ab_{33})_2 = 48.0/14 = 3.43.$$

These values, along with the estimated cell means, are in Table 11.11. The estimated cell means are plotted, along with the actual cell means, in Figure 11.4.

GENERAL PRINCIPLES

The general principles and their application to higher-way interactions should now be clear. First, SS_k values are computed for the trend in question for each combination of all of the other factors involved. These SS_k are then totaled, and from this total is subtracted the total of the sums of squares for all lower-order effects involving the numerical factor (and involving *only* factors that are in the interaction in question) for the same trend component. The result is a sum of squares whose degrees of freedom are the number of SS_k that were summed, minus the total degrees of freedom of the sums of squares that were subtracted. A simple check on the calculations derives from the fact that the resulting sums of squares and degrees of freedom must themselves sum to the sum of squares and degrees of freedom of the interaction being partitioned.

Coefficients for estimating effects are always found by dividing the C_k value for the effect by the sum of the squares of the coefficients used in finding that C value. Estimates of a trend component in a higher-order interaction implicitly include those for the same trend in all lower-order interactions and main effects involving the numerical factor (and that involve *only* factors that are involved in the higher-order interaction).

The variances of the effects and the proportions of variance accounted for are estimated by adapting the standard formulas.

Two Numerical Factors

A multifactor design may sometimes contain more than one factor with numerical values. When it does, trends on interactions can be more finely partitioned. For example, suppose that in the study in Table 11.1 the three schools had been of three different, evenly spaced sizes (say, 400, 800, and 1,200 pupils), and we had wanted to study the effects of size of school as well as student IQ. The trend analysis on the A main effect (IQ level) would then be the same as before, but we could perform a trend analysis on school size as well. The procedure is exactly the same as that for the trend analysis on Factor A, and the analysis is shown in Table 11.12.

TWO-WAY INTERACTIONS

The two-way interaction can be divided into as many trend components as there are degrees of freedom, with each trend component being tested by a planned comparison. More specifically, each trend in Table 11.3 can be divided into two components, each with one degree of freedom. The linear component on A, for example, can be divided into two components, a linear $\times$ linear and

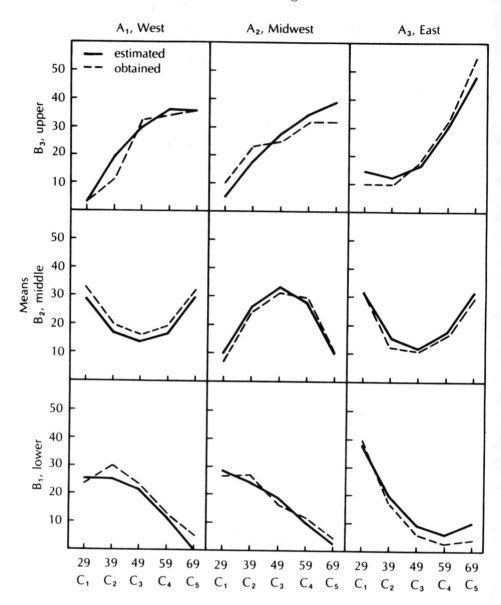

FIGURE 11.4. Estimated and obtained means, Factor C numerical; data from Table 11.6.

TABLE 11.11. Estimates of effects and cell means for data in Table 11.6, Factor C numerical.

μ	β_1	β_2	β_3	$c^*(b_1)_1$	$c^*(b_2)_1$	$c^*(b_3)_1$
20.43	-4.11	0.53	3.57	-6.60	0.12	8.24

$$c^*(ab_{ij})_2$$

	A_1	A_2	A_3
B_1	-2.14	-0.77	3.51
B_2	3.86	-5.57	5.06
B_3	-2.66	-1.29	3.43

Obtained (upper values) and estimated (lower values) cell means

		C_1	C_2	C_3	C_4	C_5
A_1	B_1	24.0	28.8	22.8	13.2	4.8
		25.2	25.1	20.6	11.9	-1.2
	B_2	31.8	20.4	15.6	19.2	30.6
		28.4	17.0	13.2	17.2	28.9
	B_3	2.4	13.2	32.4	33.6	34.8
		2.2	18.4	29.3	34.9	35.2

A_2	B_1	26.4	26.4	16.8	10.8	3.6
		28.2	23.7	17.9	10.5	1.6
	B_2	7.2	24.0	30.0	27.6	9.6
		9.6	26.4	32.1	26.6	10.1
	B_3	10.8	21.6	25.2	32.4	32.4
		4.9	17.0	26.6	33.5	37.9

A_3	B_1	37.2	18.0	6.0	2.4	3.6
		36.5	19.4	9.3	6.2	10.1
	B_2	31.2	12.0	10.8	15.6	28.8
		30.8	15.8	10.8	16.0	31.3
	B_3	9.6	9.6	18.0	31.2	52.8
		14.4	12.3	17.1	28.8	47.3

TABLE 11.12. Trend analysis on Factor B; data from Table 11.1.

	B_1	B_2	B_3
School size	400	800	1200
Mean	45.00	52.75	57.75
c_{ik} Linear	-1	0	1
Quad	1	-2	1

	C	SS	F	p	$\hat{\omega}^2$
Linear	12.75	3,251.2	81.3	<.001	.13
Quad	-2.75	50.4	1.26	.26	.00

a linear × quadratic. The first is a linear trend, across the levels of B, on the linear trend components of A; i.e., it is a tendency for the linear trend components of A to increase or decrease evenly over levels of B. The second is a quadratic trend, across levels of B, on the linear components of A.

Because A and B are numerical, however, and there is no good mathematical reason why one should take precedence over the other, the linear × linear trend can also be thought of as a linear trend, across levels of A, on the linear trend components of B. Similarly, the linear × quadratic trend can be thought of as a linear trend, across the levels of A, on the quadratic components of B. These points will be clearer when the actual analyses are described.

Carrying this partitioning across the trend components in Table 11.3, we also obtain a quadratic × linear, quadratic × quadratic, cubic × linear, and cubic × quadratic trend. There are a total of six trend components, one for each of the six degrees of freedom in the AB interaction. Each can be thought of as a trend, across one factor, on the components of a trend on the other factor.

Alternatively, these trends can be expressed in terms of an extension of the model equation. We have seen that the model equation with one factor numerical can be extended to include a term for each trend component. The model equation for the two-way analysis, with both factors numerical, can also be extended to represent each trend component. If we let V_i represent the numerical value of the ith level of Factor A, as before, and let W_j represent the numerical value of the jth level of Factor B, then the extended equation is

$$X_{ijk} = \mu + [a_1(V_i - \overline{V}) + a_2(V_i - \overline{V})^2 + a_3(V_i - \overline{X})^3 + \cdots]$$

$$+ [b_1(W_j - \overline{w}) + b_2(W_j - \overline{w})^2 + b_3(W_j - \overline{w})^3 + \cdots]$$

$$+ [ab_{11}(V_i - \overline{V})(W_j - \overline{w}) + ab_{12}(V_i - \overline{V})(W_j - \overline{w})^2 + \cdots$$

$$+ ab_{21}(V_i - \overline{V})^2(W_j - \overline{w}) + ab_{22}(V_i - \overline{V})^2(W_j - \overline{w})^2 + \cdots]$$

$$+ \epsilon_{ijk}. \tag{11.1}$$

The terms in the first set of brackets represent the A main effect; each value of a_k represents one trend component, to be tested with a planned comparison having one degree of freedom. The terms in the second set of brackets represent the B main effect; each b_l represents one B main effect trend component. In the same way, the terms in the third set of brackets are those of the AB interaction; each ab_{kl} represents one trend component of the AB interaction. The term ab_{11} represents the linear × linear component, ab_{12} represents the linear × quadratic, and so on.

Generally, Equation 11.1 is used when estimating the μ_{ij}, in much the same way that Equation 10.3 is used in the one-way trend test; those terms that are significant are kept, and those that are not significant are discarded.

Significance Tests. Each of the above interpretations of two-way trends suggests a different method of testing for them. The method suggested by Equation 11.1 is in some ways more basic, so it will be described first.

First method. Just as the rationale for the tests of one-way trend components is difficult to explain, so is the rationale for the two-way tests; we will simply describe the tests without attempting to give their rationale. However, it is well to remember that the coefficients in Table A.7 can be used only if every cell contains the same number of scores and the values of both the V_i and the W_j are equally spaced.

The procedure is illustrated in Table 11.13. With this procedure, it is necessary to construct a two-way table of coefficients for each trend component. To find the linear × linear coefficients, we write the values of the coefficients (c_{i1}) that were used to test the linear component of the A main effect in one margin of the table; in Table 11.13 the columns represent the A main effect, so these coefficients are written as column headings. In the other margin we write the coefficients for the linear trend of the B main effect. The entry in each cell is then the product of that cell's row and column values. The entry in the top left-hand cell, for example, is $(-1)(-3) = 3$. The cell entries are the coefficients for testing the linear × linear trend component. The test itself is an ordinary planned comparison, using the cell entries for the linear × linear part of Table 11.13 as coefficients for the cell means in Table 11.1. For the linear × linear trend

$$C_{11} = 3(31) + 1(43) - 1(48) - 3(58)$$
$$-3(37) - 1(59) + 1(66) + 3(69) = 17.$$

The value of SS_{11} (72.25 for these data) is then found in the usual way, using n as the multiplier, and the sum of the squares of the coefficients in the cells of the linear × linear table (40 for these data) as the divisor. The F ratio, 1.81, is not significant.

The rest of the trend components are tested in exactly the same way, but the values used as row and column values are different. For the linear × quadratic component, for example, the column values are those for the linear component of the A main effect, as before, but the row values are those for the quadratic component of the B main effect. The cell entries are the products of the row and column values, as before, and the test is a planned comparison.

Table 11.13 shows that the only significant two-way trend is the quadratic × linear, indicating that there is a linear trend in the quadratic components of A. This linear trend can be seen in Figure 11.1, which shows that the amount of curvature across levels of A increases from the nearly straight trend in school B_1 to the highly curved function in school B_3. An alternative interpretation of the quadratic × linear trend can be seen in Figure 11.5, which plots trends across levels of B for different levels of A. Here we see a quadratic trend in the slope of the function, with the greatest slope occurring for A_3 and the next greatest for A_2. (Differences in the overall heights of the curves reflect the A main effect.)

Second method. The other method of calculating the SS for each trend component is generally simpler computationally. It makes use of the fact that the linear × quadratic trend, for example, is a quadratic trend, over the levels of B, for the linear trend on A. Accordingly, to test the linear × linear and linear × quadratic trends, we first find the C values for the linear trend on A, at each level of B. These were found in Table 11.3 and are reproduced in the upper table of Table 11.14. We then test for linear and quadratic trends on these C values. To find the linear × quadratic trend, for example, we do a quadratic trend test on the linear C values: $C_{12} = 1(86) - 2(111) + 1(103) = -33$. This, as can be seen from Table 11.13, is the same as the linear × quadratic C value calculated by the previous method. The SS are then found by squaring each C value, multiplying by the number of scores in each cell, and dividing by the sum of the squares of the coefficients as before. The sum of the squares of the coefficients can be found, however, without first finding every coefficient. To find the sum of the squares of the coefficients for the linear × quadratic trend, we first find the sum of the squared coefficients for the linear trend on the A main effect (20, as can be seen from Table 11.12). We then multiply it by the sum of squared coefficients for the quadratic trend on the B main effect (6, from Table 11.2), to get $20 \times 6 = 120$. The SS for the linear × quadratic trend is thus $10(-33)^2/120 = 90.75$. The SS values for the other trends are found the same way. The complete set of tests is shown

TABLE 11.13. Trend analysis on AB interaction; data from Table 11.1, both A and B numerical.

Linear × linear		Quadratic × quadratic	

Linear × linear

A

		-3	-1	1	3
	-1	3	1	-1	3
B	0	0	0	0	0
	1	-3	-1	1	3

$C_{11} = 17,\ SS_{11} = 72.2$
$F = 1.81,\ p = .18$
$\hat{\omega}^2 = .00$

Quadratic × quadratic

A

		1	-1	-1	1
	1	1	-1	-1	1
B	-2	-2	2	2	-2
	1	1	-1	-1	1

$C_{22} = 5,\ SS_{22} = 10.4$
$F = 0.26$
$\hat{\omega}^2 = .00$

Linear × quadratic

A

		-3	-1	1	3
	1	-3	-1	1	3
B	-2	6	2	-2	-6
	1	-3	-1	1	3

$C_{12} = -33,\ SS_{12} = 90.8$
$F = 2.27,\ p = .13$
$\hat{\omega}^2 = .00$

Cubic × linear

A

		-1	3	-3	1
	-1	1	-3	3	-1
B	0	0	0	0	0
	1	-1	3	-3	1

$C_{31} = -1,\ SS_{31} = 0.2$
$F = 0.01$
$\hat{\omega}^2 = .00$

Quadratic × linear

A

		1	-1	-1	1
	-1	-1	1	1	-1
B	0	0	0	0	0
	1	1	-1	-1	1

$C_{21} = -17,\ SS_{21} = 361.2$
$F = 9.03,\ p = .003$
$\hat{\omega}^2 = .01$

Cubic × quadratic

A

		-1	3	-3	1
	1	-1	3	-3	1
B	-2	2	-6	6	-2
	1	-1	3	-3	1

$C_{32} = 29,\ SS_{32} = 70.1$
$F = 1.75,\ p = .19$
$\hat{\omega}^2 = .00$

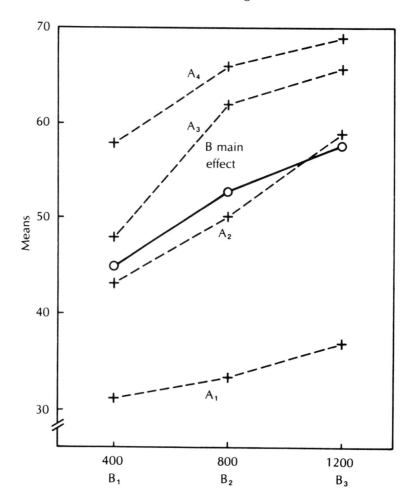

FIGURE 11.5. Cell means in Table 11.1, plotted as a function of Factor *B*.

in Table 11.14. Estimates of variances and proportions of variance accounted for follow the usual procedures for contrasts.

Estimation. The significant quadratic × linear interaction, together with the significant linear and quadratic trend in the A main effect, plus the significant linear trend in the B main effect, suggest that the data can be accounted for most economically with the equation

$$\hat{\mu}_{ij} = \hat{\mu} + \hat{a}_1^* c_{i1(a)} + \hat{a}_2^* c_{i2(a)} + \hat{b}_1^* c_{j1(b)} + \hat{ab}_{12}^* c_{ij,21(ab)}.$$

In this equation, $c_{i1(a)}$ is the coefficient of level A_i for testing the linear component of the A main effect; i.e.,

$$C_{11(a)} = -3, \quad c_{21(a)} = 1, \quad c_{31(a)} = 1, \quad c_{41(a)} = 3.$$

Similarly, $c_{i2(a)}$ is the coefficient of level A_i for testing the quadratic component of the A main effect, and $c_{j1(b)}$ is the coefficient of level B_j for testing the linear component of the B main effect. The value $c_{ij,21(ab)}$ is the coefficient of level AB_{ij} for testing the quadratic × linear component of the AB interaction; e.g.,

$$
\begin{aligned}
c_{11,21(ab)} &= 3, \; c_{12,21(ab)} = 0, \; c_{13,21(ab)} = -3, \\
c_{21,21(ab)} &= 1,
\end{aligned}
$$

and so on. Unfortunately, the subscripts make these terms look more complicated than they are. The equation for $\hat{\mu}_{11}$, for example, would be

$$\hat{\mu}_{11} = 51.83 + \hat{a}_1^*(-3) + \hat{a}_2^*(1) + \hat{b}_1^*(-1) + \hat{ab}_{12}^*(-1).$$

To estimate the effects a_1^*, a_2^*, b_1^*, and ab_{12}^*, we follow the procedure described in Chapter 10. In particular, the equation used is Equation 10.11, with the divisor being in each case the sum of the squares of the coefficients used for testing that effect. The estimates are

$$
\begin{aligned}
\hat{a}_1^* &= 100/20 = 5, \\
\hat{a}_2^* &= -11.33/4 = -2.83, \\
\hat{b}_1^* &= 12.75/2 = 6.38, \\
\hat{ab}_{12}^* &= -17/8 = -2.12.
\end{aligned}
$$

and

$$
\begin{aligned}
\hat{\mu}_{11} &= 51.83 + 5.00(-3) - 2.83(1) + 6.38(-1) - 2.12(-1) \\
&= 29.7.
\end{aligned}
$$

The other estimates, along with the obtained cell means, are shown in Table 11.15 and Figure 11.6. In Figure 11.6 the increasing quadratic trend over levels of B is especially clear.

TABLE 11.14. Alternative way to calculate AB interaction trends; data from Table 11.1, both A and B numerical.

| | A linear | | | |
	B_1	B_2	B_3	Σc^2
School size	400	800	1,200	
C_1	86	111	103	20
Linear × linear	-1	0	1	2
Linear × quad	1	-2	1	6

	C	SS	F	p	$\hat\omega^2$
Linear × linear	17	72.2	1.81	.18	.00
Linear × quad	-33	90.8	2.27	.13	.00

| | A quadratic | | | |
	B_1	B_2	B_3	Σc^2
School size	400	800	1,200	
C_2	-2	-13	-19	4
Quad × linear	-1	0	1	2
Quad × quad	1	-2	1	6

	C	SS	F	p	$\hat\omega^2$
Quad × linear	-17	361.2	9.03	.003	.01
Quad × quad	5	10.4	0.26	—	.00

| | A cubic | | | |
	B_1	B_2	B_3	Σc^2
School size	400	800	1,200	
C_3	12	-3	11	20
Cubic × linear	-1	0	1	2
Cubic × quad	1	-2	1	6

	C	SS	F	p	$\hat\omega^2$
Cubic × linear	-1	0.2	0.01	—	.00
Cubic × quad	29	70.1	1.75	.19	.00

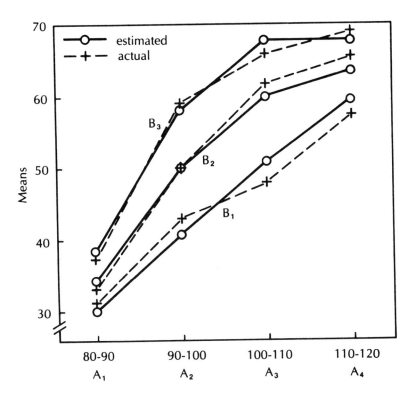

FIGURE 11.6. Estimated and obtained means, both A and B numerical; data from Table 11.1.

TABLE 11.15. Estimated effects and cell means; data from Table 11.1, both A and B numerical.

	Estimated effects			
μ	a_1^*	a_2^*	b_1^*	ab_{21}^*
51.83	5.00	-2.83	6.38	-2.12

Obtained (upper values) and estimated (lower values) cell means

	A_1	A_2	A_3	A_4
B_1	31	43	48	58
	29.7	41.2	51.2	59.7
B_2	33	50	62	66
	34.0	49.7	59.7	64.0
B_3	37	59	66	69
	38.3	58.2	68.2	68.3

TABLE 11.16. Trend analyses on B main effect and two-way interactions; data from Table 11.6.

		B main effect				
	C	Σc^2	SS	F	p	$\hat{\omega}^2$
Linear	7.68	2	2,211.8	10.1	<.001	.04
Quad	-1.60	6	32.0	0.15	—	.00

		AB interaction				
	C	df	MS	F	p	$\hat{\omega}^2$
Linear	243.3	2	121.6	0.55	—	.00
Quad	101.0	2	50.5	0.23	—	(-.01)

			BC interaction			
	C	Σc^2	MS	F	p	$\hat{\omega}^2$
Linear × linear	148.4	20	16,516.9	75.1	<.001	.24
Linear × quad	-5.2	28	14.5	0.07	—	.00
Linear × cubic	-8.8	20	58.1	0.26	—	.00
Linear × quartic	18.4	140	36.3	0.16	—	.00
Quad × linear	14.0	60	49.0	0.22	—	.00
Quad × quad	-30.8	84	169.4	0.77	—	.00
Quad × cubic	12.0	60	36.0	0.16	—	.00
Quad × quartic	0.0	420	0.0	0.00	—	.00

THREE-WAY INTERACTIONS WITH TWO NUMERICAL FACTORS

Three-way and higher-way designs in which two factors are numerical require only minor extensions of the methods discussed above. The extensions will be illustrated using the data in Table 11.6, assuming that the three income levels (Factor B) are equally spaced on some numerical economic scale.

The trends on the B main effect are tested in the usual way, and the AB and BC interactions are divided into trends exactly as in the two-way design discussed above. The results of these analyses are given in Table 11.16. Only the linear trend on the B main effect and the linear × linear trend on the BC interaction are significant.

Significance Tests. The ABC interaction can be divided into the same eight components as the BC interaction (Table 11.16), just as the AB interaction, with B numerical, is divided into the same number of components as the B main effect (Table 11.16). To make significance tests, C and SS are separately calculated for each trend component at each level of A. The three SS for a given trend are then summed and the SS for the corresponding component of the BC interaction is subtracted. In this example only the BC interaction is subtracted because it is the only lower-order effect that involves both numerical factors. The result is then treated as a sum of squares with $(I-1)$ degrees of freedom. To illustrate, with the linear × linear component, the SS for level A_1 is calculated by multiplying the cell means at A_1

	C_1	C_2	C_3	C_4	C_5
B_1	24.0	28.8	22.8	13.2	4.8
B_2	31.8	20.4	15.6	19.2	30.6
B_3	2.4	13.2	32.4	33.6	34.8

by the linear × linear coefficients:

		C_1	C_2	C_3	C_4	C_5
		-2	-1	0	1	2
B_1	-1	2	1	0	-1	-2
B_2	0	0	0	0	0	0
B_3	1	-2	-1	0	1	2

The resulting C value (139.2) is then squared, multiplied by 5 (the number of scores on which each cell mean is based), and divided by 20 (the sum of the squares of the coefficients). The result is $SS = 4,844.2$. The same procedure is followed for the other two levels of A, giving SS values of 3,317.8 for A_2 and 9,101.2 for A_3. These values are summed, and the SS for the linear × linear component of the BC interaction is subtracted to obtain the sum of squares: $4,844.2 + 3,317.8 + 9,101.2 - 16,516.9 = 746.3$, with 2 degrees of freedom. The mean square is then $746.3/2 = 373.2$, and $F = 373.2/220 = 1.70$, which is not significant.

The remaining seven tests can be made the same way. The tests are shown in Table 11.17. The quadratic × quadratic interaction is significant with $p = .02$; none of the others are significant.

The method illustrated in Table 11.14, for the data in Table 11.1, can also be used to calculate the C values in Table 11.17. We can generate the C values in Table 11.17 from those in Table 11.9 by performing linear and quadratic trend tests, over levels of B, on each level of A. The C values for the quadratic trend at A_1, for example, are -30.0, 54.0, and -37.2, in Table 11.9. To perform a linear test on these values, we use the coefficients for the linear trend on B to get $-1(-30.0) + 0(54.0) + 1(-37.2) = -7.2$, the C value for the linear × quadratic trend at A_1. The divisor used in finding SS, i.e., the sum of the squared coefficients, is the product of the sum of the squared linear coefficients on B times the sum of the squared quadratic coefficients on C; i.e., $2 \times 14 = 28$.

The remaining C and SS values can be found the same way, with a sizable saving in labor because complete tables of coefficients need not be made.

Estimation. The significant effects are now the linear component of the B main effect, the quadratic component of the AC interaction, the linear × linear

TABLE 11.17. Trend analysis on three-way interaction; data from Table 11.6, Factors B and C numerical.

	Linear × linear		Linear × quadratic		Linear × cubic		Linear × quartic	
	C	SS	C	SS	C	SS	C	SS
A_1	139.2	4,844.2	-7.2	9.3	-20.4	104.0	46.8	78.2
A_2	115.2	3,317.8	-7.2	9.3	-8.4	17.6	-3.6	0.5
A_3	190.8	9,101.2	-1.2	0.3	2.4	1.4	12.0	5.1
Sum		17,263.2		18.9		123.0		83.8
BC interaction		16,516.9		14.5		58.1		36.3
SS		746.3		4.4		64.9		47.5
df		2		2		2		2
MS		373.2		2.2		32.4		23.8
F		1.70		0.01		0.15		0.11
p		.19		—		—		—
$\hat{\omega}^2$		.00		(-.01)		(-.01)		(-.01)

	Quadratic × linear		Quadratic × quadratic		Quadratic × cubic		Quadratic × quartic	
	C	SS	C	SS	C	SS	C	SS
A_1	38.4	122.9	-175.2	1827.1	1.2.	0.1	46.8	26.1
A_2	-24.0	48.0	127.2	963.1	18.0	27.0	-20.4	5.0
A_3	27.6	63.5	-44.4	117.3	16.8	23.5	-26.4	8.3
Sum		234.4		2,907.5		50.6		39.4
BC interaction		49.0		169.4		36.0		0.0
SS		185.4		2,738.1		14.6		39.4
df		2		2		2		2
MS		92.7		1,369		7.3		19.7
F		0.42		6.22		0.03		0.09
p		—		.002		—		—
$\hat{\omega}^2$		.00		.03		(-.01)		(-.01)

component of the BC interaction, and the quadratic $\times$ quadratic component of the ABC interaction. The estimation equation is therefore

$$\hat{\mu}_{ijk} = \hat{\mu} + \hat{b}_1^* c_{j1} + \hat{c}^*(a_i)_2 c_{k2} + b\hat{c}_{11}^* c_{jk,11}$$
$$+ b\hat{c}^*(a_i)_{22} c_{jk,22}.$$

Here again we have run out of different symbols, a starred c refers to a level of the C main effect, and an unstarred c refers to a coefficient used in testing for trends. Each starred term is estimated by dividing the related C value by the sum of its squared coefficients. The complete set of estimates is in Table 11.18, and the estimated cell means are in Figure 11.7.

TABLE 11.18. Estimated effects and cell means; data from Table 11.6, B and C numerical.

		Estimated effects			
μ	b_1^*	$c^*(a_1)_2$	$c^*(a_2)_2$	$c^*(a_3)_2$	bc_{11}^*
20.43	3.84	-0.31	-2.54	4.00	7.42

$bc^*(a_1)_{22}$	$bc^*(a_2)_{22}$	$bc^*(a_3)_{22}$
-2.09	1.51	-0.53

Obtained (upper values) and estimated (lower values) cell means

		C_1	C_2	C_3	C_4	C_5
	B_1	24.0	28.8	22.8	13.2	4.8
		26.6	26.4	21.4	11.6	-3.0
A_1	B_2	31.8	20.4	15.6	19.2	30.6
		28.2	16.6	12.7	16.6	28.2
	B_3	2.4	13.2	32.4	33.6	34.8
		4.6	19.2	29.1	34.1	34.3

	B_1	26.4	26.4	16.8	10.8	3.6
		29.4	25.0	18.6	10.2	-0.3
A_2	B_2	7.2	24.0	30.0	27.6	9.6
		9.3	26.0	31.6	26.0	9.3
	B_3	10.8	21.6	25.2	32.4	32.4
		7.4	17.9	26.3	32.7	37.0

	B_1	37.2	18.0	6.0	2.4	3.6
		38.4	20.5	9.6	5.7	8.7
A_3	B_2	31.2	12.0	10.8	15.6	28.8
		30.6	15.4	10.3	15.4	30.6
	B_3	9.6	9.6	18.0	31.2	52.8
		16.4	13.4	17.3	28.2	46.0

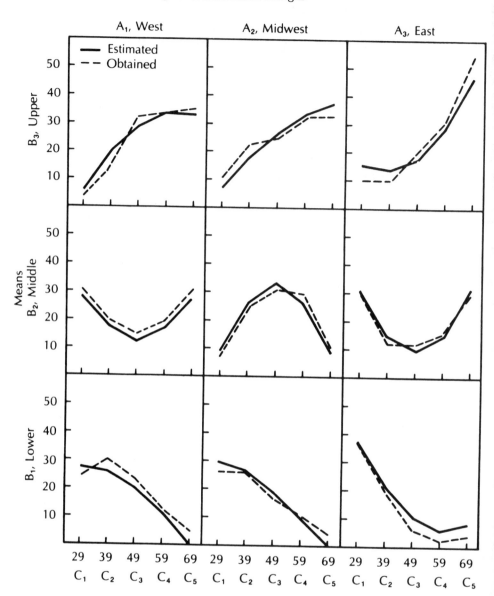

FIGURE 11.7. Estimated and obtained means, both B and C numerical; data from Table 11.6.

TABLE 11.19. Trend components of AC interaction; data from Table 11.6.

	C	Σc^2	SS	F	p	$\hat{\omega}^2$
Linear × linear	-1.2	20	1.1	0.00	—	.00
Linear × quad	60.4	28	1,954.4	8.88	.003	.03
Linear × cubic	-5.6	20	23.5	0.11	—	.00
Linear × quartic	-7.6	140	6.2	0.03	—	.00
Quad × linear	16.4	60	67.2	0.31	—	.00
Quad × quad	122.8	84	2,692.8	12.2	<.001	.04
Quad × cubic	-4.8	60	5.8	0.03	—	.00
Quad × quartic	51.6	420	95.1	0.43	—	.00

More than Two Numerical Factors

Experimental designs with more than two numerical factors are relatively rare: nevertheless, their analysis is a logical extension of the methods described in the previous sections. Main effects and two-way interactions are of course handled exactly as in the previous sections, no matter what the total number of numerical factors may be, as are all the three-way interactions in which no more than two factors are numerical. The remaining effects are three-way interactions involving three numerical factors, and interactions involving a total of more than three factors.

THREE-WAY INTERACTIONS

Three-way interactions will be illustrated with a further analysis of the data in Table 11.6. This time we will assume that the three areas of the country (Factor A) are approximately equally spaced geographically, so that we can assign to them the numerical values -1 for the west, 0 for the midwest, and +1 for the east. All three factors in that design are then numerical. For these data there is no use dividing the A main effect or the AB interaction into trend components because the sums of squares of both are so small that no single trend could be significant. (Remember, the SS for any trend cannot exceed the SS for the effect of which it is a part.) The analysis of the AC interaction is straightforward and appears in Table 11.19. The linear × quadratic and quadratic × quadratic components are significant.

The three-way interaction, with 16 degrees of freedom, can now be divided into 16 trend components. The 16 components correspond to the 16 ways of combining the two possible components (linear and quadratic) on A, the two on B, and the four on C. They are listed as the 16 rows in Table 11.20, to which we will return later.

Each trend component is tested by a planned comparison whose coefficients are the products of the corresponding coefficients for the individual factors.

TABLE 11.20. Trend tests on three-way interaction; data from Table 11.6, all factors numerical.

			Factor A linear			
			A_1	A_2	A_3	Σc^2
		c_{ik}	-1	0	1	2
B	C					
	Linear		193.2	115.2	190.8	20
Linear	Quad		-7.2	-7.2	-1.2	28
	Cubic		-20.4	-8.4	2.4	20
	Quartic		46.8	-3.6	12.0	140
	Linear		38.4	-24.0	27.6	60
Quad	Quad		-175.2	127.2	-44.4	84
	Cubic		1.2	18.0	16.8	60
	Quartic		46.8	-20.4	-26.4	420

		C	SS	F	p	$\hat{\omega}^2$
B	C					
	Linear	51.6	332.8	1.51	.22	.00
Linear	Quad	6.0	3.2	0.01	—	.00
	Cubic	22.8	65.0	0.30	—	.00
	Quartic	-34.8	21.6	0.10	—	.00
	Linear	-10.8	4.9	0.02	—	.00
Quad	Quad	130.8	509.2	2.31	.13	.00
	Cubic	15.6	10.1	0.05	—	.00
	Quartic	-73.2	31.9	0.14	—	.00

TABLE 11.20. *Continued*

		Factor A quadratic			
		A_1	A_2	A_3	Σc^2
	c_{ik}	1	-2	1	6
B	**C**				
	Linear	139.2	115.2	190.8	20
Linear	Quad	-7.2	-7.2	-1.2	28
	Cubic	-20.4	-8.4	2.4	20
	Quartic	46.8	-3.6	12.0	140
	Linear	38.4	-24.0	27.6	60
Quad	Quad	-175.2	127.2	-44.4	84
	Cubic	1.2	18.0	16.8	60
	Quartic	46.8	-20.4	-26.4	420

		C	SS	F	p	$\hat{\omega}^2$
B	**C**					
	Linear	99.6	413.3	1.88	.17	.00
Linear	Quad	6.0	1.1	0.00	—	.00
	Cubic	-1.2	0.1	0.00	—	.00
	Quartic	66.0	25.9	0.12	—	.00
	Linear	114.0	180.5	0.82	—	.00
Quad	Quad	-474.0	2,228.9	10.1	.002	.03
	Cubic	18.0	4.5	0.02	—	.00
	Quartic	61.2	7.4	0.03	—	.00

TABLE 11.21. Coefficients for testing linear $\times$ quadratic $\times$ quadratic interaction component on data in Table 11.6.

		A	B	\multicolumn{5}{c}{C Quadratic}
		Lin	Quad	2	-1	-2	-1	2
A_1	B_1	-1	1	-2	1	2	1	-2
	B_2	-1	-2	4	-2	-4	-2	4
	B_3	-1	1	-2	1	2	1	-2
A_2	B_1	0	1	0	0	0	0	0
	B_2	0	-2	0	0	0	0	0
	B_3	0	1	0	0	0	0	0
A_3	B_1	1	1	2	-1	-2	-1	2
	B_2	1	-2	-4	2	4	2	-4
	B_3	1	1	2	-1	-2	-1	2

Table 11.21 illustrates the method of finding the coefficients for the linear $\times$ quadratic $\times$ quadratic component. The component is linear in Factor A, so we use the coefficients for the linear trend on A, -1, 0, 1; the component is quadratic in B, so for B we use the quadratic coefficients 1, -2, 1; and the component is quadratic in C, so we use the quadratic components 2, -1, -2, -1, 2. The term in each cell is the product of the three corresponding coefficients for the individual factors. Thus, the coefficient in cell ABC_{111} is $(-1)(1)(2) =$ -2, the term in cell ABC_{121} is $(-1)(-2)(2) = 4$, and so on. The terms in the cells are the coefficients for calculating the linear $\times$ quadratic $\times$ quadratic interaction as a planned comparison.

However, it is tedious to calculate the full set of coefficients for all 16 trend components. A shorter way to test the trends is shown in Table 11.20. There we make use of the fact that we have already analyzed the same data, with B and C numerical, in Table 11.17. To generate the 16 trend components in Table 11.20, we simply perform a linear and a quadratic trend test on each of the eight sets of C values in Table 11.17. To find the linear $\times$ quadratic $\times$ quadratic trend, for example, we first write down the three C values for the quadratic $\times$ quadratic trend, across A: -175.2 at A_1, 127.2 at A_2, and -44.4 at A_3. The C value for a linear trend on these three values is $-1(-175.2) + 0(127.2)$ $+1(-44.4) = 130.8$. The sum of the squared coefficients, i.e., the divisor when finding the SS, is found by multiplying the sum of the squared coefficients of the linear trend on A $(=2)$ with that of the BC quadratic $\times$ quadratic trend (84), to get 168. Therefore, the SS is $5(130.8)^2/168 = 509.2$. The remaining trend components, calculated the same way, are shown in Table 11.20; only the quadratic $\times$ quadratic $\times$ quadratic trend is significant. Estimates of effects

TABLE 11.22. Estimated effects and cell means; data from Table 11.6, all factors numerical.

		Estimated effects			
μ	b_1^*	ac_{12}^*	ac_{22}^*	bc_{11}^*	abc_{222}^*
20.43	3.84	2.16	1.46	7.42	-0.94

Obtained (upper values) and estimated (lower values) cell means

		C_1	C_2	C_3	C_4	C_5
	B_1	24.0	28.8	22.8	13.2	4.8
		28.2	25.6	19.9	10.8	-1.5
A_1	B_2	31.8	20.4	15.6	19.2	30.6
		22.8	19.2	18.1	19.2	22.8
	B_3	2.4	13.2	32.4	33.6	34.8
		6.2	18.5	27.6	33.3	35.8

		C_1	C_2	C_3	C_4	C_5
	B_1	26.4	26.4	16.8	10.8	3.6
		29.4	25.0	18.7	10.2	-0.3
A_2	B_2	7.2	24.0	30.0	27.6	9.6
		7.1	27.1	33.8	27.1	7.1
	B_3	10.8	21.6	25.2	32.4	32.4
		7.4	17.9	26.4	32.7	37.0

		C_1	C_2	C_3	C_4	C_5
	B_1	37.2	18.0	6.0	2.4	3.6
		36.8	21.3	11.2	6.5	7.1
A_3	B_2	31.2	12.0	10.8	15.6	28.8
		31.4	14.9	9.4	14.9	31.4
	B_3	9.6	9.6	18.0	31.2	52.8
		14.8	14.2	18.9	29.0	44.5

and proportions of variance accounted for are made just as for lower-order trends. Estimates of ω^2 are in Table 11.20; estimates of the effects and cell means are in Table 11.22. The estimated cell means are plotted in Figure 11.8.

HIGHER-WAY INTERACTIONS

Higher-way interactions in which all factors are numerical are analyzed just like the above interaction. Each interaction can be analyzed into as many

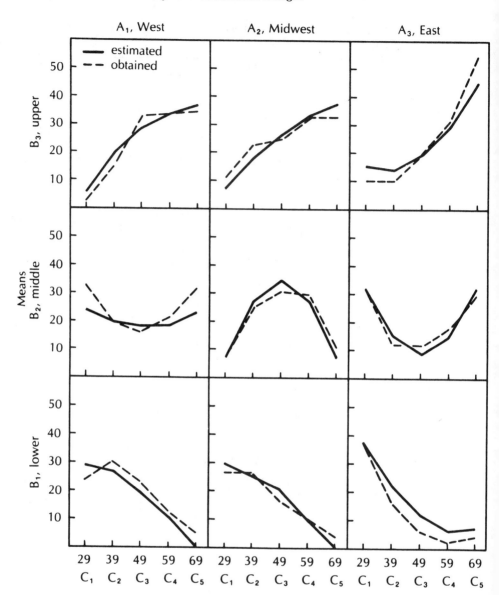

FIGURE 11.8. Estimated and obtained means, all three factors numerical; data from Table 11.6.

trends as there are degrees of freedom, with each trend being a kind of product of the possible trends on the individual factors. The coefficients of the trends on the interactions, moreover, are the products of the coefficients of the trends on the individual factors.

However, it is tedious to calculate all of the products; it is simpler to proceed by the methods used in Tables 11.3 and 11.14, and in Tables 11.9, 11.17, and 11.20. By this method, only one factor is singled out at first, and C values are calculated for all possible trends on that factor, at all possible combinations of levels of the other factors. Then a second factor is singled out, and for each trend on the first factor and all possible values of the remaining factors, C values are computed for all trends on the new factor. This second set of C values is calculated on the C values of the initial analysis. The number of new C values will be the product of the numbers of trends possible on each of the first two factors chosen. This procedure is continued until all of the numerical factors have been singled out and their trends have been calculated. The divisor for finding each SS is then found as the product of the divisors for the corresponding trends on the individual factors.

For interactions involving a mixture of numerical and nonnumerical factors, the procedure is only a little more complicated. First, the procedure described in the last paragraph is carried out, but only the numerical factors are singled out for trend calculations. The result is as many trends as there are combinations of trends on the numerical factors, and as many C values for each trend as there are combinations of levels of nonnumerical factors.

Suppose, for example, that the three-way design of Table 11.6 was expanded to a four-way design by the addition of a fourth factor, D, with four levels. If A, B, and C were all numerical, as before, the four-way interaction would be divisible into 16 trends (the 16 trends found in Table 11.20), and there would be four C values (for the four levels of D) for each trend. If only B and C were numerical, there would be eight trends (see Table 11.17), and for each trend there would be 12 C values, one for each combination of the three levels of A and the four levels of D.

Then for each C value an SS is calculated, and the SS values for a given trend are summed. From this sum is subtracted one or more other sums of squares and SS values. To determine which terms need to be subtracted, find all main effects and lower-order interactions that involve *all* of the numerical factors involved in the trend being analyzed; then subtract the SS for each. In the first example above, with A, B, and C numerical, the only term that would be subtracted would be that for the corresponding trend component of the ABC interaction. With only B and C numerical, we would subtract the corresponding trends in the ABC interaction, BCD interaction, and BC interaction. The degrees of freedom are the total number of SS that were summed, minus the sum of the degrees of freedom of the terms that were subtracted. As a check, the final degrees of freedom should sum to the degrees of freedom for the interaction as a whole.

Estimates of effects, cell means, variances, and proportions of variance accounted for then follow exactly the same procedures as were described earlier.

Factors with Unequally Spaced Numerical Values

In all of the examples above, the factors had equally spaced numerical values. As we saw in Chapter 10, equally spaced numerical values simplify trend analyses. However, trend analyses on main effects and interactions can be performed when the values are not equally spaced, so long as the sample sizes are equal. (They can also be performed with unequal sample sizes, but the problem becomes much more complicated.) Instead of using the coefficients in Table A.7, we must calculate the coefficients. The coefficients are found just as described for the one-way design in Chapter 10. In the example in Table 11.1, if the school sizes had been 400, 1,100, and 1,200, we could still have tested the linear trend on the B main effect, but the coefficients would have been proportional to -500, 200, and 300, the differences between the school sizes and their mean of 900. In the interaction, the linear × linear, quadratic × linear, and cubic × linear trends could be tested, using the above three values for the coefficients of Factor B. The quadratic trend on the B main effect, as well as the linear × quadratic, quadratic × quadratic, and cubic × quadratic trends, could also be calculated after finding the appropriate quadratic coefficients on B. (In this example, they would be 1, -8, and 7.)

Warning: Although assumptions are not explicitly mentioned in this chapter, they cannot be forgotten. All of the tests described here are planned comparisons or tests based on planned comparisons, so all of the assumptions necessary for planned comparisons are necessary for trend tests. These assumptions have been discussed in detail in previous chapters and will not be discussed again here. However, it is well to remember that planned comparisons are seldom robust with respect to the assumption of equality of variances and covariances; the special methods of handling this problem, discussed in previous chapters, apply equally well to trend analyses. Remember too that the rather elaborate tests described in this chapter do not *have* to be made just because one or more factors have numerical values. It is always permissible to proceed with the analyses as though the factors did not have numerical values. The tests described here merely provide a way of taking what would otherwise be a few overall tests and partitioning them into a larger number of more specific tests that *may* give a more accurate picture of the differences in the data.

Exercises

(1.) Reanalyze the data in Problem 2, Chapter 5, assuming that the levels of B represent different, equally spaced academic levels; i.e., do a trend analysis on the levels of Factor B. Then do a trend analysis assuming that both A and B are numerically valued factors. What conclusions do you draw about analyzing numerically valued factors with only two levels?

(2.) Do a trend analysis on the data in Problem 3, Chapter 6.

(3.) In Problem 1, Chapter 7, the logarithms of the distances are equally spaced, making a trend test with equally spaced values appropriate, so long as it is remembered that the values on the distance factor are the *logs* of the distances rather than the distances themselves. Do the trend analysis under this assumption; do significance tests and estimate cell means using effects significant at the .05 level. Then plot the actual and estimated cell means against the actual distances rather than plotting them against the logs of the distances. Discuss and interpret the results in terms of the graph.

(4.) Suppose in Table 7.9 that there was a third level of A, with values.

	C_1	C_2	A_3 C_3	C_4	C_5
B_1	11	7	10	2	14
B_2	6	-7	3	4	1
B_3	-3	-7	2	-3	1

Do complete trend tests, assuming that Factors A, B, and C, in turn, are numerically valued. Do trend tests assuming that each pair of factors, AB, AC, and BC, is numerically valued. Finally, do a trend test assuming that ABC are all numerically valued. Whenever C is assumed to be numerically valued, assume that the entire design is a fixed-effects design; in that case, use the highest trend on the highest interaction as the error term.

(5.) Do a trend test on the data in Table 8.1, assuming that Factor A is numerically valued (in equal increments). Do significance tests and estimate cell means using effects significant at the .05 level.

(6.) Do a set of trend analyses like those in Problem 4 (i.e., on all possible combinations of numerical factors) on the data in Table 8.6, with all factors assumed to be fixed whenever necessary.

(7.) Do a trend analysis on Factor A in the following repeated measures design; estimate and graph the cell means, using effects that are significant at the .05 level.

		| A 20	40	60
	S_1	11	10	15
	S_2	2	5	9
B	S_3	6	3	12
	S_4	7	4	13
	S_5	11	12	9
	S_6	7	5	12

(8.) Do a trend analysis on the latitudes main effect in Problem 3, Chapter 8.

12

Basic Matrix Algebra

The remaining material requires an elementary knowledge of matrix algebra. This chapter contains the basic concepts necessary to understand it. If you know no matrix algebra, it should teach you enough to understand the remaining chapters. If you are not sure of your knowledge of matrix algebra, you should probably at least scan this material. If you already have a basic knowledge of matrix algebra, you may skip most of this chapter, although you should read at least the last section.

Matrix algebra is very much like ordinary algebra; variables are added, subtracted, multiplied, etc., but the variables are arrays, called matrices, instead of single numbers. A matrix is a two-dimensional table of numbers. The following, for example, is a three by four (three rows by four columns) matrix:

$$G = \begin{vmatrix} 4 & 3 & -7 & 1 \\ -2 & 6 & 12 & -5 \\ 0 & 2 & 1 & 9 \end{vmatrix}.$$

A vector is a matrix that has only one row or one column. If it has only one column, it is called a column vector. If it has only one row, it is a row vector. The following is a three-dimensional row vector:

$$H = | -5 \quad 10 \quad 2 |.$$

In matrix algebra, as in ordinary algebra, variables have letter names. However, in matrix algebra a single letter represents an entire vector or matrix. Capital letters are often used to represent vectors and matrices. For example, we identified the above matrix by the letter G, meaning that the single letter G represents the entire three by four array. Similarly, we represented the vector by H. We can then perform a number of operations, such as addition, subtraction, and multiplication, on the matrices and vectors, and we can indicate these operations by the usual algebraic symbols on the identifying letters. For example, if we are adding two matrices represented by the letters A and B, we write $A + B$, just as in ordinary algebra.

Matrix Addition and Subtraction

Two matrices can be added only if both have the same number of rows and the same number of columns. To add two such matrices, we add corresponding elements. If

$$A = \begin{vmatrix} 3 & -2 & 1 \\ 4 & 1 & 0 \end{vmatrix}, \ B = \begin{vmatrix} 4 & 6 & -5 \\ -1 & 0 & 3 \end{vmatrix},$$

then

$$A + B = \begin{vmatrix} 3+4 & -2+6 & 1-5 \\ 4-1 & 1+0 & 0+3 \end{vmatrix} = \begin{vmatrix} 7 & 4 & -4 \\ 3 & 1 & 3 \end{vmatrix}.$$

Similarly, to subtract two matrices we subtract each element in the second matrix from the corresponding element in the first. Given A and B as above,

$$A - B = \begin{vmatrix} 3-4 & -2-6 & 1+5 \\ 4+1 & 1-0 & 0-3 \end{vmatrix} = \begin{vmatrix} -1 & -8 & 6 \\ 5 & 1 & -3 \end{vmatrix}.$$

Multiplication of Matrices

Multiplication is a little more complicated than addition and subtraction. To begin with, the order of multiplication is important; A times B is not the same as B times A. To multiply two matrices, the number of *columns* in the first matrix has to be equal to the number of *rows* in the second. Multiplication then proceeds in a "row by column" manner. Consider matrices C and D below:

$$C = \begin{vmatrix} 3 & -2 & 1 \\ 4 & 2 & 0 \end{vmatrix}, \ D = \begin{vmatrix} 4 & 6 & -5 \\ -1 & 0 & 3 \\ 2 & 5 & 0 \end{vmatrix}.$$

Here, multiplication of C times D is possible because C has three columns and D has the same number of rows. (Multiplication of D times C would not be possible.) We write the product as CD, just as we do for multiplication in ordinary algebra. To find the value in the first row and column of CD, we take the values in the first row of C and the first column of D, and we match them to each other

$$\begin{array}{llll} C: & 3 & -2 & 1, \\ D: & 4 & -1 & 2. \end{array}$$

Then we multiply corresponding pairs of elements and sums, getting

$$(3)(4) + (-2)(-1) + (1)(2) = 16.$$

The remaining elements in the product are found in basically the same way. The element in any given row and column of CD is found by first pairing the elements in the corresponding row of C with those in the corresponding column of D, and then by multiplying paired elements and summing. To find the element in the first row and the second column of CD, we pair the first row of C with the second column of D

$$\begin{array}{llll} C: & 3 & -2 & 1, \\ D: & 6 & 0 & 5. \end{array}$$

Multiplying and summing, we get

$$(3)(6) + (-2)(0) + (1)(5) = 23.$$

The complete matrix is

$$CD = \begin{vmatrix} 16 & 23 & -21 \\ 14 & 24 & -14 \end{vmatrix}.$$

Note that the product has the same number of rows as C (that is, 2) and the same number of columns as D (that is, 3).

IDENTITY MATRIX

The *identity matrix*, usually assigned the letter I, acts like the number one in ordinary multiplication. It is a square matrix (that is, it has the same number of columns as rows) with ones on the main diagonal leading from the upper left to the lower right, and with zeros for all of the off-diagonal elements. The exact number of rows and columns depends on the application, with the letter I usually being used regardless of the number of rows and columns. You may check for yourself that for any matrix A, $AI = A$ and $IA = A$ (assuming that I has the correct number of rows and columns in each case).

A kind of matrix division is also possible, but only in a limited way. To understand it, we must first understand some other concepts about matrices.

MATRIX TRANSPOSES

Sometimes addition, subtraction, or multiplication of matrices is possible after one or both matrices have been *transposed*. To transpose a matrix, we simply exchange rows and columns. For example, the transpose of C is

$$C^t = \begin{vmatrix} 3 & 4 \\ -2 & 2 \\ 1 & 0 \end{vmatrix}.$$

One important use of transposing is to enable the multiplication of a matrix by itself. We cannot write AA unless A is a square matrix, but we can always write $A^t A$ and AA^t. For the matrix above,

$$C^t C = \begin{vmatrix} 3 & 4 \\ -2 & 2 \\ 1 & 0 \end{vmatrix} \begin{vmatrix} 3 & -2 & 1 \\ 4 & 2 & 0 \end{vmatrix} = \begin{vmatrix} 25 & 2 & 3 \\ 2 & 8 & -2 \\ 3 & -2 & 1 \end{vmatrix}.$$

Two things are worth noting about this product. The first is that the diagonal elements are all positive; the second is that the elements above the diagonal are equal to the corresponding elements below the diagonal. The product of a matrix by its own transpose will always have both of these properties. When the values above and below the diagonal are equal, we say that the matrix is *symmetric*.

SCALAR MULTIPLICATION

It is often convenient to define a simpler multiplication, the multiplication of a matrix by a single value. In matrix algebra, a single number is called a *scalar* and is often denoted by a lowercase letter. (However, in keeping with the earlier chapters, we will still denote random variables such as X_i and V_i with uppercase letters.) To multiply a matrix by a scalar, we multiply every element of the matrix by the scalar. For example,

$$3C = \begin{vmatrix} 3(3) & 3(-2) & 3(1) \\ 3(4) & 3(2) & 3(0) \end{vmatrix} = \begin{vmatrix} 9 & -6 & 3 \\ 12 & 6 & 0 \end{vmatrix}.$$

Proper Numbers and Proper Vectors

Given any matrix A, certain vectors bear a special relationship to that matrix. Premultiplication of such a vector by the matrix gives the same result as multiplication by a scalar. That is, if X is one of these vectors, then $AX = kX$, where k is some scalar. For example, you may verify that if

$$E = \begin{vmatrix} 29 & 9 & -3 \\ 9 & 18 & -6 \\ -3 & -6 & 2 \end{vmatrix}, \; X = \begin{vmatrix} 2 \\ -3 \\ 1 \end{vmatrix},$$

then

$$EX = \begin{vmatrix} 28 \\ -42 \\ 14 \end{vmatrix} = 14X.$$

When $AX = kX$ for some k, then X is said to be a *proper vector, principal component, characteristic vector* or *eigenvector* of A (all four terms are used, depending on the preferences of a given author), and k is said to be the *proper number, principal root, characteristic root* or *eigenvalue* of A. In the above example, E is a square symmetric matrix; although proper numbers and proper vectors exist for other matrices, we will use them only in relation to square symmetric matrices.

Procedures for finding the proper numbers and proper vectors of a matrix are complicated and difficult. They are best left to computer programs that have been written by experts. Such programs are widely available. However, the following facts about proper numbers and vectors are important.

If X is a proper vector of a matrix A, and p is any nonzero scalar, then pX is also a proper vector for A. In practice, we resolve this ambiguity in either of two ways. Either we regard X and pX as the same proper vector (even though technically they are different vectors), or we choose some arbitrary value, c, and *normalize* X so that $X^t X = c$. By convention, c is usually set to either k, the proper number, or one.

A square symmetric matrix has at least as many proper vectors as it has rows and columns. Two proper vectors are said to be *orthogonal* if and only if $X^tY = 0$. (Note that such a product of two vectors is always a scalar.) If A is symmetric with n rows and columns, then we can always find exactly n proper vectors, each of which is orthogonal to all the others. For the matrix E, given above, the proper vectors are

$$X_1 = \begin{vmatrix} 2 \\ -3 \\ 1 \end{vmatrix}, \quad X_2 = \begin{vmatrix} 5 \\ 3 \\ -1 \end{vmatrix}, \quad X_3 = \begin{vmatrix} 0 \\ 1 \\ 3 \end{vmatrix}.$$

Every proper vector is associated with a single proper number. The proper numbers associated with the above proper vectors are

$$k_1 = 14, \quad k_2 = 35, \quad k_3 = 0.$$

However, two or more proper vectors may have the same proper number. Suppose we choose a set of n orthogonal proper vectors, and we consider the n proper numbers belonging to them. It is entirely possible that the same proper number will appear more than once in that set. However, it can be proved that no matter how we choose the set of n proper vectors, we will always obtain the same set of n proper numbers. Those n proper numbers, with some numbers possibly appearing more than once, are the proper numbers of the matrix. Thus the above proper numbers are the proper numbers of E. As another example, if

$$F = \begin{vmatrix} 15 & 6 & -6 \\ 6 & 6 & 12 \\ -6 & 12 & 6 \end{vmatrix},$$

then you should be able to verify that

$$X_1 = \begin{vmatrix} 4 \\ 1 \\ -1 \end{vmatrix}, \quad X_2 = \begin{vmatrix} 0 \\ 1 \\ 1 \end{vmatrix}, \quad X_3 = \begin{vmatrix} 1 \\ -2 \\ 2 \end{vmatrix}$$

are a set of three orthogonal proper vectors of F with associated proper numbers

$$k_1 = 18, \quad k_2 = 18, \quad k_3 = -9.$$

These are the proper numbers of F. Note that we say there are three proper numbers in the set even though the number 18 appears twice. We say that 18 has *multiplicity* 2.

If all of the proper numbers in the set of n are different, then the matrix has exactly n proper vectors, and every proper number is paired with exactly one proper vector. However, if they are not all different, then there are an infinite number of proper vectors, and there are an infinite number of ways to choose the set of n orthogonal vectors. In such a case, the specific choice of n orthogonal vectors usually is not important.

The determinant of a matrix can be defined in a variety of ways. The following definition suits our purposes and fits the above discussion: the determinant is the product of the set of n proper numbers. In the above examples, the determinant of E is $(14)(35)(0) = 0$; the determinant of F is $(18)(18)(-9) = -2,916$.

Special Types of Matrices

The *rank* of a matrix is the number of nonzero proper numbers in the full set of n (i.e., counting multiplicities). The rank of E in the last set of examples is 2 because there are two nonzero proper numbers (i.e., 14 and 35). The rank of F is three because there are three nonzero proper numbers (i.e., 18, 18, and -9). If the rank is less than the number of rows and columns, the matrix is said to be *singular*; if it is equal to the number of rows and columns, it is said to be *nonsingular*. The significance of this terminology will become evident in the next section. Note that we can alternatively define singularity in terms of the determinant. If the determinant is zero, the matrix is singular; if it is nonzero, the matrix is nonsingular.

If all of the proper numbers of a matrix are positive, the matrix is said to be *positive definite*. If they are all nonnegative (i.e., some may be zero), the matrix is said to be *positive semidefinite*. In particular, given any matrix, A, $A^t A$ and AA^t are both positive semidefinite. A nonsingular positive semidefinite matrix is automatically positive definite.

Matrix Inverses

The matrix operation analogous to division is inversion. It applies only to square, nonsingular matrices. The inverse of a matrix A is usually written as A^{-1}. It has the special property that $A^{-1}A = AA^{-1} = I$, where I is the identity matrix as before. It is the closest operation to division in matrix algebra, and it is used primarily to solve for unknown vectors in equations. Suppose, for example, that we want to find a vector X such that $AX = Y$, where A is a square matrix whose elements are known, and Y is a vector whose elements are also known. Then the solution is $X = A^{-1}Y$, as can be shown from the following equalities:

$$AX = Y; \quad A^{-1}AX = A^{-1}Y; \quad IX = A^{-1}Y; \quad X = A^{-1}Y.$$

In the above examples, E is singular so it has no inverse; you can verify that the inverse of F is

$$F^{-1} = \begin{vmatrix} 1/27 & 1/27 & -1/27 \\ 1/27 & -1/54 & 2/27 \\ -1/27 & 2/27 & -1/54 \end{vmatrix} = (1/54) \begin{vmatrix} 2 & 2 & -2 \\ 2 & -1 & 4 \\ -2 & 4 & -1 \end{vmatrix}.$$

If the inverse exists, it is unique. However, the calculation of an inverse tends to be rather complicated; it is best done with one of the many computer programs that have been written for that purpose.

Linear Equations and Least-Squares Solutions

One use of matrix algebra is to solve linear equations. Suppose, for example, that we have the following three linear equations in three unknowns:

$$
\begin{aligned}
15x_1 + 6x_2 - 6x_3 &= -5, \\
16x_1 + 6x_2 + 12x_3 &= 4, \\
-6x_1 + 12x_2 + 6x_3 &= 10.
\end{aligned}
$$

The coefficients in these equations are identical to those in the matrix F above. Consequently, the equations can be written in matrix form as $FX = Y$, where the elements of Y are -5, 5, -10. Using the theory above, we get

$$
X = F^{-1}Y = \begin{vmatrix} 1/3 \\ -1 \\ 2/3 \end{vmatrix}.
$$

Of course, this solution is possible only because F is a square, nonsingular matrix. If F had not been square and nonsingular, there would have been either no solution or an infinite number of possible solutions. One case with no solution is important in statistics. It is the case in which we assume that a variable, Y, depends partly on one or more other variables, but also depends partly on random error.

Suppose we are concerned about students who may enter an elementary statistics class with an inadequate background in mathematics. We give each student in the statistics class a test on general mathematical ability at the beginning of the semester. We are interested in how well we can use these scores to predict students' performances in the statistics class.

The data in Table 12.1 might result from such a study. In this table, X is a score on general math ability and Y is a measure of performance in the statistics class. Then the prediction equation can be written as $a + bX_i = Y_i + e_i$, with e_i representing random error. The entire set of equations can be written as

$$
PW = Y + E,
$$

where P is a matrix with two columns: The first column consists entirely of ones, and the second is the list of X_i values in Table 12.1; W is a vector containing the two unknown values a and b, Y is a vector consisting of the list of Y_i in Table 12.1, and E is a vector of the unknown error terms.

We cannot solve these equations because there are 26 equations and only the two unknowns, a and b, to be solved for (the e_i cannot be solved for directly). Instead we seek a *least-squares solution*: a solution that minimizes $E^t E$, the sum of the squared deviations of the predicted from the obtained test scores. The solution is known to be

$$
W = (P^t P)^{-1} P^t Y.
$$

TABLE 12.1. Hypothetical data: Math. ability (X) vs. stat. performance (Y).

X	Y	X	Y	X	Y	X	Y
6	2	9	3	8	7	6	3
8	4	10	7	9	6	8	3
9	6	9	4	11	7	9	7
7	4	9	8	8	6	7	4
9	5	7	5	9	5	9	5
8	3	8	4	9	4	9	7
8	4	9	4				

	X	Y	XY
Sum	218	172	
Mean	8.385	4.855	
Sum of squares or cross products	1,860	685	1,092
s^2 or $C(V,X)$	1.237	2.487	1.044
s or r_{VX}	1.112	1.577	.596

$$\hat{\omega}^2 \;=\; r^2 = .355, \; F_{(1,24)} = 13.2, \; p = .001$$
$$\hat{Y}_i \;=\; 4.885 + .845(X_i - 8.385) = .845X_i - 2.196$$

In this example,

$$P^tP \;=\; \begin{vmatrix} 26 & 218 \\ 218 & 1860 \end{vmatrix}, \; (P^tP)^{-1} = \begin{vmatrix} 2.225 & -.2608 \\ -.2608 & .0311 \end{vmatrix},$$

$$P^tY \;=\; \begin{vmatrix} 127 \\ 1092 \end{vmatrix}, \; W + (1/418) \begin{vmatrix} -918 \\ 353 \end{vmatrix} = \begin{vmatrix} -2.196 \\ .845 \end{vmatrix},$$

as can be verified by doing the actual calculations. Our model, then, is $-2.196 + .845X_i = Y_i + e_i$. This is identical to the result that would be obtained using the regression formulas in any elementary statistics text. However, this general method is easily extended to cover *multiple regression*, where the Y_i depend linearly on more than one other variable (e.g., statistics performance may depend on both math ability and "intellectual maturity").

Covariance Matrices

One type of square matrix is especially important in statistics. It is the *covariance matrix* and its close relative, the *correlation matrix*. Basically, the covariance matrix is just a table listing the variances and covariances of two or more variables. For the example of math scores and statistics performance,

$$s_x^2 = 1.237, \quad s_y^2 = 2.487, \quad C(X,Y) = 1.044.$$

We can represent these values by the square, symmetric matrix

$$\begin{vmatrix} 1.237 & 1.044 \\ 1.044 & 2.487 \end{vmatrix}.$$

If a third variable had been involved, the matrix would have had a third row and column, a fourth variable would have added a fourth row and column, and so on.

A correlation matrix is a covariance matrix on the z scores instead of on the raw scores of the variables. The main diagonal of a correlation matrix always contains ones, and the off-diagonal entries are Pearson product-moment correlations. In our example, $r_{xy} = .596$, so the correlation matrix is

$$\begin{vmatrix} 1 & .596 \\ .596 & 1 \end{vmatrix}.$$

Covariance and correlation matrices are not merely convenient tables. They enter into matrix formulas and equations that are fundamental to many statistical procedures.

Sum of Cross-Products Matrices

A type of matrix that plays an important role in analysis of variance is the *sum of cross-products* matrix. A sum of cross-products matrix is a kind of multidimensional sum of squares, and it plays much the same role in multivariate designs as a sum of squares in an ordinary analysis of variance.

For example, in a multivariate one-way analysis of variance, two such matrices are important: the *sum of cross products between* matrix (SC_{bet}) is the *sum of cross products within* matrix (SC_w).

To construct each matrix, we proceed in a manner analogous to that of the ordinary one-way design. First we construct *raw sum of cross-products* matrices, and then we construct the sum of cross-products matrices from them by subtraction. Each matrix has as many rows and columns as there are dependent variables. The values on the diagonal are just the raw sums of squares for the individual dependent variables. The values off the diagonal are sums of cross products.

As an example, we will construct the RC_{bet} matrix for the data in Table 12.2. These data are from three different courses given at a university. The first column in each group contains the same "mental maturity" scores as in Table 2.1. The second column contains the scores of the same subjects on a standardized IQ test (with 100 subtracted from each score to simplify calculations). The third column is the chronological age of each subject. It is reasonable to suppose not only that subjects in different classes may differ on any of these scores but that the three types of scores may be related to each other.

For these data,

$$RC_{bet} = \begin{vmatrix} 198.6 & 806.8 & 1,250.4 \\ 806.8 & 4,873.0 & 6,093.6 \\ 1,250.4 & 6,093.6 & 8,557.0 \end{vmatrix}.$$

The first value on the diagonal, 198.6, is RS_{bet} calculated on the mental maturity scores; it is the same as the value calculated in Chapter 2; the second value on the diagonal, 4,873.0, is RS_{bet} calculated on the IQ scores; the third value, 8,557.0, is RS_{bet} calculated on the chronological ages.

The first off-diagonal value, 806.8, is the sum of cross products calculated on the mental maturity scores (X_{ij}) and the IQ scores (Y_{ij}). It is calculated in exactly the same way as the sums of squares, except that cross products are taken instead of squares. More specifically, it is

$$[(20)(101) + (8)(108) + (23)(50)]/5 = 806.8.$$

A comparison of this formula with the totals in Table 12.2 should clarify the relationship between these sums of cross products and ordinary sums of squares.

The *raw sum of cross-products within* (RC_w) matrix is calculated similarly; the diagonals contain ordinary raw sums of squares within on the individual dependent variables. The off-diagonal elements are calculated in the same way, except that sums of cross products are substituted for sums of squares. For the data in Table 12.2,

$$RC_w = \begin{vmatrix} 233 & 751 & 1,296 \\ 751 & 6,323 & 6,042 \\ 1,296 & 6,042 & 8,669 \end{vmatrix}.$$

The first value, 233, is RS_w for the mental maturity scores, taken from Table 2.5. The other two values on the diagonal (6,323 and 8,669) are the RS_w for IQ and age, respectively. The off-diagonal entries are calculated by exactly the same formula, except that squares are replaced by cross products. The first off-diagonal entry is found by adding the cross products of each subject's mental maturity score with the same subject's IQ score; i.e.,

$$(5)(23) + (3)(22) + (5)(17) + \cdots + (5)(-6) + (2)(29) = 751,$$

and the remaining off-diagonal entries are found similarly.

Finally, the *raw sum of cross-products mean* (RC_m) matrix is calculated:

$$RC_m = \begin{vmatrix} 173.4 & 880.6 & 1,213.8 \\ 880.6 & 4,472.07 & 6,164.2 \\ 1,213.8 & 6,164.2 & 8,496.6 \end{vmatrix}.$$

TABLE 12.2. Scores on mental maturity (X), IQ's minus 100 (Y), and chronological ages (X), for students in three different university courses.

	A_1			A_2			A_3			Sum		
	X	Y	Z	X	Y	Z	X	Y	Z	X	Y	Z
	5	23	27	1	36	23	5	10	28			
	3	22	21	2	19	22	4	1	26			
	5	17	26	2	15	20	7	16	26			
	6	26	31	0	32	18	5	-6	24			
	1	13	23	3	6	22	2	29	20			
Sum	20	101	128	8	108	105	13	50	124	51	259	357
Mean	4.0	20.2	25.6	1.6	21.6	21.0	4.6	10.0	24.8	3.4	17.3	23.8

As before, the main diagonal contains the RS_m for the individual dependent variables, and the off-diagonal entries are calculated by the same formula with the square replaced by a cross product:

$$(51)(259)/15 = 880.6, \quad (51)(357)/15 = 1,213.8, \quad \text{etc.}$$

We can now use these raw sum of cross-products matrices to calculate the sum of cross-products matrices. The subtraction proceeds exactly as for the calculation of sums of squares from raw sums of squares, except that we subtract matrices instead of scalar values. As in the one-way design, $SC_m = RC_m$, and

$$SC_{bet} = RC_{bet} - SC_m = \begin{vmatrix} 25.2 & -73.8 & 36.6 \\ -73.8 & 400.93 & -70.6 \\ 36.6 & -70.6 & 60.4 \end{vmatrix},$$

$$SC_w = RC_w - SC_{bet} - SC_m = \begin{vmatrix} 34.4 & -55.8 & 45.6 \\ -55.8 & 1,450.0 & -51.6 \\ 45.6 & -51.6 & 112.0 \end{vmatrix}.$$

Also, as in the one-way design, we can calculate $SC_t = SC_{bet} + SC_w$.

The concept of a sum of cross-products matrix can be extended to other problems in the analysis of variance. For every F test described in the previous chapters, there will be a corresponding multivariate test (discussed in Chapter 13) and analysis of covariance (discussed in Chapter 14). The basic principle for all of these tests is that the single values that ordinarily enter into the F ratio are replaced by matrices. Every term entering into an F ratio involves the squaring of some quantity or quantities. For the matrices, we always replace the square with a cross product, but otherwise the calculations are the same.

We will illustrate this point with two more examples. The first is a multivariate planned comparison on the data of Table 12.2. We will consider the null hypothesis that $\mu_1 + \mu_3 - 2\mu_2 = 0$ for all three of the variables (the alternative hypothesis is that the equality does not hold for one or more of the variables). The second example will be an expansion on the two-way design in Table 5.2, giving scores of mental patients after taking drugs. To these data we will add a second dependent variable representing ratings, by the patients' doctors, of the severity of each patient's illness. The ratings are assumed to be on a ten-point scale, a ten representing the most severe illness. The data are given in Table 12.3.

We will consider the planned comparison first. For it, the usual univariate procedure is first to calculate C_k and then SS_k. In the multivariate case, the values on the main diagonal of the matrix will be the SS_k calculated on each of the three dependent variables. For the off-diagonal values, we note that the squaring process occurs when we calculate $SS_k = 5C_k^2/6$. Accordingly, we substitute a cross product at that point. For example, $C_k = 5.4$ for mental maturity, -13.0 for IQ, and 8.4 for age. The first off-diagonal element concerns mental maturity and IQ, so we calculate it using the cross product of the

TABLE 12.3. Improvement scores (X) and severity of illness (Y).

		B_1		B_2		B_3			
		X	Y	X	Y	X	Y		
A_1		8	5	8	7	4	3		
		4	3	10	7	6	6		
		0	4	6	4	8	6		
		—	—	—	—	—	—		
	Sum	12	12	24	18	18	15	54	45
	Mean	4.0	4.0	8.0	6.0	6.0	5.0	6.0	5.0
A_2		10	5	0	2	15	8		
		6	6	4	7	9	3		
		14	7	2	6	12	7		
		—	—	—	—	—	—		
	Sum	30	18	6	15	36	18	72	51
	Mean	10.0	6.0	2.0	5.0	12.0	6.0	8.0	5.7
	Sum	42	30	30	33	54	33	126	96
	Mean	7.0	5.0	5.0	5.5	9.0	5.5	7.0	5.3

corresponding C_k values, i.e., as $5(5.4)(-13.0)/6 = -58.5$. The complete matrix is

$$SC_k = \begin{vmatrix} 24.30 & -58.50 & 37.80 \\ -58.50 & 140.83 & -91.00 \\ 37.80 & -91.00 & 58.80 \end{vmatrix}.$$

For the two-way design in Table 12.3, we would ordinarily calculate SS_m, SS_a, SS_b, SS_{ab}, and SS_w. Analogously, we calculate the sum of cross-products matrices, SC_m, SC_a, SC_b, SC_{ab}, and SC_w. The diagonal elements of these matrices will be the ordinary sums of squares calculated on each independent variable. The off-diagonal element (there is only one because there are only two independent variables) will be calculated in basically the same way, but with cross products substituted in place of squares.

As in the one-way design, we begin by calculating the raw sum of cross-products matrices. For the A main effect, for example, the main diagonal of RC_a will contain the RS_a for the two variables. Since each RS_a is calculated by squaring row totals, the off-diagonal element of RC_a will be calculated by summing cross products of row totals; i.e.,

$$[(54)(45) + (72)(51)]/9 = 678.$$

The complete matrix is

$$RC_a = \begin{vmatrix} 900 & 678 \\ 678 & 514 \end{vmatrix}.$$

TABLE 12.4. RC and SC matrices for data in Table 12.2.

		RC		SC	
m	$=$	882	672	882	672
		672	512	672	512
a	$=$	900	678	18	6
		678	514	6	2
b	$=$	930	672	48	0
		672	513	0	1
ab	$=$	1,092	708	144	30
		708	522	30	7
w	$=$	1,198	753	106	45
		753	566	45	44
t	$=$			316	81
				81	54

The complete set of RC matrices for the data in Table 12.3 is in Table 12.4. The SC matrices are found from the RC matrices by subtraction, just as the SS are found from the RS in a univariate design. The SC matrices are also shown in Table 12.4. Their use is explained in the next two chapters.

13
Multivariate Analysis of Variance

This chapter and the next deal specifically with multivariate extensions of the analysis of variance. They are concerned with measures on more than one dependent variable. The two examples in Chapter 12 are illustrative. In Table 12.2 the experiment of Table 2.1 is extended to include measures of IQ and chronological age as well as of intellectual maturity. There are thus three dependent variables. In Table 12.3 the data in Table 5.1 are extended to include severity of illness as a second dependent variable.

Each dependent variable can be treated in either of two ways. We may be interested in studying it directly, or we may be interested in using it to "control" statistically for variables that we cannot control for experimentally. In the latter case, the dependent variable plays a role very similar to that of an additional factor in the experiment. We will consider that case in Chapter 14. In this chapter, we will study methods for testing the dependent variables directly.

Multivariate Procedures

Multivariate analysis of variance is a set of techniques for dealing with several dependent variables simultaneously. For example, given the experiment in Table 12.2, we may be interested in any or all of three null hypotheses, one for each dependent variable. All three are of the usual form for a one-way design; they all state that there are no differences among the group means.

We can combine these three null hypotheses into one general null hypothesis, that there are no differences among group means on *any* of the dependent variables. In other words, we can hypothesize that the three null hypotheses stated above are all simultaneously true. Multivariate analysis of variance provides the techniques for a single test of this combined null hypothesis.

Unfortunately, the combined null hypothesis cannot always be tested with a simple F ratio. Instead, more complicated procedures, leading to statistics for which good tables may not be readily available, have to be used. Four procedures are in general use. We will describe two at length. The first is based on the concept of a likelihood ratio; the second is based on the concept of an optimal linear combination. Each has advantages and disadvantages relative to the other, and each also has advantages and disadvantages over the practice of separately testing each dependent variable. We will first describe each procedure and then discuss the relative advantages and disadvantages. After that we will briefly cover the other two procedures. Then we will discuss the relative advantages and disadvantages of multivariate procedures as compared to some alternatives. Finally, we will show how the restrictive assumptions in mixed and random designs can be avoided by treating the random factor as the dimensions of a multivariate dependent variable.

LIKELIHOOD RATIO PROCEDURE (WILKS'S LAMBDA)

The likelihood ratio procedure, commonly called *Wilks's lambda*, is based on a very general method for obtaining parameter estimates and doing statistical tests. The general method is described in most mathematical statistics texts. All of the F tests that we have described, with the exception of some post hoc tests, are likelihood ratio tests. Likelihood ratio tests have three desirable properties: They are usually fairly easy to derive, they tend to be very powerful, and when no exact distribution can be found for a likelihood ratio statistic, approximations are readily available.

To calculate the likelihood ratio statistic for a multivariate test, we begin with the SC matrices described in Chapter 12. The statistic can then be calculated in a number of ways. Probably the simplest way is to use the determinants of the SC matrices. For any test, there will be an SC matrix corresponding to the numerator term, and an SC matrix corresponding to the denominator term, of the univariate F ratio. If we designate the "numerator" matrix by SC_n and the "denominator" matrix by SC_d, the maximum likelihood statistic is

$$\Lambda = \det(SC_d)/\det(SC_d + SC_n),$$

where "det" indicates that the determinant of the matrix is to be taken. (The "inversion," with the denominator term, SC_d, in the numerator of Λ, is intentional; it is standard practice to define Λ in this way.) If we let $SC_s = SC_d + SC_n$, then

$$\Lambda = \det(SC_d)/\det(SC_s).$$

Before discussing the distribution of Λ, we will illustrate it using the examples in Chapter 12. For the first example, we will use the data in Table 12.2, testing the very general null hypothesis that there are no differences among the groups on any of the three dependent variables. This is a one-way design, and the corresponding univariate test would be based on the ratio MS_{bet}/MS_w. Accordingly, for the multivariate test we will let $SC_n = SC_{bet}$ and $SC_d = SC_w$. We then have

$$
\begin{aligned}
\Lambda &= \det(SC_w)/\det(SC_w + SC_{bet}) \\
&= 2,393,800/5,329,900 = .4491.
\end{aligned}
$$

The second example in Chapter 12 is a contrast comparing μ_2 with the average of μ_1 and μ_3. For that contrast, the univariate test uses SS_k/MS_w.

Accordingly, SC_n is now the *matrix* SC_k, given in Chapter 12. The statistic is

$$\Lambda = \det(SC_w)\det[SC_w + SC_k]$$
$$= 2,393,760/4,178,140 = .5729.$$

Finally, for the third example, using the data in Table 12.3 and the matrices in Table 12.4, we will test the A main effect, the B main effect, and the interaction. Following the same procedures as above, we have

$$\Lambda_a = \det(SC_w)/\det(SC_w + SC_a)$$
$$= 2,639/3,103 = .8505,$$
$$\Lambda_b = \det(SC_w)/\det(SC_w + SC_b)$$
$$= 2,639/4,905 = .5380,$$
$$\Lambda_a b = \det(SC_w)\det(SC_w + SC_{ab})$$
$$= 2,639/7,125 = .3703.$$

In this last example, we assume a fixed-effects model. If one or more factors had been random, the choice of SC_d would have been different, but the principle would have been the same. Suppose, for example, that Factor A had been fixed and Factor B had been random. Then the tests on the B main effect and the interaction would have been the same, but the univariate test on the A main effect would have used MS_{ab} in the denominator (see Chapter 6). Accordingly, the multivariate test on the A main effect would have been

$$\Lambda_a = \det(SC_{ab})\det(SC_{ab} + SC_a)$$
$$= 108/162 = .6667.$$

Thus, every *planned* test described in the previous chapters has a relatively simple multivariate extension.

To test hypotheses, we must know the distribution of Λ. Unfortunately, the distribution may be complicated, depending on a number of parameters. However, good approximations are available. The simplest approximation uses the fact that the logarithm of a likelihood ratio statistic has approximately a chi-square distribution. Specifically, if we let

$$p = \text{\# of dependent variables,}$$
$$\nu_1, \nu_2 = \text{numerator and denominator degrees of freedom,}$$
$$\text{respectively, that would be appropriate}$$
$$\text{in a univariate } F \text{ test,}$$
$$m^* = \nu_2 + (\nu_1 - p - 1)/2$$

then $-m^* \log_e(\Lambda)$ has approximately a chi-square distribution with degrees of freedom $= p\nu_1$, if ν_2 is large.

TABLE 13.1. Calculations for maximum-likelihood tests of examples in text.

| | One-way design | | Two-way design | | | |
	Bet	Contrast	a	b	ab	$a^{(1)}$
p	3	3	2	2	2	2
ν_1	2	1	1	2	2	1
ν_2	12	12	2	12	12	2
Λ	.4491	.5729	.8505	.5380	.3703	.6667

			Chi-square			
m^*		11	10.5	11	11.5	11.5	1
Chi-square		8.81	5.83	1.77	7.13	11.4	.41
ν		6	3	2	4	4	2
p		.19	.12	.41	.12	.022	.82

			F			
g	2	1	1	2	2	1
F	1.64	2.48	.96	2.00	3.53	.25
ν_1	6	3	2	4	4	2
ν_2	20	10	11	22	22	1
p	.19	.13	.41	.12	.023	.82

[1] Assumes Factor B is random.

For a better but more complicated approximation, let

$$
\begin{aligned}
g &= \begin{cases} 1 & \text{if } p^2 + \nu_1^2 = 5, \\ [(p^2\nu_1^2 - 4)/(p^2 + \nu_1^2 - 5)]^{1/2} & \text{otherwise,} \end{cases} \\
h &= gm^* - (p\nu_1 - 2)/2, \\
W &= \Lambda^{(1/g)}.
\end{aligned}
$$

Then $[(1-W)/W][h/(p\nu_1)]$ has approximately an F distribution with numerator degrees of freedom $= p\nu_1$ and denominator degrees of freedom $= h$. In fact, the distribution is exact if $g = 1$ or $g = 2$. Table 13.1 summarizes all of the calculations for the tests described above. The F test is exact for all of the examples, and the chi-square approximation is very accurate.

OPTIMAL LINEAR COMBINATION (ROY'S MAXIMUM ROOT)

With this procedure, we first create a new variable that is a linear combination of the dependent variables in the experiment. In our one-way design, for example, we create a new random variable, W_{ij}, that is a linear combination of the three original scores (i.e., intellectual maturity, IQ, and chronological age). We then do an ordinary analysis of variance on the W_{ij}, calculating an

F ratio and finding its p value. Of course, the result will depend on which linear combination we use.

Roy's maximum root method finds the *best* linear combination, i.e., the linear combination that maximizes the F ratio. That linear combination is fairly difficult to calculate. The coefficients for the linear combination are the elements of a proper vector of $SC_d^{-1}SC_n$. Unfortunately, this matrix is not symmetric, so its proper vectors are not easy to find. We usually find them indirectly, by first finding a matrix C such that $SC_d = CC^t$. (Methods for doing this can be found in texts on linear algebra.) We then calculate the matrix

$$B = C^{-1}SC_n(C^t)^{-1}.$$

This is a symmetric matrix, so its proper vectors can be found by the usual methods. The particular proper vector that we want is associated with the largest proper number. If we let this proper vector be P, the coefficients we are looking for are the elements of the vector $Q = (C^t)^{-1}P$. If the associated proper number is λ, then the maximal F ratio is $\lambda\nu_2/\nu_1$.

Usually the calculations can all be performed by an existing computer program, so it is not necessary to be able to do them yourself. However, we will illustrate by doing them step-by-step for the overall test on the data of Table 12.2. The numerator matrix in this example is SC_{bet}, and the denominator matrix is SC_w. We first use a method called *Cholesky factorization* on SC_w to get

$$C = \begin{vmatrix} 5.8652 & 0 & 0 \\ -9.5138 & 36.8712 & 0 \\ 7.7747 & .6066 & 7.1544 \end{vmatrix}.$$

This *triangular* matrix is easily inverted, giving

$$C^{-1} = \begin{vmatrix} .17050 & 0 & 0 \\ .04399 & .02712 & 0 \\ -.18901 & -.00230 & .13977 \end{vmatrix}.$$

Next we calculate

$$C^{-1}SC_{bet}(C^t)^{-1} = \begin{vmatrix} .73256 & -.15224 & .08906 \\ -.15224 & .16757 & .10866 \\ .08906 & .10866 & .12979 \end{vmatrix}.$$

The largest proper number of this matrix is .7767, and the associated proper vector is $P^t = (.96959, -.22527, .09564)$. Finally $Q^t = (.13733, -.00633, .0134)$.

These are the coefficients of our linear combination, but it is usual to normalize them so the sum of their squares will equal one. Doing that and putting them into the linear combination, we have

$$W + .994X - .046Y + .097Z.$$

Nearly all of the weight is on the intellectual maturity scores, X. Basically, this tells us that any significant difference we find will be due mostly to differences in intellectual maturity rather than to differences in either IQ or chronological age.

However, these coefficients must be interpreted with some caution. First, a large coefficient is less impressive if the associated dependent variable has a small variance. To interpret the coefficients properly, we should first multiply each by the standard deviation of the associated dependent variable and then normalize them. This gives the relative weight of each variable when expressed as a z score. The standard deviations are found from the variances on the diagonal of the error matrix (SC_w in our example). For our example, the adjusted weights are (.945, -.283, .166), so the dominance of the intellectual maturity score is reduced slightly. Second, the relative sizes of the coefficients for two variables may depend on the other variables that are present. The coefficients tell us only the relative importance of each variable as a member of the entire set of variables. (A similar problem arises in multiple regression.)

We cannot find the significance level from an F table because the F ratio was not calculated on a single dependent variable. It was calculated on a linear combination chosen to make it as large as possible. Such a calculation is bound to make use of chance variations to increase the value of F. Therefore, instead of calculating F, we calculate $\theta = \lambda/(1 + \lambda)$ and look it up in a special table. The table can be found in Harris (1975). To use it we need three parameters. They are

$$
\begin{aligned}
s &= \text{the smaller of } p \text{ and } \nu_1, \\
m &= (|\nu_1 - p| - 1)/2, \\
n^* &= (\nu_2 - p - 1)/2.
\end{aligned}
$$

In our example, $p = 3, s = 2, m = 0, n^* = 4$, and

$$\theta = .7767/(1 + .7767) = .4372.$$

The table is somewhat limited; it contains values for the .05 and .01 levels only. Moreover, for $n = 4$ we must interpolate. However, the .05 critical value is approximately .616, so the result is not significant at the .05 level.

There is no good approximation for Roy's maximum root, but an upper bound on F can be calculated. To find it, let

$$
\begin{aligned}
s^* &= \text{the larger of } p \text{ and } \nu_1, \\
F &= \lambda(\nu_2 - s^* - 1)/s^*,
\end{aligned}
$$

and treat F as having an F distribution with ν_1 degrees of freedom in the numerator and $(\nu_2 + \nu_1 - s^*)$ degrees of freedom in the denominator. Remember, however, that this is an upper bound; if the test is not significant by this criterion, then the null hypothesis cannot be rejected, but if it is significant, the exact test must be performed.

Final results for all of the tests are shown in Table 13.2. Note that for three tests—the contrast in the one-way design and the two tests on the A main effect in the two-way design—an ordinary F ratio is used, and the result is

TABLE 13.2. Examples of Roy's maximum root procedure. The A_i are the coefficients.

	One-way design		Two-way design			
	Bet	Contrast	a	b	ab	$a^{(1)}$
λ	.7767	.7454	.1758	.8182	1.634	.5000
A_1	.994	.973	.949	-.689	.830	-.204
A_2	-.046	-.122	-.316	.725	-.557	.979
A_3	.097	.194				
s	2	1	1	2	2	1
m	0	1/2	0	-1/2	-1/2	-1/2
n^*	4	4	4	4	4	-1/2
θ	.4372	.4271	.1495	.4500	.6203	.3333
p	NS	.13[2]	.41[2]	NS	<.05	.82[2]

[1] Assumes Factor B random.
[2] Based on F ratio.

identical to that in Table 13.1. For these tests, $s = 1$. Whenever $s = 1$, the quantity

$$F = [(n^* + 1)/(m + 1)]\lambda$$

has an ordinary F distribution with $\nu_1 = 2m + 2$ and $\nu_2 = 2n^* + 2$. In fact, the F value obtained in this manner is identical to the F value obtained using Wilks's lambda.

COMPARISON OF MULTIVARIATE APPROACHES

Each of these two multivariate approaches has advantages and disadvantages. A major advantage of Wilks's lambda is that it is based on the likelihood ratio. The likelihood ratio is commonly used in both univariate and multivariate statistics. The t and F tests are both based on likelihood ratios. The method is commonly used because it is usually simple to derive and it usually has more power than other methods. However, for some analyses, Roy's maximum root has more power.

Another advantage of Wilks's lambda is ease of calculation. Determinants are generally easier to calculate than proper numbers. Moreover, very good, relatively simple statistics, approximately distributed as F and chi-square, are available. By contrast, Roy's maximum root depends on a table that has three parameters and is not found in most statistics texts.

The principal advantage of Roy's maximum root is that it generates the optimal linear combination. This allows us to determine which dependent variables are most responsible for a significant result, and how they interact (e.g., do they all add together or are some dependent variables subtracted from others?). Wilks's lambda does not generate any such single set of coefficients.

Roy's maximum root can use the coefficients to make post hoc comparisons in a kind of limited way. The limitation is that all such comparisons must be made on an artificial dependent variable, constructed from the linear combination obtained in the overall analysis of variance. Since the overall test was not significant in the example, no such post hoc test can be significant either. However, we will do a post hoc test to illustrate the method. The linear combination found in the overall test was

$$W = .994X - .046Y + .097Z.$$

In theory, we must calculate the value of W for each score and then do the post hoc test on those values. However, it is sufficient to find the mean value of W for each group

$$\overline{W}_{i.} = .994\overline{X}_{i.} - .046\overline{Y}_{i.} + .097\overline{Z}_{i.}.$$

We get

$$\overline{W}_{1.} = 5.53, \ \overline{W}_{2.} = 2.6338, \ \overline{W}_{3.} = 6.518$$

and

$$\begin{aligned} C_k &= 5.53 - 2(2.6338) + 6.518 = 6.7804, \\ SS_k &= 38.31. \end{aligned}$$

To find the error term, let A be the vector of coefficients used to calculate W; i.e., for this example, $A^t = (.994, -.046, .097)$. Then the error sum of squares is

$$SS_e = A^t SC_w A = 52.47.$$

The appropriate value of λ for the post hoc comparison is

$$\lambda = SS_k/SS_e = .7301.$$

Finally, we calculate $\theta = \lambda/(\lambda + 1) = .4220$, and we look this up using the same parameters as for the original overall significance test. Of course, it is not significant.

No similar post hoc test based on Wilks's lambda is available.

OTHER MULTIVARIATE METHODS

Two other multivariate methods have been proposed. The first, *Pillai's trace*, uses $V = tr[SC_n(SC_n + SC_d)^{-1}]$, where "$tr$" indicates that the trace (the sum of the values on the diagonal) of the matrix is to be taken. If we let

$$F = (2n^* + s + 1)/(2m + s + 1)V/(s - V),$$

where, as above,

$$\begin{aligned} n^* &= (\nu_2 - p - 1)/2, \\ m &= (|\, p - \nu_1\,| - 1)/2, \\ s &= \text{smaller of } p \text{ and } \nu_1, \end{aligned}$$

then F has approximately an F distribution with $s(2m + s + 1)$ degrees of freedom in the numerator and $s(2n^* + s + 1)$ degrees of freedom in the denominator.

The other, the *Hotelling–Lawley trace*, uses $U = tr(SC_d^{-1}SC_n)$. If we define m, n^*, and s as above,

$$F = 2(sn^* + 1)U/[s^2(2m + s + 1)]$$

has approximately an F distribution with $s(2m + s + 1)$ degrees of freedom in the numerator and $2(sn^* + 1)$ degrees of freedom in the denominator.

Univariate Versus Multivariate Approaches

We have described four methods for multivariate analysis of variance in this chapter. However, other approaches are possible. One is a univariate test on each dependent variable. Another is a single univariate test on a linear combination of the variables, with the particular linear combination being chosen in advance. The last method is best if we have good reasons for being interested in a particular linear combination, but that is seldom true. Thus, usually the choice is between a single multivariate test and a set of univariate tests.

UNIVARIATE VERSUS MULTIVARIATE TESTS

A comparison of the univariate and multivariate approaches is analogous to the comparison we made earlier between the overall F test and planned comparisons. The multivariate approach is kind of a "shotgun" approach; it tells us whether there are *any* differences among *any* of the means on *any* of the dependent variables. It does not tell us directly where those differences lie, although Roy's maximum root method gives us some ideas. By contrast, the univariate approach, by testing each dependent variable individually, tells us for which dependent variable(s) there are significant differences. In addition, the univariate approach is more readily adapted to the use of post hoc tests.

However, the univariate approach requires that we do more than one test. Consequently, we encounter the usual difficulties attendant on multiple tests: if only one or two significant results are obtained, were these due to chance? Moreover, unlike orthogonal contrasts, these tests are not statistically independent. Consequently, special caution should be exercised when interpreting significant results. Moreover, there are times when we do not wish to test each dependent variable individually. In some research designs the different dependent measures are regarded as different behavioral manifestations of a single underlying trait. For example, we might assess "interest" in a given activity by three measures: responses on a rating scale, amount of time spent on the activity in a free-choice situation, and some physiological measure taken while the subject is engaged in the activity. If all of these are considered to be measuring different aspects of "interest," and if our primary concern is with the underlying variable rather than with the separate behavioral measures, then a multivariate analysis is probably more appropriate than multiple univariate tests. On the other hand, if our interest is primarily in the individual behavioral measures, then univariate tests might be more appropriate.

The two approaches also differ in power, but the relative power depends on the data being analyzed. In the example of the one-way design above, most of the difference was due to the single intellectual maturity score. In such a case, a univariate test on that score will concentrate the power on the one dependent variable where the differences lie. The multivariate test loses power because it must also consider the possibility of chance differences on the other two variables. In fact, the univariate test on intellectual maturity was found to be significant in Chapter 2, but the univariate F ratio for IQ is only 1.659 ($p = .24$), and the univariate F ratio for chronological age is only 3.236 ($p = .076$).

The multivariate approach is likely to have more power when small differences exist on each of the dependent variables. The multivariate test accumulates these small differences to produce greater significance than any of the univariate tests could produce singly.

ASSUMPTIONS

The assumptions needed for multivariate tests are basically extensions of those that are needed for univariate tests. In place of the normality assumption, we assume that the dependent variables have a joint multivariate normal distribution. In place of the assumption of equal variances, we assume that all groups have identical covariance matrices. The assumptions of random sampling and independence between groups remain unchanged.

Basically, then, the multivariate tests require more assumptions than the univariate tests. The robustness of the assumptions is difficult to assess because of their complexity. However, the conclusions drawn from the univariate case can probably be generalized. Multivariate tests are probably robust to the assumption of multivariate normality if the sample sizes are large, and they are probably robust to the assumption of equal covariance matrices if the samples are equal and not too small. In addition, there is some evidence that Pillai's trace is generally more robust than the other three methods.

If the design contains random factors, the problem becomes more complex. Special assumptions have to be made about the variances and covariances of the random effects for each dependent variable, as well as for the covariances between the effects for different dependent variables (e.g., between a certain main effect for the first dependent variable and the same main effect for the second dependent variable). The assumptions are too complicated to describe easily. However, if the usual univariate assumptions hold for each dependent variable taken individually, then there will probably be no problem with the multivariate analysis.

PLANNED LINEAR COMBINATION

Occasionally we may have an a priori idea of how the dependent variables should be combined. For example, if our dependent variables are believed to be different behavioral manifestations of the same underlying variable, then a simple sum might be most appropriate. In other cases, a difference or some other linear combination might be best. If a good linear combination is chosen, then interpretation is simpler, and the test might have considerably more power. That is because a univariate test can be used instead of a multivariate test. The univariate test does not have to compensate for the possibly fortuitous choice of the linear combination. The test is also simpler. We just create a new dependent variable from the appropriate linear combination. We then use ordinary univariate procedures on this new dependent variable. Moreover, only a single test is conducted on each hypothesis instead of the multiple tests that characterize the univariate approach. Because of its simplicity and the possibility of additional power, this alternative to the multivariate test should be seriously considered.

Mixed-Effects Designs

In one case, a multivariate analysis of variance may be more valid than a univariate analysis. Earlier, we found that a univariate mixed-effects design often required special assumptions about equal variances and covariances. Moreover, the tests often were not robust to these assumptions. We proposed Box's correction as an approximate solution to the problem. However, an exact solution exists if there is only one random factor and the variables are jointly normally distributed. The problem arises whenever the random factor is crossed with one or more levels of another factor; the solution consists of regarding those levels as different dependent variables. We will illustrate this with several examples from earlier chapters.

TWO-WAY DESIGN

The first and simplest example is the data in Table 6.2. The data are from a two-way design with A fixed and B random. The test of the A main effect has two degrees of freedom, so it can be analyzed into two orthogonal contrasts. The actual choice of contrasts is arbitrary; the contrasts do not even have to be orthogonal so long as, taken together, they are equivalent to the hypothesis of no A main effect. Our first contrast will be a simple comparison of the first two means with each other. Our second contrast will compare the average of the first two means with the third. To illustrate a point to be made later, we will normalize our coefficients so that the sum of their squares is one, although again this is not necessary for the test. The coefficients are then

$$c_{11} = (1/2)^{1/2}, \quad c_{21} = -(1/2)^{1/2}, \quad c_{31} = 0,$$
$$c_{12} = (1/6)^{1/2}, \quad c_{22} = (1/6)^{1/2}, \quad c_{32} = -(2/3)^{1/2}.$$

Now instead of applying the coefficients to the marginal means, we will apply them to each row of the table (i.e., to each level of B), obtaining the twenty pairs of scores found in Table 13.3. If the null hypothesis of no A main effect is true, then the means of these scores should be zero.

If we regard A_1' and A_2' as two separate dependent variables, then this is a kind of multivariate analogue of the single-sample t test. We can also regard it as a kind of single-sample multivariate F test. We will follow the F test approach because it generalizes more easily to more complex designs.

In effect, we will do an F test on the grand mean of a set of data containing only a single group. (We regard A_1' and A_2' as two dependent variables, not two groups.) Our numerator sum of cross-products matrix will be SC_m, and our denominator matrix will be SC_w.

Because there is only one group, and each variable has only a single mean, each mean is the grand mean for that variable. We calculate the diagonal of SC_m by squaring each column total and dividing by 20, the number of scores in the column. (For this purpose, we can ignore the fact that each of the original cell means was based on 10 scores, regarding each cell mean as a single score.) We get the off-diagonal value by taking the product of the two totals and dividing by 20.

Similarly, RC_w is found by summing squares and cross-products of the two sets of scores. Finally, because SC_m represents the only "effect," $SC_w = RC_w - SC_m$. Both of these matrices are shown in Table 13.3. We can now test our null hypothesis by any of the four methods above. The associated univariate degrees of freedom are $\nu_1 = 1, \nu_2 = 19$. Since $\nu_1 = 1$, all methods give identical results. Wilks's lambda is simple to calculate; for it we just take

$$\Lambda = \det(SC_w)/\det(SC_w + SC_m) = .8293,$$
$$F_{(2,18)} = 1.853,\ p = .19.$$

This test is exact because $g = 1$. (Note that $p = 2$, the number of contrasts.) Roy's maximum root is more complicated; it gives the optimal linear combination, but that is not meaningful when arbitrary contrasts have been chosen anyway. (If the contrasts had themselves been of interest for some reason, then the optimal linear combination might have been of interest.)

By choosing our contrasts as we did, making them orthogonal and making the squares of the coefficients sum to one, we obtain a small added bonus. We can estimate K_B—Box's correction—from SC_w by a simple formula. Letting $u =$ the sum of the values on the diagonal of SC_w, $v =$ the sum of the squares of all the values in SC_w, and $r =$ the number of rows (and columns) of SC_w

$$K_B = u^2/(rv)$$
$$= 573.8^2/[(2)(166384)] = .989,$$

the same value we obtained in Chapter 6.

The same general approach can be used for more complicated designs. We construct contrasts for effects involving factors with which the random factor

TABLE 13.3. Testing the A main effect on the data in Table 6.2, treating contrasts as multivariate dependent variables.

	A_1'	A_2'
B_1	2.828	5.715
B_2	-4.243	6.532
B_3	3.536	1.225
B_4	-4.243	8.982
B_5	2.121	6.124
B_6	-2.121	-2.041
B_7	0.707	-4.491
B_8	-2.121	0.408
B_9	-5.657	-1.633
B_{10}	-9.900	0.816
B_{11}	6.364	2.041
B_{12}	1.414	-0.816
B_{13}	1.414	0.816
B_{14}	2.121	3.674
B_{15}	4.243	-1.633
B_{16}	-0.707	0.408
B_{17}	1.414	1.633
B_{18}	5.657	0.816
B_{19}	1.414	8.165
B_{20}	-0.707	-3.674
Sums	-9.192	33.068
Means	-0.460	1.653

$$SC_m \begin{vmatrix} 4.225 & -15.199 \\ -15.199 & 54.675 \end{vmatrix}$$

$$\Lambda = .8293, \quad F_{(2,18)} = 1.85, \quad p = .19$$

$$SC_w = \begin{vmatrix} 300.275 & -26.659 \\ -26.659 & 273.492 \end{vmatrix}$$

is crossed. Then we calculate the value of each contrast at each level of the random factor. The resulting "scores" are regarded as a multivariate sample, and we apply the multivariate analysis of variance. Also, just as the test on the A main effect was reduced to a test of the grand mean of the contrast, tests in later examples can be reduced to lower-order tests. Tests of main effects can be reduced to tests of grand means, tests of two-way interactions can be reduced to tests of main effects or even of grand means, and so on. In each case, K_B can also be estimated from the denominator SC matrix.

T^2 TEST

The test usually advocated for the data in Table 13.3 is called Hotelling's T^2 test. It is a multivariate generalization of the ordinary t test on a single group. However, the results are identical to those above; in fact,

$$\lambda = 1/[1 + T^2/(J - 1)],$$

so that either statistic is easily calculated from the other. We chose the approach above because it generalizes easily to more complicated designs. The generalization of the T^2 test is less apparent.

THREE-WAY DESIGN

The completely crossed three-way design with one random factor is a simple extension of the above analysis. We will illustrate it with the data in Table 7.8. The univariate test of the A main effect is valid because it has only one degree of freedom. For the test of the B main effect, we use the BC table, treating it just like the two-way table above. Again, we calculate two contrasts and test the null hypothesis that the marginal means of both contrasts are zero. We get $F_{(2,3)} = 7.21$, $p = .072$.

We use the full data table to test the interaction. The coefficients of the two contrasts we will use are as follows:

AB_{11}	AB_{12}	AB_{13}	AB_{21}	AB_{22}	AB_{23}
1/2	-1/2	0	-1/2	1/2	0
$(1/12)^{1/2}$	$(1/12)^{1/2}$	$-(1/3)^{1/2}$	$-(1/12)^{1/2}$	$-(1/12)^{1/2}$	$(1/3)^{1/2}$

The data are shown in Table 13.4. Following the same procedures as above, $F_{(2,3)} = 19.4$, $p = .020$. The ability to estimate K_B is more important here, because no simple procedure exists for estimating it directly from the original covariance matrix. If we estimate it from SC_w, $K_B = .689$, suggesting that the assumption of equal variances and covariances is questionable.

$A \times B \times C(B)$ DESIGN

In completely crossed designs, the tests are always made on simple means. The problem is more complicated in nested designs, but the principles are the same. We will illustrate it first with the $A \times B \times C(B)$ design, with C random,

TABLE 13.4. Analysis of AB interaction for data in Table 7.8.

	A'_1	A'_2
C_1	1.5	7.217
C_2	3.0	3.464
C_3	4.5	7.217
C_4	6.5	8.949
C_5	-3.0	5.196
Sums	12.5	32.043
Means	2.5	6.409

$$C_M = \begin{vmatrix} 31.250 & 80.107 \\ 80.107 & 205.350 \end{vmatrix}$$

$$\lambda = \quad .07175, \; F_{(2,3)} = 19.4, \; p = .20$$

$$SC_w = \begin{vmatrix} 51.500 & 16.167 \\ 16.167 & 17.900 \end{vmatrix}$$

in Table 8.6. In that table, if we look at each level of B separately, we find that the three nested $A \times C$ tables have the same form as a two-way, mixed-effects table. For each of these tables, we can calculate contrasts, across the levels of A, at each level of C. The coefficients for the contrasts we will use are as follows:

A_1	A_2	A_3	A_4
$(1/2)^{1/2}$	$-(1/2)^{1/2}$	0	0
$(1/6)^{1/2}$	$(1/6)^{1/2}$	$-(2/3)^{1/2}$	0
$(1/12)^{1/2}$	$(1/12)^{1/2}$	$(1/12)^{1/2}$	$-(3/4)^{1/2}$

The values of the contrasts are in Table 13.5. Clearly, we can regard the levels of A' as dependent variables, the levels of C as multivariate scores, and the levels of B as the groups in a one-way design.

In this design, the grand mean is the average, over levels of B, of the contrasts on A. Thus, the test of the grand mean in Table 13.5 is equivalent to the test of the A main effect in Table 8.6. Similarly, the test of the B main effect in Table 13.5 is a test of the null hypothesis that the contrasts on A have the same value at each level of B; this is equivalent to the test of the AB interaction in Table 8.6. Thus, here the test of the A main effect is reduced to the test of a grand mean, and the test of the AB interaction is reduced to the test of a B main effect.

TABLE 13.5. Multivariate analysis of data in Table 8.6.

	B_1			B_2			B_3		
	A'_1	A'_2	A'_3	A'_1	A'_2	A'_3	A'_1	A'_2	A'_3
C_1	-4.95	-.41	-4.62	2.12	2.04	-8.08	2.12	.41	-4.04
C_2	-2.83	4.90	0.00	1.41	4.08	-9.24	-3.54	-1.22	-2.60
C_3	-4.24	-1.63	-2.02	-2.12	1.22	-2.60	-6.36	3.67	-9.53

$$SC_m = \begin{vmatrix} 37.556 & -26.686 & 87.274 \\ -26.686 & 18.963 & -62.016 \\ 87.274 & -62.016 & 202.815 \end{vmatrix}$$

$$\Lambda = .2166, \ F_{(3,4)} = 4.82, \ p = .082$$

$$SC_{bet} = \begin{vmatrix} 31.466 & 11.290 & -28.146 \\ 11.290 & 4.481 & -8.498 \\ -28.146 & -8.498 & 31.241 \end{vmatrix}$$

$$\Lambda = .4007, \ F_{(6,8)} = .77, \ p = .62$$

$$SC_w = \begin{vmatrix} 50.000 & -.192 & 8.845 \\ -.192 & 40.889 & -15.164 \\ 8.845 & -15.164 & 62.611 \end{vmatrix}$$

The procedures for a multivariate one-way design were described above; we need not repeat them here. For the A main effect (the grand mean in Table 13.5), we get $\Lambda = .2166$, $F_{(3,4)} = 4.82$, $p = .082$. For the AB interaction (the A main effect in Table 13.5), we get $\Lambda = .4007$, $F_{(6,8)} = .77$, $p = .62$. Both tests are exact.

As above, SC_w can be used to estimate K_B; we get $K_B = .902$.

$A \times B(A) \times C(A)$ DESIGN

Our final example is the $A \times B(A) \times C(A)$ design, with C random in Table 8.8. Here, because C is crossed with B, we will calculate the contrasts across the levels of B, using the same two contrasts as in the two-way design of Table 6.2. The results are in Table 13.6; there is a separate set of values for each level of A. The data in Table 13.6 are those of a one-way multivariate analysis of variance with $I = 2$, $n = 5$, $p = 2$. The only test we are interested in is $b(a)$ because the univariate tests are valid on all of the other effects. This

TABLE 13.6. Multivariate analysis of data in Table 8.8.

	A_1		A_2	
	B_1'	B_2'	B_3'	B_4'
C_1	-3.465	-4.613	-.707	-.653
C_2	-5.869	-.449	1.556	-1.470
C_3	1.202	-1.919	1.061	-4.042
C_4	-4.738	-1.429	1.273	-4.899
C_5	-2.475	2.490	3.606	-2.654
Sums	-15.344	-5.920	6.788	-13.717
Means	-3.069	-1.184	1.358	-2.743

$$RC_{bet} = \begin{vmatrix} 56.305 & -.457 \\ -.457 & 44.640 \end{vmatrix}$$

$$\Lambda = \quad .1876, \ F_{(4,14)} = 4.58, \ p = .015$$

$$SC_w = \begin{vmatrix} 38.830 & -4.540 \\ -4.540 & 33.731 \end{vmatrix}$$

is the test of the null hypothesis that all contrasts are zero, in *each* level of A; i.e., it is a test of the null hypothesis that all of the population means, on all of the dependent variables and in both groups, in Table 13.6, are zero. This is equivalent to a simultaneous test that the grand mean is zero and that there is no A main effect. The denominator matrix is still SC_w, but the numerator matrix is now $RC_{bet} = SC_{bet} + SC_m$, combining the grand mean and the A main effect. These matrices are shown in Table 13.6. We get $F_{(4,14)} = 4.58$, $p = .015$. Also, as above, $K_B(= .980)$ is calculated from SC_w.

UNIVARIATE VERSUS MULTIVARIATE ANALYSIS

We now have a choice of conducting either a univariate or a multivariate analysis when there is a random factor. It is relevant to ask which type of analysis is better. Unfortunately, no general answer can be given to this question. For the data illustrated in this chapter, the univariate analyses are usually better. Even the most conservative univariate analysis—using Box's correction—compares well with, and frequently has a smaller p value than, the multivariate analysis. Unfortunately, these examples are not necessarily typical of real data.

We can say only one thing for certain. If $K_B = 1$, then the univariate test, without any correction on the degrees of freedom, is valid and most powerful.

When $K_B < 1$, the univariate test is no longer valid without corrections to the degrees of freedom. Then the relative power depends in complicated ways on the variances and covariances. In some cases, the least powerful univariate method—using Box's correction—has more power than the multivariate method. In others, the multivariate method has more power than the univariate analysis, even with no correction to the degrees of freedom. In the final analysis, when K_B is suspected of being smaller than 1, the choice may depend more on convenience and available computer programs than on theoretical or mathematical considerations.

14

Analysis of Covariance

In multivariate analysis of variance we test all of the dependent variables simultaneously. In *analysis of covariance* we use some dependent variables as "controls" when testing for others. If we cannot control for certain variables experimentally by making them the levels of a factor, we may be able to control for them statistically by analysis of covariance. In this chapter, we will first give a simple example. We will then describe the model for analysis of covariance, the problems of interpretation, and the assumptions that must be made. Next, we will discuss the advantages and disadvantages of analysis of covariance, as compared with other possible ways of solving the same problem. Finally, we will describe the general analysis of covariance, with examples. The reader who wants only a general understanding of analysis of covariance can skip the final section.

Example

The data in Table 14.1 are a simple example of a one-way design with one dependent variable and one covariate. They are the same as the data in Table 2.1 (scores on a test of "intellectual maturity"), with the addition of a covariate, chronological age. (The covariate is identical to the second dependent variable in Table 12.1; thus, we see already that each variable may be treated as either a dependent variable or a covariate.)

Suppose we suspect that the differences in scores might be due to differences in the ages of the students taking the courses rather than differences among the courses they are taking. One way to separate the effect of age from the effect of the course would be to use a two-way design with Factor B being age.

However, for such a design, we would have to have subjects of about the same ages in all courses. This is impossible because we cannot force students to take the courses, so we do the next best thing. We estimate the differences in "intellectual maturity" that we would have obtained had we been able to control for chronological age. That is, we control statistically for chronological age by doing an analysis of covariance.

TABLE 14.1. Data from Table 3.1, with the addition of a random, numerical factor, age. (In an ordinary analysis of variance, $\hat{\alpha}_1 = .6$, $\hat{\alpha}_2 = -1.8$, $\hat{\alpha}_3 = 1.2$.)

	A_1		A_2		A_3		Sum	
	X	Y	X	Y	X	Y	X	Y
	5	27	1	23	5	28		
	3	21	2	22	4	26		
	5	26	2	20	7	26		
	6	31	0	18	5	24		
	1	23	3	22	2	20		
Sum	20	128	8	105	23	124	51	357
Mean	4.0	25.6	1.6	21.0	4.6	24.8	3.4	23.8
$\hat{\alpha}_i$	-.13		-.66		.79			
$\hat{\theta} = .407$								

THE MODEL

The model for the analysis of covariance is like that for the two-way design with one factor (the covariate) numerical. However, we must considerably limit the model to avoid problems of calculation and interpretation. First, we limit the trend on the covariate to a linear trend and, second, we assume that there is no interaction between the two factors. Basically, these assumptions are equivalent to the assumption that the variables have a bivariate normal distribution. They are not essential to the analysis, but they simplify both the calculations and the interpretation of the results. We will briefly discuss ways of relaxing these assumptions later.

With these assumptions, the model equation is

$$X_{ij} = \mu + \alpha_i + \theta_y Y_{ij} + \epsilon_{ij}, \tag{15.1}$$

where X_{ij} is the dependent variable ("intellectual maturity" in this case) and Y_{ij} is the covariate (chronological age). In this equation, θ_y represents the effect of the covariate on the dependent variable; α_i represents the differences between groups, as usual, but now it represents the differences after the effect of the covariate has been accounted for.

INTERPRETATION

Methods of analysis will be given later. When they are applied to these data, they give $\hat{\theta}_y = .407, \hat{\alpha}_1 = -.13, \hat{\alpha}_2 = -.66, \hat{\alpha}_3 = .79$. The test of $H_0 : \theta_y = 0$ is significant ($p < .001$), but the test of the α_i is not ($p = .25$); test scores are related to chronological age, but they are not significantly related to courses when chronological age is taken into account.

TABLE 14.2. Sample data for which analysis of covariance is significant but analysis of variance is not. (In an ordinary analysis of variance, $\hat{\alpha}_1 = 0.6$, $\hat{\alpha}_2 = 0.2$, $\hat{\alpha}_3 = -0.8$.)

	A_1		A_2		A_3		Sum	
	X	Y	X	Y	X	Y	X	Y
	5	27	3	23	3	28		
	3	21	4	22	2	26		
	5	26	4	20	5	26		
	6	31	2	18	3	24		
	1	23	5	22	0	20		
Sum	20	128	18	105	13	124	51	357
Mean	4.0	25.6	3.6	21.0	2.6	24.8	3.4	23.8
$\hat{\alpha}_i$	-.13		1.34		-1.21			
$\hat{\theta} = .407$								

Interpretation of these results is complicated by the fact that the Y_{ij} and X_{ij} are not independent. In this example, the analysis of variance was significant, but the analysis of covariance was not. Table 14.2 shows an example in the analysis of variance is not significant ($F = .91$) but the analysis of covariance is ($F = 4.32$, $p = .041$). The data in this table are the same as those in Table 14.1, except that the group means are different. In Table 14.2 the $\overline{X}_{i.}$ are nearly equal, but the $\overline{Y}_{i.}$ are very different from each other. When the effects of these $\overline{Y}_{i.}$ are factored out, a sizable effect results.

Table 14.3 shows a result that is less common but can occur. Both analyses are significant for these data ($F = 4.40$, $p = .037$ for the analysis of variance; $F = 4.11$, $p = .046$ for the analysis of covariance), but the effects are in different directions, with $\hat{\alpha}_2$ smallest in the analysis of variance and largest in the analysis of covariance.

These differences occur because differences among the X_{ij} within a group are no longer regarded as due to random error. Instead, they are regarded as partly due to random error and partly due to a linear relationship with the Y_{ij}. Basically, the analysis of covariance model assumes that the X_{ij} in each group are a linear function of the Y_{ij}. It does not assume that the same linear function holds for each group, but it does assume that the linear functions all have the same slope, θ_y. Figures 14.1, 14.2, and 14.3 show the best estimates of the functions, along with their slopes, for each group. Because the functions are all assumed to have the same slope, they can differ only in their overall heights. The null hypothesis of no A main effect hypothesizes that these functions all have the same height, i.e., that in fact they are all exactly the same function.

In Table 14.1 and Figure 14.1, the analysis of variance is significant because

TABLE 14.3. Sample data for which analysis of variance and covariance are both significant but in opposite directions. (In an ordinary analysis of variance, $\hat\alpha_1 = 0.6$, $\hat\alpha_2 = -1.8$, $\hat\alpha_3 = 1.2$.)

	A_1		A_2		A_3		Sum	
	X	Y	X	Y	X	Y	X	Y
	5	25	1	13	5	40		
	3	19	2	12	4	38		
	5	24	2	10	7	38		
	6	29	0	8	5	36		
	1	21	3	12	2	32		
Sum	20	118	8	55	23	184	51	357
Mean	4.0	23.6	1.6	11.0	4.6	36.8	3.4	23.8
$\hat\alpha_i$	0.68		3.41		-4.09			
$\hat\theta = .407$								

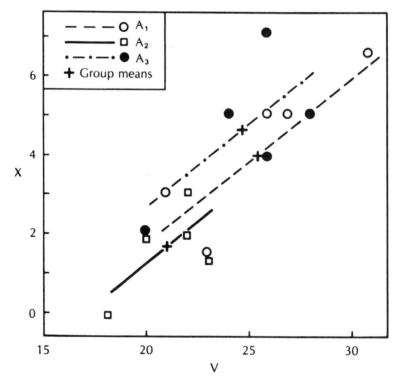

FIGURE 14.1. Data from Table 14.1 plotted against the covariate.

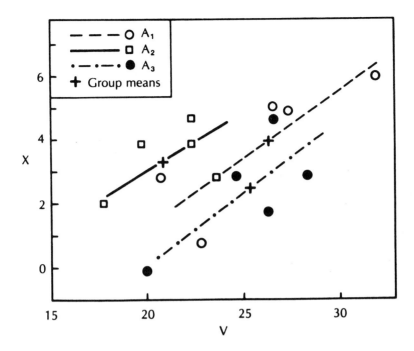

FIGURE 14.2. Data from Table 14.2 plotted against the covariate.

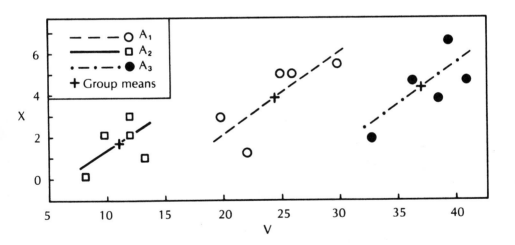

FIGURE 14.3. Data from Table 14.3 plotted against the covariate.

the mean of A_2 is much smaller than the means of A_1 and A_3. However, the mean of the covariate for A_2 is also much smaller, so the three regression lines very nearly coincide. Consequently, the analysis of covariance is not significant. We would conclude from these data that differences in test scores were not due directly to differences in the quality of the courses, but rather that students taking some courses were older than those taking other courses.

In Table 14.2 and Figure 14.2, the three group means are nearly equal, so the analysis of variance is not significant. However, the group means of the covariate are not equal, so the regression lines differ from each other significantly. From these data we would conclude that the classes did differ in how much they helped the students to improve their test scores, but that these differences were obscured in the raw data because the most helpful courses tended to be taken by younger students with lower initial scores.

In Table 14.3 and Figure 14.3, the differences among the group means are large, but the differences among the covariate means are even larger. Consequently, even though the mean of Group A_2 is lower than those of Groups A_1 and A_3, the overall height of its regression line is greater. We might conclude that the most helpful course, A_2, was so good it more than overcame a large initial handicap, due to age, on the part of those who took it.

PROBLEMS IN INTERPRETATION

Generally, interpretations require some care. In Table 14.1 and Figure 14.1, for example, although the courses seem to have no direct effect on test scores, there may be an indirect effect. Courses A_1 and A_3 obviously tend to attract older students. There may be nothing in the courses themselves that would raise the score of the average student, but higher-scoring students are found in these classes.

In a similar, classical example, suppose the data in Table 14.1 were achievement scores and the covariate was time spent studying. We would be naive to conclude from the negative results of the analysis of covariance that there were no differences among the classes. We would be wiser to try to learn why students study more in some courses than in others.

For some problems the analysis of covariance may simply be inappropriate. Suppose, for example, we were confronted with three people, each of whom had taken one of the three courses, and our problem was to guess who had the lowest score on "intellectual maturity." If all we knew about each was the course taken, the analysis of covariance would be irrelevant to our decision; no matter what the effects of age may be, it is still a fact that students in A_2 tend to have substantially lower scores than those in A_1 and A_3.

For other problems, the analysis of covariance is appropriate. If we knew the ages of the three people, then the analysis of covariance indicates that our best choice would be the youngest of the three, regardless of which course that person had taken. Similarly, a student who wished to score highly should probably be advised to take whichever course that student would enjoy most, forgetting about raising his or her test score by taking one course rather than another.

We must also avoid unwarranted extrapolations beyond the range of the data. We concluded from Table 14.2 and Figure 14.2 that A_2 was more helpful but attracted younger students. Extending the regression line, we would be tempted to conclude that A_2 would also be most helpful for older students. However, there are no older students in our sample from group A_2 so we have no way of knowing whether or not such a generalization would hold. It could be, in fact, that older students avoid A_2 because it would be harmful to them in some way.

Some interpretation problems can be eliminated by making the covariate independent of the effect being studied. For example, in a study of the aggressiveness of chickens given four different diets, the covariate might be the weights of the chickens. If the chickens have been assigned randomly to the four groups, the values of the covariate should be independent of the groups effect. Then the only effect of the covariate is to reduce the within error, increasing the power of the test and eliminating most interpretation problems. Some statisticians recommend the use of analysis of covariance only when the covariate is independent of the groups effect.

General Analysis of Covariance

The analysis of covariance can be generalized in three ways. First, it can be generalized to other than simple one-way designs. In fact, any test that can be made with an F ratio, in a fixed-effects design, can be made as an analysis of covariance. (The limitation to fixed-effects designs will be discussed later.) The model for a more complex design is just the usual linear model with an added linear term for the covariate. For example, the model for the two-way analysis is

$$X_{ijk} = \mu + \alpha_i + \beta_j + \alpha\beta_{ij} + \theta_y Y_{ij} + \epsilon_{ijk}. \tag{14.2}$$

Analyses of covariance can also be done on planned comparisons.

Second, there can be any number of covariates. The model equation then contains one linear term for each covariate. The one-way model with two covariates, Y_{ij} and Z_{ij}, would be

$$X_{ij} = \mu + \alpha_i + \theta_y Y_{ij} + \theta_z Z_{ij} + \epsilon_{ij}.$$

Third, there can be any number of dependent variables; i.e., multivariate analyses of covariance can be conducted. For multivariate analyses, there will a different slope constant (e.g., θ_y and θ_z) for each dependent variable.

RANDOM AND MIXED DESIGNS

Theoretically, analysis of covariance is a very versatile technique making it possible to control statistically for variables that are not easily controlled experimentally. However, there are limitations to its use.

In general, analyses of covariance should be limited to fixed-effects designs. Because the covariate is treated as an effect, the effects in an analysis of covariance are not all orthogonal. The nonorthogonality creates difficulties with assumptions about the variances of the random effects. In fact, in most analyses of covariance it is probably impossible to specify variances that will make the analyses valid. Consequently, it is best to avoid performing analyses of covariance on models with random effects.

ASSUMPTIONS

The analysis of covariance also requires special assumptions in addition to those of the ordinary analysis of variance. It assumes that the dependent variables and covariates are all linearly related to each other, and that the slopes of the linear functions are the same for all groups in the design. The effects of departures from these assumptions have not been thoroughly tested. For this and other reasons described below, it is probably better to avoid an analysis of covariance when an alternative analysis is feasible.

Linearity. With a single dependent variable and a single covariate, the assumption of linearity is less restrictive than it at first appears. If the relationship between the dependent variable and the covariate is not linear, for example, perhaps the relationship between the dependent variable and the square of the covariate is linear. We would then simply use the squares of the covariate values as the Y_{ij} and do the analysis as usual. The same would hold true if a linear relationship held between the dependent variable and the square root, the logarithm, or some other function of the covariate.

However, the assumption can never be completely eliminated; a transformation on the covariate simply raises the question of whether the transformed Y_{ij} are linearly related to the X_{ij}. Moreover, if there are more than one dependent variable or more than one covariate, there may not be any transformations that make all of the variables linearly related to each other.

We do not know the effects of violation of the assumption of linearity on the analysis of covariance. Earlier we saw that an analysis of variance is often robust with respect to all assumptions other than random sampling; an analysis of covariance may be robust to linearity, but we cannot be sure.

Moreover, interpretation may be more difficult without the linearity assumption. With the assumptions of linearity and equal slopes, the null hypothesis is equivalent to the assertion that the regression lines coincide. However, if the regression lines have possibly different kinds of curvature, we cannot say that they coincide even if they have the same general slope and intercept. Therefore, the practical meaning of the null hypothesis is no longer clear.

Equal Slopes. The assumption of equal slopes can be tested. However, the test is complicated, and it must be used carefully; if we accept the null hypothesis of equal slopes, we are risking an unknown probability of a Type II error.

Even if the slopes are not equal, we can still conduct the analysis of variance, using a modified test. However, interpretation of the results becomes considerably more complicated when the slopes are not equal. The original null hypothesis that the linear functions were in fact all identical must then be changed to state that the different linear functions all cross each other at a single value of the covariate. Such a null hypothesis is probably not only uninteresting in most cases, but its meaning is probably unclear in most others. The analysis is also much more complicated when equal slopes are not assumed. For these reasons, it is probably best to avoid an analysis of covariance when the slopes are not equal.

Alternatives to Analysis of Covariance

The analysis of covariance is often abused as a way to avoid the extra effort required for good experimental controls. Controlling for the values of the covariate frequently requires careful selection of subjects. The alternative of letting the values of the covariate select themselves is too tempting to resist, especially if a computer is available to do the work of analysis.

EXPERIMENTAL CONTROL OF THE COVARIATES

Nevertheless, there are a number of advantages to controlling the covariate values experimentally. Suppose, for example, that we had determined the ages of all of our potential subjects in advance. We might then have been able to divide our subjects into, say, three or four age groups and select a certain number of subjects in each age group from each course. The result, if we had used four age groups, would have been a 3×4 analysis of variance. The more complicated analysis of covariance would not have been required, nor would the assumptions of linearity and equal slope. Furthermore, if four groups did not provide a fine enough division of ages, age could still be added as a covariate. The design would then be a two-way analysis of covariance. The analysis of such a design is only a little more complicated than the one-way analysis of covariance.

The principle advantages of controlling for the values of the covariate are the reduced number of assumptions required for the analysis and the consequent increase in degrees of freedom in analyzing the data. The interaction between the covariate and the main factor, for example, can be tested and estimated. However, there are also disadvantages in controlling for the covariate. In the design of Table 14.1, controlling for the covariate would have hidden the tendency for older students to take certain courses. A separate study, using ages as data, would have been necessary to reveal that tendency. In the analysis of covariance both the relationship and the tendency are clear.

The example discussed previously, for which the covariate was time spent studying, may pose a more difficult problem. Presumably, the way to control for studying would be to have some way of making sure that every student studies for a prescribed amount of time, perhaps by specifying certain times during which the students must study and prohibiting them from studying any other time. Such a procedure would probably be highly impractical because both the very good and very poor students would resist such restrictions. Moreover, such controls might change the nature of the covariate; regulated study time might not be used as efficiently as unregulated study time by some students, whereas it may be used more efficiently by others. Thus, the variable of study time might not be the same variable in the two experiments.

Neither approach—statistical control of the covariate by analysis of covariance nor experimental control through prior selection of the covariate values—is universally the best. We need to weigh very carefully both the advantages and disadvantages of each in light of the purposes of the experiment. We should also consider the possibility that the best procedure might be to not control for the covariate at all; e.g., we might be interested simply in determining which course has the highest scoring students, regardless of the students' ages.

OTHER KINDS OF STATISTICAL CONTROL

Even when experimental control of the covariate is not desirable, there may be better methods of statistically controlling for the covariate. Frequently, for example, the actual function relating the covariate to the data is at least approximately known. A classical case is an extinction study in which all subjects are given a standard number of trials for a learning task, after which they are divided into two or more groups in which learning is extinguished under different conditions. Often an analysis of covariance is performed on the extinction data, with the covariate being the level of initial learning. However, such data may violate the assumption of equal slope. Instead, the amount of extinction may be approximately proportional to the level of initial learning, with the proportionality differing among groups. If the slopes were proportional to the amount of learning, they could not all be equal. Alternatively, we could divide each extinction score by the associated learning score and do an ordinary analysis of variance on the ratios. Such an analysis would not only be more appropriate, it would also avoid the extra problems (in both assumptions and computations) of an analysis of covariance. (With some data, the ratios might have unequal variances; then it might be better to analyze, say, the logarithms of the ratios.)

In the last example the data were assumed to be proportional to the covariate; this is by no means the only relationship that may hold. The data might be proportional, for example, to the square, the square root, the logarithm, or some other function of the covariate. Or it might be that the algebraic difference between the covariate and variate is a constant (except for random error) for a given group, with the value of that constant possibly different from one group to another. To test whether the value of the constant differs from group to group, an analysis of variance would be done on the differences between the values of the variate and the covariate. In the example in Table 14.1 and Figure 14.1, the slope of the function relating the variate to the covariate was close to one-half (actually closer to .4). If we had had good a priori reasons for believing that the slope should be one-half, we could have subtracted one-half the value of the covariate from each X_{ij} and performed an analysis of variance on the resulting differences.

The possible relations that we might choose are endless, with each reducing the problem to a simple analysis of variance instead of the more complicated analysis of covariance. The important restriction here is that we should have good a priori reasons for believing that the assumed function actually holds (at least approximately) for the data being analyzed. In the absence of such prior reasons, this alternative is less desirable.

Data Analysis

We will now describe the data analysis step-by-step. The analysis depends heavily on the concepts developed in Chapters 12 and 13. If all of your analyses are to be done by computer, you needn't know the techniques in detail. However, it is still worthwhile to understand the data analysis, both to have a better understanding of what the computer does and to understand some of the variety of tests that can be done.

After each step is described, it will be illustrated with two examples:

(a) an overall analysis of covariance on the data in Table 12.1, with X as the dependent variable and Y and Z as the covariates; and

(b) an overall analysis of covariance on the same data, with X and Y as dependent variables and Z as the covariate.

THE MODEL

We will first generalize the model to the case where there may be both multiple dependent variables and multiple covariates. The model for the one-way design is the same as in Eq. 14.1, but now any or all of the terms may be vectors or matrices. Specifically, if there are multiple dependent variables, then X_{ij}, μ, α_i, and ϵ_{ij} are all vectors. The dimension of each vector is the number of dependent variables. Similarly, if there are multiple covariates, then Y_{ij} is a vector whose dimension is the number of covariates. Finally, θ may be a matrix. The number of rows of θ is the number of dependent variables; the number of columns is the number of covariates.

The generalization to higher-way designs is straightforward. We just add the parameters for the additional effects. For example, the model for the two-way design is Eq. 14.2, with the symbols now representing vectors and matrices.

PARTITIONING THE MATRICES

For each effect being tested, we first calculate the two sum of cross-product matrices, SC_w and $SC_s = SC_n + SC_w$, as in Chapter 13. (Because we are limited to fixed-effects models, SC_d in Chapter 13 is always the same as SC_w here.) We include all variables, i.e., variates and covariates, in each matrix. Next, we *partition* SC_w and SC_s, each into four *submatrices*. If we arrange the data so that the first rows and columns of each matrix correspond to the dependent variables and the last rows and columns correspond to the covariates, we can write

$$SC_w = \begin{vmatrix} A_w & B_w \\ B_w^t & C_w \end{vmatrix}, \quad SC_s = \begin{vmatrix} A_s & B_s \\ B_s^t & C_s \end{vmatrix},$$

where A_w and A_s contain variances and covariances on the dependent variables, C_w and C_s contain variances and covariances on the covariates, and B_w and B_s contain covariances between the dependent variables and covariates. In future calculations, each of these submatrices will be regarded as a matrix in its own right.

Example a. The two matrices, as given in Chapter 12, are

$$SC_w = \begin{vmatrix} 34.4 & -55.8 & 45.6 \\ -55.8 & 1,450.0 & -51.6 \\ 45.6 & -51.6 & 112.0 \end{vmatrix},$$

$$SC_s = SC_w + SC_{bet} = \begin{vmatrix} 59.6 & -129.6 & 82.2 \\ -129.6 & 1,850.93 & -122.2 \\ 82.2 & -122.2 & 172.4 \end{vmatrix}.$$

The first row and column in each matrix correspond to the dependent variable, X, while the last two correspond to the covariates, Y and Z. Therefore,

$$A_w = \begin{vmatrix} 34.4 \end{vmatrix}, \quad B_w = \begin{vmatrix} -55.8 & 45.6 \end{vmatrix},$$

$$B_w^t = \begin{vmatrix} -55.8 \\ 45.6 \end{vmatrix}, \quad C_w = \begin{vmatrix} 1,450.0 & -51.6 \\ -51.6 & 112.0 \end{vmatrix},$$

$$A_s = \begin{vmatrix} 59.6 \end{vmatrix}, \quad B_s = \begin{vmatrix} -129.6 & 82.2 \end{vmatrix},$$

$$B_s^t = \begin{vmatrix} -129.6 \\ 82.2 \end{vmatrix}, \quad C_s = \begin{vmatrix} 1,850.93 & -122.2 \\ -122.2 & 172.4 \end{vmatrix}.$$

Example b. For this case, SC_w and SC_s are the same, but X and Y are dependent variables, and Z is the covariate. Then

$$A_w = \begin{vmatrix} 34.4 & -55.8 \\ -55.8 & 1{,}450.0 \end{vmatrix}, \quad B_w = \begin{vmatrix} 45.6 \\ -51.6 \end{vmatrix}$$

$$B_w^t = \begin{vmatrix} 45.6 & -51.6 \end{vmatrix}, \quad C_w = \begin{vmatrix} 112.0 \end{vmatrix},$$

$$A_s = \begin{vmatrix} 59.6 & -129.6 \\ -129.6 & 1{,}850.93 \end{vmatrix}, \quad C_s = \begin{vmatrix} 82.2 \\ -122.2 \end{vmatrix},$$

$$B_s^t = \begin{vmatrix} 82.2 & -122.2 \end{vmatrix}, \quad C_s = \begin{vmatrix} 172.4 \end{vmatrix}.$$

ADJUSTED SUMS OF CROSS-PRODUCTS

Two matrices are now computed; they are the *adjustment factors*, AF_w and AF_s. Each is formed in the same way from the corresponding submatrices:

$$AF_w = B_w C_w^{-1} B_w^t, \quad AF_s = B_s C_s^{-1} B_s^t.$$

The two *adjusted sum of cross-products matrices* are then

$$ASC_w = A_w - AF_w, \quad ASC_s = A_s - AF_s.$$

These take the place of SC_w and SC_s for testing the null hypotheses.

Example a. For the single dependent variable and two covariates,

$$C_w^{-1} = \begin{vmatrix} .0007012 & .0003230 \\ .0003230 & .0090774 \end{vmatrix},$$

$$AF_w = \begin{vmatrix} -55.8 & 45.6 \end{vmatrix} \begin{vmatrix} .0007012 & .0003230 \\ .0003230 & .0090774 \end{vmatrix} \begin{vmatrix} -55.8 \\ 45.6 \end{vmatrix}$$

$$= 19.4144,$$

and

$$C_s^{-1} = \begin{vmatrix} .0005668 & .0004018 \\ .0004018 & .0060852 \end{vmatrix},$$

$$AF_s = \begin{vmatrix} -129.6 & 82.2 \end{vmatrix} \begin{vmatrix} 1{,}850.93 & -122.2 \\ -122.2 & 172.4 \end{vmatrix} \begin{vmatrix} -122.2 \\ 82.2 \end{vmatrix}$$

$$= 42.0771.$$

Then

$$ASC_w = 34.4 - 19.4144 = 13.9886,$$
$$ASC_s = 59.6 - 42.0771 = 17.5229.$$

Example b. When Z is the only covariate, $C_w^{-1} = 1/112 = .008929$, and $C_s^{-1} = 1/172.4 = .005800$, so

$$
AF_w = \begin{vmatrix} 45.6 \\ -55.8 \end{vmatrix} |.008929| \begin{vmatrix} 45.6 & -51.6 \end{vmatrix}
$$

$$
= \begin{vmatrix} 18.566 & -21.009 \\ -21.009 & 23.773 \end{vmatrix},
$$

$$
AF_s = \begin{vmatrix} 82.2 \\ -122.2 \end{vmatrix} |.005800| \begin{vmatrix} 82.2 & -122.2 \end{vmatrix}
$$

$$
= \begin{vmatrix} 39.193 & -58.265 \\ -58.265 & 86.617 \end{vmatrix},
$$

and

$$
ASC_w = \begin{vmatrix} 34.4 & -55.8 \\ -55.8 & 1450.0 \end{vmatrix} - \begin{vmatrix} 18.566 & -21.009 \\ -21.009 & 23.773 \end{vmatrix}
$$

$$
= \begin{vmatrix} 15.834 & -34.791 \\ -34.791 & 1,426.2 \end{vmatrix},
$$

$$
ASC_s = \begin{vmatrix} 59.6 & -129.6 \\ -129.6 & 1850.93 \end{vmatrix} - \begin{vmatrix} 39.193 & -58.265 \\ -58.265 & 86.617 \end{vmatrix}
$$

$$
= \begin{vmatrix} 20.407 & -71.335 \\ -71.335 & 1,764.3 \end{vmatrix}.
$$

PARAMETER ESTIMATES

Let $\hat{\pi}(v)$ be the usual estimates of a certain parameter, for the variates, and $\hat{\pi}(c)$ be the estimates of the same parameter, for the covariates. Both $\hat{\pi}(v)$ and $\hat{\pi}(c)$ may be vectors, but each corresponds to what would be a single parameter estimate in a univariate analysis of variance. If we wish to estimate α_1 in a simple one-way analysis of covariance, then $\hat{\pi}(v)$ is $\hat{\alpha}_1$ for the variates, and $\hat{\pi}(c)$ is $\hat{\alpha}_1$ for the covariates. If there is only one variate, then $\hat{\pi}_{(v)}$ is a scalar; if there are two or more variates, then $\hat{\pi}_{(v)}$ is a vector. Similarly, $\hat{\pi}_{(c)}$ is either a scalar or a vector, depending on the number of covariates. In a two-way design, the $\hat{\alpha}_i$, $\hat{\beta}_j$, or $\alpha\hat{\beta}_{ij}$ may each be the $\hat{\pi}(v)$ and $\hat{\pi}(c)$, depending on what is being estimated.

Then

$$
\hat{\theta} = B_w C_w^{-1},
$$
$$
\hat{\pi} = \hat{\pi}(v) - \hat{\theta}\hat{\pi}(c),
$$

where $\hat{\pi}$ is the new set of estimates of the parameter.

These equations can be understood better when they are illustrated by examples.

Example a. For this example, B_w and C_w^{-1} are given above, and thus,

$$\hat{\theta} = \begin{vmatrix} -55.8 & 45.6 \end{vmatrix} \begin{vmatrix} .0007012 & .0003230 \\ .0003230 & .0090774 \end{vmatrix}$$

$$= \begin{vmatrix} -.0244 & .3959 \end{vmatrix}.$$

These two values can be interpreted as regression coefficients in a regression equation predicting the dependent variable from the two covariates. The first value, -.0244, is the coefficient of the first covariate, IQ; the second is the coefficient of the second covariate, age.

The parameters to be estimated in a one-way design are the α_i. Thus, to estimate, say α_1, we let $\hat{\pi}(v)$ be the usual estimate of α_1 obtained on the dependent variable. From Chapter 2, we find that $\hat{\pi}(v) = \hat{\alpha}_1 = 0.6$. Similarly, $\hat{\pi}(c)$ is the *vector* of estimates of α_1 obtained from the two covariates:

$$\hat{\pi}(c) = \begin{vmatrix} 2.933 \\ 1.800 \end{vmatrix}.$$

Consequently, from the formula given above,

$$\hat{\pi} = \hat{\alpha}_1 = .60 - \begin{vmatrix} -.0244 & .3959 \end{vmatrix} \begin{vmatrix} 2.933 \\ 1.800 \end{vmatrix} = -.42.$$

Using the same calculations on the other two parameters, $\hat{\alpha}_2 = -.588$, $\hat{\alpha}_3 = .630$.

Example b. From the above figures, we get

$$\hat{\theta} = \begin{vmatrix} 45.6 \\ -51.6 \end{vmatrix} \begin{vmatrix} .008929 \end{vmatrix} = \begin{vmatrix} .4071 \\ -.4607 \end{vmatrix}.$$

These values are interpreted as coefficients in two separate simple regression equations. The first is the coefficient for predicting intellectual maturity from chronological age; the second, -.4607, is the coefficient for predicting IQ from chronological age.

Again, the parameters to be estimated are the α_i. To estimate α_1, we let $\hat{\pi}(v)$ be the vector of estimates of α_1 on the two dependent variables:

$$\hat{\pi}(v) = \hat{\alpha}_1 = \begin{vmatrix} 0.600 \\ 2.933 \end{vmatrix}.$$

Similarly, $\hat{\pi}(c)$ is the single estimate of α_1, obtained from the covariate: $\hat{\pi}(c) = 1.8$. Using the formula above, we obtain

$$\hat{\pi} = \hat{\alpha}_1 = \begin{vmatrix} 0.600 \\ 2.933 \end{vmatrix} - \begin{vmatrix} .4071 \\ -.4607 \end{vmatrix} \begin{vmatrix} 1.8 \end{vmatrix} = \begin{vmatrix} -.133 \\ 3.763 \end{vmatrix},$$

The first value, -.133, is $\hat{\alpha}_1$ for intellectual maturity; the second value, 3.763, is $\hat{\alpha}_1$ for IQ. The other $\hat{\alpha}_i$ are

$$\hat{\alpha}_2 = \begin{vmatrix} -.660 \\ 3.043 \end{vmatrix}, \; \hat{\alpha}_3 = \begin{vmatrix} .793 \\ -6.806 \end{vmatrix}.$$

STATISTICAL TESTS

Significance tests are the same as in ordinary multivariate analysis of variance, except that adjusted SC matrices are used.

Tests of Effects. To test a given effect, ASC_s takes the place of SC_s, and ASC_w takes the place of SC_w. (For procedures that use it, $ASC_n = ASC_s - ASC_w$.) The numerator degrees of freedom, ν_1, are the same as for the associated analysis of variance; the denominator degrees of freedom, ν_2, are the usual denominator degrees of freedom, minus the number of covariates.

Example a. We will use Wilks's lambda from Chapter 13 on both examples. With one dependent variable and two covariates, $AF_w = 19.4144$, $ASC_w = 14.9856$, and $A_w = 34.4$. The determinant of a scalar is equal to the scalar, so to test $H_0 : \alpha_i = 0$ (all i), we use $ASC_w = 14.9856$, $ASC_s = 17.5229$. Then

$$
\begin{aligned}
\Lambda &= 14.9856/17.5229 = .8552, \\
\nu_1 &= 2, \; \nu_2 = 10, \\
p &= 1, \; m^* = 10, \; g = 1, \; h = 10, \\
W &= \Lambda = .4356, \; F_{(2,10)} = .847,
\end{aligned}
$$

which is obviously not significant. (Note: The usual denominator degrees of freedom for these data would be 12; with two covariates, $\nu_2 = 12 - 2 = 10$.)

Example b. With two dependent variables, we must use matrices instead of scalars. To test $H_0 : \alpha_i = 0$, we use the ASC_w and ASC_s calculated previously for Example b, and

$$
\begin{aligned}
\Lambda &= \det(ASC_w)/\det(ASC_s) = .6913, \\
\nu_1 &= 2, \; \nu_2 = 10, \\
p &= 2, \; m^* = 9.5, \; g = 2, \; h = 18, \\
W &= \Lambda^{1/2} = .8314, \; F_{(4,18)} = .91,
\end{aligned}
$$

which is not significant.

Tests on θ. The test on $H_0 : \theta = 0$ is complicated by the fact that θ is not orthogonal to the effects being tested. Consequently, the test on θ can either ignore the other effects or take them into account. The tests on effects described above take θ into account. (If they ignored θ, the tests would be exactly the same as in an ordinary analysis of variance.)

We can also test θ taking some effects into account and ignoring others. The possibilities are numerous; however, in practice only three cases are likely to arise.

(1) Testing θ, taking all effects into account. This test is not orthogonal to the analysis of covariance on the effects, so it is not recommended for routine use. However, it may be useful as a preliminary test to determine whether the covariate has enough of an effect to make an analysis of covariance worthwhile. If the test is significant, then the analysis of covariance should probably be performed. If it is not, then the covariate does not account for a significant amount of variance above that accounted for by the effects. This indicates that an ordinary analysis of variance may be as informative as an analysis of covariance (but remember the caveats about accepting a null hypothesis).

To test $H_0 : \theta = 0$, we use the procedures in Chapter 13, except that AF_w takes the place of SC_n, and ASC_w takes the place of SC_w (i.e., $AF_w + ASC_w = A_w$ takes the place of SC_s). The numerator degrees of freedom, ν_1, are the total number of coefficients (i.e., the number of dependent variables times the number of covariates); the denominator degrees of freedom, ν_2, are the number of degrees of freedom of the associated analysis of variance, minus ν_1.

Example a.

$$\Lambda = AF_w/A_w = 14.9856/34.4 = .4356,$$
$$\nu_1 = 2, \; \nu_2 = 12 - 2 = 10,$$
$$p = 1, \; m^* = 10, \; g = 1, \; h = 10,$$
$$W = \Lambda = .4356, \; F_{(2,10)} = 6.48.$$

The F ratio is exact, and $p = .016$.

Example b.

$$ASC_w = \begin{vmatrix} 15.834 & -34.791 \\ -34.791 & 23.773 \end{vmatrix},$$

$$A_w = \begin{vmatrix} 34.4 & -55.8 \\ -55.8 & 1450.0 \end{vmatrix},$$

$$\Lambda = det(ASC_w)/det(A_w) = 21,372/46,766 = .4570,$$

$$\nu_1 = 2, \; \nu_2 = 12 - 2 = 10,$$

$$p = 2, \; m^* = 9.5, \; g = 2, \; h = 18,$$

$$W = \Lambda^{1/2} = .6760, \; F_{(4,18)} = 2.16.$$

Again, this is exact, and $p = .17$.

(2) Testing θ ignoring all effects, but taking into account the grand mean. This is usually the test of choice if one is not interested in performing an analysis of covariance on the grand mean. It is orthogonal to an ordinary analysis of variance, but not to an analysis of covariance, test of the grand mean. It is equivalent to a test using the model in Equation 14.1, but using $(Y_{ij} - \overline{Y}_{..})$ in place of Y_{ij}. If this test is used, and the grand mean is also tested, the grand mean should be tested using an ordinary analysis of variance.

To perform the test, we calculate a new matrix, $SC_\theta = RC_w - SC_m$. Next we partition these matrices into the four submatrices, $A_\theta, B_\theta, C_\theta$, just as we did before. Finally we calculate $AF_\theta = B_\theta C_\theta^{-1} B_\theta^t$. AF_θ plays the same role as AF_w in the test on θ described above, i.e., $ASC_\theta = AF_\theta + ASC_w$.

Example a. RC_w and SC_m are given in Chapter 13. Their difference is

$$SC_\theta = \begin{vmatrix} 59.6 & -129.6 & 82.2 \\ -129.6 & 1,850.93 & -122.2 \\ 82.2 & -122.2 & 172.4 \end{vmatrix}.$$

Next we partition this matrix into four submatrices. The two that we will use are

$$B_\theta = \begin{vmatrix} -129.6 & 82.2 \end{vmatrix}, \quad C_\theta = \begin{vmatrix} 1,850.93 & -122.2 \\ -122.2 & 172.4 \end{vmatrix}.$$

From these we calculate

$$\begin{aligned} AF_\theta &= 19.41441746, \\ ASC_\theta &= AF_\theta + ASC_w = 57.0626, \\ \Lambda &= 14.9856/57.0626 = .2626. \end{aligned}$$

Finally, ν_1, ν_2, p, m^*, g, and h are as before, and $F_{(2,10)} = 14.04$, $p = .001$.

Example b. SC_θ is the same as in Example a. However, the partitioning is different. We now have

$$B_\theta^t = \begin{vmatrix} 82.2 & -122.2 \end{vmatrix}, \quad C_\theta = \begin{vmatrix} 172.4 \end{vmatrix}.$$

From these we calculate

$$AF_\theta = \begin{vmatrix} 39.193 & -58.265 \\ -58.265 & 86.617 \end{vmatrix}, \quad ASC_\theta = \begin{vmatrix} 55.027 & -93.056 \\ -93.056 & 1,512.82 \end{vmatrix},$$

$$\Lambda = 21,372/74,586 = .2865, \quad F_{(4,18)} = 3.907, \quad p = .019.$$

All other parameters are as in Example b above.

(3) Testing θ ignoring all other parameters, including all effects *and* the grand mean. This is the test of choice if one wants to test the grand mean after first correcting for the covariate (i.e., as part of the analysis of covariance). Essentially, it is the appropriate test using the model in Eq. 14.1 without first subtracting $\overline{Y}_{...}$. It is orthogonal to an analysis of covariance, but not to an ordinary analysis of variance, test of the grand mean. Thus, if this test is used, the grand mean should also be tested using an analysis of covariance with adjusted sum of cross-products matrices.

The test is identical to the tests that have just been described, except that for this test $SC_\theta = RC_w$. For Example a, $F_{(2,10)} = 68.31$, $p < .001$; for Example b, $F_{(4,18)} = 28.60$, $p < .001$.

Note that the F ratio is smallest for the first of these three cases and largest for the third. This will not always be true, but most often it will. For case (1), we first account for as much variance as possible using the effects and grand mean alone; then the remaining variance is attributed to the covariate. For case (3) we first account for as much variance as possible using the covariate alone; then the remaining variance is attributed to the effects and the grand mean.

15

General Linear Model

Every analysis we have studied in this text has been based on a model equation containing terms that are added to obtain the observed scores. All of these are instances of the *general linear model*. The *univariate* general linear model is most easily represented in matrix terms as

$$X = AP + E, \tag{15.1}$$

where X is a vector of N observed scores, P is a vector of m unknown parameters, A is an $N \times m$ matrix of coefficients, and E is a vector of N random errors. (In this chapter N is a scalar representing the total number of scores; n represents the number of scores in each group.)

In this chapter we first consider the univariate fixed-effects model. Then we discuss the univariate random-effects model. Finally, we show how these are generalized to multivariate models and the analysis of covariance.

Consider the one-way analysis of variance with unequal sample sizes in Table 2.6. The linear model for the one-way design is

$$X_{ij} = \mu + \alpha_i + \epsilon_{ij}. \tag{15.2}$$

(Here, as in Chapter 2, X_{ij} is a scalar.) We can represent this in terms of Equation 15.1 by letting (we write the transposes of vectors and matrices to save space)

$$X^t = \begin{vmatrix} 3 & 5 & 6 & 1 & 1 & 2 & 2 & 0 & 3 & 5 & 5 & 2 \end{vmatrix},$$

$$p^t = \begin{vmatrix} \mu & \alpha_1 & \alpha_2 & \alpha_3 \end{vmatrix},$$

$$\tag{5.13}$$

$$A^t = \begin{vmatrix} 1 & 1 & 1 & 1 & 1 & 1 & 1 & 1 & 1 & 1 & 1 & 1 \\ 1 & 1 & 1 & 1 & 0 & 0 & 0 & 0 & 0 & 0 & 0 & 0 \\ 0 & 0 & 0 & 0 & 1 & 1 & 1 & 1 & 1 & 0 & 0 & 0 \\ 0 & 0 & 0 & 0 & 0 & 0 & 0 & 0 & 0 & 1 & 1 & 1 \end{vmatrix},$$

$$E^t = \begin{vmatrix} \epsilon_{11} & \epsilon_{12} & \epsilon_{13} & \epsilon_{14} & \epsilon_{21} & \epsilon_{22} & \epsilon_{23} & \epsilon_{24} & \epsilon_{25} & \epsilon_{31} & \epsilon_{32} & \epsilon_{33} \end{vmatrix}.$$

---

*Note: References to theorems in this chapter are to the theorems in Appendix A.

If you insert these matrices into Equation 15.1, you will see how they fit the linear model for the one-way design. For example, multiplying the first row of A by P and adding the first element of E, we get

$$X_{11} = 3 = \mu + \alpha_1 + \epsilon_{11}.$$

The matrix A always contains constant coefficients. For analysis of variance problems, it nearly always contains ones and zeros, but in general it can contain any values. For regression and correlation problems it usually contains the values of continuous variables, and for analysis of covariance problems some columns contain ones and zeros, while others contain values of continuous variables.

General Solution

The basic tasks when applying the general linear model are parameter estimation and hypothesis testing. Often the problem of estimation is simple. If we let $S = A^t A$ and $Y = A^t X$, the best point estimate of P is found by solving the linear equation $S \hat{P}_0 = Y$ (Theorem F1). (The subscript on $\hat{P}_0$ will distinguish this estimate from others to be derived later.) If S is nonsingular, the solution is $\hat{P}_0 = S^{-1} Y$ (Theorem F3).

We will assume that the elements of E are independent and normally distributed with mean 0 and variance σ_e^2; i.e., $\underline{E}(E) = 0$, $\underline{V}(E) = \sigma_e^2 I$. (The $\underline{E}$ and $\underline{V}$, representing an expectation and a variance-covariance matrix, respectively, are underlined to distinguish them from vectors or matrices.) Then if S is nonsingular, $\underline{E}(\hat{P}_0) = P$, and $\underline{V}(\hat{P}_0) = \sigma_e^2 S^{-1}$ (Theorem F9). For an unbiased estimator of σ_e^2, let

$$
\begin{aligned}
SS_{e(0)} &= (X - A\hat{P}_0)^t (X - A\hat{P}_0) \\
&= X^t X - \hat{P}_0^t S \hat{P}_0 \\
&= X^t X - \hat{P}_0^t Y \\
&= X^t X - Y^t S^{-1} Y, \\
MS_e &= SS_{e(0)} / (N - m).
\end{aligned}
$$

Then

$$
\begin{aligned}
SS_{e(0)} / \sigma_e^2 &\sim \chi_{(N-m)}^2 \quad \text{(Theorem F11)}, \\
\underline{E}(MS_e) &= \sigma_e^2.
\end{aligned}
$$

These facts can be used to obtain confidence intervals for the elements of P.

There are two problems with this approach. One is that there is no explicit provision for testing null hypotheses. The other is that S may be singular; i.e., it may not have an inverse. In fact, that is true for the sample data above.

For them we have

$$S = \begin{vmatrix} 12 & 4 & 5 & 3 \\ 4 & 4 & 0 & 0 \\ 5 & 0 & 5 & 0 \\ 3 & 0 & 0 & 3 \end{vmatrix},$$

(15.4)

$$Y^t = |\ 35 \quad 15 \quad 8 \quad 12\ |.$$

It should be obvious that the rows of S are not linearly independent; the first row is the sum of the other three. Therefore, the matrix is singular and has no inverse.

We will deal with both of these problems in the following sections. We will see that there are often several ways to deal with them. All yield essentially the same results, so the choice for a given problem depends mainly on ease of analysis and interpretation. For simplicity we will at first limit ourselves to univariate, fixed-effects models.

There are basically three different approaches to the problem of a singular matrix S, but there are often variations within these approaches. We will describe the three approaches in the following sections. We will also describe methods for testing null hypotheses.

FIRST APPROACH

The basic problem in the example above is that the columns of A are dependent; the first column is the sum of the other three. We can eliminate the problem by eliminating one column of A. In general, then, one solution is to eliminate enough columns of A to make S nonsingular. However, in doing so we must usually redefine P, and we may have to modify the remaining columns of A. This often complicates the analyses.

First Variation. We can eliminate any column of A, and the subsequent analyses depend on our choice. For our first illustration, we will eliminate the first column. Because there is then no provision for the grand mean in our model, we must modify P accordingly. Our model is now

$$X_{ij} = \mu_i + \epsilon_{ij}.$$

X and E are the same as above, but now

$$P^t = |\ \mu_1 \quad \mu_2 \quad \mu_3\ |,$$

$$S = \begin{vmatrix} 4 & 0 & 0 \\ 0 & 5 & 0 \\ 0 & 0 & 3 \end{vmatrix},$$

$$Y \quad = \quad | \ 15 \ \ 8 \ \ 12 \ |,$$

$$\hat{P}_0^t \quad = \quad | \ 0.833 \ \ -1.317 \ \ 2.917 \ |, \qquad (15.5)$$

$$SS_{e(0)} \quad = \quad 143 - [(.833)15 + (-1.317)8 + (1.083)12] = 25.95.$$

S^{-1} should be obvious.

Second Variation. We can retain the distinction between the grand mean, μ, and the effects, α_i, and still eliminate one column of A. To do this we add a linear restriction on P.

For the analysis with equal sample sizes, we added the restriction $\Sigma_i \alpha_i = 0$ in Chapter 2. We could apply the same restriction to the design with unequal sample sizes. However, we can find a better one. Averaging over both i and j in Equation 15.2 and taking the expected value, we get

$$\underline{E}(\overline{X}_{..}) = \mu + (1/N)\Sigma_i n_i \alpha_i.$$

If $\overline{X}_{..}$ is to be an unbiased estimate of μ, then we must have the restriction,

$$\Sigma_i n_i \alpha_i = 0. \qquad (15.6)$$

That is the restriction we will use. It not only gives us a simple estimate of μ, it also simplifies other analyses as we will see. With this restriction,

$$\alpha_I = -\Sigma_{i=1}^{I-1}(n_i/n_I)\alpha_i.$$

For our data,

$$\alpha_3 = -(4/3)\alpha_1 - (5.3)\alpha_2.$$

This enables us to express the X_{ij} in terms of μ, α_1, and α_2 alone; i.e.,

$$\begin{aligned} X_{1j} \quad &= \quad \mu + \alpha_1 + \epsilon_{1j}, \\ X_{2j} \quad &= \quad \mu + \alpha_2 + \epsilon_{2j}, \\ X_{3j} \quad &= \quad \mu - (4/3)\alpha_1 - (5/3)\alpha_2 + \epsilon_{3j}. \end{aligned}$$

In terms of the general model, X and E are the same as above, but

$$P^t \quad = \quad | \ \mu \ \ \alpha_1 \ \ \alpha_2 \ |,$$

$$A^t \ = \ \begin{vmatrix} 1 & 1 & 1 & 1 & 1 & 1 & 1 & 1 & 1 & 1 & 1 \\ 1 & 1 & 1 & 1 & 0 & 0 & 0 & 0 & 0 & -4/3 & -4.3 & -4/3 \\ 0 & 0 & 0 & 0 & 1 & 1 & 1 & 1 & 1 & -5/3 & -5/3 & -5/3 \end{vmatrix},$$

$$S = \begin{vmatrix} 12 & 0 & 0 \\ 0 & 28/3 & 20/3 \\ 0 & 20/3 & 40/3 \end{vmatrix},$$

$$Y = \begin{vmatrix} 35 & -1 & -12 \end{vmatrix},$$

$$\hat{P}_0^t = \begin{vmatrix} 2.917 & 0.833 & -1.317 \end{vmatrix},$$

and $SS_{e(0)} = 25.95$ and $MS_e = 2.883$, as above.

SECOND APPROACH

The second approach to the problem of a singular S involves adding linear restrictions explicitly to P; i.e., we require that $C_0 P = 0$ for some appropriately chosen C_0. Applying the same restrictions to $\hat{P}_0$, we have the equation

$$\begin{vmatrix} S \\ C_0 \end{vmatrix} \hat{P}_0 = \begin{vmatrix} Y \\ 0 \end{vmatrix}.$$

As an example, we will use the restriction in Equation 15.6:

$$C_0 = \begin{vmatrix} 0 & 4 & 5 & 3 \end{vmatrix}.$$

Note, however, that this is not the only matrix we could have used to place restrictions on P. In general, C_0 can be any matrix that meets two requirements. The first is that C_0 must put enough restrictions on P to make $\begin{vmatrix} A^t & C_0^t \end{vmatrix}$ have rank m. (Equivalently, $\begin{vmatrix} S & C_0^t \end{vmatrix}$ must have rank m.) The second is that no row *or combination of rows* of C_0 can be a linear combination of the rows of A (or therefore of S). For example, we could have used either of the following:

$$C_0 = \begin{vmatrix} 0 & 1 & 1 & 1 \end{vmatrix},$$

$$C_0 = \begin{vmatrix} -1 & 1 & 1 & 1 \end{vmatrix}.$$

The first of these corresponds to the restriction $\Sigma_i \alpha_i = 0$, while the second corresponds to the restriction $\Sigma_i \alpha_i = \mu$. We chose the restriction we did because it fit the definition of α_i in our model. However, the following C_0 is not legitimate:

$$C_0 = \begin{vmatrix} 0 & 1 & -1 & 0 \end{vmatrix}.$$

This C_0 can be obtained by subtracting $1/5$ of row three of S from $1/4$ of row two; i.e., it is equal to KS where

$$K = \begin{vmatrix} 0 & 1/4 & -1/5 & 0 \end{vmatrix}.$$

In general, there will be a number of C_0 from which to choose. We want to choose that which best fits the model and/or most simplifies the calculations.

There are now several ways to estimate P.

First Variation. The best estimate of P can be obtained directly by solving the equation

$$\left| \begin{array}{c} S \\ C_0 \end{array} \right| P = \left| \begin{array}{c} y \\ 0 \end{array} \right|$$

using any of a number of methods for solving simultaneous linear equations. (E.g., we could choose m linearly independent equations and solve them directly.)

Second Variation. The best estimate of P can also be obtained directly as

$$\hat{P}_0 = (S + C_0^t C_0)^{-1} Y \quad \text{(Theorems F1 and F4).}$$

In our example we will let

$$C_0 = \left| \begin{array}{cccc} 0 & 4 & 5 & 3 \end{array} \right|,$$

so that

$$S + C_0^t C_0 \;=\; \left| \begin{array}{cccc} 12 & 4 & 5 & 3 \\ 4 & 20 & 20 & 12 \\ 4 & 20 & 30 & 15 \\ 3 & 12 & 15 & 12 \end{array} \right|,$$

$$(S + C_0^t C_0)^{-1} \;=\; (1/720) \left| \begin{array}{cccc} 65 & -5 & -5 & -5 \\ -5 & 75 & -55 & -55 \\ -5 & -55 & 89 & -55 \\ -5 & -55 & -55 & 185 \end{array} \right|,$$

$$\hat{P}_0^t \;=\; \left| \begin{array}{cccc} 3.083 & 0.833 & -1.317 & 1.083 \end{array} \right|.$$

Third Variation. The third variation uses the fact that if $C_0 D_0^t = 0$, and $P = D_0^t Q$ for some Q, then $C_0 P = 0$. Our problem then is to find a matrix D_0 such that $C_0 D_0^t = 0$, and

$$\text{rank} \; \left| \begin{array}{cc} C_0^t & D_0^t \end{array} \right| = m.$$

We can then substitute $D_0^t Q$ for P in the equation $X = AP + E$ to get

$$\begin{aligned} X &= A D_0^t Q + E, \\ D_0 S D_0^t \hat{Q}_0 &= D_0 Y, \\ \hat{Q}_0 &= (D_0 S D_0^t)^{-1} D_0 Y, \\ \hat{P}_0 &= D_0^t \hat{Q}_0 = D_0^t (D_0 S D_0^t)^{-1} D_0 Y \quad \text{(Theorems F1 and F6).} \end{aligned}$$

Methods for finding an appropriate D_0 are given in texts on matrix algebra. However, for most problems one can be found with a little trial and error. For our problem, C_0 has one row, so D_0 must have three linearly independent rows, all of which are orthogonal to C_0. One possibility is

$$D_0 = \begin{vmatrix} 1 & 0 & 0 & 0 \\ 0 & 5 & -4 & 0 \\ 0 & 3 & 0 & -4 \end{vmatrix},$$

from which we obtain

$$D_0 S D_0^t = \begin{vmatrix} 12 & 0 & 0 \\ 0 & 180 & 60 \\ 0 & 60 & 84 \end{vmatrix},$$

$$(D_0 Y)^t = \begin{vmatrix} 35 & 43 & -3 \end{vmatrix},$$

$$\hat{Q}_0^t = \begin{vmatrix} 2.9167 & 0.3292 & 0.2708 \end{vmatrix},$$

$$\hat{P}_0^t = \begin{vmatrix} 3.083 & 0.833 & -1.317 & 1.083 \end{vmatrix},$$

as with the first variation.

This second variation has a minor disadvantage over the first in that it requires that D_0 be found; however, it often has a decided advantage in that later calculations are simplified.

THIRD APPROACH

Only square nonsingular matrices have inverses, but every nonzero matrix has at least one *generalized inverse*. The generalized inverse of a matrix S is defined as a matrix G_0 such that $S C_0 S = S$. (Note that here S need not be square, let alone symmetric.) If S is square and nonsingular, then $G_0 = S^{-1}$; otherwise, S has an infinite number of generalized inverses.

It can be proved that, for *any* generalized inverse, G_0, of S, $P = G_0 Y$ is a solution to the equation $SP = Y$ (Theorem F1). The two matrices $(S + C_0^t C_0)^{-1}$ and $D_0^t (D_0 S D_0^t)^{-1} D_0$ are both examples of generalized inverses of $S = A^t A$ (Theorems F4 and F6). Therefore, any generalized inverse will give us a suitable estimate of P. For the reminder of this chapter, we will let G_0 represent *any* symmetric generalized inverse of S. (There is no problem in practice to requiring that G_0 be symmetric.)

Perhaps the simplest generalized inverse of the symmetric matrix S with rank r_0 is obtained as follows: Find r_0 rows, such that the submatrix formed from those rows and the same columns is nonsingular. For example, for S in Equation 15.4, rows (and columns) two through four form the submatrix

$$S^* = \begin{vmatrix} 3 & 0 & 0 \\ 0 & 5 & 0 \\ 0 & 0 & 4 \end{vmatrix}.$$

(Actually, for this S, any three rows and columns could have been chosen, but these lead to an especially simple inverse.)

We then invert S^* and place the values of the inverse in the corresponding places in S, replacing all other elements of S with zeros. For example, if we use S^* above,

$$G_0 = \begin{vmatrix} 0 & 0 & 0 & 0 \\ 0 & 1/3 & 0 & 0 \\ 0 & 0 & 1/5 & 0 \\ 0 & 0 & 0 & 1/4 \end{vmatrix}$$

is a generalized inverse of S (Theorem F7), giving us $\hat{P}_0 = G_0 Y$, or

$$\hat{P}_0^t = \begin{vmatrix} 0 & 5 & 1.6 & 3 \end{vmatrix}.$$

Another generalized inverse, called the *Penrose inverse*, is obtained by first performing a principle components analysis on S. If S has rank r, then we can write $S = UKU^t$ where K is an $r \times r$ diagonal matrix and U, with r rows and m columns, is such that $U^t U = I$. (Methods for calculating K and U are given in texts on matrix algebra.) Then $G_0 = UK^{-1}U^t$ is a generalized inverse of S. For our example,

$$G_0 = (1/960) \begin{vmatrix} 47 & 13 & 1 & 33 \\ 13 & 167 & -61 & -93 \\ 1 & -61 & 143 & -81 \\ 33 & -93 & -81 & 207 \end{vmatrix},$$

$$\hat{P}_0^t = \begin{vmatrix} 2.3375 & 1.4125 & -0.7375 & 1.6025 \end{vmatrix}.$$

The Penrose inverse has the form $G_0 = D_0^t (D_0 S D_0^t)^{-1} D_0$, with $D_0 = U^t$. The inverse using S^* has the same form, with $D_0 = \begin{vmatrix} S^{*-1} & 0 \end{vmatrix}$. All such inverses have the often useful property that $G_0 S G_0 = G_0$; i.e., S is a generalized inverse of G_0 (Theorem F12).

STATISTICAL PROPERTIES

In general, $\hat{P}_0$ is normally distributed with

$$\underline{E}(\hat{P}_0) = G_0 SP,$$
$$\underline{V}(\hat{P}_0) = \sigma_e^2 G_0 S G_0 \quad \text{(Theorem F8).}$$

Note that $\hat{P}_0$ is not necessarily an unbiased estimator of P. However, if we assume that $C_0 P = 0$, and we place the same restriction on $\hat{P}_0$, then $\underline{E}(\hat{P}_0) = P$, $\underline{V}(\hat{P}_0) = \sigma_e^2 G_0$ (Theorem F10).

No matter how the estimates of P are obtained, the same formula is used to estimate the error variance. We have

$$SS_{e(0)} = (X - A\hat{P}_0)^t (X - A\hat{P}_0)$$
$$= X^t X - \hat{P}_0^t S \hat{P}_0$$

$$\begin{aligned} &= \quad X^t X - \hat{P}_0^t Y \\ &= \quad X^t X - Y^t G_0 Y, \end{aligned}$$

and

$$SS_{e(0)}/\sigma_e^2 \sim \chi^2_{(N-r_0)} \quad \text{(Theorem F11)},$$

so that

$$MS_e = SS_{e(0)}/(N - r_0)$$

is an unbiased estimate of σ_e^2. All of the above formulas give the same result no matter how we estimate P (Theorem F2). In our example, $SS_{e(0)} = 25.95$, and $MS_e = 25.95/(12 - 3) = 2.883$, as in Chapter 2.

Testing Hypotheses

This section describes two general methods for testing null hypotheses. Both methods give identical results, the only difference being in the calculations. These methods will be used with only minor variations for all hypothesis testing.

The null hypotheses that we will consider are all assertions that one or more linear combinations of the parameters equal zero. For example, consider the null hypothesis of no main effect:

$$H_0(1) : \mu_1 = \mu_2 = \mu_3.$$

The way we express this in matrix form will depend on how we have formulated our model. For example, if P is as in Equation 15.5, then we can write

$$H_0(1) : \mu_1 - \mu_2 = 0, \ \mu_1 - \mu_3 = 0.$$

These, in turn, are equivalent to $H_0(1) : C_1 P = 0$, where

$$C_1 = \begin{vmatrix} 1 & -1 & 0 \\ 1 & 0 & -1 \end{vmatrix},$$

On the other hand, if we have expressed the model as in Equation 15.3, we can write

$$H_0(1) : \alpha_1 = \alpha_2 = \alpha_3 = 0,$$

for which

$$C_1 = \begin{vmatrix} 0 & 1 & 0 & 0 \\ 0 & 0 & 1 & 0 \\ 0 & 0 & 0 & 1 \end{vmatrix}. \tag{15.7}$$

ESTIMABLE FUNCTIONS

The null hypotheses we will consider will all be of the form $H_0(j) : C_j P = 0$, for some matrix C_j. (In this and other equations in this chapter, the zero is not necessarily a scalar; it may be a vector or matrix, all of whose values equal zero. The context will always make the meaning clear.)

However, not all parameters or linear combinations of parameters can be meaningfully estimated or tested. For example, in the one-way model it would be meaningless to estimate μ without some assumption about the α_i; the value of μ depends on the assumption about the α_i. As we saw above, there are a number of ways to express the parameters of most models, and given the parameters, the choice of assumptions is usually at least somewhat arbitrary.

The concept of *estimable functions* is sometimes useful when selecting parameters to estimate and tests to make. Basically, an estimable function is a function of the form $C_j P$ whose estimate, $C_j \hat{P}_0$, does not depend on our choice of $\hat{P}_0$ (e.g., it not depend on our choice of the linear restrictions, C_0). Such a function has a "meaningfulness" of its own without regard to our arbitrary choices. For example, in the one-way model it is not meaningful (in the absence of a specific restriction on the α_i) to estimate any of the α_i separately, but it is meaningful to estimate $\alpha_1 - \alpha_2$. No matter what restriction we place on the α_i, this is an estimate of the difference between the mean of Group 1 and the mean of Group 2.

Meaningful tests can be made only on functions that can be meaningfully estimated. Therefore, our tests will be meaningful only if the C_j for these tests can be written as estimable functions. It is not always necessary to *actually* express them as estimable functions, but it is essential that we *be able* to. For example, consider the one-way model discussed above. The C_1 for testing for differences among groups (Equation 15.7) asserts that all of the α_i are zero. This is not an estimable function because our estimates of the α_i obviously depend on the assumption we make about them. However, when we add the restriction in Equation 15.6, this hypothesis is equivalent to the hypothesis that all of the means are equal. This, in turn, is equivalent to the hypothesis that all contrasts have a value of zero. Therefore, we could have stated the same null hypothesis using

$$C_1 = \begin{vmatrix} 0 & 1 & -1 & 0 \\ 0 & 1 & 0 & -1 \end{vmatrix}. \tag{15.8}$$

This, as we will show later, gives us an estimable function. Any other contrasts would also have given us estimable functions.

Two questions now arise. The first is, how do we know whether a function is estimable? The second is, how do we find estimable functions? The first question is answered easily C_j is estimable if and only if $C_j = F_j A$ for some matrix F_j (Theorem F14). It is not necessary to actually *find* F_j, so long as we can prove that it exists. However, it is not always easy to prove that F_j exists. There is a simpler criterion: $C_j P$ is estimable if and only if

$$C_j G_0 S = C_j \quad \text{(Theorem F13)} \tag{15.9}$$

or, equivalently, $C_j(I - G_0 S) = 0$.

Consider the one-way design described above, with

$$G_0 = \begin{vmatrix} 1/12 & 0 & 0 & 0 \\ 0 & 1/6 & -1/12 & -1/12 \\ 0 & -1/12 & 7/60 & -1/12 \\ 0 & -1/12 & -1/12 & 1/4 \end{vmatrix},$$

$$G_0S = \begin{vmatrix} 1 & 0 & 0 & 0 \\ 1/3 & 2/3 & -1/3 & -1/3 \\ 5/12 & -5/12 & 7/12 & -5/12 \\ 1/4 & -1/4 & -1/4 & 3/4 \end{vmatrix}.$$

It is easy to show that Equation 15.9 holds for the C_1 in Equation 15.8 but not for the C_1 in Equation 15.7.

To find estimable functions, we must find C_j such that $C_j(I - G_0S) = 0$. Most textbooks on linear algebra tell how to solve such an equation. However, three more facts may also be of help. First, if C_j defines an estimable function, then each row of C_j defines an estimable function; second, if C_j and C_k define estimable functions, then any linear combination of the rows of C_j and C_k also defines an estimable function; third, if S is nonsingular, then all linear functions are estimable. Incidentally, it can be proved that for a correctly chosen C_0, $C_0G_0S = 0$, so C_0 never defines an estimable function (Theorem F5).

If C_jP is estimable, then the best, unbiased estimate of C_jP is $C_j\hat{P}_0$, which does not depend on which $\hat{P}_0$ we use (Theorems F15 and F16), and if C_jP and C_kP are two (not necessarily different) estimable functions,

$$\underline{C}(C_j\hat{P}_0, C_k\hat{P}_0) = \sigma_e^2 C_j G_0 C_k^t \quad \text{(Theorem F17)}, \tag{15.10}$$

where $\underline{C}$ indicates a matrix of covariances of the elements of $C_j\hat{P}_0$ with those of $C_k\hat{P}_0$ (C_j may or may not be the same as C_k).

Finally, if A and S have rank r_0, then there exist at most r_0 orthogonal (i.e., statistically independent) estimable functions. If each C_j has only a single row, then there are exactly r_0 orthogonal estimable functions. Otherwise, the number of linearly independent rows of all the C_j can be as large as, but no larger than, r_0 (Theorem F18).

Estimable functions are important in hypothesis testing because the only testable hypotheses are those that can be expressed as estimable functions (Theorem F19). If we use the first of the following methods for testing null hypotheses, it is not necessary that the hypotheses be expressed as estimable functions, but it is essential that they can be. If we use the second, then our hypotheses must be expressed as estimable functions.

FIRST METHOD

One general method for testing a null hypothesis is to first find a matrix D_j whose rows are orthogonal to those of both C_j and (if S is singular) C_0. That is, D_j must be such that $C_j D_j^t = 0$ and $C_0 D_j^t = 0$. Moreover, D_j must have as many linearly independent rows as possible, subject to that condition. In practice, this means that if C_0 and C_j, combined, have k independent rows, D_j must have exactly $r_j = (m - k)$ independent rows. In our example, D_1 can have only one row, and

$$D_1 = |\ 1 \quad 1 \quad 1\ |.$$

We then calculate

$$
\begin{aligned}
\hat{Q}_j &= (D_j S D_j^t)^{-1} D_j Y, \\
\hat{P}_j &= D_j^t \hat{Q}_j = D_j^t (D_j S D_j^t)^{-1} D_j Y = G_j Y,
\end{aligned}
$$

where

$$G_j = D_j^t (D_j S D_j^t)^{-1} D_j,$$

just as we did when finding general estimates of P.

$\hat{P}_j$ is our best estimate of P, subject to the restrictions imposed by C_0 and the null hypothesis. It is normally distributed with

$$\underline{E}(\hat{P}_j) = G_j S P,\ \underline{V}(\hat{P}_j) = \sigma_e^2 G_j \quad \text{(Theorem F20)}.$$

If $C_j P = 0$, then $\underline{E}(\hat{P}_j) = P$ (Theorem F21).

We can now calculate

$$
\begin{aligned}
SS_{e(j)} &= (X - A\hat{P}_j)^t (X - A\hat{P}_j) \\
&= X^t X - \hat{P}_j^t S \hat{P}_j \\
&= X^t X - \hat{P}_j^t Y \\
&= X^t X - Y^t G_j Y, \\
MS_{e(j)} &= SS_{e(j)}/(N - r_j),
\end{aligned}
$$

where r_j is the number of independent rows of D_j.

Letting $Z_j = D_j Y$, we can also write

$$
\begin{aligned}
SS_{e(j)} &= (X - D_j^t \hat{Q}_j)^t S (X - D_j^t \hat{Q}_j) \\
&= X^t X - \hat{Q}_j^t (D_j S D_j^t) \hat{Q}_j \\
&= X^t X - \hat{Q}_j^t Z_j \\
&= X^t X - Z_j^t (D_j S D_j^t)^{-1} Z_j.
\end{aligned}
$$

One of these formulas may be simpler to use if $\hat{Q}_j$ has already been calculated. (Note that we can find $\hat{Q}_j$ by solving the equation $(D_j S D_j^t)\hat{Q}_j = Z_j$.)

If the null hypothesis is true, then

$$
\begin{aligned}
SS_{e(j)}/\sigma_e^2 &\sim \chi_{(N-r_j)}^2 \quad \text{(Theorem F22)}, \\
\underline{E}(MS_{e(j)}) &= \sigma_e^2.
\end{aligned}
$$

However, $SS_{e(j)}$ is important mainly as a step in finding another sum of squares. If we define

$$SS_j = SS_{e(j)} - SS_{e(0)},$$
$$\nu_j = r_0 - r_j,$$
$$MS_j = SS_j/\nu_j,$$

then, in general, SS_j/σ_e^2 has a noncentral chi-square distribution with ν_j degrees of freedom and noncentrality parameter

$$\phi_j^2 = P^t S(G_0 - G_j)SP/(\nu_j\sigma_e^2) \quad \text{(Theorem F28).}$$

(Note that ν_j is the number of linearly independent rows of C_j that are also linearly independent of the rows of C_0.)

SS_j and $SS_{e(0)}$ are statistically independent (Theorem F31), and under the null hypothesis,

$$SS_j/\sigma_e^2 \sim \chi^2_{(\nu_j)},$$

so

$$MS_j/MS_e \sim F_{(\nu_j, N-r_0)}.$$

To test $H_0(j)$, we compare the obtained ratio with the F distribution as usual.

For the example,

$$D_1 S D_1^t = |12|, \ Z_1 = D_1 Y = |35|, \ \hat{Q}_1 = |2.917|,$$
$$\hat{P}_1^t = |2.917 \quad 2.917 \quad 2.917|,$$
$$SS_{e(1)} = 143 - (2.917)(35) = 40.92,$$
$$SS_1 = 40.92 - 25.95 = 14.97.$$

SS_1 is the same as SS_{bet} in Table 2.6. It has $\nu_j = 2$ degrees of freedom.

Note: In general, D_j must be orthogonal to both C_j and C_0. In this example, any matrix orthogonal to C_1 is automatically orthogonal to C_0. However, that is not always the case. In the next section we will give an example where we must make certain that D_1 is orthogonal to both C_1 and C_0.

SECOND METHOD

The second method uses C_j directly, but it requires that C_j be expressed as an estimable function. Therefore, in our example, we cannot use Equation 15.7, but we can use the C_1 in Equation 15.8. If the rows of C_j are linearly independent, then $C_j G_0 C_j^t$ is nonsingular (Theorem F23), and we can show that

$$\hat{P}_j = \hat{P}_0 - G_0 C_j^t (C_j G_0 C_j^t)^{-1} C_j \hat{P}_0 \quad \text{(Theorem F24),}$$
$$SS_j = \hat{P}_0^t C_j^t (C_j G_0 C_j^t)^{-1} C_j \hat{P}_0 \quad \text{(Theorem F25).}$$

This gives exactly the same result as the formula using D_j, so all of the above discussion applies. In particular, note that now the degrees of freedom, ν_j, are equal to the number of rows of C_j, and

$$\phi_j^2 = P^t C_j^t (C_j G_0 C_j^t)^{-1} C_j P / (\nu_j \sigma_e^2) \quad \text{(Theorem F29)},$$
$$SS_{e(j)} = SS_{e(0)} + SS_j.$$

Multiple Tests

So far we have limited ourselves to the test of a single null hypothesis. A new problem arises if we want to do a set of orthogonal tests. That problem can be solved in two different ways. The first way is generally preferable, but it cannot always be applied.

FIRST APPROACH

The first approach consists of formulating the null hypotheses so they are orthogonal. This is not always easily done, but with some designs the problem is simple. Two hypotheses, $H_0(j)$ and $H_0(k)$, are orthogonal if and only if $C_j \hat{P}_0$ and $C_k \hat{P}_0$ are independent, and by Equation 15.10, this is true if and only if $C_j G_0 C_k^t = 0$ *when both are expressed as estimable functions.* Alternatively, they are orthogonal if and only if $(G_0 - G_j)S(G_0 - G_k) = 0$ (Theorem F26).

If $H_0(j)$ and $H_0(k)$ are orthogonal, then SS_j and SS_k are statistically independent (Theorem F30).

To illustrate, we will test the null hypothesis that the grand mean is zero, again using the data in Table 2.6. The null hypothesis, $H_0(2) : \mu = 0$, has

$$C_2 = \begin{vmatrix} 1 & 0 & 0 & 0 \end{vmatrix},$$

which is easily shown to be an estimable function. Therefore, $H_0(2)$ is orthogonal to $H_0(1)$ if and only if $C_1 G_0 C_2^t = 0$, when $C_1 P$ is an estimable function as in Equation 15.8.

C_2 is orthogonal to C_1 in this example, so $H_0(2)$ can be tested directly by either of the above methods. The second is easier because an estimable function has already been found to test orthogonality. For our example, $SS_2 = 102.08$, which is equal to the SS_m in Table 2.8. It has one degree of freedom, and $F = SS_2 / MS_e = 102.08 / 2.883 = 35.4$.

SECOND APPROACH

It is not always easy to formulate orthogonal null hypotheses. The second approach is a method of formulating orthogonal tests even though the initial hypotheses may not be orthogonal. Moreover, it can be useful even with orthogonal hypotheses because it can lead to simpler calculations.

With this approach, we must test the hypotheses in a predetermined order. In effect, with each new test we test that part of the null hypothesis that is independent of all previously tested null hypotheses. The method is most easily described by means of an example.

We will illustrate the approach using the two-way design with unequal sample sizes in Table 5.9. (The exact analysis is in Table 5.11.) The matrices for the basic analysis are in Table 15.1. From it we obtain $SS_{e(0)} = 82$, with 10 d.f., so $MS_e = 82/10 = 8.2$ as in Table 5.11.

We first test the interaction. Although it is possible to express the interaction as an estimable function, it is simpler to use the first method above. The matrices are in Table 15.2 (remember that D_1 must be orthogonal to both C_1 and C_0); they give $SS_1 = 147.43$, $MS_1 = 73.72$, with 2 degrees of freedom, again as in Table 5.11.

Both tests of main effects are orthogonal to the test of the interaction. However, as we noted in Chapter 5, they are not orthogonal to each other. Consequently, to have orthogonal tests we must test them in a particular order. Either main effect can be tested first, depending on our experimental objectives. Here we will test the B main effect first, using

$$C_2 = \begin{vmatrix} 0 & 0 & 0 & 1 & 0 & 0 & 0 & 0 & 0 & 0 & 0 & 0 \\ 0 & 0 & 0 & 0 & 1 & 0 & 0 & 0 & 0 & 0 & 0 & 0 \\ 0 & 0 & 0 & 0 & 0 & 1 & 0 & 0 & 0 & 0 & 0 & 0 \end{vmatrix}.$$

Although $H_0(2)$ is orthogonal to $H_0(1)$, it is difficult to express C_2 as an estimable function; it is simpler to test $H_0(2)$ without assuming orthogonality.

We now combine $H_0(1)$ and $H_0(2)$, as if we were testing the null hypothesis that both are simultaneously true. We can do this in either of two ways: (1) we can find D_2 orthogonal to C_0, C_1, and C_2, and calculate $SS^*_{e(2)}$ using the first method above (the asterisk indicates that D_2 is now orthogonal to C_1 as well as to C_0 and C_2), or (2) we can find a way to express the combined functions $C_1 P$ and $C_2 P$ as a single estimable function and then calculate a sum of squares as in the second method above.

To use the second of these alternatives, we would have to first form a combined matrix containing both C_1 and C_2 and then reformulate it so that it was the matrix of an estimable function. The first method is simpler. We find

$$D_2 = \begin{vmatrix} 1 & 0 & 0 & 0 & 0 & 0 & 0 & 0 & 0 & 0 & 0 \\ 0 & 1 & -1 & 0 & 0 & 0 & 0 & 0 & 0 & 0 & 0 \end{vmatrix}.$$

Proceeding as usual, we obtain $SS^*_{e(2)} = 293.5$. Now, however, to obtain SS_2, we subtract $SS_{e(1)}$, not $SS_{e(0)}$. That is, $SS_2 = SS^*_{e(2)} - SS_{e(1)} = 293.5 - 240.4 = 53.1$, with two degrees of freedom.

TABLE 15.1. Basic matrices for unbalanced two-way design of Table 5.9. Interior dividing lines are for clarity. They are not part of the matrix.

μ	α		β			αβ						X
	1	2	1	2	3	11	12	13	21	22	23	
1	1	0	1	0	0	1	0	0	0	0	0	8
1	1	0	1	0	0	1	0	0	0	0	0	4
1	1	0	1	0	0	1	0	0	0	0	0	0
1	1	0	0	1	0	0	1	0	0	0	0	8
1	1	0	0	1	0	0	1	0	0	0	0	10
1	1	0	0	1	0	0	1	0	0	0	0	6
1	1	0	0	0	1	0	0	1	0	0	0	4
1	1	0	0	0	1	0	0	1	0	0	0	8
1	0	1	1	0	0	0	0	0	1	0	0	14
1	0	1	1	0	0	0	0	0	1	0	0	10
1	0	1	0	1	0	0	0	0	0	1	0	0
1	0	1	0	1	0	0	0	0	0	1	0	4
1	0	1	0	1	0	0	0	0	0	1	0	2
1	0	1	0	0	1	0	0	0	0	0	1	15
1	0	1	0	0	1	0	0	0	0	0	1	9
1	0	1	0	0	1	0	0	0	0	0	1	12

S												Y
16	8	8	5	6	5	3	3	2	2	3	3	114
8	8	0	3	3	2	3	3	2	0	0	0	48
8	0	8	2	3	3	0	0	0	2	3	3	66
5	3	2	5	0	0	3	0	0	2	0	0	36
6	3	3	0	6	0	0	3	0	0	3	0	30
5	2	3	0	0	5	0	0	2	0	0	3	48
3	3	0	3	0	0	3	0	0	0	0	0	12
3	3	0	0	3	0	0	3	0	0	0	0	24
2	2	0	0	0	2	0	0	2	0	0	0	12
2	0	2	2	0	0	0	0	0	2	0	0	24
3	0	3	0	3	0	0	0	0	0	3	0	6
3	0	3	0	0	3	0	0	0	0	0	3	36

TABLE 15.1. *Continued*

C_0

0	8	8	0	0	0	0	0	0	0	0	0
0	0	0	5	6	5	0	0	0	0	0	0
0	0	0	0	0	0	3	3	2	0	0	0
0	0	0	0	0	0	0	0	0	2	3	3
0	0	0	0	0	0	3	0	0	2	0	0
0	0	0	0	0	0	0	3	0	0	3	0
0	0	0	0	0	0	0	0	2	0	0	3

D_0

1	0	0	0	0	0	0	0	0	0	0	0
0	1	-1	0	0	0	0	0	0	0	0	0
0	0	0	6	-5	0	0	0	0	0	0	0
0	0	0	1	0	-1	0	0	0	0	0	0
0	0	0	0	0	0	2	-2	0	-3	2	0
0	0	0	0	0	0	2	0	-3	-3	0	2

$\hat{P}_0^t$

7.125	-11	.275	-2.125	2.275	-2.4	4.0	-2.4	3.6	-4.0	1.6

(Equivalently, we could have used the second method above to calculate SS_2^*, the sum of squares for the combined null hypotheses, $H_0(1)$ and $H_0(2)$, and then $SS_2 = SS_2^* - SS_1$.)

Note that SS_2 is identical to $SS_{b:a}$ in Table 5.11. It is the sum of squares for testing B after eliminating all possible contamination by A. Of course, the remainder of the test proceeds as for $SS_{b:a}$ in Table 5.11.

The next test is on the A main effect, using

$$C_3 = \begin{vmatrix} 0 & 1 & 0 & 0 & 0 & 0 & 0 & 0 & 0 & 0 & 0 & 0 \\ 0 & 0 & 1 & 0 & 0 & 0 & 0 & 0 & 0 & 0 & 0 & 0 \end{vmatrix}.$$

D_3 must now be orthogonal to C_0, C_1, C_2, and C_3, so

$$D_3 = \begin{vmatrix} 1 & 0 & 0 & 0 & 0 & 0 & 0 & 0 & 0 & 0 & 0 & 0 \end{vmatrix}.$$

(Equivalently, the null hypotheses $H_0(1)$, $H_0(2)$, and $H_0(3)$ could be combined into a single composite null hypothesis which could then be expressed as an estimable function.)

This gives $SS_{e(3)}^* = 313.75$ and $SS_3 = SS_{e(3)}^* - SS_{e(2)}^* = 20.25$ (or, if we use the first alternative, $SS_3 = SS_3^* - SS_2^*$) which is the same as SS_a in Table 5.11. We have just calculated the sum of squares for A ignoring B.

The general method for testing hypotheses sequentially should now be clear. For each $H_0(j)$, we first calculate either SS_j^* or $SS_{e(j)}^*$, based on hypotheses

TABLE 15.2. Matrices for testing interaction; data from Tables 5.9 and 15.1.

$$C_1$$

0	0	0	0	0	0	1	0	0	0	0	0
0	0	0	0	0	0	0	1	0	0	0	0
0	0	0	0	0	0	0	0	1	0	0	0
0	0	0	0	0	0	0	0	0	1	0	0
0	0	0	0	0	0	0	0	0	0	1	0
0	0	0	0	0	0	0	0	0	0	0	1

$$D_1$$

1	0	0	0	0	0	0	0	0	0	0	0
0	1	-1	0	0	0	0	0	0	0	0	0
0	0	0	6	-5	0	0	0	0	0	0	0
0	0	0	1	0	-1	0	0	0	0	0	0

$$\hat{P}_1^t$$

7.125	-1	1	.275	-2.125	2.275	0	0	0	0	0	0

$H_0(1)$ through $H_0(j)$. Then $SS_j = SS_j^* - SS_{j-1}^* = SS_{e(j)}^* - SS_{e(j-1)}^*$ (with $SS_1^* = SS_1$ and $SS_{e(1)}^* = SS_{e(1)}$ because no hypotheses are tested prior to $H_0(1)$). The SS_j constructed in this way are statistically independent (Theorem F31) (so by definition the tests are orthogonal), and each SS_j/σ_e^2 has a noncentral chi-square distribution with

$$\nu_j = \text{rank } (D_{j-1}) - \text{rank } (D_j) = r_{j-1} - r_j,$$
$$\phi_j^2 = P^t S(G_{j-1} - G_j) SP \quad \text{(Theorem F18)}.$$

In essence, then, we can write

$$H_0(j) : (G_{j-1} - G_j) SP = 0,$$

which expresses $H_0(j)$ as an estimable function, orthogonal to all other null hypotheses.

The last hypothesis we can test is $H_0(4) : \mu = 0$, using

$$C_4 = \begin{vmatrix} 1 & 0 & 0 & 0 & 0 & 0 & 0 & 0 & 0 & 0 & 0 & 0 \end{vmatrix}.$$

When we test it, we encounter a problem that we must eventually encounter when we continue testing hypotheses sequentially by the first alternative. There is no matrix, D_4, orthogonal to all the C_j. In this example we could note that $H_0(4)$ is orthogonal to the other null hypotheses and test it independently. However, there is a simpler solution that applies whether or not the hypothesis is orthogonal. There is no D_4 because the set of C_j from C_0 through

C_4, taken together, requires that $\hat{P}_4^* = 0$. Therefore, $SS_{e(4)}^* = X^t X = 1126$, and $SS_4 = SS_{e(4)}^* - SS_{e(3)}^* = 1126 - 313.75 = 812.25$. This corresponds, as it should, with SS_m in Table 5.11. Whenever we are testing hypotheses sequentially, and we encounter a hypothesis for which there is no D_j, we let $SS_{e(j)}^* = X^t X$. We then stop testing because there are no more orthogonal tests.

ANALYZING THE VARIANCE

Suppose we have chosen our null hypotheses so as to exhaust the possibilities in the data, i.e., so that there is no last D_j, and we have made all of our tests orthogonal. Then it is easily shown that

$$X^t X = SS_{e(0)} + \Sigma_j SS_j \qquad (15.11)$$

This *analysis* of $X^t X$ into a sum of terms is analogous to the analysis of a total sum of squares into individual sums of squares representing various errors and effects. In fact, in analysis of variance models it is customary to let one element of P be the grand mean μ, with the corresponding column of A consisting entirely of ones. In that case, the hypotheses can be set up so that one of the terms in the above sum is SS_m, the sum of squares for testing $H_0 : \mu = 0$. Subtracting that term from both sides, we can express $SS_t = X^t X - SS_m$ as the sum of the SS_j for the remaining tests.

Similarly, we can also write

$$
\begin{aligned}
E(X^t X) &= \underline{E}(SS_e + \Sigma_j SS_j) \\
&= (N - r_0)\sigma_e^2 + \sigma_e^2 \Sigma_j \nu_j (1 + \phi_j^2) \\
&= N\sigma_e^2 + \sigma_e^2 \Sigma_j \nu_j \phi_j^2 \\
&= N\sigma_e^2 + \Sigma_j P^t C_j^t (C_j G_0 C_j^t)^{-1} C_j P.
\end{aligned}
$$

If we let

$$\sigma_j^2 = P^t C_j^t (C_j G_0 C_j^t)^{-1} C_j P_j / N,$$

then

$$\underline{E}[(1/N)X^t X] = \sigma_e^2 + \Sigma_j \sigma_j^2, \qquad (15.12)$$

and this is analogous to the division of sums of squares in an analysis of variance model. If μ is an element of P, and SS_m is part of the sum in Equation 15.12, then one of the σ_j^2 will be $N\mu^2$. Subtracting $N\mu^2$ from both sides, we have

$$
\begin{aligned}
\sigma_t^2 &= \underline{E}[(1/N)X^t X] - N\mu^2 \\
&= \sigma_e^2 + \Sigma_j \sigma_j^2,
\end{aligned}
$$

where, of course, $N\mu^2$ is no longer included in the sum on the right. This is the general form of the analysis of variance.

Finally, the degrees of freedom can be similarly divided

$$N = (N - r_0) + \Sigma_j \nu_j, \tag{15.13}$$

and again, if $N\mu^2$ is in Equation 15.12, then the corresponding ν_j equals one. Subtracting that from both sides, we get

$$N - 1 = \nu_e + \Sigma_j \nu_j,$$

which is the general form for dividing degrees of freedom in an analysis of variance.

In the general linear model, μ need not be an element of P, and $N\mu^2$ need not appear in the sum in Equation 15.12. Then technically there is no analysis of variance. However, the analyses in Equations 15.11, 15.12, and 15.13 are analogous to those in analysis of variance models.

The analogy enables us to test residual effects. Suppose we have made a number of orthogonal tests but have used up fewer than the r_0 degrees of freedom available to us. Then, under the null hypothesis that there are no more systematic effects,

$$SS_{rem} = X^t X - SS_{e(0)} - \Sigma_j SS_j,$$

where the sum is now taken only over those tests that have actually been made, has a chi-square distribution with the degrees of freedom that were not used in the other tests.

Random Effects

A random or mixed model is characterized by the fact that some of the elements of P are random variables. For example, in the two-way mixed-effects design the β_j and $\alpha\beta_{ij}$ are random variables, while μ and the a_i are fixed, though unknown, parameters. The analysis is somewhat more complicated for random and mixed models, and we are more limited in the tests that we can make.

We can characterize such models most easily by supposing that P is a random vector with mean M and covariance matrix V. We can include fixed, nonrandom elements by making both their variances and their covariances with all other elements of P (both fixed and random) equal to zero. We will assume throughout that the random elements of P are normally distributed; it is usual to assume that their mean is zero, but the general linear model does not require that assumption.

Null hypotheses in the random model cannot always be expressed in the form $C_j P = 0$ because some null hypotheses concern the entire population of possible parameter values instead of just those that were sampled. (For example, in the one-way random model, the null hypothesis concerns the entire population of α_i, not just those for the groups that were sampled.) However, all of the null hypotheses that we will consider will *imply* that $C_j P = 0$ for some C_j (i.e., if $H_0(j)$ is true, then $C_j P = 0$, even though the opposite may not be true), and that is sufficient for testing purposes. (In the one-way random model, $H_0 : \sigma_{bet}^2 = 0$ states that all α_i in the population are equal, so all contrasts in the α_i must be zero.) Then all SS_j and $SS_{e(j)}$ are calculated as before. Moreover,

$$SS_{e(0)}/\sigma_e^2 \sim \chi_{(N-r_0)}^2,$$

just as in the fixed effects model (Theorem R1).

However, there are some differences. Each $\hat{P}_j$ (including $\hat{P}_0$) is now normally distributed with

$$\underline{E}(\hat{P}_j) = G_j SM, \; \underline{V}(\hat{P}_j) = G_j(\sigma_e^2 S + SVS)G_j \quad \text{(Theorem R3)}.$$

If $C_j M = 0$, then $\underline{E}(\hat{P}_j) = M$ (Theorem R4).

If $C_j P$ is estimable, then

$$\begin{aligned}
\underline{E}(C_j \hat{P}_0) &= C_j M, \\
\underline{C}(C_j \hat{P}_0, C_k \hat{P}_0) &= C_j(\sigma_e^2 G_0 + V)C_k^t \quad \text{(Theorem R5)}.
\end{aligned}$$

Each SS_j is of the form

$$SS_j = Y^t H_j Y,$$

where $H_j = G_0 - G_j$ if the hypotheses are orthogonal, and $H_j = G_{j-1} - G_j$ if they are not. Whichever test is used, the SS_j are independent of $SS_{e(0)}$ (Theorem R2). However, if SS_j and SS_k are to be statistically independent, we must have $H_j SVSH_k = 0$ *in addition to orthogonality* (Theorem R7). If $C_j P$ is estimable, and the tests are not sequential, then an equivalent condition is

$$C_j V C_k^t = 0 \quad \text{(Theorem R9)},$$

again, in addition to orthogonality.

For SS_j to have a chi-square distribution, we must have

$$H_j SVSH_j = (a_j - \sigma_e^2)H_j, \tag{15.14}$$

for some a_j (Theorem R6). An equivalent condition, when $C_j P$ is estimable, is

$$C_j V C_j^t = (a_j - \sigma_e^2)C_j G_0 C_j^t \quad \text{(Theorem R8)}.$$

These conditions are highly restrictive; for example, they usually cannot be met in analyses of covariance. In general, they limit us to standard analyses of variance with equal sample sizes.

When the second restriction is met, SS_j/a_j has a noncentral chi-square distribution with $\nu_j = \text{rank }(H_j)$ as in the fixed-effects model, and

$$\phi_j^2 = M^t SH_j SM/(a_j \nu_j) \quad \text{(Theorem R6)}.$$

Then

$$\begin{aligned}
\underline{E}(SS_j) &= a_j \nu_j(1 + \phi_j^2), \\
\underline{E}(MS_j) &= a_j(1 + \phi_j^2).
\end{aligned} \tag{15.15}$$

Two-Way Model

We will illustrate with a two-way analysis of variance with n scores per cell. However, even the simplest such analysis of any interest involves very large matrices. We can simplify the example and make it more general at the same time if we use *star products*.

Star Products. We will define the star product of two matrices as follows. Let a_{ij} be the element in the ith row and jth column of A. Then

$$A * B = \begin{vmatrix} a_{11}B & a_{12}B & a_{13}B & \cdots \\ a_{21}B & a_{22}B & a_{23}B & \cdots \\ a_{31}B & a_{32}B & a_{33}B & \cdots \\ \cdot & \cdot & \cdot & \cdots \\ \cdot & \cdot & \cdot & \cdots \\ \cdot & \cdot & \cdot & \cdots \end{vmatrix}.$$

That is, $A * B$ is calculated by multiplying each element of A by each element of B, in turn, and arranging the result as shown above. The number of rows in $A * B$ is the number of rows in A times the number of rows in B, and the number of columns in $A * B$ is the number of columns in A times the number of columns in B. For example,

$$\begin{vmatrix} 3 & 2 & -1 \\ 0 & 1 & 4 \end{vmatrix} * \begin{vmatrix} 4 & 6 \\ -2 & 3 \end{vmatrix} = \begin{vmatrix} 12 & 18 & 8 & 12 & -4 & -6 \\ -6 & 9 & -4 & 6 & 2 & -3 \\ 0 & 0 & 4 & 6 & 16 & 24 \\ 0 & 0 & -2 & 3 & -8 & 12 \end{vmatrix}.$$

Note that $A * B$ is defined for any A and B; i.e., A and B do not have to match in either number of rows or number of columns.

The following properties of star products are useful:

$$\begin{aligned} cA &= c * A = A * c \ (c \text{ any scalar}), \\ (A * B) * C &= A * (B * C), \\ A * (B + C) &= (A * B) + (A * C), \\ (A + B) * C &= (A * C) + (B * C), \\ (AB) * (CD) &= (A * C)(B * D). \end{aligned}$$

Example: Balanced Two-Way Design. Let g be the number of levels of Factor A and h be the number of levels of Factor B (I and J could be confused with matrices). We then define

U_k: a k-dimensional column vector consisting entirely of ones (note that $U_k^t U_k = k$);

I_k: a $k \times k$ identity matrix; and

TABLE 15.3. Basic matrices for two-way analysis of variance with equal sample sizes. The sample size is n, the number of levels of Factor A is g, and the number of levels of Factor B is h. Other symbols are defined in the text.

$$A = \begin{array}{|ccc|} I_g * U_h & U_g * I_h & I_g * I_h \end{array} * U_n$$

$$S = n \begin{vmatrix} gh & hU_g^t & gU_h^t & U_g^t * U_h^t \\ hU_g & hI_g & U_g * U_h^t & I_g * U_h^t \\ gU_h & U_g^t * U_h & gI_h & U_g^t * I_h \\ U_g * U_h & I_g * U_h & U_g * I_h & I_g * I_h \end{vmatrix}$$

$$C_0 = \begin{vmatrix} 0 & U_g^t & 0 & 0 \\ 0 & 0 & U_h^t & 0 \\ 0 & 0 & 0 & I_g * U_h^t \\ 0 & 0 & 0 & U_g^t * I_h \end{vmatrix} \qquad D_0 = \begin{vmatrix} 1 & 0 & 0 & 0 \\ 0 & J_g & 0 & 0 \\ 0 & 0 & J_h & 0 \\ 0 & 0 & 0 & J_g * J_h \end{vmatrix}$$

$$D_0 S D_0^t = \begin{vmatrix} gh & 0 & 0 & 0 \\ 0 & hI_{g-1} & 0 & 0 \\ 0 & 0 & gI_{h-1} & 0 \\ 0 & 0 & 0 & I_{g-1} * I_{h-1} \end{vmatrix}$$

$$(D_0 S D_0^t) = (1/n) \begin{vmatrix} 1/(gh) & 0 & 0 & 0 \\ 0 & (1/h)I_{g-1} & 0 & 0 \\ 0 & (1/g) & I_{h-1} & 0 \\ 0 & 0 & 0 & {}_{g-1} * I_{h-1} \end{vmatrix}$$

$$G_0 = (1/n) \begin{vmatrix} 1/(gh) & 0 & 0 & 0 \\ 0 & (1/h)L_g & 0 & 0 \\ 0 & 0 & (1/g)L_h & 0 \\ 0 & 0 & 0 & L_g * L_h \end{vmatrix}$$

0: a matrix of any dimensions, consisting entirely of zeros.

With this notation, the basic matrices are in Table 15.3.

The first "element" of A contains the coefficients of μ; it is just a vector of gh ones. The second contains the coefficients of the α_i; in effect it contains g submatrices, each $h \times g$, where the ith submatrix has ones in the ith row and zeros everywhere else. The third contains the coefficients of the β_j; it consists of an $h \times h$ identity matrix repeated g times. The fourth contains the coefficients of the $\alpha\beta_{ij}$; it is an identity matrix with gh rows and columns. (These correspondences between the submatrices and the parameters will be retained in all of the following matrices.) Finally, the entire matrix is multiplied by U_n, indicating that each cell contains n scores.

Following the derivation described earlier, we obtain S, whose rows and columns are partitioned in the same way as the columns of A.

The first row of C_0 is the restriction that the α_i sum to zero; the second row is the restriction that the β_j sum to zero; the third row is the restriction that the $\alpha\beta_{ij}$ sum to zero across levels of B; and the fourth row is the restriction that the $\alpha\beta_{ij}$ sum to zero across levels of A.

Before describing the other matrices, we need two more definitions. Let J_k be any $(k-1) \times k$ matrix such that $J_k J_k^t = I_{k-1}$ and $J_k U_k = 0$. In effect, J_k is a matrix of coefficients of orthogonal contrasts, normalized so that the squares of the coefficients sum to one. We note then that $J_k^t J_k = I_k - (1/g)U_k U_k^t$, and we define $L_k = J_k^t J_k$. Finally, we note that $L_k U_k = 0$, $U_k^t L_k = 0$, and $L_k^2 = L_k$.

With these definitions, we obtain D_0. The second row of D_0 is any set of normalized orthogonal contrasts in the α_i; the third row is any set of normalized orthogonal contrasts in the β_j; the fourth row is a set of normalized orthogonal contrasts in the $\alpha\beta_{ij}$. (Normalized orthogonal contrasts are not necessary but they are very convenient.) If we follow the rules for star products, it is not hard to show that D_0 is orthogonal to C_0.

Notice that G_0 contains only four submatrices, arranged along its diagonal. Each submatrix corresponds to a single set of parameters. This greatly simplifies later calculations. The H_j for each null hypothesis can be derived from G_0 by setting the submatrices for the other effects equal to zero.

For example, the matrices for testing the A main effect are in Table 15.4. C_1 requires that $\alpha_i = 0$ for all i. D_1 differs from D_0 only in that the second row of D_0 is missing from D_1. G_1 differs from G_0 in that the second row has been set to zero. Every row of H_1 except the second is zero. Similar patterns arise when testing the grand mean, the B main effect, and the AB interaction. It is not difficult to show that all of these tests are orthogonal.

For the random model (i.e., A and B both random), let

$$
V = \begin{vmatrix} 0 & 0 & 0 & 0 \\ 0 & \sigma_a^2 I_g & 0 & 0 \\ 0 & 0 & \sigma_b^2 I_h & 0 \\ 0 & 0 & 0 & \sigma_{ab}^2 I_g * I_h \end{vmatrix}.
$$

TABLE 15.4. Matrices for testing A main effect in two-way design with equal sample sizes.

$$C_1 = \begin{vmatrix} 0 & I_g & 0 & 0 \end{vmatrix}$$

$$D_1 = \begin{vmatrix} 1 & 0 & 0 & 0 \\ 0 & 0 & J_h & 0 \\ 0 & 0 & 0 & J_g * J_h \end{vmatrix}$$

$$G_1 = (1/n)\begin{vmatrix} 1/(gh) & 0 & 0 & 0 \\ 0 & 0 & 0 & 0 \\ 0 & 0 & (1/g)L_h & 0 \\ 0 & 0 & 0 & L_g * L_h \end{vmatrix}$$

$$H_1 = (1/n)\begin{vmatrix} 0 & 0 & 0 & 0 \\ 0 & (1/h)L_g & 0 & 0 \\ 0 & 0 & 0 & 0 \\ 0 & 0 & 0 & 0 \end{vmatrix}$$

(This is not the only possible covariance matrix, but it is the simplest. It corresponds to the assumption that all parameters are independent and all parameters of the same type have equal variances.)

It is now a rather tedious but straightforward task to show that $H_1 SVSH_1 = n(\sigma_{ab}^2 + h\sigma_a^2)H_1$ so $a_1 = \sigma_e^2 + n\sigma_{ab}^2 + nh\sigma_a^2$. In the random model, the first element of M is μ, and the remaining elements are zero. From this it is again tedious but straightforward to show that $\phi_1^2 = 0$. Finally, the rank of H_1 is the rank of L_g, which is easily shown to be $(g-1)$, giving

$$\nu_1 = g - 1,$$
$$\underline{E}(SS_1) = \nu_1(\sigma_e^2 + n\sigma_{ab}^2 + nh\sigma_a^2),$$
$$\underline{E}(MS_1) = \sigma_e^2 + n\sigma_{ab}^2 + nh\sigma_a^2.$$

This is, of course, $\underline{E}(MS_a)$ given in Chapter 6 for the two-way random model. The other expected mean squares are derived similarly.

The two-way mixed-effects model is somewhat more complicated. If Factor A is fixed, we must have $\Sigma_i \alpha\beta_{ij} = 0$; i.e., the $\alpha\beta_{ij}$ are random but not independent across levels of A. The requirement that the $\alpha\beta_{ij}$ sum to zero is equivalent to the requirement that the elements in each row of the covariance matrix sum to zero. In addition, we traditionally assume that the covariances of the $\alpha\beta_{ij}$ are equal, giving

$$V = \begin{vmatrix} 0 & 0 & 0 & 0 \\ 0 & 0 & 0 & 0 \\ 0 & 0 & \sigma_b^2 I_h & 0 \\ 0 & 0 & 0 & \sigma_{ab}^2[g/(g-1)]L_g * I_h \end{vmatrix}.$$

(Again, this is not the only possible covariance matrix, but it is probably the simplest.)

The second row is zero, reflecting the fact that Factor A is now fixed. The third row is the same as before. The last row contains the matrix $[g/(g-1)]L_g$, a matrix whose diagonal elements are all equal to one, and whose off-diagonal elements are all equal to $-1/(g-1)$. It is not hard to show that the elements sum to zero across both rows and columns. The star product with I_h indicates that this matrix occurs at each level of Factor B.

Again by a tedious but straightforward process we can determine that Equation 15.14 holds for the A main effect, with

$$a_1 = \sigma_e^2 + n[g/(g-1)]\sigma_{ab}^2.$$

With Factor A fixed, M contains nonzero values for both μ and the α_i. To indicate this, let $\underline{\alpha}$ be a vector whose elements are the α_i. Then

$$M^t = \begin{vmatrix} \mu & \underline{\alpha}^t & 0 & 0 \end{vmatrix}$$

and

$$M^t SH_1 SM = nh\underline{\alpha}^t L_g \underline{\alpha} = nhg\sigma_a^2. \tag{15.16}$$

The last equality holds because of a special property of every matrix L_k; for a k-dimensional vector, X, $X^t L_k X = k\sigma_x^2$.

We thus have

$$\begin{aligned} \underline{E}(SS_1) &= (g-1)\sigma_e^2 + ng\sigma_{ab}^2 + nhg\sigma_a^2, \\ \underline{E}(MS_1) &= \sigma_e^2 + n[g/(g-1)]\sigma_{ab}^2 + nh[g/(g-1)]\sigma_a^2. \end{aligned}$$

Again, this is the expected value given in Chapter 6. The expected values of the other mean squares are derived similarly.

If there is no A main effect, then from Equation 15.16 we can see that $\phi_1^2 = 0$, so MS_1 can be used to test the null hypothesis. However, the value of a_j tells us that the denominator for the test has to be the mean square for testing the interaction.

POOLED SUMS OF SQUARES AND QUASI-F RATIOS

The rules for pooling sums of squares are fairly simple. If two or more independent sums of squares have the same expected value, they can be added, and the sum will have a chi-square distribution. Its degrees of freedom will be equal to the sum of the degrees of freedom of the original sums of squares.

The rules for quasi-F ratios are also fairly simple. When no appropriate denominator can be found for testing a given effect, we may be able to construct one as a linear combination of independent mean squares, all of which have *central* chi-square distributions. Then if

$$(MS_{\text{pooled}}) = \Sigma_j c_j MS_j,$$

we have

$$
\begin{aligned}
\underline{E}(MS_{\text{pooled}}) &= \Sigma_j c_j \underline{E}(MS_j) = \Sigma_j c_j a_j, \\
\underline{V}(MS_{\text{pooled}}) &= \Sigma_j c_j^2 \underline{V}(MS_j) = 2\Sigma_j c_j^2 a_j^2/\nu_j,
\end{aligned}
\tag{15.17}
$$

where $\nu_j = \text{rank}\ (H_j)$ is the degrees of freedom of MS_j.

We construct MS_{pooled} to have the same expected value as the MS for the effect being tested. MS_{pooled} does not usually have a chi-square distribution, but its distribution is likely to be similar to that of a chi-square variable divided by its degrees of freedom, i.e., it ranges from approximately zero to infinity and is likely to have a single mode. Our problem is to determine which chi-square variable would have the most similar distribution. A simple answer is to find the chi-square variable that has the same mean and variance. If x/k is a chi-square variable with ν degrees of freedom, then $\underline{E}(x) = k\nu$ and $\underline{V}(x) = 2k^2\nu$. This will have the same mean and variance as MS_{pooled} if and only if

$$
\begin{aligned}
k &= \underline{V}(MS_{\text{pooled}})/[2\underline{E}(MS_{\text{pooled}})], \\
\nu &= 2\underline{E}(MS_{\text{pooled}})^2/\underline{V}(MS_{\text{pooled}}).
\end{aligned}
$$

Usually, we do not know either $\underline{E}(MS_{\text{pooled}})$ or $\underline{V}(MS_{\text{pooled}})$. However, from Equation 15.15, we can estimate the a_j as

$$\hat{a}_j = MS_j,$$

and then from Equation 15.17, we can get

$$
\begin{aligned}
\underline{\hat{E}}(MS_{\text{pooled}}) &= \Sigma_j c_j \hat{a}_j = MS_{\text{pooled}}, \\
\underline{\hat{V}}(MS_{\text{pooled}}) &= 2\Sigma_j c_j^2 \hat{a}_j^2/\nu_j = 2\Sigma_j c_j^2 MS_j^2/\nu_j.
\end{aligned}
$$

The estimate of $\underline{V}(MS_{\text{pooled}})$ is slightly biased, but it is less likely to be negative than the unbiased estimate, and it serves well for the purpose of estimating ν and k. Generally, it tends to be conservative, leading to a slight underestimate of ν.

Finally, then we have

$$\hat{\nu} = 2\underline{\hat{E}}(MS_{\text{pooled}})^2/\underline{\hat{V}}(MS_{\text{pooled}}) = (MS_{\text{pooled}})^2/(\Sigma_j c_j^2 MS_j^2/\nu_j).$$

Multivariate Analysis of Variance

The generalization to the multivariate general linear model is straightforward. If there are p dependent variables, then P, which has been a vector, becomes an $m \times p$ matrix, and similarly, E becomes an $n \times p$ matrix. Instead of calculating sums of squares, we calculate matrices of sums of cross-products. Instead of forming F ratios, we use the procedures described in Chapter 14. Instead of a single matrix, V, there is a separate covariance matrix for each dependent variable, and each matrix must satisfy Equation 15.14. However, no special assumptions need be made about the covariances among the elements in the same row of P.

Analysis of Covariance

The analysis of covariance can be handled by taking those columns of X that are regarded as covariates and making them columns of A. We also add a row to P for each coefficient θ. Although some of the covariates may themselves be random variables, we regard them as constants for the purpose of the analysis. There is nothing wrong with this practice; the resulting analysis is called a *conditional analysis* (conditional on the obtained values of the covariates), with the conditional probability of a Type I error being α. Because the conditional probability is α for every possible set of values of the covariates, the unconditional probability is also α. (However, for conventional analyses of covariance, the warnings given in Chapter 14 still apply.)

Appendix A

Theorems and Proofs

Background

This appendix contains basic theorems related to the assertions in Chapter 16. The proofs are sometimes sketched rather than given in full detail. They occasionally use, without proof, facts that are easily found in a test on matrix algebra. Additional material can be found in Scheffé (1959) and Searle (1971).

Definitions

The following concepts are used throughout this appendix:

A matrix, B, is *idempotent* if $B^2 = B$.

The *rank* of a matrix is equal to its number of linearly independent rows (or, equivalently, its number of linearly independent columns). The rank of B will be symbolized by $r(B)$.

The *column space* of a matrix, B, is the set of all vectors that can be expressed as a linear function of the columns of B, i.e., all X such that $X = BY$ for some Y.

The *row space* of B is the set of all vectors that can be expressed as a linear function of the rows of B, i.e., all X^t such that $X^t = Y^t B$ for some Y.

The *trace* of a matrix is the sum of its diagonal elements. The trace of B will be symbolized by $tr(B)$.

A quadratic form is any function of the form $X^t B X$. B need not be symmetric, but if it is not, we can let $C = .5(B + B^t)$, and then $X^t C X = X^t B X$ for all X.

A quadratic form is *positive definite* if $X^t B X > 0$ for all $X \neq 0$; it is *positive semidefinite* if $X^t B X \geq 0$ for all X.

A square matrix, B, is positive (semi)definite if the quadratic form $X^t B X$ is positive (semi)definite.

The *generalized inverse* of B is any matrix, G, such that $BGB = B$. (Note that in general G is not uniquely defined.)

The following theorems are in five sections. The first section contains general theorems about matrices. The second contains theorems about generalized inverses. The third contains theorems about the distribution of quadratic forms. The theorems in the first section can be found in most texts on matrix algebra. Those in the second and third sections are taken, with minor modification, from Searle (1971), Chapters 1 and 2. They are given here without proofs.

The theorems in the fourth section pertain to fixed-effects models, while those in the fifth pertain to random-effects models. Many of the theorems in the first three sections are used in the fourth and fifth sections (some are included just because they are generally useful), whose theorems verify the statements made in Chapter 16.

Preliminary Matrix Theorems

Theorem M1. $r(AB) \leq \min[r(A), r(B)]$.

Theorem M2. *If B is positive definite, then $r(A^t B A) = r(A)$. In particular, because I is positive definite, $r(A^t A) = r(A^t I A) = r(A)$.*

Theorem M3. $tr(AB) = tr(BA)$.

Theorem M4. *Any nonzero matrix, B, can be factored into three matrices, U, D, and V, such that $B = UDV^t$, where D is a diagonal matrix with $r(B)$ rows and columns, all of whose diagonal elements are nonzero, and $U^t U = V^t V = I$. If B is a square, symmetric matrix, then $U = V$.*

Theorem M5. *A matrix, B, is positive semidefinite if and only if, when B is factored as in Theorem M4, all diagonal elements of D are greater than zero. If B is also nonsingular, then $r(D) = r(B)$, and B is positive definite.*

Theorem M6. *If $B = A^t A$, for some A, then B is positive semidefinite.*

Theorem M7. *Assume B is positive semidefinite. Then $X^t B X = 0$ if and only if $BX = 0$.*

Theorem M8. *If B is idempotent, then $r(B) = tr(B)$.*

Theorems Concerning Generalized Inverses

In the following theorems, we consider solutions to equations of the form $AX = Y$, with A and Y known. In every case we assume that the equation is *consistent*, i.e., that it has at least one solution. Also, unless otherwise stated, G is any generalized inverse of A.

We add one more definition. When B is factored as in Theorem M4, the *Penrose inverse* of B is defined to be $VD^{-1}U^t$.

Theorem G1. *X is a solution to the equation $AX = Y$ if and only if $X = GY$ where G is a generalized inverse of A (i.e., $AGA = A$).*

Theorem G2. *X is a solution to the equation $AX = Y$ if and only if $X = GY + (GA - I)Z$, where G is any generalized inverse of A, and Z is an arbitrary vector. (Note that this means all possible solutions can be found after calculating any generalized inverse of A.)*

Theorem G3. *Let $H = GA$ where the rank of A is r. Then H is idempotent with rank r, $(I - H)$ is idempotent with rank $(q - r)$, and $H(I - H) = 0$.*

Theorem G4. *Let $H = GA$. Then KX is invariant to the particular solution, X, of $AX = Y$ if and only if $KH = K$.*

Theorem G5. *The Penrose inverse, and only the Penrose inverse, has all of the following properties:*

(1) $AGA = A$,

(2) $GAG = G$, *and*

(3) AG *and* GA *are both symmetric.*

Theorem G6. *Let $B = A^t A$, and let G be any generalized inverse of B. Then the following are true:*

(1) G^t *is also a generalized inverse of B.*

(2) $AGA^t A = A$ *(i.e., GA^t is a generalized inverse of A).*

(3) AGA^t *is invariant to the choice of G.*

(4) AGA^t *is symmetric whether G is or not.*

Corollary G6. *If G is a generalized inverse of $B = A^t A$, then*

(1) $AG^t A^t A = A$;

(2) $A^t AGA^t = A^t$;

(3) $A^t AG^t A^t = A^t$; *and*

(4) $AG^t A^t = AGA^t$, *and* $AG^t A^t$ *is symmetric.*

Theorem G7. *If A is nonsingular, then A^{-1} is the only generalized inverse of A.*

Preliminary Distribution Theorems

In the following theorems, we assume that X has a multivariate normal distribution with mean $\underline{E}(X) = M$ and covariance matrix $\underline{V}(X) = V$. $\underline{C}(X, Y)$ will represent the matrix of covariances between the two random variables X and Y.

Theorem D1.

$$
\begin{aligned}
\underline{E}(X^t AX) &= tr(AV) + M^t AM, \\
\underline{V}(X^t AX) &= 2\, tr[(AV)^2] + 4M^t AV AM, \\
\underline{C}(X, X^t AX) &= 2V AM.
\end{aligned}
$$

Theorem D2. $X^t AX$ has a noncentral chi-square distribution with $\nu = tr(AV)$ and $\phi^2 = M^t A^t V AM/\nu$ if and only if

(1) $VAVAV = VAV$,

(2) $M^t AV = M^t AV AV$, and

(3) $M^t AM = M^t AV AM$.

(Note that all three conditions will hold if $AV A = A$; if no specific restrictions can be placed on M, then the single condition, $AV A = A$, is necessary and sufficient.)

Corollary D2.1. If V is nonsingular, then $X^t AX$ has a noncentral chi-square distribution with ν and ϕ^2 as in Theorem D2 if and only if AV is idempotent. (Note that this applies only when V is nonsingular; if V is singular, then idempotency is neither necessary nor sufficient.)

Corollary D2.2. If $V = I$, then $X^t AX$ has a noncentral chi-square distribution with $\nu = r(A)$, $\phi^2 = M^t AM/\nu$ if and only if A is idempotent.

Theorem D3. $X^t AX$ and $X^t BX$ are independent if and only if

(1) $VAVBV = 0$,

(2) $VAVBM = VBV AM = 0$, and

(3) $M^t AV BM = 0$.

(Note that all three conditions will hold if $AV B = 0$; if no restrictions can be placed on M, then the single condition, $AV B = 0$, is necessary and sufficient.)

Corollary D3.1. If V is nonsingular, then $X^t AX$ and $X^t BX$ are independent if and only if $AV B = 0$.

Corollary D3.2. If $V = I$, then $X^t AX$ and $X^t BX$ are independent if and only if $AB = 0$.

Theorems on Fixed-Effects Models

The following theorems concern the material in Chapter 16 on fixed-effects models. Some facts are also included that are needed for the theorems; these facts are stated as lemmas. Every effort was made to make the proofs as simple as possible, but occasionally additional concepts such as Lagrange multipliers were unavoidable.

ASSUMPTIONS

Unless stated otherwise, all notation is identical to that used in Chapter 16. We assume i, j, $k \neq 0$ unless a theorem specially states otherwise. We also assume that $X = AP + E$, where A is $N \times m$ with rank r_0, and P and E are vectors.

In addition, we assume the following:

E has a multivariate normal distribution with $\underline{E}(E) = 0$, $\underline{V}(E) = \sigma_e^2 I$.

$\underline{E}(X) = AP$, $\underline{V}(X) = \sigma_e^2 I$.

$S = A^t A$ is $m \times m$ with rank r_0.

$Y = A^t X$, $\underline{E}(Y) = SP$, $\underline{V}(Y) = \sigma_e^2 S$.

P is constant, but unknown, so no specific restrictions can be placed on M when applying Theorems D2 and D3 to X or Y.

C_0 is constructed to be linearly independent of A with $r(C_0) = m - r_0$,
$$r\left(\begin{vmatrix} A^t & C_0^t \end{vmatrix}\right) = m.$$

D_0 is any $r_0 \times m$ matrix such that
$$C_0 D_0^t = 0, \; r(D_0) = r_0.$$

G_0 is any symmetric generalized inverse of S. (The requirement of symmetry causes no practical problems because if G_0 is a generalized inverse of S, then so is $.5(G_0 + G_0^t)$.)

$\hat{P}_0 = G_0 Y$.

$$
\begin{aligned}
SS_{e(0)} &= (X - A\hat{P}_0)^t(X - A\hat{P}_0) \\
&= X^t X - \hat{P}_0^t S \hat{P}_0 \\
&= X^t X - Y^t \hat{P}_0 \\
&= X^t X - Y^t G_0 Y \\
&= X^t(I - AG_0 A^t)X \\
&= \sigma_e^2 Z^t(I - AG_0 A^t)Z,
\end{aligned}
$$

where $Z = X/\sigma_e$, $\underline{E}(Z) = AP/\sigma_e$, $\underline{V}(Z) = I$.

C_j is any matrix with m columns.

D_j is any $r_j \times m$ matrix such that

$$C_0 D_j^t = 0, \ C_j D_j^t = 0, \ r(D_j) = r_j, \ r \left(\left| \begin{array}{ccc} C_0^t & C_j^t & D_j^t \end{array} \right| \right) = m.$$

If the tests are sequential, then all C_i such that $i < j$ must be included with C_j in the above equations.

$C_j P$ is estimable if and only if $C_j \hat{P}_0$ is invariant to the choice of $\hat{P}_0$.

$$G_j = D_j^t (D_j S D_j^t)^{-1} D_j.$$

$$\hat{P}_j = G_j Y.$$

$$\begin{aligned} SS_{e(j)} &= (X - A\hat{P}_j)^t (X - A\hat{P}_j) \\ &= X^t X - \hat{P}_j^t S \hat{P}_j \\ &= X^t X - Y^t \hat{P}_j \\ &= X^t X - Y^t G_j Y \\ &= X^t (I - AG_j A^t) X \\ &= \sigma_e^2 Z^t (I - AG_j A^t) Z, \end{aligned}$$

where $Z = X/\sigma_e$, $\underline{E}(Z) = AP/\sigma_e$, $\underline{V}(Z) = 0$.

$SS_j = SS_{e(j)} - SS_{e(i)}$, where generally, $i = j - 1$ if the tests are sequential, and $i = 0$ if they are not.

$H_0(j) : C_j P = 0$ and $H_0(k) : C_k P = 0$ are orthogonal if and only if $C_j G_0 C_k^t = 0$.

THEOREMS

Theorem F1. $\hat{P}_0 = G_0 Y$ *minimizes* $SS_{e(0)} = (X - A\hat{P}_0)^t (X - A\hat{P}_0)$ *if and only if* G_0 *is a generalized inverse of* S, *that is, if and only if* $S\hat{P}_0 = Y$.

Proof. By Theorem G1, $\hat{P}_0$ is a solution to the equation $SP = Y$ if and only if $\hat{P}_0 = G_0 Y$, where G_0 is a generalized inverse of S. Let $\hat{P}_0 = G_0 Y$ for some generalized inverse, G_0, of S; then

$$\begin{aligned} SS_{e(0)} &= (X - A\hat{P}_0)^t (X - A\hat{P}_0) \\ &= [(X - AP) + A(P - \hat{P}_0)]^t [(X - AP) + A(P - \hat{P}_0)] \\ &= (X - AP)^t (X - AP) + (P - \hat{P}_0)^t S(P - \hat{P}_0). \end{aligned}$$

Note that the cross-product is zero because

$$(P - \hat{P}_0)^t A^t (X - AP) = (P - \hat{P}_0)^t (Y - SP) = 0.$$

The first term above does not depend on $\hat{P}_0$; the second is a positive semidefinite quadratic form, so its minimum is achieved when $\hat{P}_0 = P$.

Now let $\hat{P}_0^* = G_0 Y + (G_0 S - I)U + V$ for some U and V (see Theorem G2). Then after some algebra we have

$$(P - \hat{P}_0^*)^t S (P - \hat{P}_0^*) = P^t S P + V^t S V.$$

Being positive semidefinite, $V^t S V \geq 0$. If it is greater than zero, then $\hat{P}_0^*$ does not minimize $SS_{e(0)}$. If it equals zero, then by Theorem M7, $SV = 0$, and

$$S\hat{P}_0 = S[G_0 Y + (G_0 S - I)U + V] = S G_0 Y = S\hat{P}_0 = Y.$$

$\square$

Theorem F2. $SS_{e(0)}$ *is invariant to the choice of* G_0 *(and, therefore, of* $\hat{P}_0$*).*

Proof. $SS_{e(0)} = X^t(I - A G_0 A^t)X$, but $A G_0 A^t$ is invariant to G_0 by Theorem G6, Pt. 3, so $I - A G_0 A^t$ is invariant. $\square$

Theorem F3. *If* S *is nonsingular, then* $\hat{P}_0 = S^{-1}Y$ *uniquely minimizes* $SS_{e(0)}$.

Proof. The proof follows immediately from Theorems G7 and F1. $\square$

Theorem F4. $G_0 = (S + C_0^t C_0)^{-1}$ *is a generalized inverse of* S.

Proof. Note first that, if $K^t = |\ A^t \quad C_0^t\ |$, then K^t has rank m (by construction), so by Theorem M2, $K^t K = S + C_0^t C_0$ has rank m and is nonsingular, guaranteeing the existence of an inverse. Now $\hat{P}_0 = G_0 Y$ is the least-squares solution to

$$\left|\begin{array}{c} X \\ 0 \end{array}\right| = \left|\begin{array}{c} A \\ C_0 \end{array}\right| \hat{P}_0 + \left|\begin{array}{c} E \\ F \end{array}\right|.$$

But C_0 is linearly independent of A so we can minimize $E^t E$ and $F^t F$ simultaneously and independently. $F^t F$ is minimized by letting $F = 0$, which is possible because $r(C_0) < m$, and $E^t E$ is minimized, by Theorem F1, if and only if $S\hat{P}_0 = Y$. Therefore, the $\hat{P}_0$ that minimizes the above function is the one that sets $C_0 \hat{P}_0 = 0$, $S\hat{P}_0 = Y$. Consequently, $\hat{P}_0 = G_0 Y$ minimizes $SS_{e(0)}$ and, by Theorem F1, G_0 is a generalized inverse of S. (For a more complete and detailed proof, see Searle, 1971, pp. 21-22.) $\square$

Theorem F5. *If* $\hat{P}_0 = G_0 Y$ *and* $C_0 \hat{P}_0 = 0$, *then* $C_0 G_0 S = 0$.

Proof. $0 = C_0 \hat{P}_0 = C_0 G_0 Y = C_0 G_0 A^t X$, and this must hold for all X, so $C_0 G_0 A^t = 0$ and $C_0 G_0 A^t A = C_0 G_0 S = 0$. $\square$

Lemma F1. $C_0 K = 0$ *if and only if* $K = G_0 S L$ *for some* L.

Proof. By construction, $r(C_0) = m - r_0$ by Theorem G3, $r(G_0 S) = r_0$; and by Theorem F5, the rows of C_0 are orthogonal to $G_0 S$. Therefore, taken together, the rows of C_0 and $(G_0 S)^t$ span the space of all m-dimensional vectors. Consequently, K is orthogonal to the rows of C_0 if and only if K is in the column space of $G_0 S$, i.e., if and only if $K = G_0 SL$ for some L.   □

<u>Lemma F2.</u> *Given the constraint, $C_0 \hat{P}_0 = 0$, $\hat{P}_0$ is invariant to G_0.*

<u>Proof.</u> Suppose $C_0 \hat{P}_0 = C_0 \hat{P}_0^* = 0$. By Theorem G2, $\hat{P}_0^* = \hat{P}_0 + (G_0 S - I)K$ for some K, and

$$C_0 \hat{P}_0^* = C_0 [\hat{P}_0 + (G_0 S - I)K] = 0.$$

But, then expanding the product and applying Theorem F5, we have $C_0 K = 0$ and, by Lemma F1, $K = G_0 SL$ for some L. Substituting back and applying Theorem G3, we have

$$\hat{P}_0^* = \hat{P}_0 + (G_0 S - I)G_0 SL = \hat{P}_0.$$

□

<u>Lemma F3.</u> $r(D_j SD_j^t) = r_j$. *This lemma applies to the case $j = 0$. (Note: This implies that $D_j SD_j^t$ is nonsingular if and only if the rows of D_j are linearly independent.)*

<u>Proof.</u> By Theorems M6 and F4, $(S + C_0^t C_0)$ is positive definite. Because D_j is orthogonal to C_0, $D_j (S + C_0^t C_0)D_j^t = D_j SD_j^t$, so by Theorem M2, $r(D_j SD_j^t) = r_j$.   □

<u>Theorem F6.</u> $G_0 = D_0^t (D_0 SD_0^t)^{-1} D_0$ *is a generalized inverse of S.*

<u>Proof.</u> By construction (see Chapter 15), $G_0 Y$ minimizes the least-squares problem $X = AP + E$, subject to $C_0 P = 0$. But because C_0 is linearly independent of A, this minimizes the unconstrained problem; i.e., $\hat{P}_0 = G_0 Y$ minimizes $SS_{e(0)}$ (see arguments in Theorem F4). Then, by Theorem F1, G_0 is a generalized inverse of S.   □

<u>Theorem F7.</u> *Let K be the $r_0 \times r_0$ submatrix composed of the first r_0 rows and columns of S. Then, if K is nonsingular,*

$$G_0 = \begin{vmatrix} K^{-1} & 0 \\ 0 & 0 \end{vmatrix}$$

is a generalized inverse of S.

(Note: If a different set of r_0 linearly independent rows and columns is used, the theorem can still be applied. First the rows and columns are permuted so that the first r_0 rows and columns are the selected ones. Then the theorem is applied and, finally, the rows and columns are repermuted.)

<u>Proof.</u> Letting

$$S = \begin{vmatrix} K & L \\ L^t & M \end{vmatrix},$$

we get

$$SG_0 S = \begin{vmatrix} K & L \\ L^t & L^t K^{-1} L \end{vmatrix}.$$

Therefore, we must prove that $L^t K^{-1} L = M$. To do this, consider

$$F = \begin{vmatrix} K^{-1} & 0 \\ -L^t K^{-1} & I \end{vmatrix} \begin{vmatrix} K & L \\ L^t & M \end{vmatrix} \begin{vmatrix} K^{-1} & -K^{-1}L \\ 0 & I \end{vmatrix} = \begin{vmatrix} K^{-1} & 0 \\ 0 & M - L^t K^{-1} L \end{vmatrix}.$$

The first r_0 rows of F are linearly independent because K^{-1} is nonsingular. Thus, either $r(F) > r_0$ or $L^t k^{-1} L = M$. But, by Theorem M1, $r(F) \le r(S) = r_0$, so $L^t K^{-1} L = M$.   □

Theorem F8. $\underline{E}(\hat{P}_0) = G_0 SP$, $\underline{V}(\hat{P}_0) = \sigma_e^2 G_0 SG_0$. (Note that, if $G_0 SG_0 = G_0$, then $\underline{V}(\hat{P}_0) = \sigma_e^2 G_0$. Note also that, in general, $\underline{E}(\hat{P}_0)$ depends on the choice of G_0.)

Proof.

$$\begin{aligned} \underline{E}(\hat{P}_0) &= \underline{E}(G_0 Y) = G_0 \underline{E}(Y) = G_0 SP, \\ \underline{V}(\hat{P}_0) &= \underline{V}(G_0 Y) = G_0 \underline{V}(Y)G_0 = \sigma_e^2 G_0 SG_0. \end{aligned}$$

□

Theorem F9. If S is nonsingular, then

$$\underline{E}(\hat{P}_0) = P, \ \underline{V}(\hat{P}_0) = \sigma_e^2 S^{-1}.$$

Proof. Follows immediately from Theorem F8.   □

Theorem F10. Given the constraints, $C_0 \hat{P}_0 = C_0 P_0 = 0$,

$$\underline{E}(\hat{P}_0) = P, \ \underline{V}(\hat{P}_0) = \sigma_e^2 G_0.$$

Proof. By Lemma F2, $\hat{P}_0$ is invariant to G_0, so we can let $G_0 = D_0^t(D_0 SD_0^t)^{-1}D_0$ and $P = D_0^t Q$ for some D_0 and Q. Then by Theorem F8,

$$\begin{aligned} \underline{E}(\hat{P}_0) &= G_0 SP = D_0^t(D_0 SD_0^t)^{-1}D_0 SD_0^t Q = D_0^t Q = P, \\ \underline{V}(\hat{P}_0) &= \sigma_e^2 G_0 SG_0 = \sigma_e^2 D_0^t(D_0 SD_0^t)^{-1}D_0 SD_0^t(D_0 SD_0^t)^{-1}D_0 \\ &= \sigma_e^2 G_0. \end{aligned}$$

□

Lemma F4. $tr(AG_j A^t) = tr(SG_j) = r(D_j) = r_j$. This lemma applies to the case $j = 0$.

Proof. By Theorem M3,

$$\begin{aligned} tr(AG_j A^t) &= tr(A^t AG_j) = tr(SG_j) \\ &= tr[SD_j^t(D_j SD_j^t)^{-1}D_j] \\ &= tr[(D_j SD_j^t)(D_j SD_j^t)^{-1}] = r(D_j). \end{aligned}$$

If $j = 0$, then by Theorem G6, Pt. 3, $AG_0 A^t$ is invariant to the choice of G_0, so we can choose G_0 as in Theorem F6.   □

Theorem F11. $SS_{e(0)}/\sigma_e^2 \sim \chi^2_{(N-r)}$.

Proof. $SS_{e(0)}/\sigma_e^2 = Z^t(I - AG_0A^t)Z$. Therefore, by Corollary D.2, we must prove that $(I - AG_0A^t)$ is idempotent. But

$$(I - AG_0A^t)^2 = (I - AG_0A^t) - (I - AG_0A^t)AG_0A^t,$$

and the last term is zero upon expanding and applying Theorem G6, Pt. 2. Applying Lemma F4, we obtain

$$
\begin{aligned}
\nu_{e(0)} &= tr(I - AG_0A^t) = N - tr(AG_0A^t) = N - r_0, \\
\phi^2_{e(0)} &= P^t A^t(I - AG_0A^t)AP/(\nu_j\sigma_e^2) \\
&= P^t(S - SG_0S)P/(\nu_j\sigma_e^2) = 0.
\end{aligned}
$$

$\square$

Theorem F12. $KSK = K$ if and only if $K = U^t(VSU^t)^{-1}V$ for some U and V. (If K is symmetric, then $U = V$.) In particular, $G_jSG_j = G_j$, and this applies to G_0 if it is constructed as in Theorem F6.

Proof. If $K = U^t(VSU^t)^{-1}V$, then

$$KSK = U^t(VSU^t)^{-1}VSU^t(VSU^t)^{-1}V = K.$$

Conversely, assume $KSK = K$ and let $K = U^t LV$ when factored as in Theorem M4. Then $U^t LVSU^t LV = U^t LV$. Premultiplying by $L^{-1}U$ and postmultiplying by $V^t L^{-1}$, we have $VSU^t = L^{-1}$, or $L = (VSU^t)^{-1}$. (The inverse exists because, by definition, L is nonsingular.) Then $K = U^t(VSU^t)^{-1}V$. $\square$

Theorem F13. $C_j P$ is estimable if and only if $C_j G_0 S = C_j$. (Note that by Theorem F5, $C_0 P$ is not estimable.)

Proof. Let $\hat{P}_0$ and $\hat{P}_0^*$ be any two solutions to $SP = Y$ (see Theorem F1). Then, by Theorem G2,

$$
\begin{aligned}
C_j \hat{P}_0^* &= C_j[\hat{P}_0 + (G_0 S - I)U] \\
&= C_j \hat{P}_0 + C_j(G_0 S - I)U,
\end{aligned}
$$

and $C_j P$ is estimable if and only if $C_j(G_0 S - I)U = 0$ for all U, i.e., if and only if $C_j G_0 S = C_j$.

$\square$

Theorem F14. $C_j P$ is estimable if and only if $C_j = F_j A$ for some matrix F_j.

Proof. By Theorem F13, $C_j P$ is estimable if and only if

$$C_j = C_j G_0 S = C_j G_0 A^t A = F_j A,$$

where $F_j = C_j G_0 A^t$.

Conversely, if $C_j = F_j A$ for some F_j, then by Theorem G6, Pt. 2,

$$C_j G_0 S = F_j A G_0 A^t A = F_j A.$$

□

Theorem F15. *If $C_j P$ is estimable, then $C_j \hat{P}_0$ is the best linear unbiased estimator of $C_j P$ (defining "best" as the estimator whose individual elements have minimum variances).*

Proof. $C_j \hat{P}_0 = C_j G_0 Y = C_j G_0 A^t X$, which is linear in X. By Theorems F8 and F14, $\underline{E}(C_j \hat{P}_0) = C_j G_0 SP = C_j P$, so the estimate is unbiased. Finally, let KX be any linear unbiased estimator of $C_j P$. Then

$$
\begin{aligned}
\underline{V}(KX) &= \underline{V}[(KX - C_j \hat{P}_0) + C_j \hat{P}_0] \\
&= \underline{V}(KX - C_j \hat{P}_0) + \underline{V}(C_j \hat{P}_0) + 2\underline{C}(KX - C_j \hat{P}_0, \ C_j \hat{P}_0).
\end{aligned}
$$

KX is unbiased, so $K\underline{E}(X) = KAP = C_j P$ for all P, and therefore, $KA = C_j$. Thus, the last term is

$$
\begin{aligned}
2\underline{C}(KX - C_j \hat{P}_0, \ C_j \hat{P}_0) &= 2\underline{C}(KX, \ C_j \hat{P}_0) - 2\underline{V}(C_j \hat{P}_0) \\
&= 2\underline{C}(KX, \ C_j G_0 A^t X) - 2\underline{V}(C_j G_0 A^t X) \\
&= 2\sigma_e^2 K A G_0 C_j^t - 2\sigma_e^2 C_j G_0 S G_0 C_j^t \\
&= 2\sigma_e^2 C_j G_0 C_j^t - 2\sigma_e^2 C_j G_0 C_j^t = 0.
\end{aligned}
$$

Therefore,

$$\underline{V}(KX) = \underline{V}(KX - C_j \hat{P}_0) + \underline{V}(C_j \hat{P}_0).$$

The variances are the elements on the diagonal; because covariance matrices are positive semidefinite, the minimum is found by making the first quantity zero, i.e., by setting $KX = C_j \hat{P}_0$. □

Theorem F16. *If $C_j P$ is estimable, then $C_j \hat{P}_0$ is invariant to the choice of $\hat{P}_0$.*

Proof. Let $\hat{P}_0$ and $\hat{P}_0^*$ be two estimates of P. Then, by Theorems G2 and F13,

$$C_j \hat{P}_0^* = C_j[\hat{P}_0 + (G_0 S - I)K] = C_j[\hat{P}_0 + (C_j G_0 S - C_j)K] = C_j \hat{P}_0$$

□

Theorem F17. *If $C_j P$ is estimable, or if $G_0 S G_0 = G_0$, then*

$$\underline{C}(C_j \hat{P}_0, \ C_k \hat{P}_0) = \sigma_e^2 C_j G_0 C_k^t.$$

(Note that this theorem includes the case $C_j = C_k$.)

Proof.

$$\begin{aligned}
\underline{C}(C_j\hat{P}_0,\ C_k\hat{P}_0) &= \underline{C}(C_jG_0A^tX,\ C_kG_0A^tX)\\
&= C_jG_0A^t\underline{V}(X)AG_0C_k^t\\
&= \sigma_e^2C_jG_0SG_0C_k^t\\
&= \sigma_e^2C_jG_0C_k^t.
\end{aligned}$$

$\square$

Theorem F18. *The number of linearly independent estimable functions is less than or equal to r_0.*

Proof. By Theorem F13, $C_j = F_jA$ for some F_j. Therefore, $r(C_j) \leq r_0$. But suppose we have a number of C_j, all linearly independent. Then we can combine them into one large, composite C, which also is an estimable function and therefore has rank less than or equal to r_0. $\square$

Theorem F19. *Only null hypotheses that can be expressed as estimable functions can be tested.*

Proof. See Searle (1971), pp. 193–195. $\square$

Theorem F20. $\hat{P}_j$ *is normally distributed with* $\underline{E}(\hat{P}_j) = G_jSP$, $\underline{V}(\hat{P}_j) = \sigma_e^2G_j$.

Proof. $\hat{P}_j$ is normally distributed because it is a linear function of X. Using Theorem F12, we get

$$\begin{aligned}
\underline{E}(\hat{P}_j) &= G_j\underline{E}(Y) = G_jSP,\\
\underline{V}(\hat{P}_j) &= G_j\underline{V}(Y)G_j = \sigma_e^2G_jSG_j = \sigma_e^2G_j.
\end{aligned}$$

$\square$

Theorem F21. *If* $C_jP = 0$, *then* $\underline{E}(\hat{P}_j) = P$.

Proof. If $C_jP = 0$, then $P = D_j^tQ_j$ for some Q_j. From Theorem F20,

$$\underline{E}(\hat{P}_j) = G_jSP = G_jSD_j^tQ_j = D_j^t(D_jSD_j^t)^{-1}D_jSD_j^tQ_j = D_j^tQ_j = P.$$

$\square$

Theorem F22. *If* $C_jP = 0$, *then* $SS_{e(j)}/\sigma_e^2 \sim \chi^2_{(N-r_j)}$.

Proof. By Corollary D2.2,

$$SS_{e(j)}/\sigma_e^2 = Z^t(I - AG_jA^t)Z$$

has a noncentral chi-square distribution if and only if $(I - AG_jA^t)$ is idempotent, but expanding the product and applying Theorem F12 yields

$$(I - AG_jA^t)^2 = (I - AG_jA^t) - (I - AG_jA^t)AG_jA^t = I - AG_jA^t.$$

Next, by Lemma F4,

$$\nu_{e(j)} = tr(I - AG_jA^t) = N - r(D_j) = N - r_j.$$

Finally, if $C_jP = 0$, then $P = D_j^tQ_j$ for some Q_j, so

$$\phi_{e(j)}^2 = Q_j^tD_j(S - SG_jS)D_j^tQ_j/(\nu_j\sigma_e^2) = 0$$

upon expanding the product and letting

$$G_j = D_j^t(D_jSD_j^t)^{-1}D_j.$$

(Note: If $P = D_j^tQ_j + M_j$, then $C_jP = C_jM_j$, and $\phi_{e(j)}^2 = M_j^t(S - SG_jS)M_j/(\nu_j\sigma_e^2)$.)

$\square$

<u>Theorem F23.</u> *If $C_jG_0S = C_j$, then $r(C_jG_0C_j^t) = r(C_j)$. (Note that if the rows of C_j are also linearly independent, then $C_jG_0C_j^t$ is nonsingular.)*

Proof. Let $S = UKU^t$ when factored as in Theorem M4. Then $r(C_jG_0U^tKU) = r(C_j)$, so by Theorem M1, $r(C_jG_0U^t) \geq r(C_j)$. But, by the same theorem, $r(C_jG_0U^t) \leq r(C_j)$. Thus, $R(C_jG_0U^t) = r(C_j)$.

Finally,

$$C_jG_0C_j^t = C_jG_0SG_0C_j^t = (C_jG_0U^t)K(C_jG_0U^t)^t,$$

and S is positive semidefinite, so by Theorem M5, K is positive definite. Therefore, by Theorem M2,

$$r(C_jG_0C_j^t) = r(C_jG_0U^t) = r(C_j).$$

$\square$

<u>Theorem F24.</u> *Let F_j be the generalized inverse of $C_jG_0C_j^t$. Then*

$$\hat{P}_j = \hat{P}_0 - G_0C_j^tF_jC_j\hat{P}_0.$$

Proof. We must minimize $(X - A\hat{P}_j)^t(X - A\hat{P}_j)$, subject to $C_j\hat{P}_j = 0$. Using a Lagrange multiplier, L, we must minimize

$$(X - A\hat{P}_j)^t(X - A\hat{P}_j) + L^tC_j\hat{P}_j.$$

Taking derivatives, we obtain the equations

$$S\hat{P}_j + C_j^tL - Y = 0, \ C_j\hat{P}_j = 0.$$

Solving the first equation, we obtain

$$S\hat{P}_j = Y - C_j^tL, \ \hat{P}_j = G_0(Y - C_j^tL) = \hat{P}_0 - G_0C_j^tL.$$

Substituting in the second yields

$$\begin{aligned} C_j\hat{P}_j &= C_j(\hat{P}_0 - G_0C_j^tL) = 0, \\ C_jG_0C_j^tL &= C_j\hat{P}_0, \\ L &= F_jC_j\hat{P}_0. \end{aligned}$$

The proof is finished by substituting this into the first equation. $\square$

Theorem F25. *Define F_j as in Theorem F24. Then*

$$SS_j = SS_{e(j)} - SS_{e(0)} = \hat{P}_0 C_j^t F_j C_j \hat{P}_0.$$

Proof. Applying Theorem F24, we get

$$\begin{aligned} SS_{e(j)} - SS_{e(0)} &= (X^t X - Y^t \hat{P}_j) - (X^t X - Y^t \hat{P}_0) \\ &= Y^t (\hat{P}_0 - \hat{P}_j) \\ &= Y^t G_0 C_j^t F_j C_j \hat{P}_0 \\ &= \hat{P}_0^t C_j^t F_j C_j \hat{P}_0. \end{aligned}$$

$\square$

Theorem F26. *Let $SS_j = Y^t(G_0 - G_j)Y$ and $SS_k = Y^t(G_0 - G_k)Y$. Then SS_j and SS_k are independent if and only if*

$$(G_0 - G_j)S(G_0 - G_k) = 0.$$

Proof. Follows directly from Theorem D3. $\square$

Lemma F5. $C_j G_j = 0.$

Proof. $C_j G_j = C_j D_j^t (D_j S D_j^t)^{-1} D_j$, and by construction, $C_j D_j^t = 0$. $\square$

Theorem F27. *Assume that either $C_j P$ is estimable or $G_0 S G_0 = G_0$. Then $C_j \hat{P}_0$ and $C_k \hat{P}_0$ are independent if and only if $(G_0 - G_j)S(G_0 - G_k) = 0$.*

Proof. If $(G_0 - G_j)S(G_0 - G_k) = 0$, then applying Lemma F5 and Theorem F13, we obtain

$$0 = C_j(G_0 - G_j)S(G_0 - G_k)C_k^t = C_j G_0 S G_0 C_k^t = C_j G_0 C_k^t,$$

and $C_j \hat{P}_0$ and $C_k \hat{P}_0$ are independent by Theorem F16.

Conversely, suppose $C_j \hat{P}_0$ and $C_k \hat{P}_0$ are independent. Then by Theorem F25, SS_j and SS_k are independent because SS_j and SS_k depend only on $C_j \hat{P}_0$ and $C_k \hat{P}_0$ respectively. However, then $(G_0 - G_j)S(G_0 - G_k) = 0$ by Theorem F26. $\square$

Lemma F6. *Given C_j, G_j is invariant to the choice of D_j.*

Proof. Let D_j be any matrix of K independent rows, such that $C_j D_j^t = 0$. Let D_j^* be another. Then $D_j^* = K D_j$ for some nonsingular matrix, K (because the row spaces of D_j and D_j^* are both orthogonal to that of C_j). Then

$$\begin{aligned} G_j^* &= D_j^{*t}(D_j^* S D_j^{*t})^{-1} D_j^* \\ &= D_j^t K^t (K D_j S D_j^t K^t)^{-1} K D_j \\ &= D_j^t K^t (K^t)^{-1} (D_j S D_j^t)^{-1} K^{-1} K D_j \\ &= G_j. \end{aligned}$$

$\square$

<u>Lemma F7.</u> *Let* $C_j D_j^t = 0$, $C_j D_k^t = 0$, $C_k D_k^t = 0$. *Then* $G_j S G_k = G_k$.

<u>Proof.</u> By Lemma F6, G_j and G_k are invariant to the choice of D_j and D_k. Let

$$D_j = \left| \begin{array}{c} D_k \\ F \end{array} \right|$$

and $G_j = G_k + F^t(FSF^t)^{-1}F$, so applying Theorem F12 yields

$$
\begin{aligned}
G_j S G_k &= [G_k + F^t(FSF^t)^{-1}F]SG_k \\
&= G_k S G_k + F^t(FSF^t)^{-1}FSD_k^t(D_k S D_k^t)^{-1}D_k \\
&= G_k S G_k = G_k.
\end{aligned}
$$

We conclude the proof by showing that F, with rank $r_j = r_k$, can be constructed. We note that $r(D_j) = r_j$, and therefore D_j is constructed to be orthogonal to a number of C_i whose combined ranks must be $m - r_j$. Let C^* be constructed so that its rows are the rows of all these C_i; that is,

$$r(C^*) = m - r_j.$$

Finally, note that $r(D_k S) \le r_k$, so

$$r(\, |C^{*t} \quad (D_k S)^t| \,) \le (m - r_j) + r_k = m - (r_j - r_k).$$

Therefore, we can construct a matrix of rank $r_j - r_k$ orthogonal to the rows of both C^* and $D_k S$. □

<u>Theorem F28.</u> *Let* $C_i D_i^t = 0$, $C_i D_j^t = 0$, $C_j D_j^t = 0$, *and* $SS_j = SS_{e(j)} - SS_{e(i)}$. *Then* SS_j/σ_e^2 *has a noncentral chi-square distribution with*

$$
\begin{aligned}
\nu_j &= r_i - r_j, \\
\phi_j^2 &= P^t S(G_i - G_j)SP/(\nu_j \sigma_e^2).
\end{aligned}
$$

This theorem includes the case $i = 0$.

<u>Proof.</u> By Corollary D2.2,

$$(SS_{e(j)} - SS_{e(i)})/\sigma_e^2 = Z^t(AG_iA^t - AG_jA^t)Z$$

has a noncentral chi-square distribution if and only if $(AG_iA^t - AG_jA^t)$ is idempotent, but applying Lemma F7 and Theorem F12, we get

$$
\begin{aligned}
(AG_iA^t - AG_jA^t)^2 &= AG_iSG_iA^t - AG_iSG_jA^t - AG_jSG_iA^t + AG_jSG_jA^t \\
&= AG_iA^t - AG_jA^t.
\end{aligned}
$$

Next, applying Lemma F4 gives

$$
\begin{aligned}
\nu_j &= tr(AG_iA^t - AG_jA^t) = tr(AG_iA^t) - tr(AG_jA^t) \\
&= r_i - r_j, \\
\phi_j^2 &= P^t A^t(AG_iA^t - AG_jA^t)AP/(\nu_j \sigma_e^2) \\
&= P^t S(G_i - G_j)SP/(\nu_j \sigma_e^2).
\end{aligned}
$$

Finally, note that if $i = 0$ then AG_0A^t is invariant to G_0, so we can assume G_0 has been constructed according to Theorem F5.   □

Theorem F29. *Define F_j as in Theorem F24, and let $SS_j = SS_{e(j)} - SS_{e(0)}$. Then if C_jP is estimable,*

$$\phi_j^2 = P^t C_j^t F_j (C_j G_0 C_j^t) F_j C_j P / (\nu_j \sigma_e^2).$$

If $F_j (C_j G_0 C_j^t) F_j = F_j$ (e.g., if $C_j G_0 C_j^t$ is nonsingular), then

$$\phi_j^2 = P^t C_j^t F_j C_j P / (\nu_j \sigma_e^2).$$

Proof. From Theorem F25,

$$
\begin{aligned}
SS_j / \sigma_e^2 &= \hat{P}_0^t C_j^t F_j C_j \hat{P}_0 / \sigma_e^2 \\
&= X^t A G_0 C_j^t F_j C_j G_0 A^t X / \sigma_e^2 \\
&= Z^t A G_0 C_j^t F_j C_j G_0 A^t Z,
\end{aligned}
$$

and the rest follows by applying Theorem D2.   □

Theorem F30. *Assume that either C_jP is estimable, or $G_0 S G_0 = G_0$. Then if $H_0(j)$ and $H_0(k)$ are orthogonal, $C_j \hat{P}_0$ and $C_k \hat{P}_0$ are independent, as are*

$$SS_j = SS_{e(j)} - SS_{e(0)} \quad and \quad SS_k = SS_{e(k)} - SS_{e(0)}.$$

Proof. From Theorem F17,

$$\underline{C}(C_j \hat{P}_0,\; C_k \hat{P}_0) = \sigma_e^2 C_j G_0 C_k^t,$$

which is zero if and only if $C_j G_0 C_k^t = 0$.

From Theorem F25, SS_j and SS_k depend entirely on $C_j \hat{P}_0$ and $C_k \hat{P}_0$, respectively. Therefore, they are independent if and only if $C_j \hat{P}_0$ and $C_k \hat{P}_0$ are independent.   □

Theorem F31. *Let $C_i D_i^t = 0$, $C_i D_j^t = 0$, $C_j D_j^t = 0$, $C_i D_k^t = 0$, $C_j D_k^t = 0$, and $C_k D_k^t = 0$. (Note: These conditions arise when hypotheses are tested sequentially.) Let*

$$
\begin{aligned}
SS_{e(i)} &= X^t (I - AG_i A^t) X, \\
SS_j &= X^t (AG_i A^t - AG_j A^t) X, \\
SS_k &= X^t (AG_j A^t - AG_k A^t) X.
\end{aligned}
$$

Then these sums of squares are all statistically independent. Moreover, this theorem includes the case $i = 0$.

Proof. By Corollary D3.2, $X^t K X$ and $X^t L X$ are independent if and only if $KL = 0$. Using Theorem M6, Pt. 2, Lemma F3, and Theorem F12, we find

(1) $SS_{e(i)}$ and SS_k are independent because

$$(I-AG_iA^t)(AG_jA^t-AG_kA^t) = AG_jA^t-AG_iSG_jA^t-AG_kA^t+AG_iSG_kA^t = 0,$$

(2) $SS_{e(i)}$ and SS_j are independent by a similar argument, and

(3) SS_j and SS_k are independent because

$$(AG_iA^t-AG_jA^t)(AG_jA^t-AG_kA^t) = AG_iSG_jA^t-AG_jSG_jA^t-AG_iSG_kA^t$$

$$+AG_jSG_kA^t = 0.$$

Finally, if $i = 0$, the arguments of Theorem F27 apply.   □

Theorems Concerning Random Parameters

ASSUMPTIONS

In the following theorems, we assume P is a random variable having a multivariate normal distribution with $\underline{E}(P) = M$, $\underline{V}(P) = V$, and E and P are independent. We also assume that M is unknown so no specific restrictions can be placed on M when applying Theorems D2 and D3. Finally, we let $SS_j = X^tAH_jAX = Y^tH_jY$, where $H_j = G_0 - G_j$ if the hypotheses are orthogonal, and $H_j = G_{j-1} - G_j$ if they are not. Consequently, all sums of squares would have noncentral chi-square distributions and be statistically independent if the parameters were fixed.

Note that when P is random,

$$\begin{aligned}\underline{E}(X) &= AM, \underline{V}(X) = \sigma_e^2I + AVA^t,\\ \underline{E}(Y) &= SM, \underline{V}(Y) = \sigma_e^2S + SVS,\\ \underline{E}(Z) &= (1/\sigma_e)AM, \underline{V}(Z) = I + (1/\sigma_e^2)AVA^t, \text{ where } Z = (1/\sigma_e)X.\end{aligned}$$

THEOREMS

<u>Theorem R1.</u> $SS_{e(0)}/\sigma_e^2 \sim \chi^2_{(N-r_0)}$.

<u>Proof.</u> We have given

$$SS_{e(0)}\sigma_e^2 = Z^t(I - AG_0A^t)Z.$$

By Theorem D2, we must prove that

$$(I - AG_0A^t)[I + (1/\sigma_e^2)AVA^t](I - AG_0A^t) = I - AG_0A^t.$$

But, by Theorem M6, Pt. 2,

$$(I - AG_0A^t)AVA^t = AVA^t - AG_0A^tAVA^t = AVA^t - AVA^t = 0.$$

Therefore,

$$(I - AG_0A^t)[I + (1/\sigma_e^2)AVA^t] = (I - AG_0A^t),$$

which, from the proof of Theorem F11, is idempotent.

Then, applying Lemma F4 and the above arguments yields

$$
\begin{aligned}
\nu_{e(0)} &= tr(I - AG_0A^t)[I + (1/\sigma_e^2)AVA^t] = tr(I - AG_0A^t) \\
&= N - tr(AG_0A^t) = N - r_0, \\
\phi_{e(0)}^2 &= M^tA^t(I - AG_0A^t)AM/(\nu_{e(0)}\sigma_e^2) \\
&= M^t(S - SG_0S)M/(\nu_{e(0)}\sigma_e^2) = 0.
\end{aligned}
$$

$\square$

Theorem R2. $SS_{e(0)}$ and SS_j are independent.

Proof. By Theorem D3, we must prove that

$$(I - AG_0A^t)(\sigma_e^2I + AVA^t)(AH_jA^t) = 0.$$

But this follows directly upon expanding the product and applying Theorem M6, Pt. 2, and Lemma F7. $\square$

Theorem R3. $\hat{P}_j$ is normally distributed with

$$\underline{E}(\hat{P}_j) = G_jSM, \quad \underline{V}(\hat{P}_j) = G_j(\sigma_e^2S + SVS)G_j.$$

This includes the case $j = 0$. (Note that, if $G_jSG_j = G_j$, then $\underline{V}(\hat{P}_j) = \sigma_e^2G_j + G_jSVSG_j$.)

Proof. $\hat{P}_j$ is normally distributed because it is a linear function of X:

$$
\begin{aligned}
\underline{E}(\hat{P}_j) &= G_j\underline{E}(Y) = G_jSM, \\
\underline{V}(\hat{P}_j) &= G_j\underline{V}(Y)G_j = G_j(\sigma_e^2S + SVS)G_j.
\end{aligned}
$$

$\square$

Theorem R4. If $C_jM = 0$, then $\underline{E}(\hat{P}_j) = M$.

Proof. The proof follows the same reasoning as in Theorem F21. $\square$

Theorem R5. If C_jP is estimable,

$$\underline{E}(C_j\hat{P}_0) = C_jM, \quad \underline{C}(C_j\hat{P}_0, C_k\hat{P}_0) = C_j(\sigma_e^2G_0 + V)C_k^t.$$

Proof. Applying Theorem F13 and R3, we get

$$
\begin{aligned}
\underline{E}(C_j\hat{P}_0) &= C_jG_0SM = C_jM, \\
\underline{C}(C_j\hat{P}_0, C_k\hat{P}_0) &= C_jG_0(\sigma_e^2S + SVS)G_0C_k^t = C_j(\sigma_e^2G_0 + V)C_k^t.
\end{aligned}
$$

$\square$

Lemma R1. *Let C_i, C_j, C_k, D_i, D_j, and D_k be as in Theorem F31. Let $H_j = G_i - G_j$, $H_k = G_j - G_k$. Then $H_j S H_j = H_j$ and $H_j S H_k = 0$.*

Proof. The proof follows directly from Lemma F7.   □

Theorem R6. *SS_j / a_j has a noncentral chi-square distribution, for some a_j, with*

$$\nu_j = r_i - r_j, \quad \phi_j^2 = M^t S H_j S M / (\nu_j a_j),$$

if and only if $H_j S V S H_j = (a_j - \sigma_e^2) H_j$.

Proof.

$$SS_j = Y^t H_j Y.$$

By Theorem D2, we must have

$$[(1/a_j) H_j](\sigma_e^2 S + S V S)[(1/a_j) H_j] = (1/a_j) H_j$$

or, equivalently,

$$H_j(\sigma_e^2 S + S V S) H_j = a_j H_j.$$

Expanding and applying Lemma R1, we obtain

$$\sigma_e^2 H_j + H_j S V S H_j = a_j H_j,$$

and the rest follows directly.

Then

$$
\begin{aligned}
\nu_j &= tr[(1/a_j) H_j(\sigma_e^2 S + S V S)] \\
&= (1/a_j)[\sigma_e^2 \, tr(H_j S) + tr(H_j S V S)].
\end{aligned}
$$

But applying Theorem M3 and Lemma R1 yields

$$tr(H_j S V S) = tr(H_j S H_j S V S) = tr(H_j S V S H_j S) = (a_j - \sigma_e^2) \, tr(H_j S).$$

Therefore, by Lemma F4,

$$\nu_j = tr(H_j S) = tr(G_i S) - tr(G_j S) = r_i - r_j.$$

Next,

$$
\begin{aligned}
\phi_j^2 &= M^t S H_j(\sigma_e^2 S + S V S) H_j S M / (\nu_j a_j^2) \\
&= M^t S(\sigma_e^2 H_j S H_j + H_j S V S H_j) S M / (\nu_j a_j^2) \\
&= M^t S(a_j H_j) S M / (\nu_j a_j^2) \\
&= M^t S H_j S M / (\nu_j a_j).
\end{aligned}
$$

Finally, note that if $i = 0$, the arguments in Theorem F28 apply.   □

Theorem R7. *SS_j and SS_k are independent if and only if $H_j S V S H_k = 0$.*

<u>Proof.</u> By Theorem D3, we must have

$$H_j(\sigma_e^2 S + SVS)H_k = \sigma_e^2 H_j SH_k + H_j SVSH_k = 0,$$

but the first term is zero by Lemma R1.

Finally, note that if $i = 0$, the arguments in Theorem F28 apply.   □

Theorem R8. *Let $H_j = G_0 - G_j$. Assume $C_j P$ is estimable. Then SS_j/a_j has a noncentral chi-square distribution with ν_j and ϕ_j^2 as in Theorem R6 if and only if*

$$C_j V C_j^t = (a_j - \sigma_e^2)C_j G_0 C_j^t.$$

<u>Proof.</u> Assume that SS_j/a_j has a noncentral chi-square distribution. Then, from Theorem R6,

$$C_j(G_0 - G_j)SVS(G_0 - G_j)C_j^t = (a_j - \sigma_e^2)C_j(G_0 - G_j)C_j^t.$$

Applying Lemma F5 and Theorem F13 to both sides, we obtain

$$C_j V C_j^t = (a_j - \sigma_e^2)C_j G_0 C_j^t.$$

Conversely, assume

$$C_j V C_j^t = (a_j - \sigma_e^2)C_j G_0 C_j^t.$$

Then, by Theorem D3, we must prove that

$$(1/a_j^2)H_j \underline{V}(Y)H_j = (1/a_j)H_j.$$

But by Lemma R1, this reduces to

$$(G_0 - G_j)SVS(G_0 - G_j) = a_j(G_0 - G_j).$$

Finally, comparing the definition of SS_j with Theorem F25, letting $\hat{P}_0 = G_0 Y = G_0 A^t X$, and noting that X can be any arbitrary vector, we get

$$A(G_0 - G_j)A^t = AG_0 C_j^t F_j C_j G_0 A^t.$$

Combining these facts, we must prove that

$$(G_0 C_j^t F_j C_j G_0)SVS(G_0 C_j^t F_j C_j G_0) = (a_j - \sigma_e^2)G_0 C_j^t F_j C_j G_0.$$

But this is readily proved by applying the identities

$$C_j V C_j^t = (a_j - \sigma_e^2)G_j G_0 C_j^t, \ C_j G_0 S = C_j.$$

□

Theorem R9. *Let $H_j = G_0 - G_j$, $H_k = G_0 - G_k$. Then SS_j and SS_k are independent if and only if*

$$C_j G_0(\sigma_e^2 S + SVS)G_0 C_k^t = 0.$$

If $C_j P$ is estimable, this reduces to

$$C_j(\sigma_e^2 G_0 + V)C_k^t = 0.$$

<u>Proof.</u> Assume $C_j G_0(\sigma_e^2 S + SVS)G_0 C_k^t = 0$. Then, by Theorem R3, $C_j \hat{P}_0$ and $C_k \hat{P}_0$ are independent. But by Theorem F25, SS_j and SS_k depend only on these quantities, so they are independent too.

Conversely, assume that SS_j and SS_k are independent. Then, applying Theorem D3, we get

$$
\begin{aligned}
0 &= C_j(G_0 - G_j)(\sigma_e^2 S + SVS)(G_0 - G_k) \\
  &= C_j G_0(\sigma_e^2 S + SVS)G_0 C_k^t.
\end{aligned}
$$

Finally, if $C_j P$ is estimable, we can apply Theorem F13 to the above formulas. □

Appendix B

Tables

TABLE B.1. Upper-tail significance levels of the standard normal distribution.

z	0	1	2	3	4	5	6	7	8	9
0.00	.5000	.4960	.4920	.4880	.4840	.4801	.4761	.4721	.4681	.4641
0.10	.4602	.4562	.4522	.4483	.4443	.4404	.4364	.4325	.4286	.4247
0.20	.4207	.4168	.4129	.4090	.4052	.4013	.3974	.3936	.3897	.3859
0.30	.3821	.3783	.3745	.3707	.3669	.3632	.3594	.3557	.3520	.3483
0.40	.3446	.3409	.3372	.3336	.3300	.3264	.3228	.3192	.3156	.3121
0.50	.3085	.3050	.3015	.2981	.2946	.2912	.2877	.2843	.2810	.2776
0.60	.2743	.2709	.2676	.2643	.2611	.2578	.2546	.2514	.2483	.2451
0.70	.2420	.2389	.2358	.2327	.2297	.2266	.2236	.2206	.2177	.2148
0.80	.2119	.2090	.2061	.2033	.2005	.1977	.1949	.1922	.1894	.1867
0.90	.1841	.1814	.1788	.1762	.1736	.1711	.1685	.1660	.1635	.1611
1.00	.1587	.1562	.1539	.1515	.1492	.1469	.1446	.1423	.1401	.1379
1.10	.1357	.1335	.1314	.1292	.1271	.1251	.1230	.1210	.1190	.1170
1.20	.1151	.1131	.1112	.1093	.1075	.1056	.1038	.1020	.1003	.0985
1.30	.0968	.0951	.0934	.0918	.0901	.0885	.0869	.0853	.0838	.0823
1.40	.0808	.0793	.0778	.0764	.0749	.0735	.0721	.0708	.0694	.0681
1.50	.0668	.0655	.0643	.0630	.0618	.0606	.0594	.0582	.0571	.0559
1.60	.0548	.0537	.0526	.0516	.0505	.0495	.0485	.0475	.0465	.0455
1.70	.0446	.0436	.0427	.0418	.0409	.0401	.0392	.0384	.0375	.0367
1.80	.0359	.0351	.0344	.0336	.0329	.0322	.0314	.0307	.0301	.0294
1.90	.0287	.0281	.0274	.0268	.0262	.0256	.0250	.0244	.0239	.0233
2.00	.0227	.0222	.0217	.0212	.0207	.0202	.0197	.0192	.0188	.0183
2.10	.0179	.0174	.0170	.0166	.0162	.0158	.0154	.0150	.0146	.0143
2.20	.0139	.0136	.0132	.0129	.0125	.0122	.0119	.0116	.0113	.0110
2.30	.0107	.0104	.0102	.0099	.0096	.0094	.0091	.0089	.0087	.0084
2.40	.0082	.0080	.0078	.0075	.0073	.0071	.0069	.0068	.0066	.0064
2.50	.0062	.0060	.0059	.0057	.0055	.0054	.0052	.0051	.0049	.0048
2.60	.0047	.0045	.0044	.0043	.0041	.0040	.0039	.0038	.0037	.0036
2.70	.0035	.0034	.0033	.0032	.0031	.0030	.0029	.0028	.0027	.0026
2.80	.0026	.0025	.0024	.0023	.0023	.0022	.0021	.0021	.0020	.0019
2.90	.0019	.0018	.0018	.0017	.0016	.0016	.0015	.0015	.0014	.0014

TABLE B.1. *Continued*

3.00	.0013	.0013	.0013	.0012	.0012	.0011	.0011	.0011	.0010	.0010
3.10	.0010	.0009	.0009	.0009	.0008	.0008	.0008	.0008	.0007	.0007
3.20	.0007	.0007	.0006	.0006	.0006	.0006	.0006	.0005	.0005	.0005
3.30	.0005	.0005	.0005	.0004	.0004	.0004	.0004	.0004	.0004	.0003
3.40	.0003	.0003	.0003	.0003	.0003	.0003	.0003	.0003	.0003	.0002
3.50	.0002	.0002	.0002	.0002	.0002	.0002	.0002	.0002	.0002	.0002
3.60	.0002	.0002	.0001	.0001	.0001	.0001	.0001	.0001	.0001	.0001
3.70	.0001	.0001	.0001	.0001	.0001	.0001	.0001	.0001	.0001	.0001
3.80	.0001	.0001	.0001	.0001	.0001	.0001	.0001	.0001	.0001	.0001

TABLE B.2. Values of the standard normal distribution for selected two-tailed significance levels.

	Significance Level									
	0	1	2	3	4	5	6	7	8	9
.000		3.891	3.719	3.616	3.540	3.481	3.432	3.390	3.353	3.320
.001	3.290	3.264	3.239	3.216	3.195	3.175	3.156	3.138	3.121	3.105
.002	3.090	3.076	3.062	3.048	3.036	3.023	3.011	3.000	2.989	2.978
.003	2.968	2.958	2.948	2.938	2.929	2.920	2.911	2.903	2.894	2.886
.004	2.878	2.870	2.863	2.855	2.848	2.841	2.834	2.827	2.820	2.814
.005	2.807	2.801	2.794	2.788	2.782	2.776	2.770	2.765	2.759	2.753
.006	2.748	2.742	2.737	2.732	2.727	2.721	2.716	2.711	2.706	2.702
.007	2.697	2.692	2.687	2.683	2.678	2.674	2.669	2.665	2.661	2.656
.008	2.652	2.648	2.644	2.640	2.636	2.632	2.628	2.624	2.620	2.616
.009	2.612	2.608	2.605	2.601	2.597	2.594	2.590	2.586	2.583	2.579
.01	2.576	2.543	2.512	2.484	2.457	2.432	2.409	2.387	2.366	2.346
.02	2.326	2.308	2.290	2.273	2.257	2.241	2.226	2.212	2.197	2.183
.03	2.170	2.157	2.144	2.132	2.120	2.108	2.097	2.086	2.075	2.064
.04	2.054	2.044	2.034	2.024	2.014	2.005	1.995	1.986	1.977	1.969
.05	1.960	1.951	1.943	1.935	1.927	1.919	1.911	1.903	1.896	1.888
.06	1.881	1.873	1.866	1.859	1.852	1.845	1.838	1.832	1.825	1.818
.07	1.812	1.805	1.799	1.793	1.787	1.780	1.774	1.768	1.762	1.757
.08	1.751	1.745	1.739	1.734	1.728	1.722	1.717	1.711	1.706	1.701
.09	1.695	1.690	1.685	1.680	1.675	1.670	1.665	1.660	1.655	1.650
.1	1.645	1.598	1.555	1.514	1.476	1.440	1.405	1.372	1.341	1.311
.2	1.282	1.254	1.227	1.200	1.175	1.150	1.126	1.103	1.080	1.058
.3	1.036	1.015	0.994	0.974	0.954	0.935	0.915	0.896	0.878	0.860
.4	0.842	0.824	0.806	0.789	0.772	0.755	0.739	0.722	0.706	0.690
.5	0.674	0.659	0.643	0.628	0.613	0.598	0.583	0.568	0.553	0.539
.6	0.524	0.510	0.496	0.482	0.468	0.454	0.440	0.426	0.412	0.399
.7	0.385	0.372	0.358	0.345	0.332	0.319	0.305	0.292	0.279	0.266
.8	0.253	0.240	0.228	0.215	0.202	0.189	0.176	0.164	0.151	0.138
.9	0.126	0.113	0.100	0.088	0.075	0.063	0.050	0.038	0.025	0.013

TABLE B.3. Values of the chi-square distribution for selected one-tailed significance levels.

Significance Levels

DF	.9990	.9975	.9950	.9900	.9750	.9500	.9000	.7500	.5000
1	.0000	.0000	.0000	.0002	.0010	.0039	.0158	.1015	.4549
2	.0020	.0050	.0100	.0201	.0506	.1026	.2107	.5754	1.386
3	.0243	.0449	.0717	.1148	.2158	.3518	.5844	1.213	2.366
4	.0908	.1449	.2070	.2971	.4844	.7107	1.064	1.923	3.357
5	.2102	.3075	.4117	.5543	.8312	1.145	1.610	2.675	4.351
6	.3811	.5266	.6757	.8721	1.237	1.635	2.204	3.455	5.348
7	.5985	.7945	.9893	1.239	1.690	2.167	2.833	4.255	6.346
8	.8571	1.104	1.344	1.646	2.180	2.733	3.490	5.071	7.344
9	1.152	1.450	1.735	2.088	2.700	3.325	4.168	5.899	8.343
10	1.479	1.827	2.156	2.558	3.247	3.940	4.865	6.737	9.342
11	1.834	2.232	2.603	3.053	3.816	4.575	5.578	7.584	10.34
12	2.214	2.661	3.074	3.571	4.404	5.226	6.304	8.438	11.34
13	2.617	3.112	3.565	4.107	5.009	5.892	7.042	9.299	12.34
14	3.041	3.582	4.075	4.660	5.629	6.571	7.790	10.17	13.34
15	3.483	4.070	4.601	5.229	6.262	7.261	8.547	11.04	14.34
16	3.942	4.573	5.142	5.812	6.908	7.962	9.312	11.91	15.34
17	4.416	5.092	5.697	6.408	7.564	8.672	10.09	12.79	16.34
18	4.905	5.623	6.265	7.015	8.231	9.390	10.86	13.68	17.34
19	5.407	6.167	6.844	7.633	8.907	10.12	11.65	14.56	18.34
20	5.921	6.723	7.434	8.260	9.591	10.85	12.44	15.45	19.34
21	6.447	7.289	8.034	8.897	10.28	11.59	13.24	16.34	20.34
22	6.983	7.865	8.643	9.542	10.98	12.34	14.04	17.24	21.34
23	7.529	8.450	9.260	10.20	11.69	13.09	14.85	18.14	22.34
24	8.085	9.044	9.886	10.86	12.40	13.85	15.66	19.04	23.34
25	8.649	9.646	10.52	11.52	13.12	14.61	16.47	19.94	24.34
26	9.222	10.26	11.16	12.20	13.84	15.38	17.29	20.84	25.34
27	9.803	10.87	11.81	12.88	14.57	16.15	18.11	21.75	26.34
28	10.39	11.50	12.46	13.56	15.31	16.93	18.94	22.66	27.34
29	10.99	12.13	13.12	14.26	16.05	17.71	19.77	23.57	28.34
30	11.59	12.76	13.79	14.95	16.79	18.49	20.60	24.48	29.34
40	17.92	19.42	20.71	22.16	24.43	26.51	29.05	33.66	39.34
60	31.74	33.79	35.53	37.48	40.48	43.19	46.46	52.29	59.33
120	77.75	81.07	83.85	86.92	91.57	95.70	100.6	109.2	119.3

TABLE B.3. *Continued*

Significance Levels

DF	.2500	.1000	.0500	.0250	.0100	.0050	.0025	.0010
1	1.323	2.706	3.841	5.024	6.635	7.879	9.141	10.83
2	2.773	4.605	5.991	7.378	9.210	10.60	11.98	13.82
3	4.108	6.251	7.815	9.348	11.34	12.84	14.32	16.27
4	5.385	7.779	9.488	11.14	13.28	14.86	16.42	18.47
5	6.626	9.236	11.07	12.83	15.09	16.75	18.39	20.52
6	7.841	10.64	12.59	14.45	16.81	18.55	20.25	22.46
7	9.037	12.02	14.07	16.01	18.48	20.28	22.04	24.32
8	10.22	13.36	15.51	17.53	20.09	21.95	23.77	26.12
9	11.39	14.68	16.92	19.02	21.67	23.59	25.46	27.88
10	12.55	15.99	18.31	20.48	23.21	25.19	27.11	29.59
11	13.70	17.28	19.68	21.92	24.72	26.76	28.73	31.26
12	14.85	18.55	21.03	23.34	26.22	28.30	30.32	32.91
13	15.98	19.81	22.36	24.74	27.69	29.82	31.88	34.53
14	17.12	21.06	23.68	26.12	29.14	31.32	33.43	36.12
15	18.25	22.31	25.00	27.49	30.58	32.80	34.95	37.70
16	19.37	23.54	26.30	28.85	32.00	34.27	36.46	39.25
17	20.49	24.77	27.59	30.19	33.41	35.72	37.95	40.79
18	21.60	25.99	28.87	31.53	34.81	37.16	39.42	42.31
19	22.72	27.20	30.14	32.85	36.19	38.58	40.88	43.82
20	23.83	28.41	31.41	34.17	37.57	40.00	42.34	45.31
21	24.93	29.62	32.67	35.48	38.93	41.40	43.78	46.80
22	26.04	30.81	33.92	36.78	40.29	42.80	45.20	48.27
23	27.14	32.01	35.17	38.08	41.64	44.18	46.62	49.73
24	28.24	33.20	36.42	39.36	42.98	45.56	48.03	51.18
25	29.34	34.38	37.65	40.65	44.31	46.93	49.44	52.62
26	30.43	35.56	38.89	41.92	45.64	48.29	50.83	54.05
27	31.53	36.74	40.11	43.19	46.96	49.64	52.22	55.48
28	32.62	37.92	41.34	44.46	48.28	50.99	53.59	56.89
29	33.71	39.09	42.56	45.72	49.59	52.34	54.97	58.30
30	34.80	40.26	43.77	46.98	50.89	53.67	56.33	59.70
40	45.62	51.81	55.76	59.34	63.69	66.77	69.70	73.40
60	66.98	74.40	79.08	83.30	88.38	91.95	95.34	99.60
120	130.1	140.2	146.6	152.2	158.9	163.6	168.1	173.6

TABLE B.4. Values of the t distribution for selected two-tailed significance levels.

Significance Levels

df	.500	.200	.100	.050	.020	.010	.005	.002	.001
1	1.000	3.078	6.314	12.71	31.82	63.66	127.3	318.3	636.6
2	.8165	1.886	2.920	4.303	6.965	9.925	14.09	22.33	31.60
3	.7649	1.638	2.353	3.182	4.541	5.841	7.453	10.21	12.92
4	.7407	1.533	2.132	2.776	3.747	4.604	5.598	7.173	8.610
5	.7267	1.476	2.015	2.571	3.365	4.032	4.773	5.893	6.869
6	.7176	1.440	1.943	2.447	3.143	3.707	4.317	5.208	5.959
7	.7111	1.415	1.895	2.365	2.998	3.499	4.029	4.785	5.408
8	.7064	1.397	1.860	2.306	2.896	3.355	3.833	4.501	5.041
9	.7027	1.383	1.833	2.262	2.821	3.250	3.690	4.297	4.781
10	.6998	1.372	1.812	2.228	2.764	3.169	3.581	4.144	4.587
11	.6974	1.363	1.796	2.201	2.718	3.106	3.497	4.025	4.437
12	.6955	1.356	1.782	2.179	2.681	3.055	3.428	3.930	4.318
13	.6938	1.350	1.771	2.160	2.650	3.012	3.372	3.852	4.221
14	.6924	1.345	1.761	2.145	2.624	2.977	3.326	3.787	4.140
15	.6912	1.341	1.753	2.131	2.602	2.947	3.286	3.733	4.073
16	.6901	1.337	1.746	2.120	2.583	2.921	3.252	3.686	4.015
17	.6892	1.333	1.740	2.110	2.567	2.898	3.222	3.646	3.965
18	.6884	1.330	1.734	2.101	2.552	2.878	3.197	3.610	3.922
19	.6876	1.328	1.729	2.093	2.539	2.861	3.174	3.579	3.883
20	.6870	1.325	1.725	2.086	2.528	2.845	3.153	3.552	3.850
21	.6864	1.323	1.721	2.080	2.518	2.831	3.135	3.527	3.819
22	.6858	1.321	1.717	2.074	2.508	2.819	3.119	3.505	3.792
23	.6853	1.319	1.714	2.069	2.500	2.807	3.104	3.485	3.768
24	.6848	1.318	1.711	2.064	2.492	2.797	3.091	3.467	3.745
25	.6844	1.316	1.708	2.060	2.485	2.787	3.078	3.450	3.725
26	.6840	1.315	1.706	2.056	2.479	2.779	3.067	3.435	3.707
27	.6837	1.314	1.703	2.052	2.473	2.771	3.057	3.421	3.690
28	.6834	1.313	1.701	2.048	2.467	2.763	3.047	3.408	3.674
29	.6830	1.311	1.699	2.045	2.462	2.756	3.038	3.396	3.659
30	.6828	1.310	1.697	2.042	2.457	2.750	3.030	3.385	3.646
40	.6807	1.303	1.684	2.021	2.423	2.704	2.971	3.307	3.551
60	.6786	1.296	1.671	2.000	2.390	2.660	2.915	3.232	3.460
120	.6765	1.289	1.658	1.980	2.358	2.617	2.860	3.160	3.373
INF	.6745	1.282	1.645	1.960	2.326	2.576	2.807	3.090	3.290

TABLE B.5. Values of the F distribution for selected one-tailed significance levels.

df						Significance Level			
Num	Den	.5000	.2500	.1000	.0500	.0250	.0100	.0050	.0025	.0010
1	1	1.000	5.828	39.86	161.4	647.8	4052	16211	64845	
1	2	.6667	2.571	8.526	18.51	38.51	98.50	198.5	398.5	998.5
1	3	.5851	2.024	5.538	10.13	17.44	34.12	55.55	89.58	167.0
1	4	.5486	1.807	4.545	7.709	12.22	21.20	31.33	45.67	74.14
1	5	.5281	1.692	4.060	6.608	10.01	16.26	22.78	31.41	47.18
1	6	.5149	1.621	3.776	5.987	8.813	13.75	18.63	24.81	35.51
1	8	.4990	1.538	3.458	5.318	7.571	11.26	14.69	18.78	25.41
1	10	.4897	1.491	3.285	4.965	6.937	10.04	12.83	16.04	21.04
1	12	.4837	1.461	3.177	4.747	6.554	9.330	11.75	14.49	18.64
1	15	.4778	1.432	3.073	4.543	6.200	8.683	10.80	13.13	16.59
1	20	.4719	1.404	2.975	4.351	5.871	8.096	9.944	11.94	14.82
1	24	.4690	1.390	2.927	4.260	5.717	7.823	9.551	11.40	14.03
1	30	.4662	1.376	2.881	4.171	5.568	7.562	9.180	10.89	13.29
1	40	.4633	1.363	2.835	4.085	5.424	7.314	8.828	10.41	12.61
1	60	.4605	1.349	2.791	4.001	5.286	7.077	8.495	9.962	11.97
1	120	.4577	1.336	2.748	3.920	5.152	6.851	8.179	9.539	11.38
1	INF	.4549	1.323	2.706	3.841	5.024	6.635	7.879	9.141	10.83
2	1	1.500	7.500	49.50	199.5	799.5	4999	20000	79999	
2	2	1.000	3.000	9.000	19.00	39.00	99.00	199.0	399.0	999.0
2	3	.8811	2.280	5.462	9.552	16.04	30.82	49.80	79.93	148.5
2	4	.8284	2.000	4.325	6.944	10.65	18.00	26.28	38.00	61.25
2	5	.7988	1.853	3.780	5.786	8.434	13.27	18.31	24.96	37.12
2	6	.7798	1.762	3.463	5.143	7.260	10.92	14.54	19.10	27.00
2	8	.7568	1.657	3.113	4.459	6.059	8.649	11.04	13.89	18.49
2	10	.7435	1.598	2.924	4.103	5.456	7.559	9.427	11.57	14.91
2	12	.7348	1.560	2.807	3.885	5.096	6.927	8.510	10.29	12.97
2	15	.7262	1.523	2.695	3.682	4.765	6.359	7.701	9.173	11.34
2	20	.7177	1.487	2.589	3.493	4.461	5.849	6.986	8.206	9.953
2	24	.7136	1.470	2.538	3.403	4.319	5.614	6.661	7.771	9.339
2	30	.7094	1.452	2.489	3.316	4.182	5.390	6.355	7.365	8.773
2	40	.7053	1.435	2.440	3.232	4.051	5.179	6.066	6.986	8.251
2	60	.7012	1.419	2.393	3.150	3.925	4.977	5.795	6.632	7.768
2	120	.6972	1.402	2.347	3.072	3.805	4.787	5.539	6.301	7.321
2	INF	.6931	1.386	2.303	2.996	3.689	4.605	5.298	5.991	6.908

TABLE B.5. *Continued*

df		Significance Level							
Num	Den	.5000	.2500	.1000	.0500	.0250	.0100	.0050	.0025	.0010
3	1	1.709	8.200	53.59	215.7	864.2	5403	21615	86460	
3	2	1.135	3.153	9.162	19.16	39.17	99.17	199.2	399.2	999.2
3	3	1.000	2.356	5.391	9.277	15.44	29.46	47.47	76.06	141.1
3	4	.9405	2.047	4.191	6.591	9.979	16.69	24.26	34.96	56.18
3	5	.9071	1.884	3.619	5.409	7.764	12.06	16.53	22.43	33.20
3	6	.8858	1.784	3.289	4.757	6.599	9.780	12.92	16.87	23.70
3	8	.8600	1.668	2.924	4.066	5.416	7.591	9.596	11.98	15.83
3	10	.8451	1.603	2.728	3.708	4.826	6.552	8.081	9.833	12.55
3	12	.8353	1.561	2.606	3.490	4.474	5.953	7.226	8.652	10.80
3	15	.8257	1.520	2.490	3.287	4.153	5.417	6.476	7.634	9.335
3	20	.8162	1.481	2.380	3.098	3.859	4.938	5.818	6.757	8.098
3	24	.8115	1.462	2.327	3.009	3.721	4.718	5.519	6.364	7.554
3	30	.8069	1.443	2.276	2.922	3.589	4.510	5.239	5.999	7.054
3	40	.8023	1.424	2.226	2.839	3.463	4.313	4.976	5.659	6.595
3	60	.7977	1.405	2.177	2.758	3.343	4.126	4.729	5.343	6.171
3	120	.7932	1.387	2.130	2.680	3.227	3.949	4.497	5.048	5.781
3	INF	.7887	1.369	2.084	2.605	3.116	3.782	4.279	4.773	5.422
4	1	1.823	8.581	55.83	224.6	899.6	5625	22500	90000	
4	2	1.207	3.232	9.243	19.25	39.25	99.25	199.2	399.2	999.2
4	3	1.063	2.390	5.343	9.117	15.10	28.71	46.19	73.95	137.1
4	4	1.000	2.064	4.107	6.388	9.605	15.98	23.15	33.30	53.44
4	5	.9646	1.893	3.520	5.192	7.388	11.39	15.56	21.05	31.09
4	6	.9419	1.787	3.181	4.534	6.227	9.148	12.03	15.65	21.92
4	8	.9146	1.664	2.806	3.838	5.053	7.006	8.805	10.94	14.39
4	10	.8988	1.595	2.605	3.478	4.468	5.994	7.343	8.888	11.28
4	12	.8885	1.550	2.480	3.259	4.121	5.412	6.521	7.762	9.633
4	15	.8783	1.507	2.361	3.056	3.804	4.893	5.803	6.796	8.253
4	20	.8683	1.465	2.249	2.866	3.515	4.431	5.174	5.967	7.096
4	24	.8633	1.445	2.195	2.776	3.379	4.218	4.890	5.596	6.589
4	30	.8584	1.424	2.142	2.690	3.250	4.018	4.623	5.253	6.125
4	40	.8536	1.404	2.091	2.606	3.126	3.828	4.374	4.934	5.698
4	60	.8487	1.385	2.041	2.525	3.008	3.649	4.140	4.637	5.307
4	120	.8439	1.365	1.992	2.447	2.894	3.480	3.921	4.362	4.947
4	INF	.8392	1.346	1.945	2.372	2.786	3.319	3.715	4.106	4.617

TABLE B.5. *Continued*

df		Significance Level							
Num	Den	.5000	.2500	.1000	.0500	.0250	.0100	.0050	.0025	.0010
5	1	1.894	8.820	57.24	230.2	921.8	5764	23056	92224	
5	2	1.252	3.280	9.293	19.30	39.30	99.30	199.3	399.3	999.3
5	3	1.102	2.409	5.309	9.013	14.88	28.24	45.39	72.62	134.6
5	4	1.037	2.072	4.051	6.256	9.364	15.52	22.46	32.26	51.71
5	5	1.000	1.895	3.453	5.050	7.146	10.97	14.94	20.18	29.75
5	6	.9765	1.785	3.108	4.387	5.988	8.746	11.46	14.88	20.80
5	8	.9483	1.658	2.726	3.687	4.817	6.632	8.302	10.28	13.48
5	10	.9319	1.585	2.522	3.326	4.236	5.636	6.872	8.288	10.48
5	12	.9212	1.539	2.394	3.106	3.891	5.064	6.071	7.196	8.892
5	15	.9107	1.494	2.273	2.901	3.576	4.556	5.372	6.263	7.567
5	20	.9004	1.450	2.158	2.711	3.289	4.103	4.762	5.463	6.461
5	24	.8953	1.428	2.103	2.621	3.155	3.895	4.486	5.106	5.977
5	30	.8902	1.407	2.049	2.534	3.026	3.699	4.228	4.776	5.534
5	40	.8852	1.386	1.997	2.449	2.904	3.514	3.986	4.470	5.128
5	60	.8802	1.366	1.946	2.368	2.786	3.339	3.760	4.185	4.757
5	120	.8752	1.345	1.896	2.290	2.674	3.174	3.548	3.922	4.416
5	INF	.8703	1.325	1.847	2.214	2.567	3.017	3.350	3.677	4.103
6	1	1.942	8.983	58.20	234.0	937.1	5859	23437	93750	
6	2	1.282	3.312	9.326	19.33	39.33	99.33	199.3	399.3	999.3
6	3	1.129	2.422	5.285	8.941	14.73	27.91	44.84	71.71	132.8
6	4	1.062	2.077	4.010	6.163	9.197	15.21	21.97	31.54	50.53
6	5	1.024	1.894	3.405	4.950	6.978	10.67	14.51	19.58	28.83
6	6	1.000	1.782	3.055	4.284	5.820	8.466	11.07	14.35	20.03
6	8	.9711	1.651	2.668	3.581	4.652	6.371	7.952	9.828	12.86
6	10	.9544	1.576	2.461	3.217	4.072	5.386	6.545	7.871	9.926
6	12	.9434	1.529	2.331	2.996	3.728	4.821	5.757	6.803	8.379
6	15	.9327	1.482	2.208	2.790	3.415	4.318	5.071	5.891	7.092
6	20	.9221	1.437	2.091	2.599	3.128	3.871	4.472	5.111	6.019
6	24	.9169	1.414	2.035	2.508	2.995	3.667	4.202	4.763	5.550
6	30	.9117	1.392	1.980	2.421	2.867	3.473	3.949	4.442	5.122
6	40	.9065	1.371	1.927	2.336	2.744	3.291	3.713	4.144	4.731
6	60	.9014	1.349	1.875	2.254	2.627	3.119	3.492	3.868	4.372
6	120	.8964	1.328	1.824	2.175	2.515	2.956	3.285	3.612	4.044
6	INF	.8914	1.307	1.774	2.099	2.408	2.802	3.091	3.375	3.743

TABLE B.5. *Continued*

df		Significance Level								
Num	Den	.5000	.2500	.1000	.0500	.0250	.0100	.0050	.0025	.0010
10	1	2.042	9.320	60.19	241.9	968.6	6056	24224	96899	
10	2	1.345	3.377	9.392	19.40	39.40	99.40	199.4	399.4	999.4
10	3	1.183	2.445	5.230	8.786	14.42	27.23	43.69	69.81	129.2
10	4	1.113	2.082	3.920	5.964	8.844	14.55	20.97	30.04	48.05
10	5	1.073	1.890	3.297	4.735	6.619	10.05	13.62	18.32	26.92
10	6	1.048	1.771	2.937	4.060	5.461	7.874	10.25	13.24	18.41
10	8	1.018	1.631	2.538	3.347	4.295	5.814	7.211	8.866	11.54
10	10	1.000	1.551	2.323	2.978	3.717	4.849	5.847	6.987	8.754
10	12	.9886	1.500	2.188	2.753	3.374	4.296	5.085	5.966	7.292
10	15	.9773	1.449	2.059	2.544	3.060	3.805	4.424	5.097	6.081
10	20	.9663	1.399	1.937	2.348	2.774	3.368	3.847	4.355	5.075
10	24	.9608	1.375	1.877	2.255	2.640	3.168	3.587	4.025	4.638
10	30	.9554	1.351	1.819	2.165	2.511	2.979	3.344	3.720	4.239
10	40	.9500	1.327	1.763	2.077	2.388	2.801	3.117	3.438	3.874
10	60	.9447	1.303	1.707	1.993	2.270	2.632	2.904	3.177	3.541
10	120	.9394	1.279	1.652	1.910	2.157	2.472	2.705	2.935	3.237
10	INF	.9342	1.255	1.599	1.831	2.048	2.321	2.519	2.711	2.959
12	1	2.067	9.406	60.71	243.9	976.7	6106	24426	97707	
12	2	1.361	3.393	9.408	19.41	39.41	99.42	199.4	399.4	999.4
12	3	1.197	2.450	5.216	8.745	14.34	27.05	43.39	69.32	128.3
12	4	1.126	2.083	3.896	5.912	8.751	14.37	20.70	29.66	47.41
12	5	1.085	1.888	3.268	4.678	6.525	9.888	13.38	17.99	26.42
12	6	1.060	1.767	2.905	4.000	5.366	7.718	10.03	12.95	17.99
12	8	1.029	1.624	2.502	3.284	4.200	5.667	7.015	8.613	11.19
12	10	1.012	1.543	2.284	2.913	3.621	4.706	5.661	6.754	8.445
12	12	1.000	1.490	2.147	2.687	3.277	4.155	4.906	5.744	7.005
12	15	.9886	1.438	2.017	2.475	2.963	3.666	4.250	4.884	5.812
12	20	.9775	1.387	1.892	2.278	2.676	3.231	3.678	4.151	4.823
12	24	.9719	1.362	1.832	2.183	2.541	3.032	3.420	3.826	4.393
12	30	.9665	1.337	1.773	2.092	2.412	2.843	3.179	3.525	4.001
12	40	.9610	1.312	1.715	2.003	2.288	2.665	2.953	3.246	3.642
12	60	.9557	1.287	1.657	1.917	2.169	2.496	2.742	2.988	3.315
12	120	.9503	1.262	1.601	1.834	2.055	2.336	2.544	2.749	3.016
12	INF	.9450	1.237	1.546	1.752	1.945	2.185	2.358	2.527	2.742

TABLE B.5. *Continued*

df		Significance Level							
Num	Den	.5000	.2500	.1000	.0500	.0250	.0100	.0050	.0025	.0010
15	1	2.093	9.493	61.22	245.9	984.9	6157	24630	98522	
15	2	1.377	3.410	9.425	19.43	39.43	99.43	199.4	399.4	999.4
15	3	1.211	2.455	5.200	8.703	14.25	26.87	43.08	68.82	127.4
15	4	1.139	2.083	3.870	5.858	8.657	14.20	20.44	29.26	46.76
15	5	1.098	1.885	3.238	4.619	6.428	9.722	13.15	17.66	25.91
15	6	1.072	1.762	2.871	3.938	5.269	7.559	9.814	12.65	17.56
15	8	1.041	1.617	2.464	3.218	4.101	5.515	6.814	8.355	10.84
15	10	1.023	1.534	2.244	2.845	3.522	4.558	5.471	6.514	8.129
15	12	1.012	1.480	2.105	2.617	3.177	4.010	4.721	5.515	6.709
15	15	1.000	1.426	1.972	2.403	2.862	3.522	4.070	4.665	5.535
15	20	.9887	1.374	1.845	2.203	2.573	3.088	3.502	3.940	4.562
15	24	.9831	1.347	1.783	2.108	2.437	2.889	3.246	3.618	4.139
15	30	.9776	1.321	1.722	2.015	2.307	2.700	3.006	3.320	3.753
15	40	.9721	1.295	1.662	1.924	2.182	2.522	2.781	3.044	3.400
15	60	.9667	1.269	1.603	1.836	2.061	2.352	2.570	2.788	3.078
15	120	.9613	1.243	1.545	1.750	1.945	2.192	2.373	2.551	2.783
15	INF	.9559	1.216	1.487	1.666	1.833	2.039	2.187	2.330	2.513
20	1	2.119	9.581	61.74	248.0	993.1	6209	24836	99345	
20	2	1.393	3.426	9.441	19.45	39.45	99.45	199.4	399.4	999.4
20	3	1.225	2.460	5.184	8.660	14.17	26.69	42.78	68.31	126.4
20	4	1.152	2.083	3.844	5.803	8.560	14.02	20.17	28.86	46.10
20	5	1.111	1.882	3.207	4.558	6.329	9.553	12.90	17.32	25.39
20	6	1.084	1.757	2.836	3.874	5.168	7.396	9.589	12.35	17.12
20	8	1.053	1.609	2.425	3.150	3.999	5.359	6.608	8.089	10.48
20	10	1.035	1.523	2.201	2.774	3.419	4.405	5.274	6.267	7.804
20	12	1.023	1.468	2.060	2.544	3.073	3.858	4.530	5.279	6.405
20	15	1.011	1.413	1.924	2.328	2.756	3.372	3.883	4.438	5.248
20	20	1.000	1.358	1.794	2.124	2.464	2.938	3.318	3.720	4.290
20	24	.9944	1.331	1.730	2.027	2.327	2.738	3.062	3.401	3.873
20	30	.9888	1.303	1.667	1.932	2.195	2.549	2.823	3.105	3.493
20	40	.9832	1.276	1.605	1.839	2.068	2.369	2.598	2.831	3.145
20	60	.9777	1.248	1.543	1.748	1.944	2.198	2.387	2.576	2.827
20	120	.9723	1.220	1.482	1.659	1.825	2.035	2.188	2.339	2.534
20	INF	.9669	1.191	1.421	1.571	1.708	1.878	2.000	2.117	2.266

TABLE B.5. *Continued*

df		Significance Level							
Num	Den	.5000	.2500	.1000	.0500	.0250	.0100	.0050	.0025	.0010
30	1	2.145	9.670	62.26	250.1	1001	6261	25044		
30	2	1.410	3.443	9.458	19.46	39.46	99.47	199.5	399.5	999.5
30	3	1.239	2.465	5.168	8.617	14.08	26.50	42.47	67.80	125.4
30	4	1.165	2.082	3.817	5.746	8.461	13.84	19.89	28.45	45.43
30	5	1.123	1.878	3.174	4.496	6.227	9.379	12.66	16.98	24.87
30	6	1.097	1.751	2.800	3.808	5.065	7.229	9.358	12.04	16.67
30	8	1.065	1.600	2.383	3.079	3.894	5.198	6.396	7.816	10.11
30	10	1.047	1.512	2.155	2.700	3.311	4.247	5.071	6.012	7.469
30	12	1.035	1.454	2.011	2.466	2.963	3.701	4.331	5.033	6.090
30	15	1.023	1.397	1.873	2.247	2.644	3.214	3.687	4.200	4.950
30	20	1.011	1.340	1.738	2.039	2.349	2.778	3.123	3.488	4.005
30	24	1.006	1.311	1.672	1.939	2.209	2.577	2.868	3.171	3.593
30	30	1.000	1.282	1.606	1.841	2.074	2.386	2.628	2.876	3.217
30	40	.9944	1.253	1.541	1.744	1.943	2.203	2.401	2.602	2.872
30	60	.9888	1.223	1.476	1.649	1.815	2.028	2.187	2.346	2.555
30	120	.9833	1.192	1.409	1.554	1.690	1.860	1.984	2.105	2.262
30	INF	.9779	1.160	1.342	1.459	1.566	1.696	1.789	1.878	1.990
60	1	2.172	9.759	62.79	252.2	1010	6313	25253		
60	2	1.426	3.459	9.475	19.48	39.48	99.48	199.5	399.5	999.5
60	3	1.254	2.470	5.151	8.572	13.99	26.32	42.15	67.28	124.5
60	4	1.178	2.082	3.790	5.688	8.360	13.65	19.61	28.03	44.75
60	5	1.136	1.874	3.140	4.431	6.123	9.202	12.40	16.62	24.33
60	6	1.109	1.744	2.762	3.740	4.959	7.057	9.122	11.72	16.21
60	8	1.077	1.589	2.339	3.005	3.784	5.032	6.177	7.535	9.727
60	10	1.059	1.499	2.107	2.621	3.198	4.082	4.859	5.748	7.122
60	12	1.046	1.439	1.960	2.384	2.848	3.535	4.123	4.778	5.762
60	15	1.034	1.380	1.817	2.160	2.524	3.047	3.480	3.951	4.638
60	20	1.023	1.319	1.677	1.946	2.223	2.608	2.916	3.242	3.703
60	24	1.017	1.289	1.607	1.842	2.080	2.403	2.658	2.924	3.295
60	30	1.011	1.257	1.538	1.740	1.940	2.208	2.415	2.628	2.920
60	40	1.006	1.225	1.467	1.637	1.803	2.019	2.184	2.350	2.574
60	60	1.000	1.191	1.395	1.534	1.667	1.836	1.962	2.087	2.252
60	120	.9944	1.156	1.320	1.429	1.530	1.656	1.747	1.836	1.950
60	INF	.9889	1.116	1.240	1.318	1.388	1.473	1.532	1.589	1.660

TABLE B.5. *Continued*

df		Significance Level								
Num	Den	.5000	.2500	.1000	.0500	.0250	.0100	.0050	.0025	.0010
INF	1	2.198	9.849	63.33	254.3	1018	6366	25464		
INF	2	1.443	3.476	9.491	19.50	39.50	99.50	199.5	399.5	999.5
INF	3	1.268	2.474	5.134	8.526	13.90	26.13	41.83	66.75	123.5
INF	4	1.192	2.081	3.761	5.628	8.257	13.46	19.32	27.61	44.05
INF	5	1.149	1.869	3.105	4.365	6.015	9.020	12.14	16.26	23.79
INF	6	1.122	1.737	2.722	3.669	4.849	6.880	8.879	11.39	15.75
INF	8	1.089	1.578	2.293	2.928	3.670	4.859	5.951	7.245	9.334
INF	10	1.070	1.484	2.055	2.538	3.080	3.909	4.639	5.472	6.762
INF	12	1.058	1.422	1.904	2.296	2.725	3.361	3.904	4.509	5.420
INF	15	1.046	1.359	1.755	2.066	2.395	2.868	3.260	3.686	4.307
INF	20	1.034	1.294	1.607	1.843	2.085	2.421	2.690	2.975	3.378
INF	24	1.028	1.261	1.533	1.733	1.935	2.211	2.428	2.654	2.969
INF	30	1.023	1.226	1.456	1.622	1.787	2.006	2.176	2.350	2.589
INF	40	1.017	1.188	1.377	1.509	1.637	1.805	1.932	2.060	2.233
INF	60	1.011	1.147	1.291	1.389	1.482	1.601	1.689	1.776	1.891
INF	120	1.006	1.099	1.193	1.254	1.310	1.381	1.431	1.480	1.543
INF	INF	1.000	1.000	1.000	1.000	1.000	1.000	1.000	1.000	1.000

TABLE B.6. Values of the Studentized-range distribution for selected one-tailed significance levels.

Significance Level

I	Nu	0.500	0.200	0.100	0.050	0.020	0.010	0.005	0.002	0.001
2	1	1.414	4.352	8.929	17.97	45.00	90.02	180.1	450.2	900.3
2	2	1.155	2.667	4.129	6.085	9.849	14.04	19.92	31.58	44.69
2	3	1.082	2.316	3.328	4.501	6.422	8.260	10.54	14.45	18.28
2	4	1.048	2.168	3.015	3.926	5.299	6.511	7.916	10.14	12.18
2	5	1.028	2.087	2.850	3.635	4.759	5.702	6.751	8.335	9.714
2	6	1.015	2.036	2.748	3.460	4.444	5.243	6.105	7.365	8.427
2	8	0.999	1.975	2.630	3.261	4.096	4.745	5.420	6.365	7.129
2	10	0.990	1.941	2.563	3.151	3.909	4.482	5.065	5.860	6.487
2	12	0.984	1.918	2.521	3.081	3.792	4.320	4.849	5.557	6.106
2	15	0.978	1.896	2.479	3.014	3.680	4.167	4.647	5.279	5.760
2	20	0.972	1.874	2.439	2.950	3.575	4.024	4.460	5.023	5.444
2	24	0.969	1.864	2.419	2.919	3.525	3.955	4.371	4.903	5.297
2	30	0.966	1.853	2.400	2.888	3.475	3.889	4.285	4.787	5.156
2	40	0.963	1.843	2.381	2.858	3.427	3.825	4.202	4.677	5.022
2	60	0.960	1.833	2.363	2.829	3.380	3.762	4.122	4.570	4.893
2	120	0.957	1.822	2.344	2.800	3.334	3.702	4.044	4.468	4.771
2	INF	0.954	1.812	2.326	2.772	3.290	3.643	3.970	4.370	4.654
3	1	2.338	6.615	13.44	26.98	67.51	135.0	270.1	675.2	1350
3	2	1.908	3.820	5.733	8.331	13.38	19.02	26.97	42.70	60.42
3	3	1.791	3.245	4.467	5.910	8.306	10.62	13.50	18.45	23.31
3	4	1.736	3.004	3.976	5.040	6.668	8.120	9.813	12.51	14.98
3	5	1.705	2.872	3.717	4.602	5.885	6.976	8.196	10.05	11.67
3	6	1.684	2.788	3.558	4.339	5.433	6.330	7.306	8.742	9.959
3	8	1.659	2.689	3.374	4.041	4.934	5.635	6.370	7.406	8.249
3	10	1.645	2.632	3.270	3.877	4.666	5.270	5.888	6.737	7.411
3	12	1.635	2.596	3.204	3.773	4.500	5.046	5.596	6.339	6.917
3	15	1.625	2.560	3.140	3.673	4.343	4.836	5.325	5.974	6.470
3	20	1.616	2.524	3.077	3.578	4.194	4.639	5.074	5.640	6.065
3	24	1.611	2.507	3.047	3.532	4.122	4.546	4.955	5.484	5.877
3	30	1.606	2.490	3.017	3.486	4.053	4.455	4.841	5.334	5.698
3	40	1.602	2.473	2.988	3.442	3.985	4.367	4.731	5.191	5.527
3	60	1.597	2.456	2.959	3.399	3.920	4.282	4.625	5.054	5.365
3	120	1.592	2.440	2.930	3.356	3.856	4.200	4.523	4.923	5.211
3	INF	1.588	2.424	2.902	3.314	3.793	4.120	4.424	4.798	5.063

TABLE B.6. *Continued*

Significance Level

I	Nu	0.500	0.200	0.100	0.050	0.020	0.010	0.005	0.002	0.001
4	1	2.918	8.075	16.36	32.82	82.12	164.3	328.5	821.3	1643
4	2	2.377	4.559	6.772	9.798	15.70	22.29	31.60	50.02	70.77
4	3	2.230	3.833	5.199	6.825	9.540	12.17	15.45	21.09	26.64
4	4	2.163	3.526	4.586	5.757	7.558	9.173	11.06	14.08	16.84
4	5	2.124	3.358	4.264	5.218	6.613	7.804	9.140	11.18	12.96
4	6	2.098	3.252	4.065	4.896	6.067	7.033	8.087	9.643	10.96
4	8	2.067	3.126	3.834	4.529	5.465	6.204	6.981	8.080	8.977
4	10	2.049	3.053	3.704	4.327	5.142	5.769	6.412	7.300	8.006
4	12	2.037	3.006	3.621	4.199	4.941	5.502	6.068	6.836	7.435
4	15	2.025	2.960	3.540	4.076	4.752	5.252	5.750	6.412	6.920
4	20	2.013	2.914	3.462	3.958	4.572	5.018	5.455	6.025	6.454
4	24	2.007	2.892	3.423	3.901	4.486	4.907	5.315	5.844	6.238
4	30	2.001	2.870	3.386	3.845	4.403	4.799	5.181	5.671	6.033
4	40	1.996	2.848	3.348	3.791	4.321	4.695	5.052	5.506	5.838
4	60	1.990	2.826	3.312	3.737	4.242	4.594	4.928	5.348	5.653
4	120	1.984	2.805	3.276	3.685	4.165	4.497	4.809	5.197	5.476
4	INF	1.978	2.784	3.240	3.633	4.090	4.403	4.694	5.053	5.309
5	1	3.335	9.138	18.49	37.08	92.78	185.6	371.2	927.9	1856
5	2	2.713	5.098	7.538	10.88	17.41	24.72	35.02	55.44	78.43
5	3	2.545	4.261	5.738	7.502	10.46	13.32	16.90	23.06	29.13
5	4	2.468	3.907	5.035	6.287	8.221	9.958	11.99	15.24	18.23
5	5	2.423	3.712	4.664	5.673	7.154	8.422	9.846	12.02	13.93
5	6	2.394	3.588	4.435	5.305	6.537	7.556	8.670	10.32	11.72
5	8	2.359	3.440	4.169	4.886	5.856	6.625	7.435	8.583	9.522
5	10	2.338	3.355	4.018	4.654	5.491	6.136	6.800	7.718	8.449
5	12	2.324	3.300	3.921	4.508	5.264	5.836	6.416	7.204	7.820
5	15	2.310	3.245	3.828	4.367	5.049	5.556	6.061	6.734	7.252
5	20	2.297	3.192	3.736	4.232	4.846	5.293	5.732	6.306	6.740
5	24	2.290	3.166	3.692	4.166	4.749	5.168	5.577	6.106	6.502
5	30	2.283	3.140	3.648	4.102	4.654	5.048	5.428	5.916	6.277
5	40	2.277	3.114	3.605	4.039	4.562	4.931	5.284	5.733	6.063
5	60	2.270	3.088	3.562	3.977	4.472	4.818	5.146	5.559	5.860
5	120	2.263	3.063	3.520	3.917	4.385	4.708	5.013	5.393	5.667
5	INF	2.257	3.037	3.478	3.858	4.300	4.603	4.886	5.235	5.484

TABLE B.6. *Continued*

Significance Level

I	Nu	0.500	0.200	0.100	0.050	0.020	0.010	0.005	0.002	0.001
6	1	3.658	9.966	20.15	40.41	101.1	202.2	404.4	1011	2022
6	2	2.973	5.521	8.139	11.73	18.76	26.63	37.73	59.72	84.48
6	3	2.789	4.597	6.162	8.037	11.18	14.24	18.06	24.63	31.10
6	4	2.704	4.205	5.388	6.706	8.747	10.58	12.73	16.18	19.34
6	5	2.655	3.988	4.979	6.033	7.583	8.913	10.41	12.70	14.70
6	6	2.623	3.850	4.726	5.628	6.910	7.972	9.135	10.86	12.32
6	8	2.584	3.686	4.431	5.167	6.166	6.959	7.796	8.985	9.957
6	10	2.561	3.590	4.264	4.912	5.767	6.428	7.109	8.052	8.804
6	12	2.546	3.528	4.156	4.750	5.519	6.101	6.693	7.497	8.127
6	15	2.531	3.467	4.052	4.595	5.283	5.796	6.307	6.991	7.517
6	20	2.516	3.407	3.950	4.445	5.061	5.510	5.951	6.529	6.966
6	24	2.508	3.377	3.900	4.373	4.954	5.373	5.783	6.314	6.711
6	30	2.501	3.348	3.851	4.301	4.850	5.242	5.621	6.108	6.469
6	40	2.494	3.318	3.802	4.232	4.749	5.114	5.465	5.912	6.240
6	60	2.486	3.289	3.755	4.163	4.650	4.991	5.316	5.724	6.022
6	120	2.479	3.260	3.707	4.096	4.554	4.872	5.172	5.546	5.815
6	INF	2.472	3.232	3.661	4.030	4.461	4.757	5.033	5.375	5.619
7	1	3.920	10.64	21.50	43.12	107.9	215.8	431.5	1079	2158
7	2	3.184	5.867	8.633	12.43	19.87	28.20	39.95	63.23	89.45
7	3	2.986	4.872	6.511	8.478	11.78	15.00	19.01	25.93	32.73
7	4	2.895	4.449	5.679	7.053	9.182	11.10	13.35	16.95	20.26
7	5	2.843	4.214	5.238	6.330	7.939	9.321	10.88	13.26	15.35
7	6	2.808	4.065	4.966	5.895	7.219	8.318	9.522	11.31	12.83
7	8	2.767	3.886	4.646	5.399	6.423	7.237	8.097	9.320	10.32
7	10	2.742	3.782	4.465	5.124	5.995	6.669	7.365	8.329	9.099
7	12	2.726	3.715	4.349	4.950	5.729	6.320	6.922	7.740	8.382
7	15	2.709	3.648	4.235	4.782	5.476	5.994	6.511	7.203	7.736
7	20	2.693	3.582	4.124	4.620	5.237	5.688	6.131	6.713	7.153
7	24	2.685	3.549	4.070	4.541	5.122	5.542	5.952	6.484	6.884
7	30	2.677	3.517	4.016	4.464	5.010	5.401	5.779	6.266	6.628
7	40	2.669	3.484	3.963	4.388	4.901	5.265	5.614	6.058	6.385
7	60	2.661	3.452	3.911	4.314	4.795	5.133	5.454	5.860	6.155
7	120	2.653	3.420	3.859	4.241	4.692	5.005	5.301	5.670	5.937
7	INF	2.645	3.389	3.808	4.170	4.592	4.882	5.154	5.490	5.730

TABLE B.6. *Continued*

Significance Level

I	Nu	0.500	0.200	0.100	0.050	0.020	0.010	0.005	0.002	0.001
8	1	4.139	11.21	22.64	45.40	113.6	227.2	454.3	1136	2272
8	2	3.361	6.158	9.049	13.03	20.81	29.53	41.83	66.21	93.66
8	3	3.152	5.104	6.806	8.852	12.29	15.64	19.82	27.03	34.12
8	4	3.055	4.655	5.926	7.347	9.553	11.54	13.88	17.61	21.04
8	5	3.000	4.405	5.458	6.582	8.242	9.669	11.28	13.74	15.90
8	6	2.964	4.246	5.168	6.122	7.482	8.612	9.852	11.69	13.26
8	8	2.919	4.055	4.829	5.596	6.641	7.474	8.354	9.606	10.63
8	10	2.893	3.944	4.636	5.304	6.189	6.875	7.584	8.566	9.352
8	12	2.876	3.871	4.511	5.119	5.907	6.507	7.117	7.948	8.601
8	15	2.859	3.800	4.390	4.940	5.640	6.162	6.685	7.384	7.924
8	20	2.842	3.729	4.271	4.768	5.386	5.839	6.285	6.870	7.313
8	24	2.833	3.694	4.213	4.684	5.264	5.685	6.096	6.630	7.031
8	30	2.825	3.659	4.155	4.601	5.146	5.536	5.914	6.401	6.763
8	40	2.816	3.624	4.099	4.521	5.030	5.392	5.739	6.182	6.509
8	60	2.808	3.589	4.042	4.441	4.918	5.253	5.571	5.974	6.268
8	120	2.799	3.554	3.986	4.363	4.808	5.118	5.410	5.775	6.039
8	INF	2.791	3.520	3.931	4.286	4.702	4.987	5.255	5.586	5.823
10	1	4.491	12.12	24.48	49.07	122.8	245.5	491.1	1228	2456
10	2	3.645	6.630	9.725	13.99	22.34	31.69	44.89	71.04	100.5
10	3	3.418	5.480	7.287	9.462	13.13	16.69	21.15	28.83	36.39
10	4	3.313	4.989	6.327	7.826	10.16	12.26	14.73	18.69	22.33
10	5	3.253	4.715	5.816	6.995	8.738	10.24	11.93	14.52	16.80
10	6	3.213	4.540	5.499	6.493	7.914	9.097	10.39	12.32	13.96
10	8	3.165	4.329	5.126	5.918	7.000	7.863	8.777	10.08	11.15
10	10	3.137	4.206	4.913	5.598	6.508	7.213	7.944	8.958	9.769
10	12	3.118	4.126	4.776	5.395	6.200	6.814	7.439	8.292	8.962
10	15	3.099	4.046	4.641	5.198	5.908	6.438	6.970	7.683	8.233
10	20	3.081	3.966	4.510	5.008	5.630	6.086	6.536	7.128	7.576
10	24	3.071	3.927	4.445	4.915	5.497	5.919	6.331	6.869	7.272
10	30	3.062	3.887	4.381	4.824	5.367	5.756	6.134	6.621	6.984
10	40	3.053	3.848	4.317	4.735	5.240	5.599	5.944	6.385	6.710
10	60	3.043	3.809	4.254	4.646	5.116	5.447	5.762	6.160	6.451
10	120	3.034	3.770	4.191	4.560	4.996	5.299	5.586	5.946	6.206
10	INF	3.024	3.730	4.129	4.474	4.878	5.157	5.418	5.741	5.973

426   Appendix B. Tables

TABLE B.6. *Continued*

Significance Level

I	Nu	0.500	0.200	0.100	0.050	0.020	0.010	0.005	0.002	0.001
12	1	4.767	12.84	25.92	51.96	130.0	260.0	519.9	1300	2600
12	2	3.867	7.002	10.26	14.75	23.54	33.40	47.30	74.87	105.9
12	3	3.626	5.778	7.667	9.946	13.79	17.53	22.20	30.26	38.20
12	4	3.515	5.253	6.645	8.208	10.64	12.84	15.42	19.56	23.36
12	5	3.451	4.960	6.100	7.323	9.135	10.70	12.46	15.16	17.53
12	6	3.409	4.773	5.762	6.789	8.259	9.485	10.83	12.83	14.53
12	8	3.358	4.546	5.362	6.175	7.287	8.176	9.117	10.46	11.56
12	10	3.328	4.414	5.134	5.833	6.762	7.485	8.233	9.274	10.11
12	12	3.308	4.327	4.986	5.615	6.434	7.060	7.697	8.568	9.253
12	15	3.288	4.240	4.841	5.403	6.122	6.660	7.200	7.924	8.483
12	20	3.268	4.154	4.699	5.199	5.825	6.285	6.738	7.335	7.788
12	24	3.258	4.111	4.628	5.099	5.682	6.105	6.520	7.060	7.467
12	30	3.248	4.068	4.559	5.001	5.542	5.932	6.310	6.798	7.161
12	40	3.238	4.025	4.490	4.904	5.406	5.764	6.108	6.548	6.872
12	60	3.227	3.982	4.421	4.808	5.273	5.601	5.913	6.309	6.598
12	120	3.217	3.938	4.353	4.714	5.144	5.443	5.726	6.081	6.338
12	INF	3.207	3.895	4.285	4.622	5.017	5.290	5.546	5.864	6.093
15	1	5.091	13.68	27.62	55.36	138.5	277.0	554.0	1385	2770
15	2	4.129	7.442	10.89	15.65	24.97	35.43	50.17	79.40	112.3
15	3	3.871	6.130	8.120	10.52	14.57	18.72	23.46	31.97	40.35
15	4	3.752	5.566	7.025	8.664	11.22	13.53	16.24	20.59	24.59
15	5	3.684	5.251	6.439	7.716	9.610	11.24	13.09	15.91	18.40
15	6	3.639	5.049	6.075	7.143	8.673	9.951	11.35	13.44	15.22
15	8	3.584	4.804	5.644	6.483	7.632	8.552	9.527	10.92	12.06
15	10	3.552	4.660	5.397	6.114	7.069	7.812	8.583	9.655	10.51
15	12	3.531	4.565	5.236	5.878	6.716	7.356	8.009	8.902	9.606
15	15	3.509	4.470	5.079	5.649	6.379	6.927	7.476	8.214	8.785
15	20	3.488	4.376	4.923	5.427	6.059	6.523	6.981	7.585	8.044
15	24	3.477	4.329	4.847	5.319	5.904	6.330	6.747	7.291	7.701
15	30	3.466	4.281	4.770	5.211	5.753	6.142	6.521	7.010	7.375
15	40	3.455	4.234	4.694	5.106	5.605	5.961	6.303	6.742	7.066
15	60	3.444	4.186	4.619	5.001	5.460	5.784	6.094	6.486	6.773
15	120	3.433	4.138	4.543	4.898	5.319	5.614	5.893	6.243	6.496
15	INF	3.422	4.089	4.468	4.796	5.182	5.448	5.699	6.011	6.234

TABLE B.7. Orthogonal polynomials.

I	deg.	1	2	3	4	5	6	7	8	9	10	11	12	Den.
3	1st	-1	0	1										2
	2nd	1	-2	1										6
4	1st	-3	-1	1	3									20
	2nd	1	-1	-1	1									4
	3rd	-1	3	-3	1									20
5	1st	-2	-1	0	1	2								10
	2nd	2	-1	-2	-1	2								14
	3rd	-1	2	0	-2	1								10
	4th	1	-4	6	-4	1								70
6	1st	-5	-3	-1	1	3	5							70
	2nd	5	-1	-4	-4	-1	5							84
	3rd	-5	7	4	-4	-7	5							180
	4th	1	-3	2	2	-3	1							28
	5th	-1	5	-10	10	-5	1							252
7	1st	-3	-2	-1	0	1	2	3						28
	2nd	5	0	-3	-4	-3	0	5						84
	3rd	-1	1	1	0	-1	-1	1						6
	4th	3	-7	1	6	1	-7	3						154
	5th	-1	4	-5	0	5	-4	1						84
	6th	1	-6	15	-20	15	-6	1						924
8	1st	-7	-5	-3	-1	1	3	5	7					168
	2nd	7	1	-3	-5	-5	-3	1	7					168
	3rd	-7	5	7	3	-3	-7	-5	7					264
	4th	7	-13	-3	9	9	-3	-13	7					616
	5th	-7	23	-17	-15	15	17	-23	7					2184
	6th	1	-5	9	-5	-5	9	-5	1					264
9	1st	-4	-3	-2	-1	0	1	2	3	4				60
	2nd	28	7	-8	-17	-20	-17	-8	7	28				2772
	3rd	-14	7	13	9	0	-9	-13	-7	14				990
	4th	14	-21	-11	9	18	9	-11	-21	14				2002
	5th	-4	11	-4	-9	0	9	4	-11	4				468
	6th	4	-17	22	1	-20	1	22	-17	4				1980

TABLE B.7. *Continued*

I	deg.	1	2	3	4	5	6	7	8	9	10	11	12	Den.
10	1st	-9	-7	-5	-3	-1	1	3	5	7	9			330
	2nd	6	2	-1	-3	-4	-4	-3	-1	2	6			132
	3rd	-42	14	35	31	12	-12	-31	-35	-14	42			8580
	4th	18	-22	-17	3	18	18	3	-17	-22	18			2860
	5th	-6	14	-1	-11	-6	6	11	1	-14	6			780
	6th	3	-11	10	6	-8	-8	6	10	-11	3			660
11	1st	-5	-4	-3	-2	-1	0	1	2	3	4	5		110
	2nd	15	6	-1	-6	-9	-10	-9	-6	-1	6	15		858
	3rd	-30	6	22	23	14	0	-14	-23	-22	-6	30		4290
	4th	6	-6	-6	-1	4	6	4	-1	-6	-6	6		286
	5th	-3	6	1	-4	-4	0	4	4	-1	-6	3		156
	6th	15	-48	29	36	-12	-40	-12	36	29	-48	15		11220
12	1st	-11	-9	-7	-5	-3	-1	1	3	5	7	9	11	572
	2nd	55	25	1	-17	-29	-35	-35	-29	-17	1	25	55	12012
	3rd	-33	3	21	25	19	7	-7	-19	-25	-21	-3	33	5148
	4th	33	-27	-33	-13	12	28	28	12	-13	-33	-27	33	8008
	5th	-33	57	21	-29	-44	-20	20	44	29	-21	-57	33	15912
	6th	11	-31	11	25	4	-20	-20	4	25	11	-31	11	4488

TABLE B.8. Possible balanced incomplete blocks designs $(I, J \leq 20)$.

I	J	g	h	λ	H	I	J	g	h	λ	H
3	3	2	2	1	0.7500	3	15	10	2	5	0.7500
4	4	3	3	2	0.8889	5	15	12	4	9	0.9375
5	5	4	4	3	0.9375	6	15	5	2	1	0.6000
3	6	4	2	2	0.7500	6	15	10	4	6	0.9000
4	6	3	2	1	0.6667	10	15	6	4	2	0.8333
6	6	5	5	4	0.9600	10	15	9	6	5	0.9259
7	7	3	3	1	0.7778	15	15	7	7	3	0.9184
7	7	4	4	2	0.8750	15	15	8	8	4	0.9375
7	7	6	6	5	0.9722	15	15	14	14	13	0.9949
4	8	6	3	4	0.8889	4	16	12	3	8	0.8889
8	8	7	7	6	0.9796	8	16	14	7	12	0.9796
3	9	6	2	3	0.7500	16	16	6	6	2	0.8889
9	9	8	8	7	0.9844	16	16	10	10	6	0.9600
5	10	4	2	1	0.6250	16	16	15	15	14	0.9956
5	10	6	3	3	0.8333	17	17	16	16	15	0.9961
5	10	8	4	6	0.9375	3	18	12	2	6	0.7500
6	10	5	3	2	0.8000	4	18	9	2	3	0.6667
10	10	9	9	8	0.9877	6	18	15	5	12	0.9600
11	11	5	5	2	0.8800	9	18	8	4	3	0.8438
11	11	6	6	3	0.9167	9	18	10	5	5	0.9000
11	11	10	10	9	0.9900	9	18	16	8	14	0.9844
3	12	8	2	4	0.7500	10	18	9	5	4	0.8889
4	12	6	2	2	0.6667	18	18	17	17	16	0.9965
4	12	9	3	6	0.8889	19	19	9	9	4	0.9383
6	12	10	5	8	0.9600	19	19	10	10	5	0.9500
9	12	4	3	1	0.7500	19	19	18	18	17	0.9969
9	12	8	6	5	0.9375	4	20	15	3	10	0.8889
12	12	11	11	10	0.9917	5	20	8	2	2	0.6250
13	13	4	4	1	0.8125	5	20	12	3	6	0.8333
13	13	9	9	6	0.9630	5	20	16	4	12	0.9375
13	13	12	12	11	0.9931	6	20	10	3	4	0.8000
7	14	6	3	2	0.7778	10	20	18	9	16	0.9877
7	14	8	4	4	0.8750	16	20	5	4	1	0.8000
7	14	12	6	10	0.9722	16	20	15	12	11	0.9778
8	14	7	4	3	0.8571	20	20	19	19	18	0.9972
14	14	13	13	12	0.9941						

Appendix C

Analysis of Variance Using SAS

SAS (Trademark SAS Institute, Inc.) is a general statistical analysis program. To use SAS, a file containing data and instructions is first prepared. Usually the file is stored with a name having the form " *.SAS ", where the " * " may be any file name that is legal on your system. (Some aspects of SAS are necessarily specific to the system it is on; for information on those aspects, you will probably need to consult with experts on your system.)

There are a great many options for preparing the file; their use depends on the purposes and sophistication of the user. We will consider only the simpler options here. We will give just enough information to enable you to enter data and analyze it, using most of the procedures described in the text. For more sophisticated applications, you are referred to the SAS manuals.

The file can be divided conveniently into four parts. The first part basically is a single line, giving a title, etc.; the second instructs SAS about the general format of the data, the third is the data, and the fourth is a set of instructions for analyzing the data. Each line of the file, except the lines containing data, starts with a " key word " and ends with a semicolon.

The first part consists of a single line containing the key word " DATA ". It may also contain a name. A single convention is followed for all names; they may have up to eight alphabetic and/or numeric characters but always start with an alphabetic character or an underscore (" _ "). In some applications, the name identifies a separate data file; in our examples, it will simply identify the file we are using.

We will consider each of the remaining parts in the contexts of specific analyses.

One-Way Analysis Fixed-Effects (Chapter 2)

INPUT

A sample input file, for the data in Table 2.1, is given in Table C.1.

The first line simply says " DATA T0201; ". The last part is optional; it identifies the data as coming from Table 2.1. The second starts with the key word " INPUT ", telling SAS that this line describes the data that will follow. The values of both the dependent and independent variables (i.e., the groups) must be entered as data; the " INPUT " statement must contain names for these variables, listed in the order in which they appear as data. Thus, looking farther down the file, you can see that each line contains both a value of X

TABLE C.1. Input file for analyzing data in Table 2.1.

```
DATA T0201;
   INPUT GROUP $ SCORE;
   CARDS;
      G1 5
      G1 3
      G1 5
      G1 6
      G1 1
      G2 1
      G2 2
      G2 2
      G2 0
      G2 3
      G3 5
      G3 4
      G3 7
      G3 5
      G3 2
PROC ANOVA;
   CLASS GROUP;
   MODEL SCORE=GROUP;
   MEANS GROUP;
TITLE 'TABLE 2.1';
```

and a label indicating the group from which that value came. The first datum on each line is the group identifier, so the first word after " INPUT " is " GROUP ". Following the word " GROUP " is the symbol " $ "; whenever this symbol follows a word, it indicates that the variable being input is a label rather than a numerical value. The last word on this line is the word " SCORE " indicating that the second value on each line will be a student's score on the test.

The next line says " cards; ", indicating that the data follow. (The key word " CARDS " goes back to the days when data actually *were* input on punched cards.) The data are then typed, one subject per line, with the group identifier listed first and the score second (according to their order in the " INPUT " statement). Any labels can be used for the groups, just so each of the three groups has a different label. In fact, we could have just used the numbers 1, 2, 3, in which case the " $ " would have been unnecessary on the " INPUT " line. Note that the data lines do not end in semicolons; SAS will assume that the next line containing a semicolon is the first line *following* the data.

Instructions for the analysis now follow: The first step in the analysis is to specify the procedure. There are two main procedures for doing analyses of variance and covariance. " ANOVA " is the simplest, but it is somewhat limited. It can be used for any one-way design, but it cannot be used for some of the analyses to be presented later. " GLM " (General Linear Model) is more general, but it is more complicated, taking both more memory and more time. For these data, " ANOVA " is adequate, so we call it with the statement " PROC ANOVA; ".

The second statement in the analysis begins with the key word " CLASS ". The word that follows tells SAS which variable is a " classification variable ," i.e., which one serves to classify the data by groups. In our example, the variable " GROUP " classifies the data, so that word follows the key word " CLASS ".

The third statement identifies the dependent and independent variables. For a simple one-way design, there is just one of each. The dependent variable is listed first, followed by " = " and then the independent variable.

A number of other statements can follow. I have used just two. We usually want to look at group means, so the next statement, " MEANS GROUP; ", tells SAS to print the means for the individual groups. The last statement tells SAS to put a title on each page of the printout. The desired title follows the key word " TITLE " in single quotes.

OUTPUT

On most systems, SAS will output two files. The first is called " *.LOG " (the log file), and the second is " *.LIS " (the listing file), where the " * " represents the same name that was used in the original " *.SAS " file. The LOG file is just a log of the activities of SAS during the analysis. If there are any errors in the input statements, these will be indicated in the LOG file. If there are errors, there may or may not be a LIS file, and the LIS file, if there is one, might be garbage. The errors should be corrected and the data resubmitted.

The input in Table C.1 will typically produce three pages of output. The first is uninteresting: It is just a restatement of the model. The third contains the group means.

The second is more complicated; it contains the results of the analysis of variance. It is reproduced in Table C.2. (It, like most of the tables in this appendix, has been edited somewhat for this text, but no basic changes have been made.) It contains two tables.

The first table has eight columns. The first column contains the labels " MODEL ", " ERROR ", and " CORRECTED TOTAL ". They correspond, respectively, to what this text calls " between ", " within ", and " total ". The next four should be self-explanatory. The sixth contains the p-value. The seventh, labeled " R-Square ", is { $\hat{\omega}_{bet}^2$, but it is the maximum likelihood estimate, which will tend to be a bit larger than the estimate advocated in this text. The maximum likelihood extimate is just SS_{bet}/SS_t. The last column, labeled " C.V ", is the " coefficient of variation ". It is 100 times the square root of MS_w, divided by the grand mean. It can be a useful measure of relative variability, but only when the data have a meaningful zero (e.g., when the data are measured on a ratio scale such as height or weight).

TABLE C.2. SAS analysis of data in Table C.1.

TABLE 2.1, PAGE 2

ANALYSIS OF VARIANCE PROCEDURE

DEPENDENT VARIABLE: SCORE

SOURCE	DF	SUM OF SQ	MEAN SQ	F VAL	PR > F	R-SQ	C.V.
MODEL	2	25.2000	12.6000	4.40	0.0370	0.4228	49.798
ERROR	12	34.4000	2.8667		ROOT MSE		SCORE MEAN
CORRECTED TOTAL	14	59.6000			1.6931		3.4000

SOURCE	DF	ANOVA SS	F VAL	PR > F
CLASS	2	25.2000	4.40	0.0370

Two other values are " tucked " into the first table, perhaps to save space. The first value, labeled " ROOT MSE ", is the estimated standard error, i.e., the square root of MS_w; the second, labeled " SCORE MEAN ", is the grand mean.

The second table lists the ANOVA(s) of interest; in the one-way design there is only one, the ANOVA on groups. The table should be self-explanatory; in fact, for the simple one-way design it just lists the same information as in the first table.

Planned Comparisons (Chapter 3)

The ANOVA procedure cannot do general planned comparisons, but the GLM procedure can. Accordingly we will use GLM to illustrate the analyses in Chapter 3.

INPUT

Table C.3 shows an input file for the data in Table 3.1. Most of the statements are the same as in Table 3.1, but the first statement after the data is " PROC GLM ", indicating that the GLM procedure, rather than ANOVA, is to be used. Following that, there are a number of changes from Table C.1; we will discuss the new statements here; the qualifiers (i.e., " REGWF ", etc.) following the " MEANS " statement pertain to Chapter 4.

Two new kinds of statements are found in this table. The first begins with the key word " ESTIMATE ". Each of these is followed by a title in single quotes. The title is required and must not be more than 20 characters long. The title is followed by the word " GROUP ", indicating that the group variable is being studied. (This is unnecessary for the one-way design because there is only one independent variable, but it is used in multifactor designs.) Following that are four coefficients: the coefficients for a planned contrast. The first contrast should be recognizable as the test of μ_1 versus μ_2; the second as the test of μ_2 and μ_3 versus μ_4; and the third as the test of μ_1 versus the average of the other three means. In this first analysis, you are limited to testing contrasts. (Technically, you are limited to testing estimable functions, as defined in Chapter 16.) To test noncontrasts, you must change the " MODEL " statement to read " MODEL SCORE=GROUP / NOINT ". You can then test both contrasts and noncontrasts. However, this removes the term for the grand mean from the model ("NOINT" stands for " no intercept "), and that will invalidate the overall analysis. Moreover, you cannot have more than one " MODEL " statement in each analysis. However, you *can* request more than one analysis in a single input file. Accordingly, we have added another " PROC GLM " statement, indicating that the GLM procedure should be performed a second time. This second analysis has the changed " MODEL " statement and a test of a noncontrast. Specifically, it tests $H_0 : \mu_2 + \mu_3 = \mu_4$.

TABLE C.3. Input file for analyzing data in Table 3.1. Most data are omitted to save space.

```
DATA T0301;
   INPUT GROUP $ SCORE;
   CARDS;
      G1   1
      .....
PROC GLM;
   CLASS GROUP;
   MODEL SCORE=GROUP;
   ESTIMATE 'G2 VS G3'          GROUP  0 -1  1  0;
   ESTIMATE 'G2 & G3 VS G4'     GROUP  0 -1 -1  2;
   ESTIMATE 'G1 VS G2 - G4'     GROUP -3  1  1  1;
   CONTRAST 'G2 - G4'           GROUP  0 -1  1  0,
                                GROUP  0 -1  0  1;
   MEANS GROUP / REGWF REGWQ ALPHA=.05;
   TITLE 'TABLE 3-1 WITH PLANNED AND POST HOC COMPARISONS';
PROC GLM;
   CLASS GROUP;
   MODEL SCORE=GROUP / NOINT;
   ESTIMATE 'G2+G3 VS G4' GROUP  0 -1 -1  1;
   TITLE 'TABLE 3.1 WITH TEST OF A NONCONTRAST';
```

The second kind of statement in Table C.3 begins with the key word " CONTRAST "; it is also followed by a title of no more than 20 characters. However, it can be followed by coefficients for more than one contrast. In the example, it is followed by coefficients for two contrasts: a test of μ_2 versus μ_3, and a test of μ_2 and μ_3 versus μ_4. The two sets of coefficients (each preceded by the word " GROUP ") are separated by a comma. They are also on separate lines in Table C.3, but that is for readability—it is not required by SAS. " CONTRAST " differs from " ESTIMATE " in that it can test the null hypothesis that two or more, not necessarily orthogonal, contrasts are all simultaneously zero. The pair of contrasts in the example are equivalent to $H_0 : \mu_2 = \mu_3 = \mu_4$. " ESTIMATE " does t tests and prints estimates of the contrasts; " CONTRAST " uses F ratios and does not print estimates. (Note that with " / NOINT " added to the " MODEL " statement, you can do combined tests of noncontrasts using the " CONTRAST " statement.)

OUTPUT

Table C.4 shows the results of the first analysis for the data in Table C.3. The first table is the same as was described for Table C.2. The second is similar, but it lists " TYPE I SS " and " TYPE III SS ". In the simple one-way design, there is no difference, so the " TYPE III SS " can be ignored.

The third table begins with the key word " CONTRAST "; it contains the results of the test defined by the " CONTRAST " statement in the input file. Here it says that μ_2, μ_3, and μ_4 are significantly different with a p-value of 0.0027.

The fourth table contains the three contrasts tested with " ESTIMATE " statements. They give essentially the same results as in Chapter 3.

Table C.5 shows the results of the second analysis. Note that the first two tables are mostly " garbage. " They are actually the results of testing H_0 : $\mu_1 = \mu_2 = \mu_3 = \mu_4 = 0$. The relevant data are in the last table, where we test H_0 : $\mu_2 + \mu_3 = \mu_4$; for this test, $t = -3.16$, $p = .0038$.

Post Hoc Tests (Chapter 4)

Neither ANOVA nor GLM can do general post hoc tests using the Scheffé or Tukey methods. However, you can do Bonferroni tests if you yourself calculate the necessary α value. Other tests can be done by hand, with SAS providing the cell means, sums of squares, and mean squares. Both ANOVA and GLM can do multiple tests on pairs of means, using any number of methods. The example below illustrates two different methods.

INPUT

The methods are listed after the slash ("/") in the " MEANS " statement in Table C.3. The first, called " REGWF " (which stands for " Ryan–Einot–Gabriel–Welsch with F "), is basically Welsch's variation of Ryan's method, using F ratios; the second, called " REGWQ ", is the same method using Studentized range tests. Some of the other methods available, none of which are recommended for pairwise testing, are " BON ", Bonferroni method; " DUNCAN ", Duncan method; " SCHEFFÉ ", Scheffé's method; " SNK ", Neuman-Keuls method; " LSD ", planned contrasts on all pairs of means; and " TUKEY ", Tukey method.

The last item in the statement specifies the alpha level.

OUTPUT

Table C.6 shows the output for the " REGWF " procedure. The output is largely self-explanatory: Group G_4 is significantly different from the other three groups, but groups G_2 and G_3 are not significantly different from each other, nor are groups G_3 and G_1. The output from the other procedures follows the same format.

TABLE C.4. SAS tests of planned contrasts on data in Table C.3.

TABLE 3.1, GLM WITH PLANNED AND POST HOC COMPARISONS, PAGE 2

GENERAL LINEAR MODELS PROCEDURE

DEPENDENT VARIABLE: SCORE

SOURCE	DF	SUM OF SQ	MEAN SQ	F VAL	PR > F	R-SQ	C.V.
MODEL	3	212.75	70.9167	9.60	0.0002	0.507	27.52
ERROR	28	206.75	7.3839		ROOT MSE		SCORE MEAN
					2.7173		9.875
CORRECTED TOTAL	31	419.50					

SOURCE	DF	TYPE I SS	F VAL	PR > F	DF	TYPE III SS	F VAL	PR > F
GROUP	3	212.75	9.60	0.0002	3	212.75	9.60	0.000

CONTRAST	DF	SS	F VAL	PR > F
G2 - G4	2	108.583	7.35	0.0027

PARAMETER	ESTIMATE	T FOR HO: PARAMETER=0	PR > |T|	STD ERROR OF ESTIMATE
G2 VS G3	-1.75000	-1.29	0.2083	1.3587
G2 & G3 VS G4	8.50000	3.61	0.0012	2.3533
G1 VS G2 - G4	12.50000	3.76	0.0008	3.3280

TABLE C.5. Second SAS analysis of data in Table C.3.

TABLE 3.1, GLM WITH TEST OF A NONCONTRAST, PAGE 2

GENERAL LINEAR MODELS PROCEDURE

DEPENDENT VARIABLE: SCORE

SOURCE	DF	SUM OF SQ	MEAN SQ	F VAL	PR > F	R-SQ	C.V.
MODEL	4	3333.25	833.3125	112.85	0.0001	0.942	27.52
ERROR	28	206.75	7.3839		ROOT MSE		SCORE MEAN
UNCORRECTED TOTAL	32	3540.00			2.7173		9.875

NOTE: NO INTERCEPT TERM IS USED. R-SQUARE IS REDEFINED.

SOURCE	DF	TYPE I SS	F VAL	PR > F	DF	TYPE III SS	F VAL	PR > F
GROUP	4	3333.25	112.85	0.0001	4	3333.25	112.85	0.0001

PARAMETER	ESTIMATE	T FOR H0: PARAMETER=0	PR > |T|	STD ERROR OF ESTIMATE
G2+G3 VS G4	-5.25	-3.16	0.0038	1.6640

TABLE C.6. SAS post hoc tests on data in Table C.3.

TABLE 3.1, GLM WITH PLANNED AND POST HOC COMPARISONS, PAGE 4

GENERAL LINEAR MODELS PROCEDURE

RYAN-EINOT-GABRIEL-WELSCH MULTIPLE F TEST FOR VARIABLE: SCORE
NOTE: THIS TEST CONTROLS THE TYPE I EXPERIMENTWISE ERROR RATE

ALPHA=0.05   DF=28   MSE=7.38393

NUMBER OF MEANS	2	3	4
CRITICAL F	5.58263	3.34039	2.94669

MEANS WITH THE SAME LETTER ARE NOT SIGNIFICANTLY DIFFERENT.

REGWF	GROUPING		MEAN	N	GROUP
		A	13.750	8	G4
	B		10.375	8	G2
	B				
	C	B	8.625	8	G3
	C				
	C		6.750	8	G1

Two-Way Analysis (Chapter 5)

The ANOVA procedure is appropriate for any balanced two-way analysis, but the GLM procedure must be used for an unbalanced analysis. (A balanced design is any design with equal or proportional sample sizes, i.e., for which the simple main effects SS_a and SS_b are orthogonal; an unbalanced design is any design for which $SS_{a:b}$ or $SS_{b:a}$ should be calculated.) I will illustrate both. Although planned and post hoc comparisons can be done (general planned comparisons can be done only with GLM), they are somewhat complicated. They are easily done by hand anyway and will not be illustrated here.

INPUT

The input file for the data in Table 5.2 is shown in Table C.7. Table 5.2 has equal sample sizes, so the design is balanced, and the ANOVA procedure can be used.

The first difference between Table C.7 and the previous input files is in the INPUT statement. In this statement, the two factors are both listed. They, and the dependent variable, must be listed in the order in which they appear on the following lines (i.e., the diagnosis factor, " DIAG ", comparing

TABLE C.7. Input file for analysis of data in Table 5.2. " DIAG " stands for diagnosis (schizophrenic vs. manic depressive, Factor A); "DRUG" stands for drug administered (Factor B).

```
DATA T0502;
  INPUT DIAG $ DRUG $ SCORE;
  CARDS;
    SCHIZ B1  8
    SCHIZ B1  4
    SCHIZ B1  0
    SCHIZ B2  8
    SCHIZ B2 10
    SCHIZ B2  6
    SCHIZ B3  4
    SCHIZ B3  6
    SCHIZ B3  8
    MD    B1 10
    MD    B1  6
    MD    B1 14
    MD    B2  0
    MD    B2  4
    MD    B2  2
    MD    B3 15
    MD    B3  9
    MD    B3 12
PROC ANOVA;
  CLASS DIAG DRUG;
  MODEL SCORE=DIAG|DRUG;
  MEANS DIAG|DRUG;
TITLE 'TABLE 5.2';
```

schizophrenics with manic depressives, comes first, then the " DRUG ", and finally the improvement score). The dollar signs following the factors again indicate that they are represented by labels rather than numbers in the data set.

The second difference is in the CLASS statement, where DIAG and DRUG are listed to indicate that they are classification variables (i.e., factors).

The third difference is in the MODEL statement. Here a special notation is used. When a factor is listed alone, it indicates that the *SS* of the main effect for that factor should be calculated. Factor names joined by asterisks ("*") represent interactions. Thus,

MODEL SCORE=DIAG DRUG tells SAS to calculate only the sums of squares for the two main effects;

MODEL SCORE=DIAG DIAG*DRUG tells SAS to calculate only the sums of squares for the DIAG main effect and the interaction;

MODEL SCORE=DIAG DRUG DIAG*DRUG tells SAS to calculate all three sums of squares.

There is a shorthand notation for the last case above: In Table C.7 the two factors are separated by a vertical line ("|"); this tells SAS to calculate both main effects and the interaction.

Note that SAS literally does not calculate any sums of squares that you do not request. Any sums of squares that are not calculated are pooled with the error sum of squares. In the first example above, the interaction SS would be pooled with SS_w; in the second example, the SS for the drug main effect would be pooled with SS_w.

Finally, the MEANS statement uses the same notation as the MODEL statement, but it tells which means to print; I have requested the means for both main effects and the interaction.

The input file for the data in Table 5.9 is in Table C.8. This is an unbalanced design, so " GLM " is used. Otherwise, it is basically the same as Table C.7. However, for unbalanced data the order in which " DIAG " and " DRUG " appear in the MODEL statement is important. If " DIAG " (Factor A) appears first, as it does in this example, then SAS will calculate SS_a and $SS_{b:a}$; if " DRUG " (Factor B) had been listed first, then SAS would have calculated SS_b and $SS_{a:b}$. Basically, SAS extracts the variance in the order in which the effects appear. It extracts the variance of the first listed effect first, ignoring all others in the list; then it extracts the variance of the second, taking into account the first, but ignoring all others, etc.

OUTPUT

Table C.9 shows the analysis of variance for the data in Table C.7. The first table is practically identical to those of other analyses. This first table always analyzes the data as though it were a simple one-way design. The second table contains the information of interest. That table lists the data for the individual effects. The column headings should be self-explanatory.

Table C.10 shows the analysis of variance for the data in Table C.3. It has the same format as Table C.9, except that it has both " TYPE I SS " and " TYPE III SS ". The column labeled " TYPE I SS " is the relevant column. Note that the SS for " DIAG " is SS_a, while the SS for " DRUG " is $SS_{b:a}$, as was explained above.

The GLM procedure actually can calculate four types of sums of squares, although only two, " TYPE I SS " and " TYPE III SS ", are printed by default. If we had asked for " TYPE II SS ", SAS would have printed $SS_{a:b}$ and $SS_{b:a}$; the other two types are more complicated; they are advocated by some statisticians but have no simple, straightforward interpretation. All four are explained at length in the SAS manual.

TABLE C.8. Input file for data in
Table 5.9.

```
DATA T0509;
  INPUT DIAG $ DRUG $ SCORE;
  CARDS;
      SCHIZ B1   8
      SCHIZ B1   4
      SCHIZ B1   0
      SCHIZ B2   8
      SCHIZ B2  10
      SCHIZ B2   6
      SCHIZ B3   4
      SCHIZ B3   8
      MD    B1  10
      MD    B1  14
      MD    B2   0
      MD    B2   4
      MD    B2   2
      MD    B3  15
      MD    B3   9
      MD    B3  12
PROC GLM;
  CLASS DIAG DRUG;
  MODEL SCORE=DIAG|DRUG;
  MEANS DIAG|DRUG;
  TITLE 'TABLE 5.9';
```

Models with Random Factors (Chapter 6)

SAS cannot analyze random and mixed-models directly. However, you can tell it which MS to use for each hypothesis test performed. Therefore, it can do the correct analyses if you specify the design correctly.

INPUT

Table C.11 shows the input file for the data in Table 6.2. (Because Table 6.2 actually contains only one value in each cell, we will treat it here as though there were only one score in each cell.) This file differs from previous input files in the addition of a new statement labeled " TEST ". The statement says that the factor " EXAM " is to be tested using the " EXAM*COLLEGE " interaction in the denominator of the F ratio. In general, " H= " is followed by a list of one or more effects. Each effect is to be tested using the error term following " E= ". (Note that, because we are treating this as a design with only one score per cell, there is no MS_w.)

Table C.12 shows the input file for the data in Table 6.3. In it, both " UNIV " and " COURSE " follow " H= ", because both are tested against the " UNIV*COURSE " interaction.

TABLE C.9. SAS analysis of data in Table C.7.

TABLE 5.2, PAGE 2

ANALYSIS OF VARIANCE PROCEDURE

DEPENDENT VARIABLE: SCORE

SOURCE	DF	SUM OF SQ	MEAN SQ	F VAL	PR > F	R-SQ	C.V.
MODEL	5	210.000	42.0000	4.75	0.0126	0.6646	42.459
ERROR	12	106.000	8.8333		ROOT MSE		SCORE MEAN
CORRECTED TOTAL	17	316.000			2.9721		7.0000

SOURCE	DF	ANOVA SS	F VAL	PR > F
DIAG	1	18.0000	2.04	0.1789
DRUG	2	48.0000	2.72	0.1063
DIAG*DRUG	2	144.0000	8.15	0.0058

TABLE C.10. SAS analysis of data in Table C.8.

TABLE 5.9, GLM, PAGE 2

GENERAL LINEAR MODELS PROCEDURE

DEPENDENT VARIABLE: SCORE

SOURCE	DF	SUM OF SQ	MEAN SQ	F VAL	PR > F	R-SQ	C.V.
MODEL	5	231.750	46.3500	5.65	0.0099	0.7386	40.190
ERROR	10	82.000	8.2000		ROOT MSE		SCORE MEAN
					2.8636		7.1250
CORRECTED TOTAL	15	313.750					

SOURCE	DF	TYPE I SS	F VAL	PR > F	DF	TYPE III SS	F VAL	PR > F
DIAG	1	20.250	2.47	0.1471	1	27.42857	3.34	0.0973
DRUG	2	53.100	3.24	0.0824	2	47.63077	2.90	0.1013
DIAG*DRUG	2	158.400	9.66	0.0046	2	158.40000	9.66	0.0046

TABLE C.11. Input file for analyzing data in Table 6.2. Most data are omitted to save space.

```
DATA T0602;
  INPUT EXAM $ COLLEGE $ SCORE;
  CARDS;
    U1  C1  22
    . . . . . . . . .
PROC ANOVA;
  CLASS EXAM COLLEGE;
  MODEL SCORE=EXAM¦COLLEGE;
  MEANS EXAM¦COLLEGE;
  TEST H=EXAM E=EXAM*COLLEGE;
TITLE 'TABLE 6.2';
```

TABLE C.12. Input file for analyzing data in Table 6.3. Most data are omitted to save space.

```
DATA T0603;
  INPUT UNIV $ COURSE $ SCORE;
  CARDS;
    U1  C1  20
    . . . . . . . . .
PROC ANOVA;
  CLASS UNIV COURSE;
  MODEL SCORE=UNIV¦COURSE;
  MEANS UNIV¦COURSE;
  TEST H=UNIV COURSE E=UNIV*COURSE;
  TITLE 'TABLE 6.3';
```

OUTPUT

Table C.13 shows the analysis of variance for the data in Table C.11. Note that in the second table no F values are given. In that table, SAS assumes that it is a fixed-effects model. With only one score per cell there is no error term, so no F ratios can be constructed. However, in the third table, we see the test of interest. There the correct F value of 1.95 is given for the main effect of the exam.

Table C.14 shows the analysis of variance for the data in Table C.12. It is basically the same as Table C.13, except that now two tests are listed for each main effect. The tests of main effects in the second table should be ignored; they are made as though it were a fixed-effects analysis (i.e., using MS_w as the error term). The tests in the third table are correct. The test of the interaction in the second table is correct because MS_w is the correct error term for testing the interaction.

Multiple Factors (Chapter 7)

The extension to multiple factors is straightforward. We will illustrate it with the data in Table 7.9, with Factor C random.

INPUT

Table C.15 shows the input file. Most of the changes should by now be obvious. The " INPUT " statement lists the three factors as well as the dependent variable. The " CLASS " statement lists the three factors in shorthand form. The symbols " A--C "(with *two* dashes) indicate that all of the variables from " A " through " C ", as listed in the " INPUT " statement, are factors. The " MODEL " statement is now " MODEL SCORE=A|B|C; ". The vertical lines indicate that all main effects and interactions involving those three factors should be analyzed. They make the statement equivalent to the much longer statement:

" MODEL SCORE=A B C A*B A*C B*C A*B*C ";

The statement " MEANS A|B; " instructs SAS to calculate means for the A and B main effects and the AB interaction; that is, SAS will calculate the means for fixed effects only. Finally, the A main effect, B main effect, and AB interaction all have to be tested with different denominators. This necessitates the three " TEST " statements.

OUTPUT

The output has the same format as in previous tables. Its interpretation should by now be straightforward.

TABLE C.13. SAS analysis of data in Table C.11.

TABLE 6.2, PAGE 2

ANALYSIS OF VARIANCE PROCEDURE

DEPENDENT VARIABLE: SCORE

SOURCE	DF	SUM OF SQ	MEAN SQ	F VAL	PR > F	R-SQ	C.V.
MODEL	59	921.000	15.6102	.	.	1.0000	0.0000
ERROR	0	0.000	0.0000		ROOT MSE		SCORE MEAN
CORRECTED TOTAL	59	921.000			0.0000		20.500

SOURCE	DF	ANOVA SS	F VAL	PR > F
EXAM	2	58.9000	.	.
COLLEGE	19	288.3333	.	.
EXAM*COLLEGE	38	573.7667	.	.

TESTS OF HYPOTHESES USING THE ANOVA MS FOR EXAM*COLLEGE AS AN ERROR TERM

SOURCE	DF	ANOVA SS	F VAL	PR > F
EXAM	2	58.9000	1.95	0.1562

TABLE C.14. SAS analysis of data in Table C.12.

TABLE 6.3, PAGE 2

ANALYSIS OF VARIANCE PROCEDURE

DEPENDENT VARIABLE: SCORE

SOURCE	DF	SUM OF SQ	MEAN SQ	F VAL	PR > F	R-SQ	C.V.
MODEL	23	1220.81	53.0788	2.56	0.0132	0.7101	37.395
ERROR	24	498.50	20.7708		ROOT MSE		SCORE MEAN
CORRECTED TOTAL	47	1719.31			4.5575		12.1875

SOURCE	DF	ANOVA SS	F VAL	PR > F
UNIV	3	218.2292	3.50	0.0308
COURSE	5	750.9375	7.23	0.0003
UNIV*COURSE	15	251.6458	0.81	0.6601

TESTS OF HYPOTHESES USING THE ANOVA MS FOR UNIV*COURSE AS AN ERROR TERM

SOURCE	DF	ANOVA SS	F VAL	PR > F
UNIV	3	218.2292	4.34	0.0217
COURSE	5	750.9375	8.95	0.0004

TABLE C.15. Input file for analyzing data in Table 7.9. Most data are omitted to save space.

```
DATA T0709;
   INPUT A $ B $ C $ SCORE;
   CARDS;
      A1 B1 C1  11
      . . . . . . . . . . .
PROC ANOVA;
   CLASS A B C;
   MODEL SCORE=A|B|C;
   MEANS A|B;
   TEST H=A E=A*C;
   TEST H=B E=B*C;
   TEST H=A*B E=A*B*C;
TITLE 'TABLE 7.9';
```

Nested Factors (Chapter 8)

The main difference with nested factors is in the way that nesting is specified. We will illustrate it with the data in Table 8.6.

INPUT

Table C.16 shows the input file for the data in Table 8.6. Designs with nested factors are balanced so long as the cells that contain data all have the same sample size. Thus, we can use the ANOVA procedure. The only change with nested designs is in the way that nested factors are specified. Moreover, in SAS that should be no problem because SAS specifies nested factors with parentheses exactly as in this text. Thus, our model statement now indicates that the SUBJ (subjects) factor is nested in INST (instructions). In this design, the NONS (nonsense syllable) main effect and the NONS*INST interaction are both tested with the same denominator, so they are both listed in a single " TEST " statement.

The notation for nesting is almost exactly as in this text. When specifying factors, each factor name is followed by parentheses containing the name(s) of the factor(s) it is nested in. If NONS had been nested in INST, the MODEL statement would have been

MODEL SCORE=INST|NONS(INST)|SUBJ(INST);

If INST had been nested in NONS (necessitating that SUBJ be nested in NONS as well), the statement would have been

MODEL SCORE=NONS|INST(NONS)|SUBJ(NONS INST);

TABLE C.16. Input file for analyzing data in Table 8.6. " NONS " stands for nonsense syllable, "INST" stands for "instruction", and " SUBJ " stands for "subject". Most data are omitted to save space.

```
DATA T0806;
   INPUT NONS $ INST $ SUBJ $ SCORE;
   CARDS;
      N1  I1 S1   8
      . . . . . . . . . . .
PROC ANOVA;
   CLASS NONS--SUBJ;
   MODEL SCORE=NONS¦INST¦SUBJ(INST);
   MEANS NONS¦INST;
   TEST H=NONS NONS*INST E=NONS*SUBJ(INST);
   TEST H=INST E=SUBJ(INST);
   TITLE 'TABLE 8.6';
```

(Note that the two factors in parentheses are separated by a space, not an asterisk.)

When specifying effects, the nested factors all appear within a single set of parentheses at the end. If we had a four-factor, $A \times B(A) \times C \times D(C)$, design, with D random, one TEST statement would read

TEST H=B(A) A*C B*C(A) E=B*D(A C);

instructing SAS to test the B main effect, the AC interaction, and the $BC(A)$ interaction using $MS_{bd(ac)}$ in the denominator.

OUTPUT

The output for these data is not shown. It is basically the same as in previous tables except that nested factors are noted explicitly.

Other Incomplete Designs (Chapter 9)

Specialized designs such as those in Chapter 9 often involve calculating sums of squares such as $SS_{a:b}$. Such designs are unbalanced and must be analyzed using the GLM procedure. Some, such as Latin squares, are balanced and can be analyzed using ANOVA. When in doubt, GLM is the safer alternative. We will give two examples: the balanced incomplete blocks design in Table 9.4 (which requires use of GLM) and the Latin square design in Table 9.11 (which allows use of ANOVA).

TABLE C.17. Input file
for analyzing data in Table
9.4. Most data are omitted
to save space.

```
DATA T0904;
  INPUT A $ B $ SCORE;
  CARDS;
    A1 B1  19
    . . . . . . . .
PROC GLM;
  CLASS A B;
  MODEL SCORE=A B;
  MEANS A B;
TITLE 'TABLE 9.4';
```

INPUT

The input file for Table 9.4 is in Table C.17; the input file for Table 9.11 is in Table C.18. The only important differences are in the " MODEL " and " MEANS " statements. We do not want SAS to analyze interactions. SAS would if we asked it to, but then there would be no degrees of freedom left for error. When we tell SAS not to analyze interactions, the remainder, after analyzing the main effects, is left as error variance. Thus, we list the factors individually, separated by spaces, to indicate that only the main effects should be analyzed.

OUTPUT

The outputs for these two designs are not shown. They do not differ from previous examples in any important ways.

Trend Analyses (Chapter 10)

The ANOVA and GLM procedures will not do trend analyses directly. GLM can be made to do trend analyses as planned comparisons if the coefficients are entered in an " ESTIMATE " or " CONTRAST " statement. Another procedure, " REG ", will also do trend analyses, but it requires more extensive input. We will illustrate the use of REG. The REG procedure is a multiple regression procedure, but the trend analysis is fundamentally the same as a

TABLE C.18. Input file for analyzing data in Table 9.11. Most data are omitted to save space.

```
DATA T0911;
   INPUT A $ B $ C $ SCORE;
   CARDS;
      A1 B1 C1 10
      . . . . . . . . . . .
PROC ANOVA;
   CLASS A--C;
   MODEL SCORE=A B C;
   MEANS A B C;
TITLE 'TABLE 9.11';
```

polynomial regression of the dependent variable on the independent variable. Three other procedures, " RSQUARE ", " RSREG ", and " STEPWISE ", might also be useful, but we will not illustrate them here.

Trend analyses can also be done on multifactor models as in Chapter 11. However, these, like other contrasts in multifactor models, are difficult to specify in SAS. We will limit ourselves to the one-way model in Chapter 10.

INPUT

Table C.19 shows the input file for the data in Table 10.1. Note that we must specify six variables in the INPUT statement. G1 is the number of learning trials, divided by 10 to make the numbers more manageable. G2 is the square of G1, G3 is the cube, and so on. We go up to G5 because that is the highest possible trend with six groups. The sixth variable is the dependent variable, SCORE.

The INPUT statement specifies six variables, so for each subject we must now input six values. The first five are the first five powers of the number of learning trials, again divided by 10, and the sixth is the subject's score. (SAS provides a number of devices for inputting the data without typing in all six variables; however, we have typed in all the data to show what is needed as input to the REG procedure. The SAS manual tells how to create columns two through five from column one.)

The statement immediately following the data tells SAS to use the REG procedure. There are no classification variables in multiple regression, so there is no " CLASS " statement. The model statement is the same as for the ANOVA and GLM procedures, except that we must now add SS1, following a slash, telling SAS to calculate the Type I sums of squares.

TABLE C.19. Input file for analyzing data in
Table 10.1. Variables "G1" through "G5" are
the linear through quintic coefficients. Some
data are omitted to save space.

```
DATA T1001;
  INPUT G1 G2 G3 G4 G5 SCORE;
  CARDS;
    1    1    1    1    1    13
    1    1    1    1    1     5
    ........................
    2    4    8   16   32    12
    2    4    8   16   32     3
    ........................
    3    9   27   81  243    13
    3    9   27   81  243    10
    ........................
    4   16   64  256 1024    14
    4   16   64  256 1024    13
    ........................
    5   25  125  625 3125    18
    5   25  125  625 3125    18
    ........................
    6   36  216 1296 7776    17
    6   36  216 1296 7776    19
    ........................
PROC REG;
  MODEL SCORE=G1 G2 G3 G4 G5 / SS1;
TITLE 'TABLE 10.1';
```

OUTPUT

The output is shown in Table C.20. Most of the information in the first table is
basically the same as for the ANOVA and GLM procedures, but it is arranged
differently. " C TOTAL " is short for " CORRECTED TOTAL "; " DEP
MEAN " is the grand mean of the scores; " R-SQUARE " is the squared
correlation coefficient, i.e., the proportion of variance accounted for; " ADJ
R-SQ " is also the proportion of variance accounted for, but it is adjusted
downward because the unadjusted estimate tends to have a positive bias.

The next table contains the trend analysis, but it must be interpreted carefully. The estimates and their associated standard errors are valid only if the
full model, with all five powers of the independent variable, is used. The t test
and its associated p-value are based on these estimates and their standard
errors. Thus, they test the effect of removing each term, individually, from
the full model. That is why the intercept and linear terms are not significant.
If we removed them but kept all higher-order terms, the fit would not be significantly worse. Because we usually want to remove the higher-order terms
and keep the lower-order terms, these tests are not very meaningful.

Appendix C. Analysis of Variance Using SAS 455

TABLE C.20. SAS analysis of data in Table 10.1.

TABLE 10.1, PAGE 1

DEP VARIABLE: SCORE

ANALYSIS OF VARIANCE

SOURCE	DF	SUM OF SQUARES	MEAN SQUARE	F VAL	PROB>F
MODEL	5	890.60	178.1200	19.39	0.0001
ERROR	54	496.00	9.1852		
C TOTAL	59	1386.60			

ROOT MSE	3.030707	R-SQUARE	0.6423
DEP MEAN	14.3	ADJ R-SQ	0.6092
C.V.	21.19376		

PARAMETER ESTIMATES

VARIABLE	DF	PARAMETER ESTIMATE	STANDARD ERROR	T FOR H0: PARAM=0	PROB > |T|	TYPE I SS
INTERCEP	1	1.00000	29.1169	0.034	0.9727	12269.400
G1	1	14.07500	60.6128	0.232	0.8173	803.571
G2	1	-8.42917	44.2183	-0.191	0.8495	20.743
G3	1	2.41667	14.5854	0.166	0.8690	53.389
G4	1	-0.27083	2.2212	-0.122	0.9034	12.857
G5	1	0.00833	0.1268	0.066	0.9478	0.040

The last column contains the meaningful values. These are the sums of squares associated with each term. They are identical to those given in Chapter 10. Unfortunately, no F tests are reported; they have to be calculated by hand, using the ERROR mean square in the table above as the denominator term.

If, say, we chose to retain just the first three terms, we could reanalyze the data, using only those three. We could then get estimates of the coefficients. However, they would not be the same as the coefficients that were calculated in Chapter 10. Instead, they would be the coefficients for the values that entered into the model; for example, the linear coefficient would multiply G1, and the quadratic coefficient would multiply G2.

Multiway Trend Analyses (Chapter 11)

Multiway trend analyses are more difficult than one-way trend analyses using SAS. It is probably simplest to do the basic analysis using SAS, obtaining the cell means in the analysis, and then to do the trend analyses by hand.

Multivariate Analysis of Variance (Chapter 13)

The multivariate analysis of variance is a fairly straightforward extension of the univariate analysis. We will illustrate it with the data in Table 12.2.

INPUT

Table C.21 shows the input file for the data in Table 12.2. For these data the " INPUT " statement specifies two factors, labeled A and B, and two dependent variables, labeled X and Y. These are simply listed in the order in which they will appear in the data statements, since at this point SAS neither knows nor cares what kinds of variables are being input.

We will use the ANOVA procedure because, although this is a multivariate design, it is still balanced. The " CLASS " statement specifies that A and B are factors, as usual. The " MODEL " statement has both X and Y to the left of the equality—all variables to the left of the equality are assumed to be dependent variables—and has the usual shorthand notation for the effects to the right.

The new statement has the key word " MANOVA ". It can be followed by a notation like that used in " TEST " statements. The symbol " H= " is followed by a list of effects to be tested, and this is optionally followed by " E= " and the effect to be treated as error. In the first " MANOVA " statement, we assume a fixed-effects model, so all of the effects will be tested against within variance. The key word " _ALL_ " (the underline characters at the beginning and end are necessary) is a shorthand way of saying that all effects should be tested. With no " E= ", SAS will assume that the effects are to be tested against within cells error.

Just as more than one " TEST " statement can be used, so can more than one " MANOVA " statement. If factor B were random, we would have used the following two statements:

MANOVA H=A E=A*B;
MANOVA H=B A*B;

The first of these specifies that the A main effect is to be tested using the AB interaction as the error term; the second specifies that the B main effect and AB interaction are to be tested using the within cells variance.

TABLE C.21. Input file for multivariate analysis of
variance on data in Table 12.2.

```
DATA T1202;
  INPUT A $ B $ X Y;
  CARDS;
    A1 B1  8  5
    A1 B1  4  3
    A1 B1  0  4
    A1 B2  8  7
    A1 B2 10  7
    A1 B2  6  4
    A1 B3  4  3
    A1 B3  6  6
    A1 B3  8  6
    A2 B1 10  5
    A2 B1  6  6
    A2 B1 14  7
    A2 B2  0  2
    A2 B2  4  7
    A2 B2  2  6
    A2 B3 15  8
    A2 B3  9  3
    A2 B3 12  7
PROC ANOVA;
  CLASS A B;
  MODEL X Y=A¦B;
  MEANS A¦B;
  MANOVA H=_ALL_;
TITLE 'TABLE 12.2, MANOVA (CHAPTER 13)';
```

OUTPUT

SAS first shows the results of univariate tests of the two dependent variables taken individually. They are not shown here because they are identical to those shown in previous tables. The next table shows the means. This is followed by the eigenvalues and eigenvectors, which are primarily of technical interest.

Table C.22 shows the multivariate analyses. Four of them are given. The first is Wilks's lambda, based on the maximum likelihood. The next two, Pillai's trace and the Hotelling–Lawley trace, are mentioned only briefly in Chapter 13. Neither has a direct intuitive interpretation, but Pillai's trace may be useful in that it is believed to be the most robust of the four. The fourth is Roy's maximum root, based on the optimal linear combination.

The contents of Table C.22 are straightforward. The table begins with a list of definitions of terms that will be used in the actual results. Then it gives the results of the four tests. The first line for each result gives both a formula for and the value of the criterion, the second line gives the formula for an exact

TABLE C.22. SAS multivariate analysis of variance on data in Table 12.2.

TABLE 13-2, MANCOVA (CHAPTER 13), PAGE 6

ANALYSIS OF VARIANCE PROCEDURE

MANOVA TEST CRITERIA FOR THE HYPOTHESIS OF NO OVERALL A EFFECT

```
        H = ANOVA SS&CP MATRIX FOR: A
        E = ERROR SS&CP MATRIX
        P = RANK OF (H+E)              =    2
        Q = HYPOTHESIS DF             =    1
       NE= DF OF E                    =   12
        S = MIN(P,Q)                  =    1
        M = .5(ABS(P-Q)-1)           =   0.0
        N = .5(NE-P)                  =   5.0
```

WILKS' CRITERION   L = DET(E)/DET(H+E) = 0.85046729
        EXACT F = (1-L)/L*(NE+Q-P)/P WITH P AND NE+Q-P DF   (SEE RAO 1973 P 555)
                F(2,11) = 0.97   PROB > F = 0.4103

PILLAI'S TRACE   V = TR(H*INV(H+E)) = 0.14953271   (SEE PILLAI'S TABLE #2)
    F APPR = (2N+S)/(2M+S+1) * V/(S-V) WITH S(2M+S+1) AND S(2N+S) DF
                F(2,11) = 0.97   PROB > F = 0.4103

HOTELLING-LAWLEY TRACE = TR(E**-1*H) = 0.17582418   (SEE PILLAI'S TABLE #3)
    F APPR = (2S*N-S+2)*TR(E**-1*H)/(S*S*(2M+S+1)) WITH S(2M+S+1) AND 2S*N-S+2 DF
                F(2,11) = 0.97   PROB > F = 0.4103

ROY'S MAXIMUM ROOT CRITERION = 0.17582418   (SEE AMS VOL 31 P 625)
        FIRST CANONICAL VARIABLE YIELDS AN F UPPER BOUND
                F(2,11) = 0.97   PROB > F = 0.4103
```
```

F equivalent (e.g., for Wilks's criterion), an approximate F (e.g., for Pillai's trace), or an upper bound on F (e.g., Roy's maximum root). The last line gives the F ratio and its associated p-value.

Repeated Measures Designs (Chapter 13)

Repeated measures designs can be analyzed by the techniques described in Chapter 13. However, to analyze them in this way, the format for input is somewhat different.

Table C.23 shows the input file for the data in Table 6.2. Note that now all the data for a given college (colleges are the random factor) are on the same line. This fact must be reflected in the " INPUT " statement, which indicates that the three values are for " TEST1 ", " TEST2 ", and " TEST3 ". The " MODEL " statement indicates that all three variables are now dependent variables; there are no independent variables in the data statements. (The notation " TEST1--TEST3 " is shorthand for " TEST1 TEST2 TEST3 ".) Moreover, nothing follows the equals sign because as yet no independent variable has been specified. The " REPEATED " statement actually specifies the design; in this example, it states that each column of scores is for a different level of the independent variable " TESTS ".

Table C.24 shows the input file for the data in Table 7.9. Here C is random, and each line of the data statements contains the six values at a given level of C. The " INPUT " statement must reflect the fact that there are six values on a line, so we list six variables; the names of the variables can be chosen arbitrarily as they will not be used explicitly in the analyses we are interested in. The model statement again reflects the fact that there are six dependent variables on a line. The " REPEATED " statement again specifies the model. However, here there are two factors that must be specified. In this connection note that the order of the values is $AB_{11}, AB_{12}, AB_{13}, AB_{21}, \ldots$; that is, the slowest-moving index is for Factor A. With the data in this order, Factor A must be listed first in the " REPEATED " statement. Moreover, SAS has no way of guessing the number of levels of each factor, so that information must be provided. It is provided by following each factor name with its number of levels. Any number of factors can be specified in this way, so long as they are listed in the order in which the indices change (slowest-moving index first, fastest-moving last) and each factor name is followed by its number of levels. In addition, the numbers of levels must total the number of dependent variables on each line of the data statements (in this example, they must total six).

The last example is from Table 8.6. This is somewhat more complicated because there are both crossed and nested factors. It is shown in Table C.25. Because subjects are nested in the instructions factor, each line of the data statements must indicate which instructions were being used for those data. Thus, each line is preceded by B1, B2, or B3. Then the scores for the four lists of nonsense syllables follow. Again the " INPUT " statement indicates that the first variable (" INST ") on each line is categorical, while the remaining

TABLE C.23. Input file for repeated measures analysis of data in Table 6.2.

```
DATA T0602;
   INPUT TEST1 TEST2 TEST3;
   CARDS;
      22 18 13
      24 30 19
      26 21 22
      21 27 13
      26 23 17
      18 21 22
      16 15 21
      22 25 23
      14 22 20
      11 25 17
      24 15 17
      21 19 21
      20 22 20
      22 25 19
      25 19 24
      23 24 23
      22 24 21
      22 14 17
      21 23 12
      17 18 22
PROC ANOVA;
   MODEL TEST1--TEST3=;
   REPEATED TESTS;
   TITLE 'TABLE 6.2 WITH USE OF REPEATED';
```

four (" S1 " through " S4 ") are numerical. Because one of the variables in the data is a classification variable (i.e., a factor), we must include the " CLASS " statement. In the " MODEL " statement we also note that the variable " INST " is an independent variable. We want to print the means for different instructions, so we add a " MEANS " statement in the input file. Finally, each of the four columns of dependent variables (i.e., S1 through S4) represents a different level of the single factor of nonsense syllable list (" SYL "), and that fact is specified in the " REPEATED " statement.

In general, if the random factor in a repeated measures design is nested in one or more other factors, those factors are specified in the data statements and are generally treated as in nonrepeated measures designs. Factors with which the random factor is crossed are specified in the " REPEATED " statement, and all data from different levels of these factors are entered on the same line in the data statements.

TABLE C.24. Input file for repeated measures analysis of data in Table 7.9.

```
DATA T0709;
  INPUT C1 C2 C3 C4 C5 C6;
  CARDS;
     11  6 -7  1 -1 -3
     10 -2 -7  7  1 -1
     10  3 -7  2  4  2
     14  8 -3 -1  6  4
      4  3 -1  0 -7  1
PROC ANOVA;
  MODEL C1--C6=;
  REPEATED A 2, B 3;
  TITLE 'TABLE 7.9 WITH REPEATED';
```

TABLE C.25. Input file for repeated measures analysis of data in Table 8.6. " INST " is "instructions" and " NONS " is "nonsense syllable."

```
DATA T0806;
  INPUT INST $ S1 S2 S3 S4;
  CARDS;
     B1  8 15 12 17
     B1 20 24 16 20
     B1 14 20 19 20
     B2 21 18 17 28
     B2 23 21 17 31
     B2 26 29 26 30
     B3 15 12 13 18
     B3  6 11 10 12
     B3  9 18  9 23
PROC ANOVA;
  CLASS INST;
  MODEL S1--S4=INST;
  MEANS INST;
  REPEATED NONS;
  TITLE 'TABLE 8.6 USING SAS WITH REPEATED';
```

The output of a repeated measures analysis can be voluminous: up to eleven pages for the example above. (SAS allows you to save a few trees by suppressing uninteresting output; see the manual for details.) It is not reproduced here because the format is already familiar from the analyses in previous chapters. Moreover, much of it is usually of little or no interest.

The first tables show one-way analyses, in the same format as Table C.10, for each column of dependent variables in the data statements. For example, there are six such tables for the data in Table C.24 because there are six columns of dependent variables. Ordinarily, the test is on main effects and interactions involving factors in which the random factor is nested. In Table C.25 this would be " INST ". However, for Tables C.23 and C.24 there is no nesting, so for them SAS tests the null hypothesis that the grand mean of the values in each column is zero. If cell means are requested, they follow these initial analyses.

The pages that follow then give the analyses of main effects and interactions. These may be multivariate or univariate as appropriate; they follow the formats described earlier.

Finally, the last page gives an ordinary ANOVA as described in Chapters 6, 7, and 8 of this text. Two estimates of Box's correction are also given. The first, called the "Greenhouse–Geisser Epsilon," is the estimate described in this text. It tends to be somewhat conservative. The second, called the "Huynh–Feldt Epsilon," is less biased, but may occasionally be larger than one.

Analysis of Covariance (Chapter 14)

The analyses of variance described above are easily modified for analyses of covariance. We will illustrate with the two analyses described in Chapter 15. Both are on the one-way data in Table 12.1. The first analysis uses mental maturity (MM) as the dependent variable, with IQ and chronological age (CA) as the covariates; the second analysis uses mental maturity and IQ as the dependent variables, with chronological age as the covariate.

INPUT

The input file for an analysis of covariance is almost identical to that for a multivariate analysis of variance. Table C.26 shows the input for the analysis with IQ and chronological age as covariates. The GLM procedure is used for analyses of covariance. The basic difference between this and the multivariate analysis is that the covariates, " IQ " and " CA ", have been declared independent variables (i.e., they appear to the right of the " = " sign) in the " MODEL " statement.

Note that the covariates appear before the " GROUP " variable in the

TABLE C.26. Input file for analysis of covariance of data in Table 12.1 with IQ and chronological age (CA) as covariates. Most data have been eliminated to save space.

```
DATA T1201;
  INPUT GROUP $ MM IQ CA;
  CARDS;
    G1 5 23 27
    . . . . . . . . . .
PROC GLM;
  CLASS GROUP;
  MODEL MM=IQ CA GROUP;
  MEANS GROUP;
TITLE 'TABLE 12.1, GLM MANOVA WITH TWO COVARIATES (CHAPTER 14)';
```

TABLE C.27. Input file for data in Table 12.1 with mental maturity (MM) and IQ as dependent variables, and chronological age (CA) as covariate. Most data have been eliminated to save space.

```
DATA T1201;
  INPUT GROUP $ MM IQ CA;
  CARDS;
    A1 5 23 27
    . . . . . . . . . .
PROC GLM;
  CLASS GROUP;
  MODEL MM IQ=CA GROUP;
  MEANS GROUP;
  MANOVA H=_ALL_;
TITLE 'TABLE 12.1, , GLM MANOVA WITH ONE COVARIATE (CHAPTER 14)';
```

" MODEL " statement. Remember that SAS removes all the variance that can be attributed to the first variable before testing the second, etc. In an analysis of covariance, we remove all variance that can be attributed to the covariates before testing the effects.

Table C.27 shows the input file for the analysis with " CA " as the only covariate. Now " MM " and " IQ " are declared as dependent variables (to the left of the " = " sign), while " CA " is declared as an independent variable. The " MANOVA " statement also appears here because there are two dependent variables.

OUTPUT

The output, like that of a multivariate or repeated measures analysis, can be voluminous. It is not reproduced here because the format is identical to analyses that were previously discussed. The analysis of covariance follows the format for a univariate or a multivariate analysis of variance, depending on the number of dependent variables. The main difference is that the covariates are tested just like the other effects. If there is more than one covariate, they will be tested individually, and sequentially, just as the effects. Thus, there is no overall test of the covariates, and the order in which the effects are listed can make a difference. In a univariate analysis, this is not a problem – the " TYPE I " sums of squares for the covariates can be pooled for an overall test. In a multivariate analysis, this is not so easy to do.

Appendix D

Analyses Using SPSS

SPSS (a registered trademark of SPSS, Inc.) is a program that performs a variety of statistical procedures. The version described here is SPSS-X. To use SPSS, a file containing data and instructions is first prepared. Usually the file is stored with a name having the form "*.SPS", where the "*" may be any file name that is legal on your system. (Some aspects of SPSS are necessarily specific to the system it is on; for information on those aspects, you will probably need to consult with experts on your system.)

There are a great many options for preparing the file; their use depends on the purposes and sophistication of the user. We will consider only the simplest aspects here. We will give just enough information to enable you to enter data and analyze it, using most of the procedures described in the text. For more sophisticated applications, you are referred to the SPSS manuals.

The file can be divided conveniently into four parts. The first part basically is a single line, giving a title, etc.; the second instructs SPSS about the general format of the data, the third is a set of instructions for analyzing the data, and the fourth is the data. Each line of the file, except the lines containing data, starts with one or more "key words."

The first part is optional. It consists of a single line containing the key word "TITLE" followed by whatever title you choose.

We will consider each of the remaining parts in the contexts of specific analyses.

One-Way Analysis Fixed-Effects (Chapter 2)

INPUT

A sample SPSS file, for the data in Table 2.1, is in Table D.1. The first line simply says "TITLE 'TABLE 2.1' ". The word "TITLE" is the key word, indicating that "TABLE 2.1" should be used as a title on the printout. The second starts with the key words "DATA LIST", telling SPSS that this line describes the data that will follow. The third word on this line is also "LIST", indicating that the data will appear in a list format. SPSS allows the data to be in any of three formats. The first, default format is "FIXED", in which each datum must be in a specific column, and the columns must be specified. "FREE" and "LIST" are simpler. "FREE" format allows the data values to be input in any columns whatever, so long as they are in the correct order. This

TABLE D.1. Input file for analyzing data in Table 2.1.

```
TITLE 'TABLE 2.1'
DATA LIST LIST / GROUP SCORE
ONEWAY SCORE BY GROUP (1,3) / STATISTICS=DESCRIPTIVES
BEGIN DATA
    1 5
    1 3
    1 5
    1 6
    1 1
    2 1
    2 2
    2 2
    2 0
    2 3
    3 5
    3 4
    3 7
    3 5
    3 2
END DATA
```

is the most convenient format, but it encourages the user to input the data in a form that is hard to read and/or correct. The third format is "LIST"; it is like "FREE" except that each line must contain a single record (for the data in Table 2.1, each subject is a record). That is the format used here. A slash follows the word "LIST", indicating that the names of the variables will be entered next. The names of both the dependent and independent variables follow the slash. The variables must be entered in the order in which they will appear as data. Thus, looking farther down the file, you can see that each line contains both a value of X and a label indicating the group from which that value came. The first datum on each line is the group identifier, so the first word after the slash is "GROUP". The last word on this line is the word "SCORE", indicating that the second datum on each line will be a student's score on the tests. Whenever a list such as this one is typed, the items can be separated by either spaces or commas. This text will use spaces most often, with commas used when they make the data easier to read.

The key word on the next line is "ONEWAY", indicating that a one-way analysis of variance is to be performed. (Other procedures are available that will also do a one-way analysis of variance; however, ONEWAY has a number of features that the other procedures lack.) The dependent variable, "SCORE", is then listed, followed by the word "BY" and the independent variable, "GROUP". The independent variable must be followed by two integers, in parentheses, giving the lower and upper bound on the values that the independent variable can take. Here there are three groups, and we have labeled them "1", "2", and "3", so the lower bound is "1" and the upper bound is "3". Note that all variables must take on numerical values, and the values of the independent variable must be consecutive integers for the analysis to be performed properly.

The list of variables can contain more than one dependent variable. For example, if there were two dependent variables, labeled "SCORE1" and "SCORE2", we could write

ONEWAY SCORE1 SCORE2 BY GROUP (1,3)

However, there cannot be more than one independent variable, and ONEWAY cannot perform a multivariate analysis of variance. Instead it would do two univariate analyses, one on "SCORE1" and the other on "SCORE2".

Lines that specify analyses, such as ONEWAY, also allow subcommands. The subcommands follow a slash and have the form "X=Y". A number of subcommands will be used for the analyses in Chapters 3 and 4, but here we specify only one, "STATISTICS=DESCRIPTIVES". This tells SPSS to print descriptive statistics such as group means. For other options available, consult the SPSS-X User's Guide.

The fourth line contains the key words "BEGIN DATA", telling SPSS that the next lines contain data. SPSS continues reading, assuming that each line contains data, until it reads the last line, "END DATA". As stated above, all data must be numerical, and the values for the independent variable must be consecutive integers. We have chosen "1", "2", and "3" to label the respective groups.

OUTPUT

By default, SPSS-X outputs its data to the screen. However, it can save the output in a file. (The specific procedure for doing this may be system-dependent. Contact someone who knows the operation of SPSS on your system to find out how to save the output in a file.) Much of the output consists of "bookkeeping" information, i.e., a listing of the input along with information on how much time and memory were needed, etc. The relevant output is reproduced in Table D.2. (The output has been modified slightly to fit conveniently into the text. Some values are given to fewer decimals, and some headings have been shortened. However, it is essentially identical to the output of SPSS.)

The output consists of two tables. The first table gives the output of the analysis of variance. The second gives the descriptive statistics that were requested in the "STATISTICS=DESCRIPTIVES" command. Both tables should be largely self-explanatory.

TABLE D.2. SPSS output for data in Table D.1.

TABLE 2.1								Page 2

- - - - - - - - - - - - - - - - O N E W A Y - - - - - - - - - - - - - -

Variable SCORE
By Variable GROUP

ANALYSIS OF VARIANCE

SOURCE	D.F.	SUM OF SQUARES	MEAN SQUARES	F RATIO	F PROB.
BETWEEN GROUPS	2	25.2000	12.6000	4.3953	.0370
WITHIN GROUPS	12	34.4000	2.8667		
TOTAL	14	59.6000			

GROUP	COUNT	MEAN	STANDARD DEVIATION	STANDARD ERROR	95 PCT CONF INT FOR MEAN		
Grp 1	5	4.0000	2.0000	.8944	1.5167	TO	6.4833
Grp 2	5	1.6000	1.1402	.5099	.1843	TO	3.0157
Grp 3	5	4.6000	1.8166	.8124	2.3444	TO	6.8556
TOTAL	15	3.4000	2.0633	.5327	2.2574	TO	4.5426

Planned Comparisons (Chapter 3)

Contrasts are easy to test in SPSS. However, contrasts must be tested one at a time; they cannot be combined for testing. There is a limit of 10 contrasts in one analysis.

INPUT

Table D.3 shows an SPSS input file for the data in Table 3.1. (Some of the data values are missing due to space considerations.) The first two statements have the same format as in Table D.1. The third statement differs in several ways. First, it says that the independent variable ranges from 1 to 4 because there are four groups. Second, it is followed by a number of subcommands, each preceded by a slash. Although these subcommands are placed on different lines, SPSS regards them all as part of the "ONEWAY" command. The first four subcommands are pertinent to Chapter 3; they specify tests of planned comparisons. The fifth and sixth will be discussed in conjunction with the post hoc tests described in Chapter 4. The last is familiar; it asks for descriptive statistics as for the data in Chapter 2.

For each planned comparison there is a separate subcommand, preceded by a slash. The key word for this subcommand is "CONTRAST", and the

TABLE D.3. SPSS input for analyzing
data in Table 3.1. Most of the data have
been omitted to save space.

```
TITLE 'TABLE 3.1'
DATA LIST LIST / GROUP SCORE
ONEWAY SCORE BY GROUP (1,4)
    / CONTRAST   0 -1   1   0
    / CONTRAST   0 -1  -1   2
    / CONTRAST   3 -1  -1  -1
    / CONTRAST   1  1   1   1
    / RANGES TUKEY
    / RANGES 3.33 3.68 3.86
    / STATISTICS = DESCRIPTIVES
BEGIN DATA
   1   1
   . . . .
   4  14
END DATA
```

key word is followed by the coefficients for the contrast to be tested. Note
that all four coefficients are listed, even though some of these coefficients
are zero. Thus, the first contrast compares Group 2 with Group 3, the second
compares Group 4 with the average of Groups 2 and 3, and the third compares
Group 1 with the average of Groups 2, 3, and 4. (Fractional coefficients could
have been listed, indicating that averages are in fact being tested; however,
they are not necessary.) Noncontrasts, with coefficients that do not sum to
zero, can be tested using the same "CONTRAST" subcommand; SPSS will
analyze noncontrasts, but will issue a warning statement with the analysis.
The fourth planned comparison is a noncontrast; its coefficients are all equal
to one, indicating that the grand mean is being tested.

OUTPUT

SPSS first does an overall analysis of variance. Its format is identical to that
in Table D.1 and is not shown here
   Table D.4 shows the output for the four planned comparisons. There are
three sets of columns. The first column, labeled "VALUE", gives the value of
each contrast; the next four columns give the results of tests assuming equal
variances (and thus using pooled variances); the last four columns give the
results of an approximate t-test, essentially the same as that described in
Chapter 3, that does not assume equal variances. In each table, the column
labeled "S. ERR" gives the standard deviation of the contrast; the column
labeled "T" gives the t-value; the column labeled DF gives the degrees of
freedom; and the column labeled "T PR." gives the p-value. Note that for the

fourth planned comparison, the test of the grand mean, there is a message warning that that is not a contrast.

Post Hoc Tests (Chapter 4)

SPSS can do a limited variety of post hoc tests. These include the Scheffé and Tukey tests, along with the Newman–Keuls and the Duncan. However, no matter which tests are chosen, SPSS is limited to testing all pairs of means, and only pairs of means. No test more complicated than a simple pair comparison can be performed. Unfortunately, we do not recommend any of these tests for comparing all pairs of means. The Scheffé and Tukey tests lack power, and the other tests are not valid.

In addition, for tests based on the Studentized-range statistic, only one α level, .05, is permitted. However, it is possible to devise one's own tests, based on the Studentized-range statistic, by specifying the critical values of the statistic.

INPUT

Two tests are illustrated in Table D.3. Both involve the subcommand "RANGES", preceded by a slash. The first is a simple Tukey test on all pairs of means, with the default $\alpha = .05$; the second is Ryan's test. It contains only the key word "RANGES" followed by a set of critical values. These critical values are taken directly from the Studentized range table. Their meaning should be obvious if you compare them with the critical values for the Ryan test in Table 4.2. For the Welsch test we would have used 3.33, 3.50, and 3.86.

If we had chosen to do Scheffé tests, we could have specified the significance level by putting it in parentheses after the key word. For example, if we had wanted to do the tests at $\alpha = .01$, we would have written

/ RANGES=SCHEFFÉ (.01)

OUTPUT

Table D.5 shows the output for the Tukey test on the data in Table D.3. (Note that in the table it is called the "TUKEY-HSD PROCEDURE". The "HSD" stands for "honestly significant differences." By inference, the other available Tukey, which SPSS calls Tukey B, does not give honestly significant differences. In fact, it resembles the Newman–Keuls test and is invalid for basically the same reasons.) The output for Ryan's test is identical in form.

The table begins with a helpful explanation of just how the tests were conducted. However, after that, relatively little information is provided. First, SPSS generates a table like Table 4.2. The asterisks denote pairs of groups that are significant at the chosen significance level (.05 by default for the Tukey

TABLE D.4. SPSS tests of contrasts for data in Table D.3.

TABLE 3.1   Page 3

- - - - - - - - - - - - - - O N E W A Y - - - - - - - - - - - - - - - - -

			POOLED VARIANCE ESTIMATE				SEPARATE VARIANCE ESTIMATE		
	VALUE	S. ERR	T VAL	DF	T PR	S. ERR	T VAL	DF	T PR
CONTRAST 1	-1.75	1.36	-1.288	28	0.208	1.52	-1.150	14	0.270
CONTRAST 2	8.50	2.35	3.612	28	0.001	1.93	4.413	21	0.000
CONTRAST 3	-12.50	3.33	-3.756	28	0.001	3.45	-3.623	11	0.004
CONTRAST 4*	39.50	1.92	20.557	28	0.000	1.92	20.557	25	0.000

* ABOVE INDICATES SUM OF COEFFICIENTS IS NOT ZERO.

TABLE D.5. SPSS output of Tukey tests on data in Table D.3.

TABLE 3.1   Page 4

- - - - - - - - - - - - - - - - O N E W A Y - - - - - - - - - - - - - - - -

        Variable SCORE
    By Variable GROUP

MULTIPLE RANGE TEST

TUKEY-HSD PROCEDURE
RANGES FOR THE 0.050 LEVEL -

        3.86   3.86   3.86
THE RANGES ABOVE ARE TABLE RANGES.
THE VALUE ACTUALLY COMPARED WITH MEAN(J)-MEAN(I) IS..
      1.9214 * RANGE * DSQRT(1/N(I) + 1/N(J))
    (*) DENOTES PAIRS OF GROUPS SIGNIFICANTLY DIFFERENT AT THE 0.050 LEVEL

                                G G G G
                                r r r r
                                p p p p

        Mean   Group   1 3 2 4

       6.7500   Grp 1
       8.6250   Grp 3
      10.3750   Grp 2
      13.7500   Grp 4   * *

    HOMOGENEOUS SUBSETS   (SUBSETS OF GROUPS, WHOSE HIGHEST AND LOWEST MEANS
                            DO NOT DIFFER BY MORE THAN THE SHORTEST
                            SIGNIFICANT RANGE FOR A SUBSET OF THAT SIZE)

SUBSET   1

GROUP   Grp 1   Grp 3   Grp 2
MEAN   6.7500   8.6250   10.3750
- -
SUBSET   2

GROUP   Grp 2   Grp 4
MEAN   10.3750   13.7500

tests). Below that are sets of groups that are not significantly different by that test. For example, for Tukey's test, Groups 1, 2, and 3 are not significantly different, nor are Groups 2 and 4. However, Group 4 is significantly different from Groups 1 and 3.

Two-Way Analysis (Chapter 5)

The fixed-effects two-way analysis of variance uses the ANOVA procedure instead of ONEWAY. Otherwise the analysis is very similar. Although limited contrasts can be tested with the ANOVA procedure, the procedure is somewhat complicated. The tests are easily done by hand, using the output from the overall analysis, and contrasts will not be illustrated here.

INPUT

Table D.6 shows the input for the analysis of the data in Table 5.2. Table D.7 shows the input for the data in Table 5.9. In both cases, the first line contains the title to be printed at the top of each page. The second line contains a data statement similar to those in previous tables. The difference is that now there are a total of three variables—two independent variables and one dependent variable—instead of two. For example, in Tables D.6 and D.7, both factors, diagnosis ("DIAG") and drug, are regarded as independent variables. Note that they are also listed in the data. Each score is preceded by its level on factor A (diagnosis) and Factor B (drug). These levels must be consecutive integer values.

The third line has the key word "ANOVA", indicating that procedure ANOVA is to be used. For the most part, the rest is the same as before. The "STATISTICS" subcommand asks SPSS to print out the cell means for all main effects and interactions. The key word "STATISTICS" can be followed by other commands requesting that different statistics be printed. However, cell means are sufficient in most analyses.

Another subcommand that was not used here, but might be useful in some cases, is "MAXORDERS". If the subcommand "MAXORDERS=NONE" is added to the input, then the analysis is performed assuming no interaction. That is, the interaction term is pooled with the error variance. This is useful primarily when there is only one score per cell. Then the analysis will be performed using the interaction as the error term.

For unbalanced designs (i.e., designs with unequal cell sizes), another subcommand should be used. This is the subcommand "METHOD=HIERARCHICAL". This instructs SPSS to make the sums of squares orthogonal.(See the SPSS-X manual for other commands that will lead to nonorthogonal sums of squares.) In addition, the order in which the factors are listed following the key word "ANOVA" is important. By listing Factor A before Factor B, we are asking SPSS to analyze the A main effect ignoring B (i.e., to calculate SS_a), and to analyze B taking A into account (i.e., to calculate $SS_{b:a}$). If we had

TABLE D.6. SPSS input for analyzing data in Table 5.2.
"DIAG" is "diagnosis" (schizophrenic vs. manic depressive,
Factor A), and "DRUG" is "drug administered" (Factor B).

```
TITLE 'TABLE 5.2'
DATA LIST LIST / DIAG DRUG SCORE
ANOVA SCORE BY DIAG (1,2) DRUG (1,3) / STATISTICS=MCA
BEGIN DATA
   1 1  8
   1 1  4
   1 1  0
   1 2  8
   1 2 10
   1 2  6
   1 3  4
   1 3  6
   1 3  8
   2 1 10
   2 1  6
   2 1 14
   2 2  0
   2 2  4
   2 2  2
   2 3 15
   2 3  9
   2 3 12
END DATA
```

wanted to calculate SS_b and $SS_{a:b}$, we would have had to change the order of
the factors; i.e., we would have had to write "ANOVA SCORE BY DRUG(1,
3) DIAG(1,2)".

OUTPUT

Table D.8 shows the analysis of variance output for the data in Table D.7. The
output for the data in Table D.6 has the same format. It should not require
explanation. Note that the main-effects sums of squares are SS_a and $SS_{b:a}$,
as explained above.

Models with Random Factors (Chapter 6)

SPSS cannot analyze random and mixed models directly. However, you can
tell it which MS to use for each hypothesis test performed. Therefore, it can
do the correct analyses, provided you have specified the design.

TABLE D.7. SPSS input for analyzing data in Table 5.9.

```
TITLE 'TABLE 5.9'
DATA LIST LIST / DIAG DRUG SCORE
ANOVA SCORE BY DIAG (1,2) DRUG (1,3) / STATISTICS
BEGIN DATA
  1 1  8
  1 1  4
  1 1  0
  1 2  8
  1 2 10
  1 2  6
  1 3  4
  1 3  8
  2 1 10
  2 1 14
  2 2  0
  2 2  4
  2 2  2
  2 3 15
  2 3  9
  2 3 12
END DATA
```

TABLE D.8. SPSS analysis of variance output on data in Table D.7.

TABLE 5.9   Page 3

         * * *   A N A L Y S I S   O F   V A R I A N C E   * * *

Source of Variation	Sum of Squares	DF	Mean Square	F	Sig of F
Main Effects	73.350	3	24.450	2.982	.083
DIAG	20.250	1	20.250	2.470	.147
DRUG	53.100	2	26.550	3.238	.082
2-Way Interactions	158.400	2	79.200	9.659	.005
DIAG DRUG	158.400	2	79.200	9.659	.005
Explained	231.750	5	46.350	5.652	.010
Residual	82.000	10	8.200		
Total	313.750	15	20.917		

TABLE D.9. SPSS input for data in Table
6.2. Most of the data have been omitted to
save space.

```
TITLE 'TABLE 6.2'
DATA LIST LIST / TEST SCHOOL SCORE
MANOVA SCORE BY TEST(1,3) SCHOOL(1,20)
  / OMEANS=TABLES(TEST)
  / DESIGN=CONSTANT VS 1
             SCHOOL = 1
             TEST VS 2
             TEST BY SCHOOL=2
BEGIN DATA
   1  1 22
   .......
   3 20 22
END DATA
```

INPUT

Table D.9 shows the input file for the data in Table 6.2. It uses a different procedure, called "MANOVA", for the analysis. MANOVA is probably the most versatile, and complex, of all the SPSS procedures. The "MANOVA" command uses the same syntax as the "ANOVA" command. The dependent variable is listed first followed by the word "BY"; then the factors are listed. However, the subcommands that follow are different. Although a great many commands are possible, the ones shown are sufficient for a basic analysis of variance. The first subcommand, "OMEANS=TABLES(TEST)", tells SPSS to print the marginal means for the test main effect. The next subcommand, with the key word "DESIGN", actually specifies the design to be analyzed. Each main effect and interaction that should be calculated must be specified. Denominator terms have to be equated to numbers before they are used. For example, the third part of the DESIGN subcommand says "TEST VS 2", indicating that the SS for the test main effect is to be calculated and tested against error term number two. The fourth part of the subcommand, "TEST BY SCHOOL=2", tells SPSS that error term number two is the test by school interaction. Accordingly, SPSS will test the test main effect, using the test by school interaction as an error term. The first part of the "DESIGN" subcommand reads "CONSTANT VS 1". The word "CONSTANT" in SPSS refers to the grand mean. Thus, this line and the second line, "SCHOOL=1", tell SPSS to test the grand mean against the school main effect.

Table D.10 shows the SPSS input for the data in Table 6.3. The "OMEANS" command differs from that in Table D.9 because here we have chosen to print out means for both main effects and the interaction.

TABLE D.10. SPSS input for data in Table 6.3. Most of the data have been omitted to save space.

```
TITLE 'TABLE 6.3'
DATA LIST LIST / UNIV COURSE SCORE
MANOVA SCORE BY UNIV(1,4) COURSE(1,6)
   / OMEANS=TABLE(UNIV, COURSE, UNIV BY COURSE)
   / DESIGN=UNIV VS 2
              COURSE VS 2
              UNIV BY COURSE=2 VS WITHIN
   / DESIGN=CONSTANT + UNIV BY COURSE VS 1
              UNIV + COURSE = 1
BEGIN DATA
   .......
END DATA
```

Table 2.10 also differs from Table D.9 in that there are two "DESIGN" subcommands. The first analyzes the main effects and interactions. The first two lines, "UNIV VS 2" and "COURSE VS 2", tell SPSS to test the university and course main effects against error term two. The third line tells SPSS that error term two is the university by course interactions; it also tells SPSS to test the university by course interaction against MS_w.

In general, each effect to be tested or to be used as a denominator term should be listed. If an effect is to be used as a denominator, the equals sign must be used to assign it an integer value. If it is to be tested, it must be followed by the key word "VS" and a denominator term. The denominator term may be one of the integers assigned to an effect or it may be one of three predefined error terms: "WITHIN" (which can be abbreviated "W"), which is MS_w; "RESIDUAL" (which can be abbreviated "R"), which is a pooled error term consisting of all effects that have not been specified in the "DESIGN" subcommand (there are none in the examples given here); and "WR" (or "RW"), which pools the residual error terms with SS_w.

The second "DESIGN" subcommand illustrates more of the flexibility of SPSS. The grand mean can be tested with a linear combination of sums of squares in the denominator, but the linear combination includes a term (the interaction) that is subtracted (see Chapter 6 for details). Alternatively, the grand mean can be pooled with the interaction and then tested against a pooled sum of squares consisting of the two main effects. Subtraction of sums of squares is not permitted in SPSS, but pooling of sums of squares by addition is permitted. The plus sign indicates that sums of squares are to be pooled. Thus, the first line, "CONSTANT + UNIV BY COURSE VS 1", tells SPSS to pool the grand mean with the interaction and test it against error term number one; the second line, "UNIV + COURSE = 1", tells SPSS that error term number 1 is to be formed by pooling the two main effects.

Note that MANOVA allows multiple design subcommands. This is fortunate because two subcommands were necessary for this complete analysis. If we had combined all of the analyses into a single "DESIGN", the analysis would not have been successful. MANOVA takes all the effects specified under a single "DESIGN" command and treats them as separate independent effects. Thus, if all the analyses had been combined into a single "DESIGN", the effects tested would have been the university main effect, with 3 degrees of freedom, the course main effect, with 5 degrees of freedom, the interaction, with 15 degrees of freedom, and the pooled grand mean plus interaction term, with 16 degrees of freedom. However, this sums to 39 degrees of freedom for tested effects, and there are only 24 degrees of freedom available in the design. MANOVA mistakenly assumes that these effects are all independent and issues an error message saying there are too many effects. Then it continues with the analysis, but the results are mostly garbage. By specifying two different designs, we get two separate analyses, and the problem is avoided.

Note, finally, that when subcommands other than "DESIGN" are used, they should precede the "DESIGN" subcommands. The other subcommands help to specify the nature of the analysis, but the "DESIGN" subcommand actually triggers the analysis itself. In Table D.10, the "OMEANS" subcomand precedes the first "DESIGN" subcommand, so MANOVA associates it with the first analysis. If it had followed the first "DESIGN" subcommand, it would have been associated with the second analysis. If it had followed the second "DESIGN" subcommand, it would not have been associated with any analysis, and the means would not have been printed out.

OUTPUT

In Table D.9 we asked for cell means for the main effect of TEST. For each level of TEST, SPSS prints two cell means, a weighted and unweighted mean. The weighted means are calculated from the marginal totals. The unweighted means are calculated by first calculating the mean for each cell and then averaging the cell means. With unequal cell sizes, these two means will be different. With equal cell sizes, as in our examples here, there is no difference. The output of cell means from the data in Table 6.3 is similar, but there are three tables: one for each main effect, and one for the interaction.

Table D.11 shows the SPSS analysis of variance output for the data in Table D.10. The tests are divided into sets according to their error terms, with the data for each error term listed first, and then the data for the term(s) being tested. Note that Table D.11 contains two analyses of variance because there were two "DESIGN" subcommands in the input file. The word "CONSTANT" in this table again refers to the grand mean.

TABLE D.11. SPSS analysis of variance output on data in Table D.7.

```
TABLE 6.3                                              Page 5

* * * A N A L Y S I S   O F   V A R I A N C E -- DESIGN  1 * * *

Tests of Significance for SCORE using UNIQUE sums of squares
Source of Variation              SS    DF      MS      F   Sig of F

WITHIN CELLS                  498.50   24    20.77
UNIV BY COURSE (ERROR 2)      251.65   15    16.78    .81     .660

Error 2                       251.65   15    16.78
UNIV                          218.23    3    72.74   4.34     .022
COURSE                        750.94    5   150.19   8.95     .000
```

Multiple Factors (Chapter 7)

The extension to multiple factors is straightforward. We will illustrate it with the data in Table 7.1, with Factor C random.

INPUT

Table D.12 shows the SPSS input file. The statements should be familiar. In this design some effects are to be tested and are also to be used as denominator mean squares. The syntax for specifying this is found in the fourth line of the "DESIGN" subcommand. The line reads "C = 1 VS WITHIN", indicating that the C main effect is error term number one (for testing the grand mean), and that it, in turn, is to be tested using the WITHIN error term. Similarly, a later line states that the A by C interaction is error term number two, and that it also is to be tested against MS_w.

OUTPUT

The output follows the same format as in the previous designs. Its interpretation should by now be straightforward.

Nested Factors (Chapter 8)

The main difference with nested factors is in the way that nesting is specified. We will illustrate it with the data in Tables 8.1 and 8.6.

TABLE D.12. SPSS input for data in Table 7.1. Most of the data have been omitted to save space.

```
TITLE 'TABLE 7.1'
DATA LIST LIST / A B C SCORE
MANOVA SCORE BY A(1,2) B(1,2) C(1,3)
  / OMEANS=TABLE(A, B, A BY B)
  / DESIGN=CONSTANT VS 1
             A VS 2,
             B VS 3,
             C = 1 VS WITHIN
             A BY B VS 4
             A BY C = 2 VS WITHIN
             B BY C = 3 VS WITHIN
             A BY B BY C = 4 VS WITHIN
BEGIN DATA
   1 1 1   2
   . . . . . . . .
   2 2 3   6
END DATA
```

INPUT

Table D.13 shows the SPSS input file for the data in Table 8.1. The "DESIGN" subcommand shows that the word "WITHIN" is used to specify nesting. This differs from the use of parentheses as in this text, but the use of "WITHIN" is like ordinary English, so there should be no problem with the notation.

However, the "OMEANS" subcommand illustrates a problem that can occur when printing out cell and marginal means. We want to print out cell means for teachers nested in schools, but we cannot specify "TEACHER WITHIN SCHOOL" in the "OMEANS" subcommand because only the "DESIGN" subcommand recognizes the "WITHIN" key word. Should we then ask for the means for "TEACHER" or for "TEACHER BY SCHOOL"? The answer is found by noting that the "OMEANS" subcommand cannot recognize nesting. If we specified "TEACHER", it would print out marginal means as though teachers were crossed with schools. The correct set of cell means is obtained by specifying "TEACHER BY SCHOOL".

Similarly, Table D.14 shows the SPSS input file for the data in Table 8.6. The main difference is that there are multiple error terms.

Similarly, Table D.15 shows the data for an $A \times B \times C(AB)$ design to illustrate the syntax when one factor is nested in two or more others. Note that the same syntax is used here as for an interaction. That is, in the fourth line of the "DESIGN" subcommand, "C WITHIN A BY B = 1 VS WITHIN", C is said to be nested in "A BY B". Note also that in this design, with more than one score per cell, the word "WITHIN" has two meanings. The first use

TABLE D.13. SPSS input file for data in Table 8.1.

```
TITLE 'TABLE 8.1'
DATA LIST LIST / SCHOOL TEACHER SCORE
MANOVA SCORE BY SCHOOL(1,3) TEACHER(1,2)
  / OMEANS=TABLE(SCHOOL, TEACHER BY SCHOOL)
  / DESIGN=CONSTANT, SCHOOL, TEACHER WITHIN SCHOOL
BEGIN DATA
  1 1 20
  1 1 18
  1 1 14
  1 2 19
  1 2 20
  1 2 20
  2 1 14
  2 1 18
  2 1 14
  2 2 12
  2 2 12
  2 2  9
  3 1 13
  3 1 16
  3 1 13
  3 2  9
  3 2  4
  3 2  4
END DATA
```

TABLE D.14. SPSS input for data in Table 8.6.
"NONS" is "nonsense syllable," "INST" is "instructions," and "SUBJ" is "subject." Most of the data
have been omitted to save space.

```
TITLE 'TABLE 8.6'
DATA LIST LIST / NONS INST SUBJ SCORE
MANOVA SCORE BY NONS(1,4) INST(1,3) SUBJ(1,3)
  / OMEANS=TABLE(NONS, INST, NONS BY INST)
  / DESIGN=NONS VS 1,
           NONS BY INST VS 1,
           NONS BY SUBJ WITHIN INST = 1,
           INST VS 2,
           SUBJ WITHIN INST = 2
BEGIN DATA
  1 1 1  8
  . . . . . . . .
  4 3 3 23
END DATA
```

TABLE D.15. SPSS input for data from an $A \times B \times C(AB)$ design with C random. The data have been omitted to save space.

```
TITLE 'TABLE 8.X'
DATA LIST LIST / A B C SCORE
MANOVA SCORE BY A(1,2) B(1,3) C(1,2)
   / OMEANS=TABLE(A, B, A BY B)
   / DESIGN=A VS 1,
             B VS 1,
             A BY B VS 1,
             C WITHIN A BY B=1 VS WITHIN
BEGIN DATA
  . . . . . . . .
END DATA
```

of the word indicates nesting; the second indicates the error term to be used. However, there need not be any confusion because the word "WITHIN" to the left of the word "VS" always indicates nesting, while the word "WITHIN" to the right of the word "VS" always indicates the error term.

OUTPUT

The outputs are not shown here. Their interpretation should be straightforward.

Other Incomplete Designs (Chapter 9)

Specialized designs such as those in Chapter 9 often involve calculating sums of squares such as $SS_{a:b}$. Such designs are unbalanced and must be analyzed using special instructions as with the unbalanced design in Chapter 5. Others, such as Latin squares, are balanced and can be analyzed in the usual way. When in doubt, it is probably safer to assume the design is unbalanced. We will give one example, the balanced incomplete blocks design in Table 9.9, which is unbalanced.

INPUT

The SPSS file for Table 9.9 is in Table D.16. The following differences from most previous input files should be noticed: First, we have added the subcommand "METHOD=SSTYPE(SEQUENTIAL)". This requests that the effects be analyzed sequentially as in the unbalanced design in Chapter 5. When the factors are not all orthogonal, the sequential instruction guarantees that the sums of squares will be orthogonal.

Second, under the "DESIGN" subcommand, only main effects are analyzed. Generally, with designs of the type found in Chapter 9, interactions cannot be analyzed, so we request only the grand mean and the three main effects.

Third, each effect is tested "VS RESIDUAL". MANOVA pools any effects that are not specified in the design card (in this example, that would be all of the interaction effects) into a single "residual effect." In this design, the residual is the correct error term. If we had had more than one score per cell, there would also have been a within-cells effect. Then the correct error term would have been found by pooling the residual sum of squares with the within-cells sum of squares. MANOVA allows this to be done also. The key word "WR" (or, equivalently, "RW") can be used to specify the pooled error term. Therefore, if we had had more than one score per cell, each effect would have been tested "VS WR" (or "VS RW") instead of "VS RESIDUAL".

Fourth, Factor B is listed before Factor A after the "DESIGN" subcommand. We have asked MANOVA to analyze the effects sequentially. MANOVA therefore proceeds to test each effect, taking the previously tested effects into account but ignoring effects that have not yet been tested. Because B is tested before A, we will obtain SS_b, the sum of squares for B ignoring A, and then obtain $SS_{a:b}$, the sum of squares for A taking B into account. If we had put A before B, we would have obtained SS_a and $SS_{b:a}$. The placement of Factor C is immaterial in this design because C is orthogonal to both A and B. However, if it were not, we would have obtained $SS_{c:a,b}$, the main effect of C, taking both A and B into account.

OUTPUT

The cell means, obtained because the "OMEANS" subcommand was used, for the data in Table D.16 are not shown. Note, however, that the printed means for the A main effect are ordinary means, taken by simply averaging over the other two factors. They are not the values used in Chapter 9 to estimate the A main effects.

The analysis of variance for the data in Table D.16 is shown in Table D.17. The only differences from previous tables are in the last four rows. The first of these is labeled "(Corrected Model)"; it gives the total sum of squares for all of the tested effects. The second, labeled "(Corrected Total)", gives the total sum of squares for the experiment. The third, labeled "R-Squared =", gives an estimate of the total proportion of variance accounted for by all of the effects. It is just the ratio of the sums of squares in the two previous rows. The last row, labeled "Adjusted R-Squared =", gives a more conservative, and probably better, estimate of the total proportion of variance accounted for.

Trend Analyses (Chapter 10)

One-way trend analyses are easy with the ONEWAY procedure in SPSS.

TABLE D.16. SPSS input for data in Table 9.9.

```
TITLE 'TABLE 9.9'
DATA LIST LIST / A B C SCORE
MANOVA SCORE BY A(1,4) B(1,4) C(1,3)
   / CMEANS=TABLE(A B C)
   / METHOD=SSTYPE(SEQUENTIAL)
   / DESIGN=CONSTANT VS RESIDUAL
             B VS RESIDUAL
             A VS RESIDUAL
             C VS RESIDUAL
BEGIN DATA
   1 1 1 15
   2 2 1  7
   3 3 1 14
   4 4 1  8
   2 1 2 17
   3 2 2  6
   4 3 2  8
   1 4 2  7
   3 1 3 22
   4 2 3 14
   1 3 3 11
   2 4 3 15
END DATA
```

TABLE D.17. SPSS analysis of variance output on data in Table D.16.

TABLE 9.9   Page 4

* * * A N A L Y S I S   O F   V A R I A N C E -- DESIGN 1 * * *

Tests of Significance for SCORE using SEQUENTIAL Sums of Squares

Source of Variation	SS	DF	MS	F	Sig of F
RESIDUAL	21.00	3	7.00		
CONSTANT	1728.00	1	1728.00	246.86	.001
B	150.00	3	50.00	7.14	.070
A	21.00	3	7.00	1.00	.500
C	78.00	2	39.00	5.57	.098
(Corrected Model)	249.00	8	31.13	4.45	.124
(Corrected Total)	270.00	11	24.55		

R-Squared =   .922
Adjusted R-Squared = .715

TABLE D.18. SPSS input for data
in Table 10.1. Most of the data have
been omitted to save space.

```
TITLE 'TABLE 10.1'
DATA LIST LIST / GROUP SCORE
ONEWAY SCORE BY GROUP(1,6)
    / STATISTICS=DESCRIPTIVES
    / POLYNOMIAL=5
BEGIN DATA
    1  13
    . . . .
    6  17
END DATA
```

INPUT

Table D.18 shows the SPSS input for the data in Table 10.1. The only difference from the input for an ordinary one-way analysis of variance is in the
addition of the "POLYNOMIAL=5" subcommand. The five indicates that a
fifth-degree polynomial is to be fitted to the data. A number smaller than five
would request a polynomial with fewer terms. The largest term that can be
specified is five. (For polynomials higher than the fifth degree other procedures would have to be used. For example, contrasts could be requested with
the coefficients specified as in the data from Chapter 3.)

OUTPUT

The output is shown in Table D.19. The table is mostly self-explanatory.
The data for the overall test are given first, followed by the data for the
individual trends. Note that data are given for each term, as well as for the
deviations from those terms. For example, the row labeled "DEVIATION
FROM QUADRATIC" gives sums of squares, etc., for the remainder after
the linear and quadratic terms have been factored out of SS_{bet}. Note also that
the F ratios use MS_w as the error term. If residuals were to be used as error
terms, the F ratios would have to be calculated by hand. However, with the
extensive data supplied in the output, the calculations would not be difficult.

Multifactor Trend Analyses (Chapter 11)

Multifactor trend analyses such as those in Chapter 11 are relatively straightforward, but cumbersome, in SPSS. Because a special analysis on multiple
factors is involved, MANOVA must be used.

TABLE D.19. SPSS analysis of variance output for data in Table D.18.

TABLE 10.1   Page 2

- - - - - - - - - - - O N E W A Y - - - - - - - - - - -

    Variable   SCORE
    By Variable   GROUP

ANALYSIS OF VARIANCE

SOURCE	D.F.	SUM OF SQUARES	MEAN SQUARES	F RATIO	F PROB.
BETWEEN GROUPS	5	890.6000	178.1200	19.3921	.0000
LINEAR TERM	1	803.5714	803.5714	87.4856	.0000
DEVIATION FROM LINEAR	4	87.0286	21.7571	2.3687	.0640
QUAD. TERM	1	20.7429	20.7429	2.2583	.1387
DEVIATION FROM QUAD.	3	66.2857	22.0952	2.4055	.0773
CUBIC TERM	1	53.3889	53.3889	5.8125	.0193
DEVIATION FROM CUBIC	2	12.8968	6.4484	.7020	.5000
QUARTIC TERM	1	12.8571	12.8571	1.3998	.2419
DEVIATION FROM QUARTIC	1	.0397	.0397	.0043	.9478
QUINTIC TERM	1	.0397	.0397	.0043	.9478
WITHIN GROUPS	54	496.0000	9.1852		
TOTAL	59	1386.6000			

GROUP	COUNT	MEAN	ST DEV	ST ERROR	95 PCT CONF INT FOR MEAN			MIN	MAX
Grp 1	10	8.8	3.0478	.9638	6.6198	TO	10.9802	3.0	13.0
Grp 2	10	10.7	3.5917	1.1358	8.1307	TO	13.2693	3.0	16.0
Grp 3	10	12.7	2.0575	.6506	11.2281	TO	14.1719	10.0	17.0
Grp 4	10	16.3	3.3350	1.0546	13.9143	TO	18.6857	12.0	22.0
Grp 5	10	19.5	3.2059	1.0138	17.2066	TO	21.7934	14.0	26.0
Grp 6	10	17.8	2.6998	.8537	15.8687	TO	19.7313	12.0	21.0
TOTAL	60	14.3	4.8479	.6259	13.0477	TO	15.5523	3.0	26.0

TABLE D.20. SPSS input for data in Table 11.6 with Factors B and C numerically valued. The data have been omitted to save space.

```
TITLE 'TABLE 11.6'
DATA LIST LIST / A B C SCORE
MANOVA SCORE BY A(1,3) B(1,3) C(1,5)
    / CONTRAST(B)=POLYNOMIAL
    / CONTRAST(C)=POLYNOMIAL
    / PARTITION(B)
    / PARTITION(C)
    / DESIGN=A VS RESIDUAL
            B(1) VS RESIDUAL
            B(2) VS RESIDUAL
            C(1) VS RESIDUAL
            C(2) VS RESIDUAL
            C(3) VS RESIDUAL
            C(4) VS RESIDUAL
            A BY B(1) VS RESIDUAL
            A BY B(2) VS RESIDUAL
            A BY C(1) VS RESIDUAL
            A BY C(2) VS RESIDUAL
            A BY C(3) VS RESIDUAL
            A BY C(4) VS RESIDUAL
            B(1) BY C(1) VS RESIDUAL
            B(1) BY C(2) VS RESIDUAL
            ......................
            B(2) BY C(3) VS RESIDUAL
            B(2) BY C(4) VS RESIDUAL
BEGIN DATA
    ..........
END DATA
```

INPUT

Table D.20 shows the input for an analysis of variance on the data in Table 11.6, with trend analyses on Factors B and C. The first subcommand under the "MANOVA" command is "CONTRAST(B)=POLYNOMIAL". This tells MANOVA that contrasts are to be computed on Factor B. The second subcommand specifies the same treatment for Factor C. A number of kinds of contrasts are possible; the key word "POLYNOMIAL" indicates that we want a polynomial fit (i.e., a trend analysis).

The "CONTRAST" subcommand can also be used for general trend tests. For example, the subcommand

"CONTRAST(B)=SPECIAL(1 1 1, 1 -1 0, 1 1 -2)"

would tell MANOVA to divide the Factor B into three comparisons; the first comparison would test the grand mean, the second would compare the first two levels, and the third would compare the average of the first two levels against the third. In general, the key word "SPECIAL" must be followed by coefficients for as many comparisons as there are levels of B. Other key words may also follow the "CONTRAST" subcommand, but they are likely to be of less general use. They are described in the SPSS manual.

The next subcommand, "PARTITION(B)", tells MANOVA that B is to be partitioned into the previously specified contrasts when the design is analyzed. (For some reason the creators of SPSS chose to require two statements here; the first, "CONTRAST", tells *how* the factor is to be divided, and the second tells MANOVA to actually *perform* the division.) The command that follows it does the same for Factor C.

The "DESIGN" subcommand follows. Note that contrasts are specified by placing the number of the contrast in parentheses after the factor name. Thus, the quadratic trend in Factor C is labeled $C(2)$, etc.

In this list, we have omitted the three-way interactions. The data in Table 11.6 contain only cell means; given only these, MANOVA cannot compute MS_w. Accordingly, we are here treating the design as having only one score per cell. MANOVA will treat the three-way interaction as a residual error term, and we have specified that each effect is to be tested against it.

MANOVA requires the user to specify every effect to be tested; thus, the long list of tests following the "DESIGN" subcommand. The list would have been longer if we had included the three-way interactions.

OUTPUT

The MANOVA output for the data in Table D.20 is not shown. The interpretation should be straightforward.

Multivariate Analysis of Variance (Chapter 13)

The multivariate analysis of variance is a straightforward extension of the univariate analysis. We will illustrate it with the data in Table 12.2.

INPUT

Table D.21 shows the SPSS input file for the data in Table 12.2. For these data, the "DATA LIST" statement shows that there are four variables, the two independent variables (factors) of "DIAG" and "DRUG", and the two dependent variables, "X" and "Y", on each line. The only difference in the "MANOVA" statement is the addition of Y along with X as a dependent variable.

OUTPUT

The "OMEANS" subcommand in the input file requested cell means as well as results of the analysis of variance. The output (not shown) is the same as in previous analyses, except that now there are two tables—one for X and one for Y—for each effect whose means are to be calculated.

TABLE D.21. SPSS input for multivariate analysis of data in Table 12.2.

```
TITLE 'TABLE 12.2'
DATA LIST LIST / DIAG DRUG X Y
MANOVA X Y BY DIAG (1,2) DRUG (1,3)
   / OMEANS=TABLES(DIAG, DRUG, DIAG BY DRUG)
   / DESIGN=DIAG, DRUG, DIAG BY DRUG
BEGIN DATA
    1  1   8   5
    1  1   4   3
    1  1   0   4
    1  2   8   7
    1  2  10   7
    1  2   6   4
    1  3   4   3
    1  3   6   6
    1  3   8   6
    2  1  10   5
    2  1   6   6
    2  1  14   7
    2  2   0   2
    2  2   4   7
    2  2   2   6
    2  3  15   8
    2  3   9   3
    2  3  12   7
END DATA
```

Table D.22 shows the analysis of variance output for the drug main effect and "DIAG BY DRUG" interaction. MANOVA does four different multivariate tests. The first two, Pillai's trace and the Hotelling–Lawley trace, are mentioned only briefly in Chapter 14. Neither has a direct intuitive interpretation, but Pillai's trace may be useful in that it is believed to be the most robust of the four. The third is Wilks's lambda, based on the maximum likelihood, and the fourth is Roy's maximum root, based on the optimal linear combination. For each test, MANOVA prints the value of the statistic, the approximate F ratio (except for Roy's maximum root for which no really good approximation exists), the numerator degrees of freedom (in the column labeled "Hypoth. DF") and the denominator degrees of freedom (in the column labeled "Error DF") of the approximate F, and the p-value (in the column labeled "Sig. of F").

These are followed by the results of univariate F tests on the same effect. The columns labeled "Hypoth. SS" and "Hypoth. MS" give the SS and MS for the effect being tested; the "Error SS" and "Error MS" give the denominator SS and MS (SS_w and MS_w in this design). The last two columns give the F ratios and p-values.

TABLE D.22. Partial SPSS analysis of variance output on data in Table D.21.

TABLE 13.2   Page 5

* * * * * * A N A L Y S I S   O F   V A R I A N C E -- DESIGN 1 * * * * * * *

EFFECT .. DIAG BY DRUG
Multivariate Tests of Significance (S = 2, M = -1/2, N = 4 1/2)

Test Name	Value	Approx. F	Hypoth. DF	Error DF	Sig. of F
Pillais	.64477	2.85460	4.00	24.00	.046
Hotellings	1.65896	4.14740	4.00	20.00	.013
Wilks	.37039	3.53723	4.00	22.00	.023
Roys	.62034				

Note.. F statistic for WILK'S Lambda is exact.

- -

EFFECT .. DIAG BY DRUG (Cont.)
Univariate F-tests with (2,12) D. F.

Variable	Hypoth. SS	Error SS	Hypoth. MS	Error MS	F	Sig. of F
X	144.00000	106.00000	72.00000	8.83333	8.15094	.006
Y	7.00000	44.00000	3.50000	3.66667	.95455	.412

TABLE D.22. *Continued*

EFFECT .. DRUG
Multivariate Tests of Significance (S = 2, M = -1/2, N = 4 1/2)

Test Name	Value	Approx. F	Hypoth. DF	Error DF	Sig. of F
Pillais	.47176	1.85219	4.00	24.00	.152
Hotellings	.84047	2.10117	4.00	20.00	.119
Wilks	.53802	1.99829	4.00	22.00	.130
Roys	.45002				

Note.. F statistic for WILK'S Lambda is exact.

- -

EFFECT .. DRUG (Cont.)
Univariate F-tests with (2,12) D. F.

Variable	Hypoth. SS	Error SS	Hypoth. MS	Error MS	F	Sig. of F
X	48.00000	106.00000	24.00000	8.83333	2.71698	.106
Y	1.00000	44.00000	.50000	3.66667	.13636	.874

TABLE D.23. SPSS input for multivariate
analysis of data in Table 6.2.

```
TITLE 'MANOVA ON TABLE 6.2'
DATA LIST LIST / TEST1 TO TEST3
MANOVA TEST1 TO TEST3
  / WSFACTORS=TEST(3)
  / OMEANS
  / WSDESIGN=TEST
BEGIN DATA
  22 18 13
  24 30 19
  26 21 22
  21 27 13
  26 23 17
  18 21 22
  16 15 21
  22 25 23
  14 22 20
  11 25 17
  24 15 17
  21 19 21
  20 22 20
  22 25 19
  25 19 24
  23 24 23
  22 24 21
  22 14 17
  21 23 12
  17 18 22
END DATA
```

Repeated Measures Designs (Chapter 13)

Repeated measures designs can be analyzed by the multivariate techniques
described in Chapter 13. However, to analyze them in this way, the format
for input is somewhat different.

Table D.23 shows the SPSS input file for the data in Table 6.2. Note that
now all the data for a given college (colleges are the random factor) are on
the same line. This fact must be reflected in the DATA statement, which
indicates that the three values are for TEST1, TEST2, and TEST3. (The
notation "TEST1 TO TEST3" is shorthand for "TEST1 TEST2 TEST3".)
The "MANOVA" statement lists only these three variables, indicating that all
three variables are now dependent variables; initially, there are no independent
variables in the design.

The statement "WSFACTORS=TEST(3)" actually specifies that the three
dependent variables are to be regarded as the three levels of a factor in
a repeated measures design. This statement *must* immediately follow the
"MANOVA" statement. Finally, the statement "WSDESIGN=TEST" in-

TABLE D.24. SPSS input for multivariate analysis of data in Table 7.9.

```
TITLE 'MANOVA ON TABLE 7.9'
DATA LIST LIST / X1 TO X6
MANOVA X1 TO X6
   / WSFACTORS=A(2) B(3)
   / OMEANS
   / WSDESIGN=A, B, A BY B
BEGIN DATA
      11   6 -7   1 -1 -3
      10  -2 -7   7   1 -1
      10   3 -7   2   4   2
      14   8 -3  -1   6   4
       4   3 -1   0 -7   1
END DATA
```

structs MANOVA to use "TEST" as a factor in the actual analysis. The "OMEANS" command is added to illustrate that it can also be used for this kind of analysis. However, now it can be used only to obtain means for the three tests. Because the tests are regarded as dependent variables (OMEANS ignores the "WSFACTORS" statement), no list of factors is appropriate.

Table D.24 shows the SPSS input file for the data in Table 7.9. Here C is random, and each line of the data statements contains the six values at a given level of C. The "DATA" statement must reflect the fact that there are six values on a line, so we list six variables; the names of the variables can be chosen arbitrarily as they will not be used explicitly in the analyses we are interested in. The "MANOVA" statement again reflects the fact that there are six dependent variables in the design.

The "WSFACTORS" statement again specifies the model. However, here there are two factors that must be specified. In this connection note that the order of the values is AB_{11}, AB_{12}, AB_{13}, AB_{21},...; i.e., the slowest moving index is for Factor A. With the data in this order, Factor A must be listed first in the "WSFACTORS" statement. Moreover, SPSS has no way of guessing the number of levels of each factor, so that information must be provided. It is provided by following each factor name with its number of levels. Any number of factors can be specified in this way, so long as they are listed in the order in which the indices change (slowest moving index first, fastest moving last) and each factor name is followed by its number of levels. In addition, the number of levels must equal the number of dependent variables listed in the "MANOVA" statement (in this example, they must be equal to six).

The statement "WSDESIGN=A, B, A BY B" specifies the "within-subjects" effects (i.e., the effects involving the factors specified in the "WSFACTORS"

TABLE D.25. SPSS input for multivariate analysis of data in Table 8.6.

```
TITLE 'MANOVA ON TABLE 8.6'
DATA LIST LIST / INST X1 TO X4
MANOVA X1 TO X4 BY INST(1,3)
  / WSFACTORS=LIST(4)
  / OMEANS=TABLES(INST)
  / WSDESIGN
  / DESIGN=INST
BEGIN DATA
      1   8  15  12  17
      1  20  24  16  20
      1  14  20  19  20
      2  21  18  17  28
      2  23  21  17  31
      2  26  29  26  30
      3  15  12  13  18
      3   6  11  10  12
      3   9  18   9  23
END DATA
```

statement) just as the "DESIGN" statement specifies effects in an ordinary analysis of variance. For example, if we did not want to analyze the "A BY B" interaction, we could leave it off the list.

The "OMEANS" command will again produce tables of means. However, there will be a separate table for each of the six dependent variables; "OMEANS" does not recognize the "WSFACTORS" statement, so it will not produce marginal means for Factors A and B. Note that if multiple analyses were to be used, the "WSDESIGN" subcommand would have to be followed by a "DESIGN" subcommand. The "WDESIGN" subcommand by itself does not trigger an analysis. The "DESIGN" subcommand would just consist of the single word "DESIGN". In the examples given here, the analyses are triggered by the "BEGIN DATA" statements.

The last example, in Table D.25, is from Table 8.6. This is somewhat more complicated because there are both crossed and nested factors. Because subjects are nested in the instructions factor, each line of the data statements must indicate whether the first, second, or third set of instructions were being used for those data. Thus, the first datum in each line is 1, 2, or 3. Then the scores for the four lists of nonsense syllables follow. Again the "DATA" statement lists all the variables on a line; in this case the "INST" variable followed by the four dependent variables ("X1 TO X4").

The "MANOVA" command uses the standard convention for listing dependent and independent variables (factors). The four dependent variables are listed first, followed by the key word "BY", followed by the independent variable with its lowest and highest possible values in parentheses.

The "WSDESIGN" command lists no variables at all. This is an option when there is only one "within subjects" factor. Alternatively, we could have written "WSFACTORS=LIST(4)", using the complete format as in Table D.23.

The statement "OMEANS=TABLES(INST)" now takes its familiar longer form. INST is not a "within subjects" factor, so it can be listed. A separate table will be printed out for each level of "INST".

Finally, we must add the statement "DESIGN=INST" to tell MANOVA that there is one "between-subjects" factor in the design. If there were more than one such factor, we would list all of them, plus all of the interactions we wanted MANOVA to calculate, using the format already described for previous designs.

Note that the order of the statements is important here. The "WSFAC-TORS" statement *must* be the first statement following the "MANOVA" command; the "DESIGN" statement must be the last for that analysis because MANOVA always begins the analysis immediately after reading the "DE-SIGN" statement. If additional analyses were to be done, the statements for them could follow the "DESIGN" statement.

In general, if the random factor in a repeated measures design is nested in one or more other factors, those factors are specified in the "DATA", "MANOVA", and "DESIGN" statements and are generally treated as in nonrepeated-measures designs. Factors with which the random factor is crossed are specified in the "WSFACTORS" and "WSDESIGN" statements, and all data from different levels of these factors are entered on the same line in the data list.

OUTPUT

The output of a repeated measures analysis can be voluminous (up to eleven pages for the examples above), but much of it is usually of little or no interest, and much of the remainder is already familiar from the analyses in previous chapters. (SPSS does allow you to save a few trees by suppressing uninteresting output; see the manual for details.) The most relevant parts of the output for the data in Table D.25 are shown in Table D.26.

The first table lists tests for "between-subjects" effects, i.e., effects in which the random factor is nested. For this design there is only one such effect: the main effect of instructions. An ordinary F test is used and the results are identical to those obtained in Chapter 8.

The remaining tables involve "within-subjects" effects: effects that are crossed with the random factor. The first of these tables gives data relevant to the appropriateness of the univariate tests. The estimate of Box's correction, as given in Chapters 6 to 8, is listed here as the "Greenhouse–Geisser Epsilon." This estimated correction is somewhat biased downward. The "Huynh–Feldt Epsilon" is less biased, but it can be greater than one. (In this example, it actually *is* greater than one, but MANOVA printed it as one.) The last estimate is a lower bound on Box's correction. Above that is the "Mauchly sphericity test." This is a test of the null hypothesis that epsilon equals one, that is,

TABLE D.26. SPSS multivariate analysis of variance output for data in Table D.25.

```
MANOVA ON TABLE 8.6                                          Page 4

* * * * * * A N A L Y S I S   O F   V A R I A N C E -- DESIGN   1 * *

Tests of Between-Subjects Effects.

Tests of Significance for T1 using UNIQUE sums of squares
Source of Variation            SS       DF       MS        F  Sig of F

WITHIN CELLS               265.83        6     44.31
INST                       730.17        2    365.08      8.24     .019

Tests involving 'LIST' Within-Subject Effect.

Mauchly sphericity test, W =         .84624
Chi-square approx. =                 .78840 with 5 D. F.
Significance =                       .978

Greenhouse-Geisser Epsilon =         .90189
Huynh-Feldt Epsilon =               1.00000
Lower-bound Epsilon =                .33333

AVERAGED Tests of Significance that follow multivariate tests are
equivalent to univariate or split-plot or mixed-model approach to
repeated measures.
Epsilons may be used to adjust d.f. for the AVERAGED results.

- - - - - - - - - - - - - - - - - - - - - - - - - - - - - - - - - - -
EFFECT .. INST BY LIST
Multivariate Tests of Significance (S = 2, M = 0, N = 1 )

Test Name        Value   Approx. F Hypoth. DF   Error DF  Sig. of F

Pillais         .64916    .80094      6.00        10.00      .591
Hotellings     1.37087    .68544      6.00         6.00      .671
Wilks           .40074    .77291      6.00         8.00      .613
Roys            .56006
Note.. F statistic for WILK'S Lambda is exact.

EFFECT .. LIST
Multivariate Tests of Significance (S = 1, M = 1/2, N = 1 )

Test Name        Value   Exact F  Hypoth. DF   Error DF  Sig. of F

Pillais         .78344   4.82358      3.00         4.00      .081
Hotellings     3.61769   4.82358      3.00         4.00      .081
Wilks           .21656   4.82358      3.00         4.00      .081
Roys            .78344
Note.. F statistics are exact.

AVERAGED Tests of Significance for L using UNIQUE sums of squares
Source of Variation            SS       DF       MS        F  Sig of F

WITHIN CELLS               153.50       18      8.53
LIST                       259.33        3     86.44     10.14      .000
INST BY LIST                67.17        6     11.19      1.31      .302
```

the null hypothesis that the univariate tests are valid. Presumably, if this test were significant, the univariate tests would be invalid unless Box's correction were applied. If the test were not significant, the ordinary univariate test, without Box's correction, would be valid. Unfortunately, this test suffers from the same deficiencies as other tests on variances. It is risky to accept the null hypothesis with an unknown probability of Type II error, and the test is not robust to the assumption of normality (multivariate normality in this case). Therefore, I do not recommend that the test be taken too seriously.

The next two tables give the multivariate tests on the INST BY LIST interaction and the LIST main effect. They are the same as those described above for multivariate tests. The last table gives the "AVERAGED" tests of significance for the "within-subjects" effects. These are simply the ordinary univariate tests described in Chapter 8. When a multivariate test is conducted, the univariate results are given also as a kind of "bonus."

Analysis of Covariance (Chapter 14)

An analysis of variance is easily generalized to a standard analysis of covariance in MANOVA. The two examples in Chapter 14, based on the data in Table 12.1, will both be used here.

INPUT

In the data list, covariates are listed just as any other variables. They are also treated just like other variables in the "DATA LIST" statement. They are defined as covariates in the list of variables following the "MANOVA" command. In that list, the dependent variables are listed first as usual, then the word "BY" is inserted, and the factors are listed. Finally, the word "WITH" is inserted and it is followed by the covariates. This is illustrated in Table D.27, which analyzes the one-way design of Table 12.1, with mental maturity ("MM") as the dependent variable, and IQ and chronological age ("CA") as covariates; and in Table D.28, which analyzes the same design with MM and IQ as dependent variables, and CA as the covariate. In both tables, the only difference from an ordinary multivariate analysis is in the "MANOVA" statement. For an ordinary multivariate analysis, that statement would read

MANOVA MM, IQ, CA BY GROUP(1,3)

For the analysis of covariance, the covariates are listed at the end, following the key word "WITH".

There is one other new statement in the analysis, the "PMEANS" statement. The "OMEANS" statement, which is also included for completeness, calculates simple cell means of all variables, regardless of whether they are dependent variables or covariates. The "PMEANS" statement calculates "predicted means", i.e., the group means adjusted for the effect of the covariate.

TABLE D.27. SPSS input for analysis of covariance on data in Table 12.1, with mental maturity (MM) as the dependent variable, and IQ and chronological age (CA) as covariate. Most of the data have been omitted to save space.

```
TITLE 'TABLE 12.1 WITH IQ AND CA AS COVARIATES'
DATA LIST LIST / GROUP MM IQ CA
MANOVA MM BY GROUP (1,3) WITH IQ CA
   / OMEANS=TABLES(GROUP)
   / PMEANS
   / DESIGN=GROUP
BEGIN DATA
     1 5 23 27
     . . . . . . . . .
     3 2 29 20
END DATA
```

TABLE D.28. SPSS input for analysis of covariance on data in Table 12.1, with mental maturity (MM) and IQ as dependent variables, and chronological age (CA) as covariate. Most of the data have been omitted to save space.

```
TITLE 'TABLE 12.1 WITH CA AS COVARIATE'
DATA LIST LIST / GROUP MM IQ CA
MANOVA MM IQ BY GROUP (1,3) WITH CA
   / OMEANS=TABLES(GROUP)
   / PMEANS
   / DESIGN=GROUP
BEGIN DATA
     . . . . . . . . .
END DATA
```

OUTPUT

The analyses of covariance for the two examples are found in Tables D.29, for the data in Table D.27, and Table D.30, for the data in Table D.28.

The analysis in Table D.29 is a univariate analysis with two covariates. Accordingly, a univariate test is presented, with ordinary F ratios. The first row, labeled "WITHIN CELLS", is the usual within-cells error term; the second, labeled "REGRESSION", tests the regression coefficient, θ; the third tests the differences between groups.

The second and third tables focus on the relation between the dependent variable and the two covariates. The second table gives the correlation between MM and each covariate. The third gives the squares of the correlations, along with a more complete regression analysis. The column labeled "B" gives the regression coefficients, θ; the remaining columns give the results of statistical tests on these coefficients. The tests given here differ from the test listed under "REGRESSION" in the first table. In the first table the two coefficients were tested together; in this table they are tested separately. Here we see that there is a significant relationship between MM and CA ($p = .007$) but not between MM and IQ ($p = .469$). The last two rows of this table give 95% confidence intervals on the values of θ (i.e., the values here labeled "B").

The fourth table lists the group means. The second and fourth columns just list the obtained group means, and the last two columns are obviously meaningless for these data. The third column lists estimated group means after eliminating the effects of the covariates. Basically, the values in column three are $\mu + \alpha_i$, where both μ and α_i are estimated as in Chapter 15.

Table D.30 shows a multivariate analysis of covariance. It is somewhat longer and more complicated.

The first table lists the results of four multivariate overall tests on the regression coefficients, θ. These tests are significant, indicating that there is a linear relationship between the dependent variables and the covariate. The second table tests each regression coefficient separately. For these data, the "mul.R" designates ordinary correlation coefficients; if there had been more than one covariate, they would have been multiple correlation coefficients. Here we see that the relationship between MM and CA is significant ($p = .004$), but the relationship between IQ and CA is not ($p = .677$). The third table lists the regression coefficients, θ (in the column labeled "B"). If there were more than one covariate, the third table would also break down these tests into tests on the individual covariates; as it is, this table just repeats the tests in the second table. The table also give 95% confidence limits on the values of θ.

The fourth and fifth tables give the tests on differences between groups. The fourth table gives the four multivariate tests, while the fifth gives univariate tests on each dependent variable separately.

Finally, the sixth and seventh tables (not shown) give the cell means for MM and IQ. They are interpreted in the same way as the fourth table in Table D.29.

TABLE D.29. SPSS analysis of variance output on data in Table D.27.

```
TABLE 12.1 WITH IQ AND CA AS COVARIATES                      Page 4

* * * A N A L Y S I S    O F    V A R I A N C E -- DESIGN    1 * * *

Tests of Significance for MM using UNIQUE sums of squares
Source of Variation       SS    DF    MS      F   Sig of F

WITHIN CELLS            14.99   10   1.50
REGRESSION              19.41    2   9.71   6.48     .016
GROUP                    2.54    2   1.27    .85     .457

- - - - - - - - - - - - - - - - - - - - - - - - - - - - - - - -
Correlations between Covariates and Predicted Dependent Variable
            COVARIATE

VARIABLE      IQ         CA

MM          -.333       .978

- - - - - - - - - - - - - - - - - - - - - - - - - - - - - - - -
Averaged Squared Correlations between Covariates and Predicted
Dependent Variable

VARIABLE   AVER. R-SQ

IQ             .111
CA             .956
Regression analysis for WITHIN CELLS error term
--- Individual Univariate .9500 confidence intervals
Dependent variable .. MM

COVARIATE       B      Beta    Std. Err.   t-Value   Sig. of t

IQ          -.02439  -.15838       .032    -.753       .469
CA           .39590   .71436       .117    3.394       .007

COVARIATE   Lower -95%  CL- Upper

IQ           -.097        .048
CA            .136        .656

Adjusted and Estimated Means
Variable .. MM
CELL   Obs. Mean   Adj. Mean   Est. Mean  Raw Resid. Std. Resid.

  1     4.000       3.359       4.000       .000       .000
  2     1.600       2.814       1.600       .000       .000
  3     4.600       4.027       4.600       .000       .000
```

TABLE D.30. SPSS analysis of variance output on data in Table D.28.

| TABLE 12.1 WITH CA AS COVARIATE | | | | | Page 4 |

＊ ＊ ＊ ＊ A N A L Y S I S   O F   V A R I A N C E -- DESIGN 1 ＊ ＊ ＊ ＊

EFFECT .. WITHIN CELLS Regression
Multivariate Tests of Significance (S = 1, M = 0, N = 4)

Test Name	Value	Exact F	Hypoth. DF	Error DF	Sig. of F
Pillais	.54299	5.94060	2.00	10.00	.020
Hotellings	1.18812	5.94060	2.00	10.00	.020
Wilks	.45701	5.94060	2.00	10.00	.020
Roys	.54299				

Note.. F statistics are exact.

- -

EFFECT .. WITHIN CELLS Regression (Cont.)
Univariate F-tests with (1,11) D. F.

Variable	Sq. Mul. R	Mul. R	Adj. R-sq.	Hypoth. MS	Error MS
MM	.53970	.73464	.49786	18.56571	1.43948
IQ	.01640	.12804	.00000	23.77286	129.65701

Variable	F	Sig. of F
MM	12.89751	.004
IQ	.18335	.677

- -

Regression analysis for WITHIN CELLS error term
--- Individual Univariate .9500 confidence intervals
Dependent variable .. MM

COVARIATE	B	Beta	Std. Err.	t-Value	Sig. of t
CA	.40714	.73464	.113	3.591	.004

COVARIATE	Lower -95% CL- Upper	
CA	.158	.657

Dependent variable .. IQ

COVARIATE	B	Beta	Std. Err.	t-Value	Sig. of t
CA	-.46071	-.12804	1.076	-.428	.677

COVARIATE	Lower -95% CL- Upper	
CA	-2.829	1.907

TABLE D.30. *Continued*

```
EFFECT .. GROUP
Multivariate Tests of Significance (S = 2, M = -1/2, N = 4 )

Test Name     Value  Approx. F Hypoth. DF   Error DF  Sig. of F

Pillais      .31549   1.03010      4.00       22.00     .414
Hotellings   .43666    .98248      4.00       18.00     .442
Wilks        .69132   1.01355      4.00       20.00     .424
Roys         .29218
Note.. F statistic for WILK'S Lambda is exact.

- - - - - - - - - - - - - - - - - - - - - - - - - - - - - - - - -

EFFECT .. GROUP (Cont.)
Univariate F-tests with (2,11) D. F.

Variable  Hypoth. SS  Error SS Hypoth. MS  Error MS     F  Sig. of F

MM          4.57291   15.83429   2.28645   1.43948   1.588     .248
IQ        338.08879 1426.22714 169.04439 129.65701   1.304     .310

- - - - - - - - - - - - - - - - - - - - - - - - - - - - - - - - -
Adjusted and Estimated Means
Variable .. MM
CELL     Obs. Mean  Adj. Mean  Est. Mean Raw Resid. Std. Resid.

  1        4.000      3.267      4.000      .000        .000
  2        1.600      2.740      1.600      .000        .000
  3        4.600      4.193      4.600      .000        .000

- - - - - - - - - - - - - - - - - - - - - - - - - - - - - - - - -
Adjusted and Estimated Means (Cont.)
Variable .. IQ
CELL     Obs. Mean  Adj. Mean  Est. Mean Raw Resid. Std. Resid.

  1       20.200     21.029     20.200      .000        .000
  2       21.600     20.310     21.600      .000        .000
  3       10.000     10.461     10.000      .000        .000
```

Symbol Table

$a, b, c, \ldots$	Subscripts indicating effects of Factors A, B, C, $\ldots$, respectively; e.g., MS_{ac} is the mean square for the AC interaction. Use of parentheses indicates nesting; e.g., $b(a)$ means B is nested in A.
$a_k, b_k \ldots$	(In trend analyses.) Coefficients for kth trend on A, B, $\ldots$.
$\hat{a}_k^*, \hat{b}_k^*, \hat{c}_k^*$	(In trend analyses.) The coefficient used for estimating cell means.
$A, B, C, \ldots$	Factors of an experiment. May also indicate effects, e.g., an AC interaction. Also (in Chapter 23 ff.) symbols used for matrices.
$A^t, B^t, C^t, \ldots$	(In Chapter 13 ff.) Symbols used for transposes of matrices.
$A^{-1}, B^{-1}, C^{-1}, \ldots$	(In Chapter 13 ff.) Symbols used for inverses of matrices.
$A_i, B_j, C_k, \ldots$	Specific levels of Factors A, B, C, $\ldots$.
$A_i, AC_{ik}, \ldots$	Labels for individual cells in ANOVA design.
AF	(With a subscript, in analyses of covariance.) The adjustment factor for calculating ASC matrices.
ASC	(With a subscript, in analyses of covariance.) Adjusted sum of cross-products matrices.

ASX	(With a subscript, in analyses of covariance.) Adjusted sum of squares on X.
$b(a)$, $cd(ab)$, ...	(Used in subscripts.) Letters outside parentheses indicate factors that are nested in factors represented by letters inside parentheses.
bet	(Used as a subscript.) Between groups.
$B(A)$, $CD(AB)$,...	Letters outside parentheses are factors that are nested in factors inside parentheses.
c_{ik}	Coefficient used in a planned or post hoc comparison.
C_k	Sample value of planned or post hoc comparison.
$C(X, Y)$	Covariance between X and Y.
C'_k	C_k adjusted to have a standard normal distribution.
C''_k	Value for testing post hoc comparisons using Tukey test.
d_k	Estimated degrees of freedom for approximate test of a planned comparison.
$det(A)$	Determinant of matrix A.
D_k	Denominator term for approximate test of a planned comparison.
df	Degrees of freedom; also written DF.
e	Error variance; used as a subscript.
$E(X)$	Expected value of X.
F	F ratio: The ratio of two mean squares.

$F_{\alpha(\nu_1,\nu_2)}$	F-distributed random variable with ν_1 degrees of freedom in the numerator and ν_2 degrees of freedom in the denominator.
F'	F ratio for Scheffé method of testing post hoc comparisons.
F^*	Symbol for a quasi-F test.
g	Subscript indicating any general effect, e.g., ϕ_g^2; also (in balanced incomplete blocks designs.), the number of levels of B with which each A_i is paired; also (in multivariate designs), term used in converting multiway statistics, such as Wilks's lambda, to chi-square or F.
G	Value used in calculating confidence intervals for ω^2 with random factors.
h	(In balanced incomplete block designs.) The number of levels of A with which each B_j is paired; also (in multivariate designs) term used in converting multiway statistics, such as Wilks's lambda, to chi-square or F.
H	Value used in calculating confidence intervals for ω^2 with random factors. Also, the efficiency factor in balanced incomplete blocks designs.
$i, j, k, \ldots$	Subscripts indicating specific levels of Factors A, B, C, $\ldots$; use of parentheses indicates nesting, e.g., $i(j)$ means J is nested in I. Also used as subscripts to indicate trend coefficients.
$I, J, K, \ldots$	Numbers of levels of Factors A, B, C, $\ldots$.
$j(i), k1(ij), \ldots$	(Used in subscripts.) Letters outside parentheses indicate factors that are nested in factors represented by letters inside parentheses.

k	When not a subscript, symbolizes a coefficient in a linear combination of mean squares.
K_B	Box's correction: Used in multifactor mixed and random models.
Ku	Kurtosis.
m	(Used as subscript.) Grand mean, e.g., SS_m is the sum of squares for the grand mean; also (in multivariate designs), term used in Roy's maximum root test.
m^*	(In multivariate designs.) Term used in converting multiway statistics, such as Wilks's lambda, to chi-square or F.
MX	(In analysis of covariance.) Mean square of X; usually has subscripts.
MS	Mean square; always has subscripts $(a, b, \ldots)$ indicating the specific effect.
$MS_{a:b}$	Mean square for testing effect a, taking effect b into account.
n	Number of subjects in a group; sometimes has subscripts $(i, j, \ldots)$ indicating a specific group.
n^*	(In multivariate designs.) Term used in Roy's maximum root test.
$\overline{n}$	Geometric mean of sample sizes; used for approximate tests in designs with unequal sample sizes.

N	Total number of observations in experiment.
$N_{(\mu,\sigma^2)}$	Normally distributed random variable with mean (μ) and variance (σ^2).
p	"p-value" of a test (i.e., smallest value of α for which the result could be declared significant); also (in multivariate designs) number of dependent variables.
P	Probability; may have subscripts.
q	(In Youden squares.) The number of times each level of A is paired with a level of C.
Q_{ij}	(In partially nested designs.) 1 if cell AB_{ij} contains data; 0 if it does not.
r	Pearson product-moment correlation; r_{xy} is the correlation between X and Y.
rem	Used as subscript; e.g., SS_{rem} is remaining sum of squares after specific effects have been removed from total sum of squares.
rs	(With subscripts.) "Raw sum of squares" calculated on coefficients for a comparison rather than on data, used when finding the appropriate denominator for a comparison.
RC	(With a subscript, in multivariate designs.) Raw sum of cross-products matrix, analogous to RS in univariate designs.
RS	(With a subscript.) Raw sum of squares; used in intermediate step when calculating sums of squares.

s, s^*	(In multivariate designs.) Terms used in Roy's maximum root test.
s^2	Sample variance; sometimes has subscripts $(i, j, \ldots)$ indicating subgroup on which variance is calculated.
ss	(With subscripts.) "Sum of squares" calculated on coefficients for a comparison rather than on data; used when finding the appropriate denominator for a comparison.
S	Quantity used for finding confidence interval for ψ, the population value of a planned comparison; S^2 is occasionally used to symbolize a variance calculated as an intermediate step in a derivation.
S^2	See S.
SC	(With a subscript, in multivariate designs.) Sum of cross-products matrix; similar to SS in univariate designs.
Sk	Skewness.
SS	Sum of squares; always has subscripts $(a, b, c, \ldots)$ indicating the specific effect.
$SS_{a:b}$	Sum of squares for testing effect a, taking effect b into account.
t	(Used as a subscript.) Total. Also (with subscripts $a, b, \ldots$), sample total for a specific group.
$t_{(\nu)}$	t-distributed random variable with ν degrees of freedom.

t^*	An intermediate value in calculating estimates of effects in the balanced incomplete blocks design.
$t'_{(k,\nu)}$	Studentized-range distribution with k groups and ν degrees of freedom.
T	Grand total of all scores in the experiment; also (when used as a column label in a table), total number of values squared and summed to obtain RS for a given effect.
$T_a, T_{ab}, T_{acd} \ldots$	Number of scores in each level of a factor or combination of factors, used in calculating degrees of freedom.
T'	Quantity used for finding confidence interval for the population value (ψ) in the Tukey method of post hoc comparisons.
$tr(A)$	Trace of A: The sum of the values on the diagonal of A.
u	Used as a subscript to indicate an unweighted average of group variances.
V	Numerical value of factor levels used in trend tests: usually has subscripts $(i, j, \ldots)$.
$V(X)$	Variance of X.
w	(Used as a subscript.) Within variables.
W	Same as V; also (in multivariate designs), term used in converting multiway statistics, such as Wilks's lambda, to chi-square or F.

x, y, z	(Used as subscripts in multivariate designs.) Indicate a particular dependent variable.
X, Y, Z	Dependent variables; usually have subscripts $(i, j, \ldots)$ indicating a specific value or set of values. Y may be used to represent a transformation on X.
α	Without subscripts, indicates α level of significance.
$\alpha, \beta, \gamma, \ldots$	With subscripts $(i, j, \ldots)$, indicate population values for specific effects.
$\alpha\,\beta_{ij}, \alpha\gamma\delta_{ikl}, \ldots$	Interactions: Associations between Greek letters and factors are the same as for main effects.
ϵ	Random error. Usually has subscripts $(i, j, \ldots)$ indicating a specific subgroup.
θ	Slope coefficient in analysis of covariance, usually with subscripts $(y, z, \ldots)$, but may appear as a subscript; also (in multivariate designs) term used in Roy's maximum root test; also (in multivariate designs) $\lambda/(1 + \lambda)$.
λ	(In balanced incomplete blocks designs.) The number of times that any two levels of Factor A are both paired with the same level of Factor B; also (in multivariate designs) the largest proper number of a matrix; used in doing significance tests.
Λ	(In multivariate designs.) Value of Wilks's lambda, for tests of hypotheses; sometimes has subscripts $(a, b, \ldots)$.

μ	Population mean; usually has subscripts $(i, j, \ldots)$ indicating the subgroup over which a mean has been taken. Without subscripts, the population grand mean.
μ^*	Population mean value specified by a null hypothesis.
μ_r^*	rth moment about the mean.
ν	Degrees of freedom; sometimes has subscripts.
$\hat{\pi}$	(In analysis of covariance.) Usual estimate of a certain parameter.
ρ	Population value of Pearson product-moment correlation; ρ_{xy} is the correlation between x and y.
σ^2	Population variance; usually has subscripts $(a, b, \ldots)$ indicating the type of variance.
τ^2	Quantity similar to σ^2 and having the same subscripts; more convenient to use than σ^2 for some purposes.
ϕ	Noncentrality parameter of chi-square, F, and t distributions.
ϕ'	Noncentrality parameter, similar to ϕ but uaffected by sample size; $\phi' = \phi/n^{1/2}$.
$\chi^2_{(\nu)}$	Chi square distributed random variable with ν degrees of freedom.
ψ	Population value of planned comparison; usually has identifying subscripts $(k, \ldots)$.
ω^2	Proportion of variance accounted for; may have subscripts $(a, b, \ldots)$ indicating a specific effect.
	Subscript; replaces subscript over which a sum or a mean was taken.

$\sim$	"Is distributed as."
$-$	When placed over a symbol, indicates a sample mean.
$\hat{}$	When placed over a Greek symbol, indicates an estimate of that parameter.
Σ	Summation sign.

Bibliography

Aitkin, M.A. (1969). Multiple comparisons in psychological experiments. *British Journal of Mathematical and Statistical Psychology*, **22**, 193-198.

Anderson, R.L., and Houseman, E.E. (1942). Tables of orthogonal polynomial values extended to $N = 104$. *Iowa Agricultural Experimental Station Research Bulletin*, #297.

Anderson, T.W. (1958). *An Introduction to Multivariate Statistical Analysis*. New York: Wiley.

Begun, J.M., and Gabriel, K.R. (1981). Closure of the Newman–Keuls multiple comparisons procedure. *Journal of the American Statistical Association*, **76**, 241–245.

Box, G.E.P. (1953). Non-normality and tests on variances. *Biometrika*, **40**, 318–335.

Box, G.E.P. (1954a). Some theorems on quadratic forms applied in the study of analysis of variance problems: I. Effect of inequality of variance in the one-way classification. *Annals of Mathematical Statistics*, **25**, 290–302.

Box, G.E.P. (1954b). Some theorems on quadratic forms applied in the study of analysis of variance problems: II. Effect of inequality of variance and of correlation of errors in the two-way classification. *Annals of Mathematical Statistics*, **25**, 484–498.

Box, G.E.P., and Anderson, S.L. (1955). Permutation theory in the derivation of robust criteria and the study of departures from assumptions. *Journal of the Royal Statistical Society*, Series B, **17**, 1–34.

Brown, M.B., and Forsythe, A.B. (1974). The ANOVA and multiple comparisons for data with heterogeneous variances. *Biometrics*, **30**, 719–724.

Cochran, W.G. (1956). The distribution of the largest of a set of estimated variances as a fraction of their total. *Annals of Eugenics*, **11**, 47–52.

Cochran, W.G. (1974). Some consequences when the assumptions for the analysis of variance are not satisfied. *Biometrics*, **3**, 39–52.

Cochran, W.G., and Cox, G.M. (1957). *Experimental Designs*. New York: Wiley.

Cohen, A. (1968). A note on the admissibility of pooling in the analysis of variance. *Annals of Mathematical Statistics*, **39**, 1744–1746.

Cox, D.R. (1958). *Planning of Experiments.* New York: Wiley.

David, F.N., and Johnson, N.L. (1951). The effects of non-normality on the power function of the F-test in the analysis of variance. *Biometrika,* **38**, 43–47.

Davidson, M.L. (1972). Univariate vs. multivariate tests in repeated-measures experiments. *Psychological Bulletin,* **77**, 446–452.

Davis, D.J. (1969). Flexibility and power in comparisons among means. *Psychological Bulletin,* **71**, 441–444.

Delury, D.B. (1948). The analysis of covariance. *Biometrics,* **4**, 153–170.

Donaldson, T.S. (1968). Robustness of the F-test to errors of both kinds and the correlation between the numerator and denominator of the F-ratio. *Journal of the American Statistical Association,* **63**, 660–676.

Duncan, D.B. (1952). On the properties of the multiple comparison test. *Virginia Journal of Science,* **3**, 49–67.

Duncan, D.B. (1955). Multiple range and multiple F-tests. *Biometrics,* **11**, 1–42.

Duncan, D.B. (1957). Multiple range tests for correlated and heteroscedastic means. *Biometrics,* **13**, 164–176.

Duncan, D.B. (1965). A Bayesian approach to multiple comparisons. *Technometrics,* **7**, 171–222.

Duncan, O.J. (1961). Multiple comparisons among means. *Journal of the American Statistical Association,* **56**, 52–64.

Dunnett, C.W. (1955). A multiple comparisons procedure for comparing several treatments with a control. *Journal of the American Statistical Association,* **50**, 1096–1121.

Dunnett, C.W. (1964). New tables for multiple comparisons with a control. *Biometrics,* **20**, 482–491.

Einot, I., and Gabriel, K.R. (1975). A study of the powers of several methods of multiple comparisons. *Journal of the American Statistical Association,* **70**, 574–583.

Eisenhart, C. (1974). The assumptions underlying the analysis of variance. *Biometrics,* **3**, 1–21.

Elashoff, J.D. (1969). Analysis of covariance: A delicate instrument. *American Educational Research Journal,* **6**, 383–401.

Evans, S.H., and Anastasio, E.J. (1968). Misuse of analysis of covariance when treatment effect and covariate are confounded. *Psychological Bulletin,* **69**, 225–234.

Federer, W.T. (1955). *Experimental Design.* New York: MacMillan.

Federer, W.T., and Zelen, M. (1966). Analysis of multifactor classifications with unequal numbers of observations. *Biometrics*, **22**, 525–552.

Feldt, L.S., and Mahmoud, M.W. (1958). Power function charts for specification of sample size in analysis of variance. *Psychometrika*, **23**, 101–210.

Finney, D.J. (1955). *Experimental Design and Its Statistical Basis*. Chicago: University of Chicago Press.

Fisher, R.A. (1947). *The Design of Experiments*. Edinburgh: Oliver and Boyd.

Fleiss, J.L. (1969). Estimating the magnitude of experimental effects. *Psychological Bulletin*, **72**, 273–276.

Fox, M. (1956). Charts of the power of the F-test. *Annals of Mathematica Statistics*, **27**, 484–497.

Games, P.A. (1971a). Inverse relation between the risks of Type I and Type II errors and suggestions for the unequal n case in multiple comparisons. *Psychological Bulletin*, **75**, 97–102.

Games, P.A. (1971b). Multiple comparisons of means. *American Educational Research Journal*, **8**, 531–565.

Gaylon, D.W., and Hopper, F.N. (1969). Estimating the degrees of freedom for linear combinations of mean squares by Satterthwaite's formula. *Technometrics*, **11**, 691–706.

Grant, D.A. (1961). *An Introduction to Linear Statistical Models*. New York: McGraw-Hill.

Harris, R.J. (1975). *A Primer of Multivariate Statistics*. New York: Academic Press.

Harris, M., Howitz, D.G., and Mood, A.M. (1948). On the determination of sample sizes in designing experiments. *Journal of the American Statistical Association*, **43**, 391–402.

Harter, H.L. (1957). Error rates and sample sizes for range tests in multiple comparisons. *Biometrics*, **13**, 511–536.

Heilizer, F. (1964). A note on variance heterogeneity in the analysis of variance. *Psychological Reports*, **14**, 532–534.

Horsnell, G. (1953). The effect of unequal group variances in the F-test for the homogeneity of group means. *Biometrika*, **40**, 128–136.

Hsu, P.L. (1938). Contributions to the theory of student's t test as applied to the problem of two samples. *Statistical Research Memoirs*, **2**, 1–24.

Hummel, T.J., and Sligo, J.R. (1971). Empirical comparison of univariate and multivariate analysis of variance procedures. *Psychological Bulletin*, **76**, 49–57.

Kempthorne, O. (1952). *The Design and Analysis of Experiments*. New York: Wiley.

Kesselman, H.J., and Toothaker, L.E. (1973). Error rates for multiple comparison methods: Some evidence concerning the misleading conclusions of Petrinovich and Hardyck. *Psychological Bulletin*, **80**, 31–32.

Kramer, C.Y. (1956). Extension of multiple range tests to group means with unequal numbers of replications. *Biometrics*, **12**, 307–310.

Kramer, C.Y. (1957). Extension of multiple range tests to group correlated adjusted means. *Biometrics*, **13**, 13–18.

Lehmann, E.L. (1963). Robust estimation in analysis of variance. *Annals of Mathematical Statistics*, **34**, 957–966.

Lippman, L.G., and Taylor, C.J. (1972). Multiple comparisons in complex ANOVA designs. *Journal of General Psychology*, **86**, 221–223.

Miller, R.G. (1966). *Simultaneous Statistical Inference*. New York: McGraw-Hill.

Newman, D. (1939). The distribution of the range in samples from a normal population, expressed in terms of an independent estimate of standard deviation. *Biometrika*, **31**, 20–30.

Norusis, M.J. (1985). *SPSS-X Advanced Statistics Guide*. Chicago: SPSS, Inc.

Overall, J.E., and Dalal, S.N. (1968). Empirical formulae for estimating appropriate sample sizes for analysis of variance designs. *Perceptual and Motor Skills*, **27**, 363–367.

Patnaik, P.B. (1949). The noncental X^2 and F-distributions and their approximations. *Biometrika*, **36**, 202–232.

Paull, A.E. (1950). On preliminary tests for pooling mean squares in the analysis of variance. *Annals of Mathematical Statistics*, **21**, 539–556.

Petrinovich, L.F., and Hardyck, C.D. (1969). Error rates for multiple comparison methods: Some evidence concerning the frequency of erroneous conclusions. *Psychological Bulletin*, **71**, 43–54.

Ramsey, P.H. (1978). Power differences between pairwise multiple comparisons. *Journal of the American Statistical Association*, **73**, 479–487.

Robson, D.S. (1959). A simple method for construction of orthogonal polynomials when the independent variable is unequally spaced. *Biometrics*, **15**, 187–191.

Ryan, T.A. (1960). Significance tests for multiple comparison of proportions, variances and other statistics. *Psychological Bulletin*, **57**, 318–328.

Sas Introductory Guide, 3rd Ed. (1985). Cary, NC: SAS Institute, Inc.

Sas User's Guide: Basics, Version 5 Ed. (1985). Cary, NC: SAS Institute, Inc.

Sas User's Guide: Statistics, Version 5 Ed. (1985). Car, NC: SAS Institute, Inc.

Scheffé, H. (1953a). A method for judging all contrasts in the analysis of variance. *Biometrika*, **40**, 87–104.

Scheffé, H. (1953b). A "mixed model" for the analysis of variance. *Annals of Mathematical Statistics*, **27**, 23–36.

Scheffé, H. (1956). Alternative models for the analysis of variance. *Annals of Mathematical Statistics*, **27**, 251–271.

Scheffé, H. (1959). *The Analysis of Variance*. New York: Wiley.

Searle, S.R. (1971). *Linear Models*. New York: Wiley.

Searle, S.R. (1987). *Linear Models for Unbalanced Data*. New York: Wiley.

Smith, R.A. (1971). The effect of unequal group size on Tukey's HSD procedure. *Psychometrika*, **36**, 31–34.

Spjotvoll, E. (1972). On the optimality of some multiple comparison procedures. *Annals of Mathematical Statistics*, **43**, 398–411.

SPSS User's Guide, 3rd Ed. (1988). Chicago: SPSS, Inc.

Tabachnik, B.G., and Fidell, L.S. (1989). *Using Multivariate Statistics*, 2nd Ed. Cambridge: Harper and Row.

Tiku, M.L. (1967). Tables of the power of the *F*-test. *Journal of the American Statistical Association*, **62**, 525–539.

Tukey, J.W. (1949a). Comparing individual means in the analysis of variance. *Biometrics*, **5**, 99–114.

Tukey, J.W. (1949a). One degree of freedom for non-additivity. *Biometrics*, **5**, 232–242.

Tukey, J.W. (1956). Variances of variance components: I. Balanced designs. *Annals of Mathematical Statistics*, **27**, 722–736.

Tukey, J.W. (1957). Variances of variance components: II. The unbalanced single classification. *Annals of Mathematical Statistics*, **28**, 43–56.

Welsch, R.E. (1977). Stepwise multiple comparison procedures. *Journal of the American Statistical Association*, **72**, 566–575.

Welsh, B.L. (1951). On the comparison of several mean values: An alternative approach. *Biometrika*, **38**, 330–336.

Wilk, M.B., and Kempthorne, O. (1955). Fixed, mixed, and random models. *Journal of the American Statistical Association*, **50**, 1144–1167.

Wilk, M.B., and Kempthorne, O. (1957). Nonaddativities in a Latin square design. *Journal of the American Statistical Association*, **52**, 218–236.

Winer, B.J. (1971). *Statistical Principles in Experimental Design*. New York: McGraw-Hill.

Yates, F. (1936). Incomplete randomized blocks. *Annals of Eugenics*, **7**, 121–140.

Yates, F. (1940). The recovery of inter-block information in balanced incomplete block designs. *Annals of Eugenics*, **10**, 137–325.

Index

Springer Texts in Statistics *(continued from p. ii)*